Handbook of Modern
Ion Beam Materials Analysis

Second Edition

APPENDICES

Handbook of Modern Ion Beam Materials Analysis

Second Edition

APPENDICES

EDITORS:

Yongqiang Wang and Michael Nastasi

Materials Research Society
Warrendale, Pennsylvania

CAMBRIDGE
UNIVERSITY PRESS

University Printing House, Cambridge CB2 8BS, United Kingdom

One Liberty Plaza, 20th Floor, New York, NY 10006, USA

477 Williamstown Road, Port Melbourne, VIC 3207, Australia

314-321, 3rd Floor, Plot 3, Splendor Forum, Jasola District Centre, New Delhi - 110025, India

79 Anson Road, #06-04/06, Singapore 079906

Cambridge University Press is part of the University of Cambridge.

It furthers the University's mission by disseminating knowledge in the pursuit of education, learning and research at the highest international levels of excellence.

www.cambridge.org
Information on this title: www.cambridge.org/9781605112169

CODEN: MRSPDH

Published by:

Materials Research Society
506 Keystone Drive
Warrendale, PA 15086
Telephone (724) 779-3003
Fax (724) 779-83 13
Web site: http://www.rnrs.org/

A catalogue record for this publication is available from the British Library

Library of Congress Cataloging in Publication data
Handbook of modern ion beam materials analysis : appendices / editors, Yongqiang Wang and Michael A. Nastasi. -- 2nd ed.
     p. cm.
     Includes index.
     ISBN 978-1-60511-216-9
          1. Ion bombardment--Handbooks, manuals, etc. 2. Solids--Effect of radiation on--Handbooks, manuals, etc. 3. Materials--Analysis--Handbooks, manuals, etc. I. Wang, Yongqiang.
II. Nastasi, Michael Anthony
QC702.7.B65H36 2009b
620.1′127--dc22
                         2009045715

ISBN 978-1-605-11216-9 Hardback

Disclaimer of Liability

Neither the editors, authors, contributors, their affiliated institutions, nor the Materials Research Society makes any warranty or representation, express or implied, including the warranties or merchantability and fitness for a particular purpose, or assumes any legal liability or responsibility for the accuracy, completeness, or usefulness of any information, apparatus, product, or process disclosed, or represents that its use would not infringe privately owned rights.

CONTENTS

ELEMENTS

Compiled by

M. Nastasi

Los Alamos National Laboratory, Los Alamos, New Mexico, USA

J. C. Barbour

Sandia National Laboratories, Albuquerque, New Mexico, USA

J. R. Tesmer

Los Alamos National Laboratory, Los Alamos, New Mexico, USA

Table A1 lists physical properties of the elements from hydrogen (Z = 1) through americium (Z = 95), including the mass and relative abundance of each isotope. Radioactive, unstable isotopes of atomic number 20 or less that can be produced by nuclear reactions with analysis beams are indicated by asterisks. Data were obtained from Walker, F.W., Parrington, J.R., and Feiner, F. (eds.) (1989), *Nuclides and Isotopes*, 14th Ed., General Electric Company, San Jose, CA.

Table A1. The elements.

Element	Atomic number (Z)	Isotopic mass (amu)	Relative abundance	Atomic weight (amu)	Atomic density (atoms/cm³)	Mass density (grams/cm³)
H	1	1.0078 2.0141 3.0160*	99.985 0.015	1.00794		
He	2	3.0160 4.0026	0.0001 99.9999	4.002602		
Li	3	6.0151 7.0160	7.5 92.5	6.941	4.60E+22	0.53
Be	4	7.0169* 9.0122 10.0135*	100	9.01282	1.24E+23	1.85
B	5	10.0129 11.0093	19.9 80.1	10.811	1.30E+23	2.34
C	6	12.0000 13.0034 14.0032*	98.9 1.1	12.011	1.31E+23	2.62
N	7	14.0031 15.0001	99.63 0.37	14.00674	5.38E+22	1.251

Table A1. The elements (continued).

Element	Atomic number (Z)	Isotopic mass (amu)	Relative abundance	Atomic weight (amu)	Atomic density (atoms/cm³)	Mass density (grams/cm³)
O	8	15.9949	99.762	15.9994	5.38E+22	1.429
		16.9991	0.038			
		17.9992	0.2			
F	9	18.9984	100	18.9984032	5.38E+22	1.696
Ne	10	19.9924	90.51	20.1797	2.69E+22	0.901
		20.9938	0.27			
		21.9914	9.22			
Na	11	21.9944*				
		22.9898	100	22.989768	2.54E+22	0.97
Mg	12	23.9850	78.99	24.305	4.31E+22	1.74
		24.9858	10.			
		25.9826	11.01			
Al	13	25.9869*				
		26.9815	100	26.981539	6.03E+22	2.70
Si	14	27.9769	92.23	28.0855	5.00E+22	2.33
		28.9765	4.67			
		29.9738	3.1			
		31.9741*				
P	15	30.9738	100	30.973762	3.54E+22	1.82
		31.9739*				
		32.9717*				
S	16	31.9721	95.02	32.066	3.89E+22	2.07
		32.9715	0.75			
		33.9679	4.21			
		34.9690*				
		35.9671	0.02			
Cl	17	34.9689	75.77	35.4527	5.38E+22	3.17
		35.9683*				
		36.9659	24.23			
Ar	18	35.9675	0.337	39.948	2.69E+22	1.784
		36.9668*				
		37.9627	0.063			
		38.9643*				
		39.9624	99.6			
K	19	38.9637	93.2581	39.0983	1.32E+22	0.86
		39.9624	0.0117			
		40.9618	6.7302			
Ca	20	39.9626	96.941	40.078	2.33E+22	1.55
		40.9623*				
		41.9586	0.647			
		42.9588	0.135			
		43.9555	2.086			
		44.9562*				
		45.9537	0.004			
		46.9546*				
		47.9525	0.187			
Sc	21	44.9559	100	44.95591	4.02E+22	3.0

Table A1. The elements (continued).

Element	Atomic number (Z)	Isotopic mass (amu)	Relative abundance	Atomic weight (amu)	Atomic density (atoms/cm³)	Mass density (grams/cm³)
Ti	22	45.9526	8.0	47.88	5.66E+22	4.50
		46.9518	7.3			
		47.9479	73.8			
		48.9479	5.5			
		49.9448	5.4			
V	23	49.9472	0.25	50.9415	6.86E+22	5.8
		50.9440	99.75			
Cr	24	49.9461	4.35	51.9961	8.33E+22	7.19
		51.9405	83.79			
		52.9407	9.5			
		53.9389	2.36			
Mn	25	54.9380	100	54.93805	8.14E+22	7.43
Fe	26	53.9396	4.35	55.847	8.44E+22	7.83
		55.9349	83.79			
		56.9354	9.5			
		57.9333	2.36			
Co	27	58.9332	100	58.9332	9.09E+22	8.90
Ni	28	57.9353	68.27	58.69	9.13E+22	8.90
		59.9308	26.1			
		60.9311	1.13			
		61.9283	3.59			
		63.9280	0.91			
Cu	29	62.9296	69.17	63.546	8.49E+22	8.96
		64.9278	30.83			
Zn	30	63.9291	48.6	65.39	6.58E+22	7.14
		65.9260	27.9			
		66.9271	4.1			
		67.9248	18.8			
		69.9253	0.6			
Ga	31	68.9256	60.1	69.723	5.10E+22	5.91
		70.9247	39.9			
Ge	32	69.9243	20.5	72.61	4.41E+22	5.32
		71.9221	27.4			
		72.9235	7.8			
		73.9212	36.5			
		75.9214	7.8			
As	33	74.9216	100	74.92159	4.60E+22	5.72
Se	34	73.9225	0.9	78.96	3.66E+22	4.80
		75.9192	9.			
		76.9199	7.6			
		77.9173	23.5			
		79.9165	49.6			
		81.9167	9.4			
Br	35	78.9183	50.69	79.904	2.35E+22	3.12
		80.9163	49.31			

Table A1. The elements (continued).

Element	Atomic number (Z)	Isotopic mass (amu)	Relative abundance	Atomic weight (amu)	Atomic density (atoms/cm³)	Mass density (grams/cm³)
Kr	36	77.9204	0.35	83.8	2.69E+22	3.74
		79.9164	2.25			
		81.9135	11.6			
		82.9141	11.5			
		83.9115	57.			
		85.9106	17.3			
Rb	37	84.9118	72.17	85.4678	1.08E+22	1.53
		86.9092	27.83			
Sr	38	83.9134	0.56	87.62	1.79E+22	2.6
		85.9093	9.86			
		86.9089	7.00			
		87.9056	82.58			
Y	39	88.9059	100	88.90585	3.05E+22	4.5
Zr	40	89.9047	51.45	91.224	4.28E+22	6.49
		90.9056	11.27			
		91.9050	17.17			
		93.9063	17.33			
		95.9083	2.78			
Nb	41	92.9064	100	92.90635	5.54E+22	8.55
Mo	42	91.9068	14.84	95.94	6.40E+22	10.2
		93.9051	9.25			
		94.9058	15.92			
		95.9047	16.68			
		96.9060	9.55			
		97.9054	24.13			
		99.9075	9.63			
Tc	43	98	100			11.5
Ru	44	95.9076	5.52	101.07	7.27E+22	12.2
		97.9053	1.88			
		98.5059	12.7			
		99.9042	12.6			
		100.9056	17.0			
		101.9043	31.6			
		103.9054	18.7			
Rh	45	102.9055	100	102.9055	7.26E+22	12.4
Pd	46	101.9056	1.02	106.42	6.79E+22	12.0
		103.9040	11.14			
		104.9051	22.33			
		105.9035	27.33			
		107.9039	26.46			
		109.9052	11.72			
Ag	47	106.9051	51.84	107.8682	5.86E+22	10.5
		108.9048	48.16			

Table A1. The elements (continued).

Element	Atomic number (Z)	Isotopic mass (amu)	Relative abundance	Atomic weight (amu)	Atomic density (atoms/cm³)	Mass density (grams/cm³)
Cd	48	105.9065	1.25	112.411	4.63E+22	8.65
		107.9042	0.89			
		109.9030	12.49			
		110.9042	12.80			
		111.9028	24.13			
		112.9044	12.22			
		113.9034	28.73			
		115.9048	7.49			
In	49	112.9041	4.3	114.82	3.83E+22	7.31
		114.9039	95.7			
Sn	50	111.9048	1.			
		113.9028	0.7	118.71	3.70E+22	7.30
		114.9033	0.4			
		115.9017	14.7			
		116.9030	7.7			
		117.6016	24.3			
		118.9033	8.6			
		119.9022	32.4			
		121.9034	4.6			
		123.9053	5.6			
Sb	51	120.9038	57.3	121.75	3.30E+22	6.68
		122.9042	42.7			
Te	52	119.9041	0.096	127.6	2.94E+22	6.24
		121.9031	2.6			
		122.9043	0.908			
		123.9028	4.816			
		124.9044	7.14			
		125.9033	18.95			
		127.9045	31.69			
		129.9062	33.8			
I	53	126.9045	100	126.90447	2.33E+22	4.92
Xe	54	123.9061	0.10	131.29	2.70E+22	5.89
		125.9043	0.09			
		127.9035	1.91			
		128.9048	26.40			
		129.9035	4.10			
		130.9051	21.20			
		131.9041	26.90			
		133.9054	10.40			
		135.9072	8.90			
Cs	55	132.9054	100	132.90543	8.47E+21	1.87
Ba	56	129.9063	0.106	137.327	1.53E+22	3.5
		131.9050	0.101			
		133.9045	2.417			
		134.9057	6.592			
		135.9046	7.854			
		136.9058	11.230			
		137.9052	71.700			
La	57	137.9071	0.09	138.9055	2.90E+22	6.7
		138.9063	99.91			

Table A1. The elements (continued).

Element	Atomic number (Z)	Isotopic mass (amu)	Relative abundance	Atomic weight (amu)	Atomic density (atoms/cm³)	Mass density (grams/cm³)
Ce	58	135.9071 137.9060 139.9054 141.9092	0.19 0.25 88.48 11.08	140.115	2.91E+22	6.78
Pr	59	140.9077	100	140.90765	2.89E+22	6.77
Nd	60	141.9077 142.9098 143.9101 144.9126 145.9131 147.9169 149.9209	27.13 12.18 23.80 8.30 17.19 5.76 5.64	144.24	2.92E+22	7.00
Pm	61	145	100	(145)	2.69E+22	6.475
Sm	62	143.9120 146.9149 147.9148 148.9172 149.9173 151.9197 153.9222	3.1 15.0 11.3 13.8 7.4 26.7 22.7	150.36	3.02E+22	7.54
Eu	63	150.9198 152.9212	47.8 52.2	151.965	2.08E+22	5.26
Gd	64	151.9198 153.9209 154.9226 155.9221 156.9240 157.9241 159.9271	0.2 2.18 14.8 20.47 15.65 24.84 21.86	157.25	3.02E+22	7.89
Tb	65	158.9253	100	158.92534	3.13E+22	8.27
Dy	66	155.9243 157.9244 159.9252 160.9269 161.9268 162.9287 163.9292	0.06 0.1 2.34 18.9 25.5 24.9 28.2	162.5	3.16E+22	8.53
Ho	67	164.9303	100	164.93032	3.21E+22	8.80
Er	68	161.9288 163.9292 165.9303 166.9320 167.9324 169.9355	0.14 1.61 33.6 22.95 26.8 14.9	167.26	3.26E+22	9.05
Tm	69	168.9342	100	168.93421	3.33E+22	9.33

Table A1. The elements (continued).

Element	Atomic number (Z)	Isotopic mass (amu)	Relative abundance	Atomic weight (amu)	Atomic density (atoms/cm^3)	Mass density (grams/cm^3)
Yb	70	167.9339	0.13	173.04	2.43E+22	6.98
		169.9348	3.05			
		170.9363	14.30			
		171.9364	21.90			
		172.9382	16.12			
		173.9389	31.80			
		175.9426	12.70			
Lu	71	174.9408	97.4	174.967	3.39E+22	9.84
		175.9427	2.6			
Hf	72	173.9400	0.16	178.49	4.42E+22	13.1
		175.9414	5.20			
		176.9432	18.60			
		177.9437	27.10			
		178.9458	13.74			
		179.9465	35.20			
Ta	73	179.9475	0.012	180.9479	5.52E+22	16.6
		180.9480	99.988			
W	74	179.9467	0.13	183.85	6.32E+22	19.3
		181.9482	26.30			
		182.9502	14.30			
		183.9509	30.67			
		185.9544	28.60			
Re	75	184.9530	37.4	186.207	6.79E+22	21.0
		186.9557	62.6			
Os	76	183.9525	0.02	190.2	7.09E+22	22.4
		185.9538	1.58			
		186.9557	1.6			
		187.9558	13.3			
		188.9581	16.1			
		189.9584	26.4			
		191.9615	41.0			
Ir	77	190.9606	37.3	192.22	7.05E+22	22.5
		192.9629	62.7			
Pt	78	189.9599	0.01	195.08	6.61E+22	21.4
		191.9610	0.79			
		193.9627	32.9			
		194.9650	33.8			
		195.9650	25.3			
		197.9679	7.2			
Au	79	196.9666	100	196.96654	5.90E+22	19.3
Hg	80	195.9658	0.15	200.59	4.06E+22	13.53
		197.9668	10.1			
		198.9683	17.			
		199.9683	23.1			
		200.9703	13.2			
		201.9706	29.65			
		203.9735	6.8			
Tl	81	202.9723	29.524	204.3833	3.49E+22	11.85
		204.9744	70.476			

Table A1. The elements (continued).

Element	Atomic number (Z)	Isotopic mass (amu)	Relative abundance	Atomic weight (amu)	Atomic density (atoms/cm^3)	Mass density (grams/cm^3)
Pb	82	203.9730 205.9745 206.9759 207.9767	1.4 24.1 22.1 52.4	207.2	3.31E+22	11.4
Bi	83	208.9804	100	208.98037	2.82E+22	9.8
Po	84	208.9824	100	(209)	2.71E+22	9.4
At	85	210	100	(210)	0.00E+00	
Rn	86	222.0176	100	(222)	2.69E+22	9.91
Fr	87	223	100	(223)	0.00E+00	
Ra	88	226.0254	100	226.0254	1.33E+22	5.
Ac	89	227	100	227.0278	2.67E+22	10.07
Th	90	232.0381	100	232.0381	3.04E+22	11.7
Pa	91	231.0359	100	231.0359	4.01E+22	15.4
U	92	234.0410 235.0439 238.0400	0.0055 0.7200 99.2745	238.0289	4.78E+22	18.9
Np	93	237.0482	100	237.0482	5.18E+22	20.4
Pu	94	244.0642	100	(244)	4.89E+22	19.8
Am	95	243.0614	100	(243)	3.37E+22	13.6

*Radioactive, unstable istopes of atomic number 20 or less which may be produced by nuclear reactions with analysis beams.

Data from *Nuclides and Isotopes*, revised 1989 by F.W. Walker, J.R. Parrington, and F. Feiner, published by the General Electric Company.

PHYSICAL CONSTANTS, CONVERSIONS, AND USEFUL COMBINATIONS

Compiled by

J. R. Tesmer and Y. Q. Wang

Los Alamos National Laboratory, Los Alamos, New Mexico, USA

Physical Constants

Avogadro constant	$N_A = 6.022 \times 10^{23}$	atoms (molecules)/mol
Boltzmann constant	$k = 8.617 \times 10^{-5}$	eV/K
Gas constant	$R = 8.314$	J/(mol K)
Fundamental charge	$e = 1.602 \times 10^{-19}$	C
Gravitational constant	$G = 6.673 \times 10^{-11}$	N m^2/kg^2
Magnetic constant	$\mu_0 = 4\pi \times 10^{-7}$	N/A^2
Rest mass		
of electron	$M_e = 9.109 \times 10^{-31}$	kg
	$= 0.511$	MeV/c^2
of proton	$M_p = 1.673 \times 10^{-27}$	kg
	$= 938.3$	MeV/c^2
of neutron	$M_n = 1.675 \times 10^{-27}$	kg
	$= 939.6$	MeV/c^2
of deuteron	$M_d = 3.344 \times 10^{-27}$	kg
	$= 1.876 \times 10^3$	MeV/c^2
of alpha	$M_\alpha = 6.645 \times 10^{-27}$	kg
	$= 3.727 \times 10^3$	MeV/c^2
Unified mass unit	$M_u = 1.661 \times 10^{-27}$	kg
	$= 931.5$	MeV/c^2
Speed of light in a vacuum	$c = 2.998 \times 10^8$	m/s
Planck constant	$h = 6.626 \times 10^{-34}$	J s
	$= 4.136 \times 10^{-21}$	MeV s
h/2π	$\hbar = 6.582 \times 10^{-16}$	eV s

Combinations

Bohr magneton	$m_B = e\hbar/2m_e = 9.274 \times 10^{-24}$	J/T
Bohr radius	$a_0 = \hbar^2/2m_e^2 = 5.292 \times 10^{-2}$	nm
Bohr velocity	$v_0 = e^2/\hbar = 2.188 \times 10^6$	m/s
Classical electron radius	$r_e = e^2/mc^2 = 2.818 \times 10^{-6}$	nm
Compton wavelength	$\lambda_c = h/2e = 2.426 \times 10^{-12}$	m
Coulomb constant	$k = 1/(4\pi\varepsilon_0) = 8.988 \times 10^9$	N m^2/C^2
Electronic charge	$e^2 = 1.440$	eV nm
	$= 1.440 \times 10^{-13}$	MeV cm
Fine structure constant	$\alpha = e^2/\hbar c = 7.297 \times 10^{-3}$	
Hydrogen binding energy	$e^2/2a_0 = 13.606$	eV

Conversions

1 Å	$= 10^{-8}$ cm	1 T	$= 10^4$ G	
	$= 10^{-1}$ nm	1 Ci	$= 3.700 \times 10^{10}$ Bq	
1 μm	$= 10^{-4}$ cm		$= 3.700 \times 10^{10}$ decays/s	
	$= 1 \times 10^4$ Å	1 R	$= 2.58 \times 10^{-4}$ C/kg (air)	
1 barn	$= 1 \times 10^{-24}$ cm^2	1 rad	$= 1$ cGy	
1 atm	$= 101.3$ kPa		$= 1 \times 10^{-2}$ Gy	
	$= 1.013$ bar	1 rem	$= 1$ cSv	
	$= 760$ mmHg		$= 1 \times 10^{-2}$ Sv	
	$= 14.7$ psi	1 mrem	$= 10$ μSv	
1 Torr	$= 1$ mmHg			
	$= 1.333$ mbar	Speed		
1 year	$= 365.26$ days	of 1 MeV alpha	$= 6.94 \times 10^6$ m/s	
	$= 3.156 \times 10^7$ s	of 1 MeV electron	$= 2.82 \times 10^8$ m/s	
1 Coulomb	$= 6.242 \times 10^{18}$ electrons	of 1 MeVproton	$= 1.38 \times 10^7$ m/s	
1 eV	$= 1.602 \times 10^{-19}$ J	of 1 MeV deuteron	$= 9.78 \times 10^6$ m/s	
1 eV/particle	$= 23.06$ kcal/mol			
	$= 96.53$ kJ/mol			

APPENDIX
3
ENERGY-LOSS DATA

Compiled by

E. Rauhala

University of Helsinki, Helsinki, Finland

J.F. Ziegler

United States Naval Academy, Annapolis, Maryland, USA

CONTENTS

A3.1 COMPUTER PROGRAMS FOR STOPPING CALCULATIONS

The following summary of computer programs for stopping calculations was current as of July 2009.

ASTAR (M.J. Berger, J.S. Coursey, M.A. Zucker and J. Chang): http://www.exphys.uni-linz.ac.at/stopping/
ATIMA (H. Geissel, C. Scheidenberger, P. Malzacher, J. Kunzendorf, H. Weick): http://www-linux.gsi.de/~weick/atima/
CASP (P.L. Grande and G. Schiwietz): http://www.hmi.de/people/schiwietz/casp.html

GEANT4 (Geant4 collaboration): http://geant4.web.cern.ch/geant4/

MSTAR (H.Paul and A.Schinner): http://www.exphys.uni-linz.ac.at/stopping/
PSTAR (M.J. Berger, J.S. Coursey, M.A. Zucker and J. Chang): http://www.exphys.uni-linz.ac.at/stopping/
SRIM (J.F. Ziegler): http://www.srim.org/

Please note that there are terms and conditions for using each of the programs; see their respective Web sites for details.

A3.1.1 ASTAR and PSTAR

The databases ASTAR and PSTAR calculate stopping-power and range tables for helium ions and protons, respectively, according to methods described in ICRU Report 49 (ICRU, 1993). Stopping-power and range tables can be calculated for helium ions and protons in 74 materials. ASTAR and PSTAR include a Web interface for calculations and stopping plots. In the program package, the program ESTAR for electron stopping and range is also included.

The ASTAR program calculates stopping-power and range tables for helium ions in the energy range from 0.001 MeV to 1000 MeV.

The PSTAR program calculates stopping-power and range tables for protons in various materials in the energy range from 0.001 MeV to 10,000 MeV.

A3.1.2 ATIMA

ATIMA is a program developed at GSI (Gesellschaft für Schwerionenforschung, Darmstadt, Germany) that calculates various physical quantities characterizing the slowing-down of protons and heavy ions in matter for specific kinetic energies ranging from 1 keV/u to 450 GeV/u; the calculated quantities include stopping power, energy loss, energy-loss straggling, angular straggling, range, range straggling, and beam parameters (such as magnetic rigidity, time of flight, velocity). ATIMA includes a Web interface for calculations.

A3.1.3 CASP

In CasP, fast numerical calculations of the mean electronic energy transfer Q_e (due to excitation and ionization of target atoms) are performed for each individual impact parameter b in a collision. The total electronic energy-loss cross-section S_e (equivalent to the stopping power) is subsequently calculated from $Q_e(b)$. The computation of Q_e and S_e accounts for a selected predefined projectile-screening function. By selecting a proper screening function, it is possible to treat nonequilibrium energy-loss phenomena. Furthermore, the unitary convolution approximation (UCA; default selection) is a nonlinear theory, as it includes the Bloch terms. Therefore, even for bare projectiles, the results do not scale with the square of the projectile charge (as most other quantum theories do).

A3.1.4 GEANT

Geant4 is a toolkit for the simulation of the passage of particles through matter. Its areas of application include high-energy, nuclear, and accelerator physics, as well as studies in medical and space science. The two main reference works for Geant4 are those by the GEANT4 collaboration (2003) and Allison et al. (2006).

A3.1.5 MSTAR

MSTAR is a program that calculates electronic stopping powers for heavy ions (^{3}Li to ^{18}Ar). It is mainly based on the alpha stopping powers contained in ICRU Report 49 (ICRU, 1993). It can calculate the stopping of all of the elements, mixtures, and compounds contained in that report (and also of B, Ni, Zr, Gd, and Ta), for ions from ^{3}Li to ^{18}Ar (Paul and Schinner, 2001, 2002, 2003).

A3.1.6 SRIM

SRIM is a large Web site providing information on particle interactions with matter. It contains a collection of software packages that calculate many features of the transport of ions in matter. Typical applications include th following:
- Ion stopping and range in targets: Most aspects of the energy loss of ions in matter are calculated in SRIM, The Stopping and Range of Ions in Matter. SRIM includes quick calculations that produce tables of stopping powers, ranges, and straggling distributions for any ion at any energy in any elemental target. More elaborate calculations include targets with complex multilayer configurations.
- Ion implantation: Ion beams are used to modify samples by injecting atoms to change the target's chemical and electronic properties. The ion beam also causes damage to solid targets by atom displacement. Most of the kinetic effects associated with the physics of these types of interactions are found in the SRIM package.
- Sputtering: The ion beam can knock out target atoms, a process called ion sputtering. The calculation of sputtering,

by any ion at any energy, is included in the SRIM package.

Ion transmission: Ion beams can be followed through mixed gas/solid target layers, such as are found in ionization chambers or in energy degrader blocks used to reduce ion beam energies.

Ion beam therapy: Ion beams are widely used in medical therapy, especially in radiation oncology. Typical applications are included in SRIM.

A3.3 DESCRIPTION AND APPLICATION OF SR MODULE.EXE

The SRIM module is a program in the downloadable SRIM package. The following description is from the "Help-SR Module.rtf" file.

"SR Module.exe" is a program that can run under the control of other programs that needs stopping powers and ranges. When it is executed, it reads a control file and then outputs a new file containing stopping powers and ranges. This output can then be used by the other programs. Because it has no user interface, SR Module.exe is invisible to the user.

SR Module.exe uses the same controls as the "Tables of Stopping and Ranges" that can be activated from the SRIM Title Page. When "Tables of Stopping and Ranges" is run, it creates a file called "SR.IN" in the SRIM directory. It then calculates the stopping powers and ranges and places them in a separate file in the directory "/SRIM Outputs".

The SRIM Web site also includes reviews of stopping theory and history. In the SRIM database, more than 25,000 data points from 2416 publications are included. Plots of stopping powers can be found at http://www.srim.org/ SRIM/SRIMPICS/STOPPLOTS.htm.

Printed references to the SRIM package are Ziegler (2004), Ziegler *et al.* (1985), and Ziegler *et al.* (2008), as well as the references therein.

A3.2 COMPARISON OF STOPPING TABLES AND PROGRAMS

Table A3.1. Tables and computer programs for stopping power. From ICRU Report 73 (ICRU, 2005). (Grande and Schwietz 2000, 2001 refer to CasP versions 1.1 and 1.2, respectively, available from http://www.hmi.de/people/schiwietz/casp.html451, Paul 2003 refers to http://www.exphys.uni-linz.ac.at/stopping/ and Ziegler 2003 refers to SRIM version 2003, http://www.srim.org/).

Reference	Type[a] name	(E/A_1)/MeV	Z_1	Targets	Ranges?	Remarks
Benton and Henke (1969)	F, P	0.1–1200	Any ion	Any solid	Yes	Stopping power obtained as derivative of range
Brice (1972)	F	0.1–2.8	16	Gases	No	3-parameter formula
CERN (2001)	P Geant4	Wide range	All particles	All targets		See Tables 5.2–5.3
Diwan *et al.* (2001)	F	0.5–2.5	3–35	16 solids	No	Adaptation of Hubert formula to lower energy
Grande and Schiwietz (2000, 2001)	P CasP	0.001–200	≥ 1	$1 \leq Z_2 \leq 92$	No	See Table 5.3
Hiraoka and Bichsel (2000)	T	1–1000	6, 7, 10, 14, 18	72 materials	Yes	See Table 5.3
Hubert *et al.* (1980)	F, T	2.5–100	2–103	18 solid elements	Yes	
Hubert *et al.* (1989, 1990)	F, T	2.5–500	2–103	36 solid elements	Yes	See Tables 5.2–5.7
ICRU 49	T, P PSTAR, ASTAR	0.001–10000, 0.00025–250	1, 2	25 elements, 48 mixtures and compounds	Yes	
Konac *et al.* (1998)	F	0.01–100	1–100	C, Si	No	See Table 5.3
Mukherji and Nayak (1979)	F	≤ 10	6–18	Various		
Northcliffe and Schilling (1970)	T	0.0125–12	1–103	24 materials	Yes	See Tables 5.2–5.6
Paul (2003)	P MSTAR	0.001–250	3–18	Same as ICRU 49 + Ni	No	See Tables 5.2–5.5
Sigmund and Schinner (2002)	P PASS	>0.025 (recommended)	3–92 (recommended)	Potentially all elements and compounds	No	See Tables 5.2–5.7
Steward (1968)	T		6, 10, 18, 92	Al, U, H$_2$O	Yes	
	C	0.01–500	1, 2, 6, 10, 18, 54, 86	H$_2$O, Al, Cu, Ag, Pb, U		
	P		all	all	Yes	
Xia and Tan (1986)	F	0.00025–0.2	3–82 (many)	Solids	No	Relative to p stopping by Andersen and Ziegler (1977)
Ziegler (1980)	T	0.2–1000	1–92	All elements	Yes	
Ziegler (2003)	P TRIM, SRIM	Energy = 1 eV–2 GeV	1–92	All elements, many other targets	Yes	See Tables 5.2–5.6

[a]F = formula; T = table; P = program; C = curve.

15

To use SR Module.exe, first look in its subdirectory "/SR Module". You will find the program "SR Module.exe", along with the data files: SCOEF03.dat, SNUC03.dat, and VERSION. The first two files are data files for the stopping and range program, and the third contains the current version of SRIM.

1   ---Stopping/Range Input Data (Number-format: Period = Decimal Point)
2   ---Output File Name
3   "Hydrogen in Water_Liquid"
4   ---Ion(Z), Ion Mass(u)
5   1 1.008
6   ---Target Data: (Solid=0,Gas=1), Density(g/cm3), Compound Corr.
7   0 1 1
8   ---Number of Target Elements
9   2
10   ---Target Elements: (Z), Target name, Stoich, Target Mass(u)
11   1 "Hydrogen" 2 1.008
12   8 "Oxygen" 1 15.999
13   ---Output Stopping Units (1-8)
14   5
15   ---Ion Energy : E-Min(keV), E-Max(keV)
16   10 10000

SR Module.exe reads this file and generates a file called "Hydrogen in Water_Liquid" containing the stopping and range of H ions into water from 10 keV to 10 MeV.

A3.4. TABLES OF STATISTICAL COMPARISONS OF DATA AND PROGRAM CALCULATIONS

The tables in this section list quantitative comparisons between experimental data and program calculations reported by by Paul and coworkers. The quantities included are mean normalized differences $\Delta = \langle\delta\rangle = (\langle S_{exp}/S_{fit}\rangle - 1) \times 100$ between data and program fits (table values) and standard deviations $\sigma = (\langle\delta^2\rangle - \langle\delta\rangle^2)^{1/2} \times 100$ for protons, alpha particles, and heavy ions in elements and compounds (from Paul and Schinner, 2005, 2006);Paul, 2006,. Note also the following additional references: NS = Northcliffe and Schilling (1970), AZ = Andersen and Ziegler (1977), J = Janni (1982), and ZBL = Ziegler et al. (1985).

Table A3.2a. (From Paul and Schinner, 2005).
Mean normalized difference $\Delta \pm \sigma$ for H ions in Al

E/A_1 (MeV)	0.001–0.01	0.01–0.1	0.1–1.0	1–10	10–100	0.001–100
No. of points	44	263	284	171	31	793
NS 70 [7]		17 ± 10.9	−4.2 ± 7.3	−0.6 ± 2.6	−1.0 ± 0.8	4.3 ± 13
AZ 77 [23]	3.9 ± 10.6	1.1 ± 8.1	−1.1 ± 6.6	−1.0 ± 2.4	−0.5 ± 0.5	0.0 ± 6.8
J 82 [8]	18.5 ± 9.2	6.9 ± 7.4	1.9 ± 6.4	−0.7 ± 2.3	−0.5 ± 0.5	3.8 ± 7.8
ZBL 85 [24]	7.0 ± 10.2	1.6 ± 7.8	−1.1 ± 6.5	−0.3 ± 2.5	2.4 ± 1.2	0.6 ± 6.8
ICRU [11]	3.9 ± 10.6	1.1 ± 8.1	−0.7 ± 6.6	0.0 ± 2.3	0.4 ± 0.5	0.4 ± 6.8
PASS [12]	−7.1 ± 12.2	−8.7 ± 8.9	−9.6 ± 8.0	1.1 ± 3.2	3.5 ± 0.7	−6.4 ± 9.0
CasP [16–19]		2.8 ± 8.8	4.3 ± 6.4	−0.1 ± 3.1	0.4 ± 1.2	−2.5 ± 6.9
SRIM 2003 [20]	5.9 ± 10.2	2.4 ± 7.9	−0.6 ± 6.5	−1.1 ± 2.4	−0.1 ± 0.5	0.7 ± 6.8

Table A3.2b. (From Paul and Schinner, 2005).

Mean normalized difference $\Delta \pm \sigma$ for H ions in 55 solid elements

E/A_1 (MeV)	0.001–0.01	0.01–0.1	0.1–1.0	1–10	10–100	0.001–100
No. of points	272	1847	3516	1661	259	7555
AZ 77 [23]	7.6 ± 14	0.1 ± 14	−3.2 ± 8.5	−0.4 ± 5.5	−0.7 ± 0.6	−1.3 ± 9.9
J 82 [8]	13 ± 13	1.7 ± 13	−1.5 ± 7.7	−0.2 ± 5.1	−0.2 ± 0.8	0.1 ± 9.4
ZBL 85 [24]	−3.5 ± 24	−1.2 ± 13	−2.9 ± 8.1	0.2 ± 5.6	0.0 ± 2.1	−1.7 ± 10.2
SRIM 2003 [20]	5.3 ± 12	0.9 ± 11	−0.9 ± 7.0	0.3 ± 5.4	0.0 ± 0.6	0.0 ± 8.1

Table A3.2c. (From Paul and Schinner, 2006).

Mean normalized difference $\Delta \pm \sigma$ (%) for H and He ions in 23 compounds covered by ICRU 49

E/A_1 (MeV)	0–0.03	0.03–0.3	0.3–3.0	3–30	0–30
No. of points	116	1036	1237	135	2524
ICRU Rep. 49	0.2 ± 8.9	1.4 ± 5.9	1.3 ± 5.2	1.0 ± 4.4	1.3 ± 5.7
SRIM 2003_26	-7.8 ± 12	-1.0 ± 6.4	0.4 ± 5.6	-0.6 ± 4.0	-0.6 ± 6.6

Table A3.2d. (From Paul, 2006).

Mean normalized difference $\Delta \pm \sigma$ (in %) for ions from $_3$Li to $_{18}$Ar in the elemental solids covered by PASS

E/A_1 (MeV)	0.025–0.1	0.1–1	1–10	10–100	100–1000	0.025–1000
No. of points	1399	3452	1262	175	11	6299
MSTAR v.3, mode b	2.5 ± 9.9	0.1 ± 7.3	0.8 ± 5.5	0.1 ± 2.2	0.7 ± 1.4	0.8 ± 7.6
SRIM 2003.26	1.3 ± 9.7	-0.9 ± 7.0	-0.3 ± 5.6	-1.6 ± 2.9	-0.1 ± 1.6	-0.3 ± 7.4
PASS	-11.4 ± 20	-6.8 ± 12	-3.0 ± 6.6	-0.8 ± 3.0	-0.8 ± 1.9	-6.9 ± 13

Table A3.2e. (From Paul, 2006).

Mean normalized difference $\Delta \pm \sigma$ (in %) for ions from $_3$Li to $_{18}$Ar in aluminum oxide, kapton polyimide, polycarbonate (makrolon), polyethylene, polyethylene terephthalate (mylar), polypropylene, polyvinyl chloride, silicon dioxide, toluene and water (liquid)

E/A_1 (MeV)	0.025–0.1	0.1–1	1–10	10–100	0.025–100
No. of points	133	586	368	13	1100
MSTAR v.3, mode b	6.6 ± 10.4	1.6 ± 6.3	5.2 ± 4.0	0.0 ± 1.3	3.4 ± 6.6
SRIM 2003.26	-0.8 ± 8.3	-0.1 ± 5.2	-0.4 ± 5.0	-2.3 ± 1.7	-0.3 ± 5.6
PASS	-11 ± 12	-2.1 ± 7.4	-1.0 ± 5.1	-0.5 ± 1.4	-2.8 ± 8.1

A3.5 FIGURES AND TABLES OF HELIUM-ION ENERGY-LOSS DATA IN SILICON AND FITS TO DATA FROM SEMIEMPIRICAL CALCULATIONS

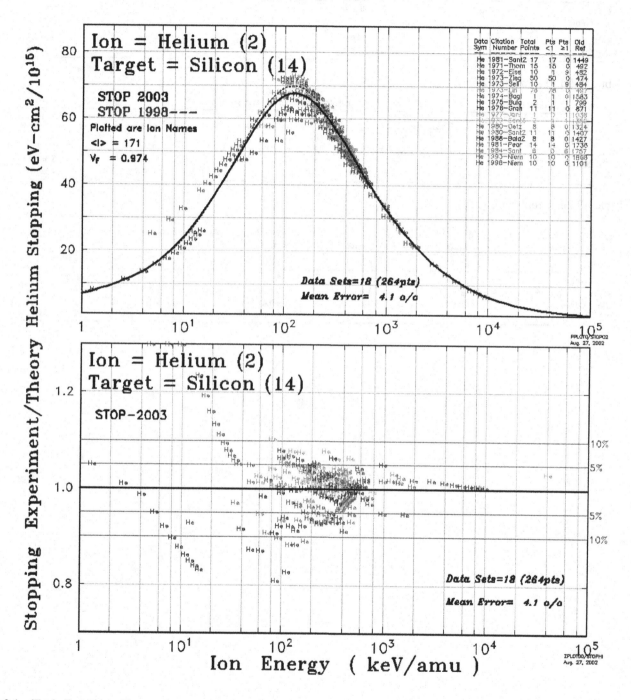

FIG. A3.1. (Top) Experimental stopping power data from 18 data sets and 264 data points for ^{4}He ions and (bottom) a best semiempirical fit in silicon and the deviation of data from the fit (from www.SRIM.org). (Multiply units of keV/amu by 4 for He energy in keV; for example, 10^3 keV/amu corresponds to 4000 keV He.)

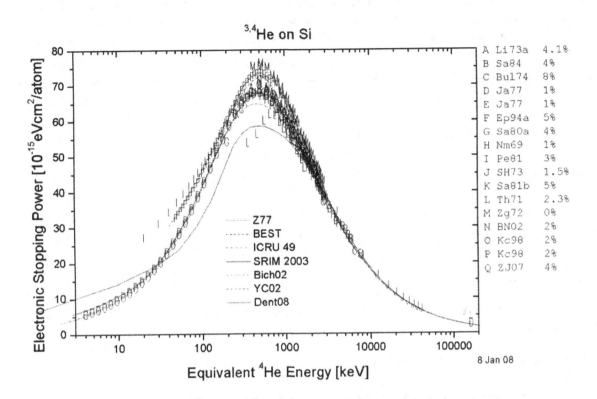

FIG. A3.2. Experimental and semi-empirical stopping powers for [3]He and [4]He ions in silicon. The curves correspond to different fitted semiempirical curves from the literature (from http://www.exphys.uni-linz.ac.at/stopping).

Table A3.3. Experimental surface Rutherford backscattering spectrometry (RBS) yield (H_0) for Si as a function of energy and calculations using different semiempirical fits (from Bianconi *et al.*, 2000. RUMP refers to http://www.genplot.com/, SRIM2000 to http://www.srim.org/ and KKKNS refers to Konac *et al.*, 1998a).

	H_0 counts/(μC msr keV)				
	1.0 MeV	1.5 MeV	2.0 MeV	2.5 MeV	3.0 Mev
Experiment	58.5	28.8	18.3	12.7	9.47
RUMP	57.2	27.7	17.1	12	9.06
SRIM2000	57.6	28.1	17.3	12.1	9.09
KKKNS	60.5	29.5	18.3	12.8	9.69

Table A3.4. Stopping cross sections from several recent publications compared with semiempirical fits (from Barradas *et al.*, 2002, with the stopping cross sections from ASTAR and SRIM-2008 added. "This work" refers to Barradas *et al.*, (2002), Lulli to Lulli *et al.*, (2000), Hoshino to Hoshino *et al.*, (2000), ZBL85 to Ziegler *et al.*, (1985), KKKNS to Konac *et al.*, (1998a)). The measurements of Barradas *et al.* (2002) agree with those of Bianconi *et al.* (2000) within about 3%, and the calculations also reproduce the data to within about 3%.

E_0 (MeV)	S_g (eV/10^{15} at/cm^2)									
	0.5	1	1.1	1.5	1.6	2	2.1	2.5	2.6	3
This work	70.9 ± 3.9	62.10 ± 41	60.43 ± 39	54.20 ± 34	52.75 ± 32	47.51 ± 37	46.32 ± 40	42.0 ± 38	41.07 ± 63	37.65 ± 78
This work[a]	68.2 ± 1.4	62.31 ± 31	60.69 ± 25	54.33 ± 15	52.81 ± 17	47.33 ± 24	46.08 ± 27	41.61 ± 35	40.62 ± 36	37.06 ± 44
Lulli et al. [19]	68.76	61.15	59.35	52.88	51.46	46.48	45.39	41.55	40.70	37.66
Hoshino et al. [43]	–	--	55.3 ± 1.2	–	50.3 ± 1.1	–	45.4 ± 1.0	–	41.5 ± 9	--
ZBL85 [23]	71.8	63.4	61.5	54.9	53.5	48.4	47.3	43.3	42.4	39.2
KKKNS [7]	67.41	59.95	58.19	51.84	50.45	45.57	44.5	40.74	39.90	36.92
ASTAR	66.2	60.4	59.0	53.5	52.3	47.8		43.1		39.3
SRIM-2008		61.2	59.5	53.2	51.8	46.9		42.0		38.1

A3.6 RATIO OF THE ENERGY LOSS FOR ^{4}HE IONS IN THE AXIAL DIRECTION TO THE ENERGY LOSS IN AMORPHOUS SILICON

Table A3.5. Ratio of the energy loss for ^{4}He ions in the axial direction to the energy loss in amorphous silicon. The values have been read from published graphs.

Energy (MeV)	Axis	Stopping ratio	Reference
0.96	⟨100⟩	0.78	Azevedo *et al.*(1998)
1.0	⟨100⟩	0.80	Shao *et al.*(2006)
1.0	⟨100⟩	0.84*	Bernardi *et al.*(2006)
1.0	⟨100⟩	0.88	Yamamoto *et al.*(1999)
1.0	⟨100⟩	0.89	dos Santos *et al.*(1995)
2.0	⟨100⟩	0.71	Shao *et al.*(2006)
2.0	⟨100⟩	0.90*	Greco *et al.*(2007)
2.0	⟨100⟩	0.85*	Bernardi *et al.*(2006)
2.0	⟨100⟩	0.74	Yamamoto *et al.*(1999)
2.0	⟨100⟩	0.64	Azevedo *et al.*(1998)
2.0	⟨100⟩	0.76	dos Santos *et al.*(1995)
1.0	⟨110⟩	0.54	Shao *et al.*(2006)
1.2	⟨110⟩	0.57	Azevedo *et al.* 2001)
1.5	⟨110⟩	0.53	Shao *et al.* (2006)
2.0	⟨110⟩	0.52	Shao *et al.* (2006)
2.0	⟨110⟩	0.48	Azevedo et. (2001)
1.0	⟨111⟩	0.84	Shao *et al.* (2006)
1.2	⟨111⟩	0.72	Azevedo *et al.* (2001)
1.5	⟨111⟩	0.71	Shao *et al.* (2006)

*Absolute stopping data given, ratio calculated assuming KKKNS stopping (Konac *et al.*, 1998, 19988a).

REFERENCES

Allison, J., Amako, K., Apostolakis, J., Araujo, H., Dubois, P.A., Asai, M., Barrand, G., Capra, R., Chauvie, S., Chytracek, R., Cirrone, G.A.P., Cooperman, G., Cosmo, G., Cuttone, G., Daquino, G.G., Donszelmann, M., Dressel, M., Folger, G., Foppiano, F., Generowicz, J., Grichine, V., Guatelli, S., Gumplinger, P., Heikkinen, A., Hrivnacova, I., Howard, A., Incerti, S., Ivanchenko, V., Johnson, T., Jones, F., Koi, T., Kokoulin, R., Kossov, M., Kurashige, H., Lara, V., Larsson, S., Lei, F., Link, O., Longo, F., Maire, M., Mantero, A., Mascialino, B., McLaren, I., Lorenzo, P.M., Minamimoto, K., Murakami, K., Nieminen, P., Pandola, L., Parlati, S., Peralta, L., Perl, J., Pfeiffer, A., Pia, M.G., Ribon, A., Rodrigues, P., Russo, G., Sadilov, S., Santin, G., Sasaki, T., Smith, D., Starkov, N., Tanaka, S., Tcherniaev, E., Tome, B., Trindade, A., Truscott, P., Urban, L., Verderi, M., Walkden, A., Wellisch, J.P., Williams, D.C., Wright, D., and Yoshida, H. (2006), *IEEE Trans. Nucl. Sci.* **53**, 270.

Andersen, H.H., and Ziegler, J.F. (1977), *The Stopping and Ranges of Ions in Matter, Vol. 3,*, Pergamon, New York.

Azevedo, G. de M., Behar, M., Dias, J.F., Grande, P.L., dos Santos, J.H.R., Stoll, R., Klatt, Chr., and Kalbitzer, S. (1998), *Nucl. Instrum. Methods* **B136–138**, 132.

Azevedo, G. de M., Dias, J.F., Behar, M., Grande, P.L., and dos Santos, J.H.R., (2001), *Nucl. Instrum. Methods* **B174**, 407.

Barradas, N.P., Jeynes, C., Webb, R.P., and Wendler, E. (2002), *Nucl. Instrum. Methods* **B194**, 15.

Benton, E.V., and Henke, R.P. ,(1969), *Nucl. Instrum. Methods* **67**, 87.

Bernardi, F., Araújo, L.L., Behar, M., and Dias, J.F. (2006), *Nucl. Instrum. Methods* **B249**, 69.

Bianconi, M., Abel, F., Banks, J.C., Climent Font, A., Cohen, C., Doyle, B.L., Lotti, R., Lulli, G., Nipoti, R., Vickridge, I., Walsh, D., and Wendler, E. (2000), *Nucl. Instrum. Methods* **B161–163**, 293.

Brice, D.K. (1972), *Phys. Rev.* **A6**, 1791.

Diwan, P.K., Scharma, A., and Kumar, S., (2001), *Nucl. Instrum. Methods* **B174**, 267.

dos Santos, J.H.R., Grande, P.L., Boudinov, H., Behar, M., Stoll, R., Klatt, Chr., Kalbitzer, S. (1995), *Nucl. Instrum. Methods* **B106**, 51.

Geant4 collaboration: Agostinelli, S., Allison, J., Amako, K., Apostolakis, J., Araújo, H., Arce, P., Asai, M., Axen, D., Banerjee, S., Barrand, G., Behner, F., Bellagamba, L., Boudreau, J., Broglia, L., Brunengo, A., Burkhardt, H., Chauvie, S., Chuma, J., Chytracek, R., Cooperman, G., Cosmo, G., Degtyarenko, P., Dell'Acqua, A., Depaola, G., Dietrich, D., Enami, R., Feliciello, A., Ferguson, C.,

Fesefeldt, H., Folger, G., Foppiano, F., Forti, A., Garelli, S., Giani, S., Giannitrapani, R., Gibin, D., Gómez-Cadenas, J.J., González, I., Gracía-Abríl, G., Greeniaus, L.G., Greiner, W., Grichine, V., Grossheim, A., Guatelli, S., Gumplinger, P., Hamatsu, R., Hashimoto, K., Hasui, H., Heikkinen, A., Howard, A., Ivanchenko, V., Johnson, A., Jones, F.W., Kallenbach, J., Kanaya, N., Kawabata, M., Kawabata, Y., Kawaguti, M., Kelner, S., Kent, P., Kimura, A., Kodama, T., Kokoulin, R., Kossov, M., Kurashige, H., Lamanna, E., Lampen, T., Lara, V., Lefébure, V., Lei, F., Liendl, M., Lockman, W., Longo, F., Magni, S., Maire, M., Medernach, E., Minamimoto, K., Mora de Freitas, P., Morita, Y., Murakami, K., Nagamatu, M., Nartallo, R., Nieminen, P., Nishimura, T., Ohtsubo, K., Okamura, M., O'Neale, S., O'Ohata, Y., Paech, K., Perl, J., Pfeiffer, A., Pia, M.G., Ranjard, F., Rybin, A., Sadilov, S., Di Salvo, E., Santin, G., Sasaki, T., Savvas, N., Sawada, Y., Scherer, S., Sei, S., Sirotenko, V., Smith, D., Starkov, N., Stöcker, H., Sulkimo, J., Takahata, M., Tanaka, S., Chernyaev, E., Safai-Tehrani, E., Tropeano, M., Truscott, P., Uno, H., Urbàn, L., Urban, P., Verderi, M., Walkden, A., Wander, W., Weber, H., Wellisch, J.P., Wenaus, T., Williams, D.C., Wright, D., Yamada, T., Yoshida, H., and Zschiesche, D. (2003), *Nucl. Instrum. Methods* **A506**, 250.

Greco, R., Luce, A., Wang, Y., and Shao, L. (2007), *Nucl. Instrum. Methods* **B261**, 538.

Hiraoka, T., and Bichsel, H. (2000), *Medical Standard Dose*, Society of Medical Standard Dose, Chiba, Japan, vol. 5 (Suppl. 1), p. 1.

Hoshino, Y., Okazawa, T., Nishii, T., Nishimura, T., Kido, Y., (2000) *Nucl. Instrum. Methods* **B171**, 409.

Hubert, F., Fleury, A., Bimbot, R., and Gardes, D. (1980), *Ann. Phys. (France)* **5S**, 3.

Hubert, F., Bimbot, R., and Gauvin, H. (1989), *Nucl. Instrum. Methods* **B36**, 357.

Hubert, F., Bimbot, R., and Gauvin, H. (1990), *At. Data Nucl. Data Tables* **46**, 1.

ICRU (1993), *Stopping Powers and Ranges for Protons and Alpha Particles*, ICRU Report 49 ICRU, Bethesda, MD.

ICRU (2005), *Stopping Powers of Ions Heavier than Helium*, ICRU Report 73, Oxford University Press, New York; *J. ICRU* **5** (1), 1.

Janni, J.F. (1982), *At. Data Nucl. Data Tables* **27**, 147.

Konac, G., Klatt, C., and Kalbizer, S. (1998), *Nucl. Instrum. Methods* **B146**, 106.

Konac, G., Kalbizer, S., Klatt, C., Niemann, D., and Stoll, R. (1998a), *Nucl. Instrum. Methods* **B138**, 159.

Lulli, G., Albertazzi, E., Bianconi, M., Bentini, G.G., Nipoti, R.,

Lotti, R. (2000), *Nucl. Instrum. Methods* **B170**, 1.

Mukherji, S., and Nayak, A. (1979), *Nucl. Instrum. Methods* **B159**, 421.

Northcliffe, L.C., and Schilling, R.F. (1970), *Nucl. Data Tables* **7**, 233.

Paul, H., and Schinner, A. (2001), *Nucl. Instrum. Methods* **B179**, 299.

Paul, H., and Schinner, A. (2002), *Nucl. Instrum. Methods* **B195**, 166.

Paul, H., and Schinner, A. (2003), *Nucl. Instrum. Methods* **B209**, 252.

Paul, H., and Schinner, A. (2005), *Nucl. Instrum. Methods* **B227**, 461.

Paul, H., and Schinner, A. (2006), *Nucl. Instrum. Methods* **B249**, 1.

Paul, H., (2006), *Nucl. Instrum. Methods* **B247**, 166.

Shao, L., Wang, Y.Q., Nastasi, M., and Mayer, J.W. (2006), *Nucl. Instrum. Methods* **B249**, 51.

Sigmund, P., and Schinner, A. (2002), *Nucl. Instrum. Methods* **B195**, 64.

Steward, P., (1968). PhD thesis, Univ, of California, Berkeley, uCRL-18127.

Xia, Y.Y., and Tan,C. (1986), *Nucl. Instrum. Methods* **B13**, 100.

Yamamoto, Y., Ikeda, A., Yoneda, T., Kajiyama, K., and Kido, Y. (1999), *Nucl. Instrum. Methods* **B153**, 10.

Ziegler, J.F. (1980), *Handbook of Stopping Cross-Sections for Energetic Ions in All Elements*, Pergamon, New York.

Ziegler, J.F., Biersack, J.P., and Littmark, U. (1985), *The Stopping and Range of Ions in Solids*, Pergamon, New York, vol. 1.

Ziegler, J.F. (2004), *Nucl. Instrum. Methods* **B219–220**, 102.

Ziegler, J.F., Biersack, J.P., and Ziegler, M.D. (2008), *The Stopping and Range of Ions in Matter*, LuLu Press, Morrisville, NC. See http://www.srim.org/SRIM%20Book.htm.

SCATTERING AND REACTION KINEMATICS

R. A. Weller

Vanderbilt University, Nashville, Tennessee, USA

CONTENTS

A4.1 INTRODUCTION

This appendix is a compendium of formulas for the analysis of particle scattering and reactions. The forms that have been chosen for the equations are those that are most useful for the analysis of backscattering, elastic recoil, and nuclear reaction experiments performed for materials analysis. The tables are organized in two sections. Table A4.1 contains nonrelativistic expressions for the analysis of elastic scattering data and should suffice for reducing data from most Rutherford backscattering and elastic recoil experiments. Table A4.2 again presents nonrelativistic equations, but these are for the analysis of nuclear reactions where the final particles are not necessarily the same as the initial ones or where processes such as Coulomb excitation of a nucleus significantly change the final kinetic energy of the system.

For high-precision work, relativistic kinematics should be used. A discussion of relativistic kinematics can be found in Hagedorn (1963).

A4.2 ELASTIC COLLISIONS

This section should be used for the analysis of nonrelativistic elastic collisions between a projectile and a target that is at rest in the laboratory. A good working definition of such collisions is that no particle's speed exceeds 10% of the speed of light, c (c = 3.0×10^8 m/s) and no nuclear reactions occur.

Figure A4.1 shows the scattering geometry and illustrates the various angles and energies. Table A4.1 contains the expressions for elastic scattering.

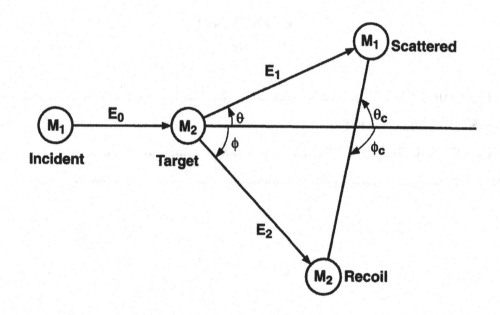

FIG. A4.1. Schematic scattering event as seen in the laboratory and center-of-mass coordinate systems, illustrating the angles, energies, and velocities for nonrelativistic elastic collisions.

E_0 = Energy of the incident projectile.
E = Total kinetic energy in the center-of-mass system.
M_1 = Mass of the incident projectile.
M_2 = Mass of the target particle.
θ = Laboratory angle of the scattered projectile.
θ_c = Center-of-mass angle of the scattered projectile.
ϕ = Laboratory angle of the recoiling target.
ϕ_c = Center-of-mass angle of the recoiling target.
E_1 = Laboratory energy of the scattered projectile.
K = Backscattering kinematic factor, E_1/E_0.
E_2 = Laboratory energy of the recoiling target.

x = Mass ratio M_1/M_2.
$\sigma(\theta)$ = Laboratory differential scattering cross section. $\sigma(\theta)\, d\Omega$ is the cross section to deflect the incident projectile by an angle θ into the solid angle $d\Omega$. $\sigma(\theta)$ is often symbolized by $d\sigma/d\Omega$.
$\sigma_c(\theta_c)$ = Differential scattering cross section in the center-of-mass system.

$\sigma_R(\phi)$ = Laboratory differential recoil cross section. Cross section per unit solid angle for the target nucleus to recoil at an angle ϕ with respect to the direction of the incident particle.

Table A4.1. Kinematic expressions for elastic scattering.

Quantity calculated	Expression	Equation number
Center-of-mass energy	$$E = \frac{M_2 E_0}{M_1 + M_2} = \frac{E_0}{1+x} \ .$$	(A4.1)
Laboratory energy of the scattered projectile for $M_1 \leq M_2$	$$K = \frac{E_1}{E_0} = \frac{\left\{ x\cos(\theta) + [1 - x^2 \sin^2(\theta)]^{1/2} \right\}^2}{(1+x)^2}$$ $$\text{when } M_1 = M_2, \theta \leq \frac{\pi}{2} \ .$$	(A4.2)
Laboratory energies of the scattered projectile for $M_1 > M_2$	$$\frac{E_1}{E_0} = \frac{\left\{ x\cos(\theta) \pm [1 - x^2 \sin^2(\theta)]^{1/2} \right\}^2}{(1+x)^2}$$ $$\text{with } \theta \leq \sin^{-1}(x^{-1}) \ .$$	(A4.3)
Laboratory energy of the recoil nucleus	$$\frac{E_2}{E_0} = 1 - \frac{E_1}{E_0} = \frac{4M_1 M_2}{(M_1 + M_2)^2} \cos^2(\phi)$$ $$= \frac{4x}{(1+x)^2} \cos^2(\phi) = \frac{4x}{(1+x)^2} \sin^2\left(\frac{\theta_c}{2}\right),$$ $$\text{where } \phi \leq \frac{\pi}{2} \ .$$	(A4.4)
Laboratory angle of the recoil nucleus	$$\phi = \frac{\pi - \theta_c}{2} = \frac{\phi_c}{2},$$ $$\sin(\phi) = \left(\frac{M_1 E_1}{M_2 E_2} \right)^{1/2} \sin(\theta) \ .$$	(A4.5)
Center-of-mass angle of the scattered projectile	$$\theta_c = \pi - 2\phi = \pi - \phi_c \ .$$ When $M_1 \leq M_2 \Rightarrow x \leq 1, \theta_c$ is defined for all $\theta \leq \pi$, and $$\theta_c = \theta + \sin^{-1}[x\sin(\theta)] \ .$$ When $M_1 > M_2 \Rightarrow x > 1, \theta_c$ is double-valued, and the laboratory scattering angle is limited to the range $\theta \leq \sin^{-1}(x^{-1})$. In this case, $$\theta_c = \theta + \sin^{-1}[x\sin(\theta)]$$ or $$\theta_c = \pi + \theta - \sin^{-1}[x\sin(\theta)] \ .$$	(A4.6)
Laboratory scattering cross section in terms of the center-of-mass cross section	$$\sigma(\theta) = \frac{\sigma_c(\theta_c)\sin^2(\theta_c)}{\sin^2(\theta)\cos(\theta_c - \theta)}$$ $$= \sigma_c[\theta_c(\theta)] \frac{\left\{ x\cos(\theta) + [1 - x^2 \sin^2(\theta)]^{1/2} \right\}^2}{[1 - x^2 \sin^2(\theta)]^{1/2}} \ .$$	(A4.7)
Laboratory recoil cross section	$$\sigma_R(\phi) = 4\sigma_c[\theta_c(\phi)]\cos(\phi) = 4\sigma_c(\pi - 2\phi)\cos(\phi) \ .$$	(A4.8)

EXAMPLE A4.1. Laboratory scattering cross sections.

Cross sections such as the Rutherford scattering cross section are ordinarily computed in the center-of-mass coordinate system where they are expressed as functions of E and θ_c, but are needed for the evaluation of experiments in the laboratory coordinate system where the relevant parameters are E_0, θ, and the particle masses. To make the transformation proceed as follows.

Given E_0, M_1, M_2, and θ:

1) Compute $x = M_1/M_2$.

2) Compute $E = E_0/(1 + x)$.

3) Compute the center-of-mass scattering angle

$$\theta_c = \theta + \sin^{-1}[x \sin(\theta)] \ .$$

4) Compute the cross section in the center-of-mass frame for the scattering event of interest. For the Rutherford scattering cross section

$$\sigma_c(\theta_c) = \frac{Z_1^2\, Z_2^2\, e^4}{16\, E^2 \sin^4(\theta_c/2)} \ . \qquad (A4.9)$$

Here Z_1 and Z_2 are the atomic numbers of the colliding atoms and e is the electron's charge (note, $e^2 = 1.44$ eV-nm).

5) Obtain the desired result

$$\sigma(\theta) = \frac{\sigma_c(\theta_c) \sin^2(\theta_c)}{\sin^2(\theta) \cos(\theta_c - \theta)} \ . \qquad (A4.10)$$

There is a relatively compact closed-form expression for the Rutherford cross section in the laboratory coordinate system (Chu *et al.*, 1978). It is

$$\sigma(\theta) = \left(\frac{Z_1 Z_2 e^2}{2E_0}\right)^2$$

$$\times \frac{\left\{\cos(\theta) + [1 - x^2 \sin^2(\theta)]^{1/2}\right\}^2}{\sin^4(\theta)\, [1 - x^2 \sin^2(\theta)]^{1/2}} \qquad (A4.11)$$

EXAMPLE A4.2. Laboratory recoil cross sections.

In elastic recoil experiments, one ordinarily measures the number of recoiling target atoms observed in a detector with fixed angle ϕ and a known solid angle. The cross section for these events is related to the center-of-mass, scattering cross section by Eq. (A4.8). The procedure for obtaining this cross section is analogous to that outlined in Example A4.1.

Given E_0, M_1, M_2, and ϕ:

1) Compute x.

2) Compute E.

3) Compute $\theta_c = \pi - 2\phi$.

4) Compute the center-of-mass scattering cross section [Note that $\sin(\theta_c/2) = \cos(\phi)$].

5) Obtain the desired result

$$\sigma_R(\phi) = 4\, \sigma_c(\theta_c) \cos(\phi) \ . \qquad (A4.12)$$

For Rutherford scattering, this result is especially simple

$$\sigma_R(\phi) = \left(\frac{Z_1 Z_2 e^2}{2E_0}\right)^2 \frac{(1 + x)^2}{\cos^3(\phi)} \ . \qquad (A4.13)$$

A4.3 INELASTIC COLLISIONS

The nuclear reactions that are most commonly used for materials analysis involve light ions. These reactions can usually be analyzed without invoking relativistic kinematics. This section, which is adapted from the compilation by Marion and Young (1968), contains nonrelativistic equations for the analysis of two kinds of scattering events:

- when there is a net energy change (a nonzero "Q" value) or

- when nucleons are transferred, resulting in reaction products with atomic masses that differ from those of the reactants (and where, in general, $Q \neq 0$ as well).

Table A4.2 contains expressions for the case where a particle of mass M_1 is incident upon another of mass M_2 that is at rest.

Figure A4.2 shows the scattering geometry and illustrates the various angles and energies.

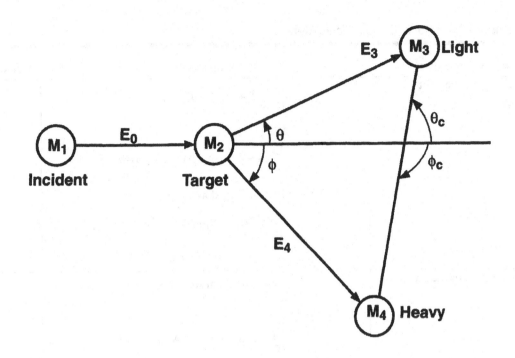

FIG. A4.2. Schematic scattering event as seen in the laboratory and center-of-mass coordinate systems, illustrating the angles, energies, and velocities for nonrelativistic inelastic collisions.

E_0 = Energy of the incident projectile.

M_1 = Mass of the incident projectile.

M_2 = Mass of the target particle.

M_3 = Mass of the lighter reaction product.

M_4 = Mass of the heavier reaction product.

Q = Energy released by the reaction = $(M_1 + M_2 - M_3 - M_4)c^2$. $Q < 0 \Rightarrow$ endothermic reaction.

θ = Laboratory angle of M_3.

θ_c = Center-of-mass angle of M_3.

ϕ = Laboratory angle of M_4.

ϕ_c = Center-of-mass angle of M_4.

E_3 = Laboratory energy of M_3.

E_{3c} = Center-of-mass energy of M_3.

E_4 = Laboratory energy of M_4.

E_{4c} = Center-of-mass energy of M_4.

$\sigma_3(\theta)$ = Laboratory differential scattering cross section for production of the light product.

$\sigma_{3c}(\theta_c)$ = Center-of-mass differential cross section for the production of the light product.

$\sigma_4(\phi)$ = Laboratory differential cross section for the production of the heavy product.

$\sigma_{4c}(\phi_c)$ = Center-of-mass differential cross section for the production of the heavy product.

$E_0 \equiv E_0 + Q = E_3 + E_4$.

$$A \equiv \frac{M_1 M_4 (E_0/E_T)}{(M_1 + M_2)(M_3 + M_4)} .$$

$$B \equiv \frac{M_1 M_3 (E_0/E_T)}{(M_1 + M_2)(M_3 + M_4)} .$$

$$C \equiv \frac{M_2 M_3}{(M_1 + M_2)(M_3 + M_4)}\left(1 + \frac{M_1 Q}{M_2 E_T}\right) = \frac{E_{4c}}{E_T} .$$

$$D \equiv \frac{M_2 M_4}{(M_1 + M_2)(M_3 + M_4)}\left(1 + \frac{M_1 Q}{M_2 E_T}\right) = \frac{E_{3c}}{E_T} .$$

The quantities A, B, C, and D obey the constraints $A + B + C + D = 1$ and $AC = BD$.

27

Table A4.2. Kinematic expressions for inelastic scattering.

Quantity calculated	Expression	Equation number
Laboratory energy of the light reaction product	$\dfrac{E_3}{E_T} = B + D + 2(AC)^{1/2}\cos(\theta_c)$. When $B \leq D$, $\dfrac{E_3}{E_T} = B\{\cos(\theta) + [D/B - \sin^2(\theta)]^{1/2}\}^2$. When $B > D$, E_3 is double-valued, $\dfrac{E_4}{E_T} = B\{\cos(\theta) \pm [D/B - \sin^2(\theta)]^{1/2}\}^2$, and $\theta \leq \sin^{-1}[(D/B)^{1/2}]$.	(A4.14)
Laboratory energy of the heavy reaction product	$\dfrac{E_4}{E_T} = A + C + 2(AC)^{1/2}\cos(\phi_c)$. When $A \leq C$, $\dfrac{E_4}{E_T} = A\{\cos(\phi) + [C/A - \sin^2(\phi)]^{1/2}\}^2$. When $A > C$, E_4 is double-valued, $\dfrac{E_3}{E_T} = A\{\cos(\phi) \pm [C/A - \sin^2(\phi)]^{1/2}\}^2$, and $\phi \leq \sin^{-1}[(C/A)^{1/2}]$.	(A4.15)
Laboratory angle of the heavy product	$\sin(\phi) = \left(\dfrac{M_3 E_3}{M_4 E_4}\right)^{1/2}\sin(\theta)$.	(A4.16)
Center-of-mass angle of the light product	$\sin(\theta_c) = \left(\dfrac{E_3/E_T}{D}\right)^{1/2}\sin(\theta)$.	(A4.17)
Cross-section transformation between the laboratory and center-of-mass systems for the light product	$\dfrac{\sigma_3(\theta)}{\sigma_{3c}(\theta_c)} = \dfrac{\sin(\theta_c)}{\sin(\theta)}\dfrac{d\theta_c}{d\theta} = \dfrac{\sin^2(\theta_c)}{\sin^2(\theta)\cos(\theta_c - \theta)}$ $= \dfrac{E_3/E_T}{(AC)^{1/2}[D/B - \sin^2(\theta)]^{1/2}}$.	(A4.18)
Cross-section transformation between the laboratory and center-of-mass systems for the heavy product	$\dfrac{\sigma_4(\phi)}{\sigma_{4c}(\phi_c)} = \dfrac{\sin(\phi_c)}{\sin(\phi)}\dfrac{d\phi_c}{d\phi} = \dfrac{\sin^2(\phi_c)}{\sin^2(\phi)\cos(\phi_c - \phi)}$ $= \dfrac{E_4/E_T}{(AC)^{1/2}[C/A - \sin^2(\phi)]^{1/2}}$.	(A4.19)
Cross-section ratio for associated particles in the laboratory system	$\dfrac{\sigma_3(\theta)}{\sigma_4(\phi)} = \dfrac{\sin(\phi)}{\sin(\theta)}\dfrac{d\phi}{d\theta} = \dfrac{\sin^2(\phi)\cos(\phi_c - \phi)}{\sin^2(\theta)\cos(\theta_c - \theta)}$.	(A4.20)

REFERENCES

Chu, W.-K., Mayer, J.W., and Nicolet, M.-A. (1979), *Backscattering Spectrometry*, Academic Press, New York, pp. 320–322.

Hagedorn, R. (1963), *Relativistic Kinematics*, W.A. Benjamin, New York, pp. 1–44.

Marion, J.B., and Young, F.C. (1968), *Nuclear Reaction Analysis*, North-Holland, Amsterdam, pp. 140–142.

APPENDIX

5

K FACTORS FOR RBS

Compiled by

J. C. Barbour

Sandia National Laboratories, Albuquerque, New Mexico, USA

CONTENTS

Table A5.1. Rutherford backscattering spectrometry kinematic factors for H as a projectile.

This table gives the RBS kinematic factors K_{M_2}, defined by Eq. (A4.2), for H as a projectile ($M_1 = 1.0078$ amu) and for the isotopic masses (M_2) of the elements. The kinematic factors are given as a function of scattering angle from 180° to 90° measured in the laboratory frame of reference. The first row for each element gives the average atomic weight of that element and the average kine‑matic factor (Average K_{M_2}) for that element. The subsequent rows give the isotopic masses for that element and the K_{M_2} for those isotopic masses. The average K_{M_2} is calculated as the weighted average of the K_{M_2} in which the relative abundances (ϖ_{M_2}) for the M_2 are given in Appendix 1:

$$\text{Average } K_{M_2} = \Sigma\, \varpi_{M_2} K_{M_2}, \text{ summed over the masses } M_2.$$

Atomic no. (Z)	El.	Isotopic mass (M2) (amu)	180°	170°	160°	150°	140°	130°	120°	110°	100°	90°
3	Li	6.941	0.5568	0.5592	0.5666	0.5789	0.5960	0.6179	0.6443	0.6748	0.7090	0.7461
		6.015	0.5084	0.5110	0.5187	0.5317	0.5500	0.5733	0.6017	0.6348	0.6721	0.7130
		7.016	0.5607	0.5632	0.5705	0.5827	0.5998	0.6215	0.6477	0.6781	0.7120	0.7488
4	Be	9.012	0.6381	0.6403	0.6468	0.6576	0.6725	0.6913	0.7139	0.7397	0.7682	0.7988
5	B	10.811	0.6877	0.6897	0.6955	0.7051	0.7184	0.7352	0.7551	0.7778	0.8027	0.8293
		10.013	0.6677	0.6697	0.6758	0.6859	0.6999	0.7175	0.7385	0.7625	0.7889	0.8171
		11.009	0.6927	0.6946	0.7004	0.7099	0.7230	0.7396	0.7592	0.7816	0.8061	0.8323
6	C	12.011	0.7143	0.7161	0.7216	0.7306	0.7429	0.7585	0.7769	0.7979	0.8208	0.8452
		12.000	0.7141	0.7159	0.7214	0.7304	0.7427	0.7583	0.7768	0.7977	0.8207	0.8450
		13.003	0.7330	0.7347	0.7399	0.7484	0.7601	0.7748	0.7921	0.8118	0.8333	0.8561
7	N	14.007	0.7495	0.7512	0.7561	0.7641	0.7752	0.7891	0.8055	0.8241	0.8443	0.8658
		14.003	0.7495	0.7511	0.7560	0.7641	0.7752	0.7891	0.8055	0.8240	0.8443	0.8657
		15.000	0.7640	0.7656	0.7702	0.7779	0.7884	0.8016	0.8172	0.8347	0.8539	0.8741
8	O	15.999	0.7770	0.7785	0.7829	0.7902	0.8003	0.8128	0.8276	0.8442	0.8624	0.8815
		15.995	0.7770	0.7785	0.7829	0.7902	0.8002	0.8128	0.8275	0.8442	0.8623	0.8815
		16.999	0.7887	0.7901	0.7943	0.8013	0.8109	0.8228	0.8369	0.8527	0.8699	0.8881
		17.999	0.7992	0.8005	0.8048	0.8112	0.8204	0.8318	0.8452	0.8603	0.8767	0.8940
9	F	18.998	0.8087	0.8100	0.8139	0.8202	0.8290	0.8399	0.8527	0.8672	0.8828	0.8993
10	Ne	20.179	0.8187	0.8199	0.8236	0.8297	0.8380	0.8484	0.8606	0.8744	0.8892	0.9048
		19.992	0.8172	0.8185	0.8222	0.8284	0.8368	0.8472	0.8595	0.8733	0.8883	0.9040
		20.994	0.8252	0.8264	0.8300	0.8359	0.8439	0.8540	0.8658	0.8790	0.8934	0.9084
		21.991	0.8324	0.8336	0.8370	0.8427	0.8504	0.8601	0.8715	0.8842	0.8979	0.9124
11	Na	22.990	0.8391	0.8403	0.8435	0.8490	0.8565	0.8658	0.8767	0.8889	0.9022	0.9160
12	Mg	24.305	0.8470	0.8479	0.8512	0.8563	0.8634	0.8725	0.8829	0.8945	0.9072	0.9203
		23.985	0.8452	0.8461	0.8494	0.8546	0.8618	0.8710	0.8815	0.8933	0.9060	0.9194
		24.986	0.8509	0.8520	0.8551	0.8602	0.8671	0.8758	0.8860	0.8973	0.9099	0.9225
		25.983	0.8562	0.8574	0.8602	0.8652	0.8719	0.8803	0.8901	0.9011	0.9129	0.9253
13	Al	26.982	0.8612	0.8621	0.8650	0.8698	0.8763	0.8844	0.8939	0.9046	0.9160	0.9280
14	Si	28.086	0.8662	0.8672	0.8700	0.8746	0.8809	0.8887	0.8979	0.9081	0.9192	0.9307
		27.977	0.8658	0.8667	0.8695	0.8741	0.8805	0.8883	0.8975	0.9081	0.9189	0.9305
		28.976	0.8701	0.8710	0.8737	0.8782	0.8844	0.8920	0.9009	0.9108	0.9216	0.9328
		29.974	0.8741	0.8750	0.8777	0.8820	0.8880	0.8954	0.9040	0.9137	0.9241	0.9349
15	P	30.974	0.8779	0.8788	0.8814	0.8856	0.8914	0.8986	0.9070	0.9163	0.9264	0.9370
16	S	32.064	0.8815	0.8826	0.8852	0.8893	0.8949	0.9018	0.9100	0.9191	0.9288	0.9390
		31.972	0.8815	0.8823	0.8849	0.8890	0.8946	0.9016	0.9097	0.9188	0.9287	0.9389
		32.972	0.8849	0.8857	0.8882	0.8922	0.8976	0.9044	0.9124	0.9212	0.9307	0.9407
		33.968	0.8881	0.8889	0.8912	0.8952	0.9005	0.9071	0.9148	0.9234	0.9327	0.9424
		35.967	0.8939	0.8947	0.8970	0.9007	0.9057	0.9120	0.9194	0.9275	0.9363	0.9455

Table A5.1. Rutherford backscattering spectrometry kinematic factors for H as a projectile (continued).

Atomic no. (Z)	El.	Isotopic mass (M₂) (amu)	180°	170°	160°	150°	140°	130°	120°	110°	100°	90°
17	Cl	35.453	0.8924	0.8932	0.8955	0.8993	0.9044	0.9108	0.9182	0.9265	0.9354	0.9447
		34.969	0.8911	0.8919	0.8942	0.8980	0.9032	0.9096	0.9171	0.9255	0.9346	0.9440
		36.966	0.8967	0.8974	0.8996	0.9032	0.9082	0.9143	0.9214	0.9294	0.9380	0.9469
18	Ar	39.948	0.9040	0.9047	0.9067	0.9101	0.9147	0.9204	0.9271	0.9345	0.9425	0.9508
		35.967	0.8939	0.8947	0.8970	0.9007	0.9057	0.9120	0.9194	0.9275	0.9363	0.9455
		37.963	0.8992	0.9000	0.9021	0.9057	0.9105	0.9165	0.9234	0.9312	0.9396	0.9483
		39.962	0.9040	0.9047	0.9068	0.9102	0.9148	0.9205	0.9271	0.9345	0.9425	0.9508
19	K	39.098	0.9020	0.9027	0.9048	0.9083	0.9129	0.9188	0.9256	0.9331	0.9413	0.9497
		38.964	0.9017	0.9024	0.9045	0.9080	0.9127	0.9185	0.9253	0.9329	0.9411	0.9496
		40.000	0.9041	0.9048	0.9069	0.9102	0.9148	0.9205	0.9272	0.9346	0.9426	0.9508
		40.962	0.9063	0.9069	0.9089	0.9123	0.9167	0.9223	0.9288	0.9361	0.9439	0.9520
20	Ca	40.078	0.9043	0.9050	0.9070	0.9104	0.9150	0.9207	0.9273	0.9347	0.9427	0.9509
		39.963	0.9040	0.9047	0.9068	0.9102	0.9148	0.9205	0.9271	0.9345	0.9425	0.9508
		41.959	0.9084	0.9090	0.9110	0.9142	0.9186	0.9241	0.9305	0.9376	0.9452	0.9531
		42.959	0.9104	0.9111	0.9130	0.9162	0.9205	0.9258	0.9320	0.9390	0.9464	0.9542
		43.956	0.9124	0.9130	0.9149	0.9180	0.9222	0.9274	0.9335	0.9403	0.9476	0.9552
		45.954	0.9160	0.9166	0.9184	0.9214	0.9255	0.9305	0.9363	0.9428	0.9498	0.9571
		47.952	0.9194	0.9199	0.9217	0.9246	0.9284	0.9333	0.9389	0.9451	0.9519	0.9588
21	Sc	44.956	0.9142	0.9148	0.9167	0.9197	0.9239	0.9290	0.9349	0.9416	0.9487	0.9561
22	Ti	47.878	0.9192	0.9198	0.9216	0.9244	0.9283	0.9332	0.9388	0.9450	0.9518	0.9588
		45.953	0.9160	0.9166	0.9184	0.9214	0.9254	0.9305	0.9363	0.9428	0.9498	0.9571
		46.952	0.9177	0.9183	0.9201	0.9230	0.9270	0.9319	0.9376	0.9440	0.9509	0.9580
		47.948	0.9194	0.9199	0.9217	0.9245	0.9284	0.9333	0.9389	0.9451	0.9519	0.9588
		48.948	0.9209	0.9215	0.9232	0.9260	0.9298	0.9346	0.9401	0.9462	0.9528	0.9597
		49.945	0.9224	0.9230	0.9247	0.9274	0.9312	0.9358	0.9413	0.9473	0.9537	0.9604
23	V	50.942	0.9239	0.9245	0.9261	0.9288	0.9325	0.9371	0.9424	0.9483	0.9546	0.9612
		49.947	0.9225	0.9230	0.9247	0.9275	0.9312	0.9358	0.9413	0.9473	0.9537	0.9604
		50.944	0.9239	0.9245	0.9261	0.9288	0.9325	0.9371	0.9424	0.9483	0.9546	0.9612
24	Cr	51.996	0.9254	0.9259	0.9275	0.9302	0.9338	0.9383	0.9435	0.9493	0.9555	0.9620
		49.946	0.9225	0.9230	0.9247	0.9275	0.9312	0.9358	0.9413	0.9473	0.9537	0.9604
		51.940	0.9253	0.9259	0.9275	0.9301	0.9338	0.9382	0.9434	0.9492	0.9555	0.9619
		52.941	0.9267	0.9272	0.9288	0.9314	0.9350	0.9394	0.9445	0.9502	0.9563	0.9626
		53.939	0.9280	0.9285	0.9301	0.9326	0.9361	0.9405	0.9455	0.9511	0.9571	0.9633
25	Mn	54.938	0.9292	0.9298	0.9313	0.9338	0.9373	0.9415	0.9464	0.9519	0.9578	0.9640
26	Fe	55.847	0.9303	0.9309	0.9324	0.9349	0.9382	0.9424	0.9473	0.9527	0.9585	0.9645
		53.940	0.9280	0.9285	0.9301	0.9326	0.9361	0.9405	0.9455	0.9511	0.9571	0.9633
		55.935	0.9305	0.9310	0.9325	0.9350	0.9383	0.9425	0.9474	0.9528	0.9586	0.9646
		56.935	0.9316	0.9321	0.9336	0.9361	0.9394	0.9435	0.9483	0.9536	0.9593	0.9652
		57.933	0.9328	0.9333	0.9347	0.9371	0.9404	0.9444	0.9491	0.9544	0.9600	0.9658
27	Co	58.933	0.9339	0.9344	0.9358	0.9382	0.9414	0.9454	0.9500	0.9551	0.9607	0.9664
28	Ni	58.688	0.9336	0.9341	0.9355	0.9379	0.9411	0.9451	0.9498	0.9549	0.9605	0.9662
		57.935	0.9328	0.9333	0.9347	0.9371	0.9404	0.9444	0.9491	0.9544	0.9600	0.9658
		59.931	0.9349	0.9354	0.9368	0.9392	0.9423	0.9462	0.9508	0.9559	0.9613	0.9669
		60.931	0.9360	0.9364	0.9378	0.9401	0.9432	0.9471	0.9516	0.9566	0.9619	0.9675
		61.928	0.9370	0.9374	0.9388	0.9411	0.9441	0.9479	0.9523	0.9573	0.9625	0.9680
		63.928	0.9389	0.9393	0.9407	0.9429	0.9458	0.9495	0.9538	0.9586	0.9637	0.9690
29	Cu	63.546	0.9385	0.9390	0.9403	0.9425	0.9455	0.9492	0.9535	0.9583	0.9634	0.9688
		62.930	0.9379	0.9384	0.9398	0.9420	0.9450	0.9487	0.9531	0.9579	0.9631	0.9685
		64.928	0.9398	0.9402	0.9416	0.9437	0.9466	0.9503	0.9545	0.9592	0.9642	0.9694

Table A5.1. Rutherford backscattering spectrometry kinematic factors for H as a projectile (continued).

Atomic no. (Z)	El.	Isotopic mass (M₂) (amu)	180°	170°	160°	150°	140°	130°	120°	110°	100°	90°
30	Zn	65.396	0.9402	0.9406	0.9419	0.9441	0.9470	0.9506	0.9548	0.9595	0.9644	0.9696
		63.929	0.9389	0.9393	0.9407	0.9429	0.9458	0.9495	0.9538	0.9586	0.9637	0.9690
		65.926	0.9407	0.9411	0.9424	0.9445	0.9474	0.9510	0.9552	0.9598	0.9648	0.9699
		66.927	0.9415	0.9420	0.9433	0.9453	0.9482	0.9517	0.9558	0.9604	0.9653	0.9703
		67.925	0.9424	0.9428	0.9441	0.9461	0.9489	0.9524	0.9565	0.9610	0.9658	0.9708
		69.925	0.9440	0.9444	0.9456	0.9476	0.9504	0.9537	0.9577	0.9621	0.9667	0.9716
31	Ga	69.723	0.9438	0.9442	0.9455	0.9475	0.9502	0.9536	0.9576	0.9619	0.9666	0.9715
		68.926	0.9432	0.9436	0.9449	0.9469	0.9497	0.9531	0.9571	0.9615	0.9663	0.9712
		70.925	0.9447	0.9452	0.9464	0.9483	0.9510	0.9544	0.9583	0.9626	0.9672	0.9720
32	Ge	72.632	0.9460	0.9464	0.9476	0.9495	0.9521	0.9554	0.9592	0.9634	0.9679	0.9726
		69.924	0.9440	0.9444	0.9456	0.9476	0.9504	0.9537	0.9577	0.9621	0.9667	0.9716
		71.922	0.9455	0.9459	0.9471	0.9490	0.9517	0.9550	0.9588	0.9631	0.9676	0.9724
		72.924	0.9462	0.9466	0.9478	0.9497	0.9524	0.9556	0.9594	0.9636	0.9681	0.9727
		73.921	0.9469	0.9473	0.9485	0.9504	0.9530	0.9562	0.9599	0.9641	0.9685	0.9731
		75.921	0.9483	0.9487	0.9498	0.9517	0.9542	0.9573	0.9610	0.9650	0.9693	0.9738
33	As	74.922	0.9476	0.9480	0.9492	0.9510	0.9536	0.9568	0.9604	0.9645	0.9689	0.9735
34	Se	78.993	0.9502	0.9506	0.9517	0.9535	0.9559	0.9589	0.9624	0.9663	0.9705	0.9748
		73.923	0.9469	0.9473	0.9485	0.9504	0.9530	0.9562	0.9599	0.9641	0.9685	0.9731
		75.919	0.9483	0.9487	0.9498	0.9517	0.9542	0.9573	0.9610	0.9650	0.9693	0.9738
		76.920	0.9489	0.9493	0.9504	0.9523	0.9548	0.9579	0.9615	0.9654	0.9697	0.9741
		77.917	0.9496	0.9499	0.9511	0.9529	0.9553	0.9584	0.9619	0.9659	0.9701	0.9745
		79.916	0.9508	0.9512	0.9523	0.9540	0.9564	0.9594	0.9629	0.9667	0.9708	0.9751
		81.917	0.9520	0.9523	0.9534	0.9551	0.9575	0.9604	0.9638	0.9675	0.9715	0.9757
35	Br	79.904	0.9508	0.9512	0.9522	0.9540	0.9564	0.9594	0.9629	0.9667	0.9708	0.9751
		78.918	0.9502	0.9506	0.9517	0.9535	0.9559	0.9589	0.9624	0.9663	0.9705	0.9748
		80.916	0.9514	0.9518	0.9528	0.9546	0.9570	0.9599	0.9633	0.9671	0.9712	0.9754
36	Kr	83.8	0.9530	0.9534	0.9544	0.9561	0.9584	0.9612	0.9646	0.9682	0.9722	0.9762
		77.920	0.9496	0.9500	0.9511	0.9529	0.9553	0.9584	0.9619	0.9659	0.9701	0.9745
		79.916	0.9508	0.9512	0.9523	0.9540	0.9564	0.9594	0.9629	0.9667	0.9708	0.9751
		81.913	0.9520	0.9523	0.9534	0.9551	0.9575	0.9604	0.9638	0.9675	0.9715	0.9757
		82.914	0.9525	0.9529	0.9539	0.9556	0.9580	0.9608	0.9642	0.9679	0.9719	0.9760
		83.911	0.9531	0.9534	0.9545	0.9562	0.9585	0.9613	0.9646	0.9683	0.9722	0.9763
		85.911	0.9542	0.9545	0.9555	0.9572	0.9594	0.9622	0.9654	0.9690	0.9728	0.9768
37	Rb	85.468	0.9539	0.9543	0.9553	0.9569	0.9592	0.9620	0.9652	0.9688	0.9727	0.9767
		84.912	0.9536	0.9540	0.9550	0.9567	0.9589	0.9618	0.9650	0.9686	0.9725	0.9765
		86.909	0.9547	0.9550	0.9560	0.9576	0.9599	0.9626	0.9658	0.9694	0.9731	0.9771
38	Sr	87.617	0.9550	0.9554	0.9564	0.9580	0.9602	0.9629	0.9661	0.9696	0.9734	0.9773
		83.913	0.9531	0.9534	0.9545	0.9562	0.9585	0.9613	0.9646	0.9683	0.9722	0.9763
		85.909	0.9542	0.9545	0.9555	0.9572	0.9594	0.9622	0.9654	0.9690	0.9728	0.9768
		86.909	0.9547	0.9550	0.9560	0.9576	0.9599	0.9626	0.9658	0.9694	0.9731	0.9771
		87.906	0.9552	0.9555	0.9565	0.9581	0.9603	0.9630	0.9662	0.9697	0.9734	0.9773
39	Y	88.906	0.9557	0.9560	0.9570	0.9586	0.9608	0.9634	0.9666	0.9700	0.9737	0.9776
40	Zr	91.221	0.9568	0.9571	0.9580	0.9596	0.9617	0.9643	0.9674	0.9708	0.9744	0.9781
		89.905	0.9561	0.9565	0.9574	0.9590	0.9612	0.9638	0.9669	0.9704	0.9740	0.9778
		90.906	0.9566	0.9569	0.9579	0.9595	0.9616	0.9642	0.9673	0.9707	0.9743	0.9781
		91.905	0.9571	0.9574	0.9584	0.9599	0.9620	0.9646	0.9676	0.9710	0.9746	0.9783
		93.906	0.9580	0.9583	0.9592	0.9607	0.9628	0.9654	0.9683	0.9716	0.9751	0.9788
		95.908	0.9588	0.9591	0.9601	0.9615	0.9636	0.9661	0.9690	0.9722	0.9756	0.9792

Table A5.1. Rutherford backscattering spectrometry kinematic factors for H as a projectile (continued).

Atomic no. (Z)	El.	Isotopic mass (M_2) (amu)	90°	100°	110°	120°	130°	140°	150°	160°	170°	180°
41	Nb	92.906	0.9785	0.9749	0.9713	0.9680	0.9650	0.9624	0.9603	0.9588	0.9579	0.9575
42	Mo	95.931	0.9792	0.9756	0.9722	0.9690	0.9661	0.9636	0.9615	0.9600	0.9591	0.9588
		91.907	0.9783	0.9746	0.9710	0.9676	0.9646	0.9620	0.9599	0.9584	0.9574	0.9571
		93.905	0.9788	0.9751	0.9716	0.9683	0.9654	0.9628	0.9607	0.9592	0.9583	0.9580
		94.906	0.9790	0.9754	0.9719	0.9686	0.9657	0.9632	0.9611	0.9596	0.9587	0.9584
		95.905	0.9792	0.9756	0.9722	0.9690	0.9661	0.9636	0.9615	0.9601	0.9591	0.9588
		96.906	0.9794	0.9759	0.9725	0.9693	0.9664	0.9639	0.9619	0.9605	0.9596	0.9593
		97.905	0.9796	0.9761	0.9727	0.9696	0.9667	0.9643	0.9623	0.9609	0.9600	0.9597
		99.908	0.9800	0.9766	0.9733	0.9702	0.9674	0.9650	0.9631	0.9616	0.9607	0.9605
43	Tc	98.000	0.9796	0.9761	0.9728	0.9696	0.9668	0.9643	0.9623	0.9609	0.9600	0.9597
44	Ru	101.019	0.9802	0.9768	0.9736	0.9705	0.9677	0.9654	0.9634	0.9620	0.9612	0.9609
		95.908	0.9792	0.9756	0.9722	0.9690	0.9661	0.9636	0.9615	0.9601	0.9591	0.9588
		97.905	0.9796	0.9761	0.9727	0.9696	0.9667	0.9643	0.9623	0.9609	0.9600	0.9597
		98.506	0.9797	0.9763	0.9729	0.9698	0.9669	0.9645	0.9625	0.9611	0.9602	0.9599
		99.904	0.9800	0.9766	0.9733	0.9702	0.9674	0.9650	0.9631	0.9616	0.9607	0.9605
		100.906	0.9802	0.9768	0.9735	0.9705	0.9677	0.9653	0.9634	0.9620	0.9611	0.9608
		101.904	0.9804	0.9771	0.9738	0.9708	0.9680	0.9657	0.9638	0.9624	0.9615	0.9612
		103.905	0.9806	0.9775	0.9743	0.9710	0.9686	0.9663	0.9644	0.9631	0.9622	0.9619
45	Rh	102.906	0.9812	0.9773	0.9741	0.9707	0.9683	0.9660	0.9641	0.9627	0.9619	0.9616
46	Pd	106.415	0.9804	0.9780	0.9749	0.9720	0.9694	0.9671	0.9653	0.9639	0.9631	0.9628
		101.906	0.9808	0.9771	0.9738	0.9708	0.9680	0.9657	0.9638	0.9624	0.9615	0.9612
		103.904	0.9810	0.9775	0.9743	0.9713	0.9686	0.9663	0.9644	0.9631	0.9622	0.9619
		104.905	0.9811	0.9777	0.9745	0.9716	0.9689	0.9666	0.9648	0.9634	0.9626	0.9623
		105.904	0.9815	0.9779	0.9748	0.9719	0.9692	0.9669	0.9651	0.9638	0.9629	0.9626
		107.904	0.9818	0.9783	0.9752	0.9724	0.9698	0.9675	0.9657	0.9644	0.9636	0.9633
		109.905	0.9815	0.9787	0.9757	0.9729	0.9703	0.9681	0.9664	0.9651	0.9643	0.9640
47	Ag	107.868	0.9813	0.9783	0.9752	0.9724	0.9698	0.9675	0.9657	0.9644	0.9636	0.9633
		106.905	0.9817	0.9781	0.9750	0.9721	0.9695	0.9673	0.9654	0.9641	0.9633	0.9630
		108.905	0.9822	0.9785	0.9755	0.9726	0.9701	0.9678	0.9661	0.9647	0.9639	0.9637
48	Cd	112.412	0.9811	0.9792	0.9762	0.9735	0.9710	0.9688	0.9671	0.9658	0.9650	0.9648
		105.907	0.9815	0.9779	0.9748	0.9719	0.9692	0.9669	0.9651	0.9638	0.9629	0.9627
		107.904	0.9818	0.9783	0.9752	0.9724	0.9698	0.9675	0.9657	0.9644	0.9636	0.9633
		109.903	0.9820	0.9787	0.9757	0.9729	0.9703	0.9681	0.9664	0.9651	0.9643	0.9640
		110.904	0.9821	0.9789	0.9759	0.9731	0.9706	0.9684	0.9667	0.9654	0.9646	0.9643
		111.903	0.9823	0.9791	0.9761	0.9733	0.9708	0.9687	0.9669	0.9657	0.9649	0.9646
		112.904	0.9826	0.9793	0.9763	0.9736	0.9711	0.9690	0.9672	0.9660	0.9652	0.9649
		113.903	0.9825	0.9794	0.9765	0.9738	0.9713	0.9692	0.9675	0.9663	0.9655	0.9652
		115.905	0.9828	0.9798	0.9769	0.9743	0.9718	0.9698	0.9681	0.9668	0.9661	0.9658
49	In	114.818	0.9826	0.9796	0.9767	0.9740	0.9716	0.9695	0.9678	0.9665	0.9658	0.9655
		112.904	0.9823	0.9793	0.9763	0.9736	0.9711	0.9690	0.9672	0.9660	0.9652	0.9649
		114.904	0.9826	0.9796	0.9767	0.9740	0.9716	0.9695	0.9678	0.9665	0.9658	0.9655
50	Sn	118.613	0.9831	0.9802	0.9774	0.9748	0.9725	0.9704	0.9688	0.9676	0.9668	0.9666
		111.905	0.9821	0.9791	0.9761	0.9733	0.9708	0.9687	0.9669	0.9657	0.9649	0.9646
		113.903	0.9825	0.9794	0.9765	0.9738	0.9713	0.9692	0.9675	0.9663	0.9655	0.9652
		114.903	0.9826	0.9796	0.9767	0.9740	0.9716	0.9695	0.9678	0.9665	0.9658	0.9655
		115.902	0.9828	0.9798	0.9769	0.9743	0.9718	0.9698	0.9681	0.9668	0.9661	0.9658
		116.903	0.9829	0.9800	0.9771	0.9745	0.9721	0.9700	0.9683	0.9671	0.9664	0.9661

Table A5.1. Rutherford backscattering spectrometry kinematic factors for H as a projectile (continued).

Atomic no. (Z)	El.	Isotopic mass (M₂) (amu)	180°	170°	160°	150°	140°	130°	120°	110°	100°	90°
		117.602	0.9663	0.9666	0.9673	0.9685	0.9702	0.9722	0.9746	0.9773	0.9801	0.9830
		118.903	0.9667	0.9669	0.9677	0.9689	0.9705	0.9725	0.9749	0.9775	0.9803	0.9832
		119.902	0.9669	0.9672	0.9679	0.9691	0.9707	0.9728	0.9751	0.9777	0.9805	0.9833
		121.903	0.9675	0.9677	0.9684	0.9696	0.9712	0.9732	0.9755	0.9781	0.9808	0.9836
		123.905	0.9680	0.9682	0.9689	0.9701	0.9717	0.9736	0.9759	0.9784	0.9811	0.9839
51	Sb	121.758	0.9674	0.9677	0.9684	0.9696	0.9712	0.9732	0.9755	0.9780	0.9808	0.9836
		120.904	0.9672	0.9675	0.9682	0.9694	0.9710	0.9730	0.9753	0.9779	0.9806	0.9835
		122.904	0.9677	0.9680	0.9687	0.9699	0.9715	0.9734	0.9757	0.9782	0.9809	0.9837
52	Te	127.586	0.9689	0.9691	0.9698	0.9709	0.9725	0.9744	0.9766	0.9790	0.9816	0.9843
		119.904	0.9669	0.9672	0.9679	0.9691	0.9707	0.9728	0.9751	0.9777	0.9805	0.9833
		121.903	0.9675	0.9677	0.9684	0.9696	0.9712	0.9732	0.9755	0.9781	0.9808	0.9836
		122.904	0.9677	0.9680	0.9687	0.9699	0.9715	0.9734	0.9757	0.9782	0.9809	0.9837
		123.903	0.9680	0.9682	0.9689	0.9701	0.9717	0.9736	0.9759	0.9784	0.9811	0.9839
		124.904	0.9682	0.9685	0.9692	0.9703	0.9719	0.9738	0.9761	0.9786	0.9812	0.9840
		125.903	0.9685	0.9687	0.9694	0.9706	0.9721	0.9740	0.9763	0.9787	0.9814	0.9841
		127.905	0.9690	0.9692	0.9699	0.9710	0.9726	0.9744	0.9766	0.9791	0.9817	0.9844
		129.906	0.9694	0.9697	0.9704	0.9715	0.9730	0.9748	0.9770	0.9794	0.9820	0.9846
53	I	126.905	0.9687	0.9690	0.9697	0.9708	0.9723	0.9742	0.9765	0.9789	0.9815	0.9842
54	Xe	131.293	0.9698	0.9700	0.9707	0.9718	0.9732	0.9751	0.9772	0.9796	0.9821	0.9848
		123.906	0.9680	0.9682	0.9689	0.9701	0.9717	0.9736	0.9759	0.9784	0.9811	0.9839
		125.904	0.9685	0.9687	0.9694	0.9706	0.9721	0.9740	0.9763	0.9787	0.9814	0.9841
		127.904	0.9690	0.9692	0.9699	0.9710	0.9726	0.9744	0.9766	0.9791	0.9817	0.9844
		128.905	0.9692	0.9694	0.9701	0.9712	0.9728	0.9746	0.9768	0.9792	0.9818	0.9845
		129.904	0.9694	0.9697	0.9704	0.9715	0.9730	0.9748	0.9770	0.9794	0.9820	0.9846
		130.905	0.9697	0.9699	0.9706	0.9717	0.9732	0.9750	0.9772	0.9795	0.9821	0.9847
		131.904	0.9699	0.9701	0.9708	0.9719	0.9734	0.9752	0.9773	0.9797	0.9822	0.9848
		133.905	0.9703	0.9706	0.9712	0.9723	0.9738	0.9756	0.9777	0.9800	0.9825	0.9851
		135.907	0.9708	0.9710	0.9716	0.9727	0.9741	0.9759	0.9780	0.9803	0.9827	0.9853
55	Cs	132.905	0.9701	0.9703	0.9710	0.9721	0.9736	0.9754	0.9775	0.9799	0.9824	0.9849
56	Ba	137.327	0.9711	0.9713	0.9719	0.9730	0.9744	0.9762	0.9782	0.9805	0.9829	0.9854
		129.906	0.9694	0.9697	0.9704	0.9715	0.9730	0.9748	0.9770	0.9794	0.9820	0.9846
		131.905	0.9699	0.9701	0.9708	0.9719	0.9734	0.9752	0.9773	0.9797	0.9822	0.9848
		133.904	0.9703	0.9706	0.9712	0.9723	0.9738	0.9756	0.9777	0.9800	0.9825	0.9851
		134.906	0.9706	0.9708	0.9714	0.9725	0.9740	0.9758	0.9778	0.9801	0.9826	0.9852
		135.905	0.9708	0.9710	0.9716	0.9727	0.9741	0.9759	0.9780	0.9803	0.9827	0.9853
		136.906	0.9710	0.9712	0.9718	0.9729	0.9743	0.9761	0.9782	0.9804	0.9829	0.9854
		137.905	0.9712	0.9714	0.9720	0.9731	0.9745	0.9763	0.9783	0.9806	0.9830	0.9855
57	La	138.905	0.9714	0.9716	0.9722	0.9733	0.9747	0.9764	0.9785	0.9807	0.9831	0.9856
		137.907	0.9712	0.9714	0.9720	0.9731	0.9745	0.9763	0.9783	0.9806	0.9830	0.9855
		138.906	0.9714	0.9716	0.9722	0.9733	0.9747	0.9764	0.9785	0.9807	0.9831	0.9856
58	Ce	140.115	0.9716	0.9719	0.9725	0.9735	0.9749	0.9766	0.9787	0.9809	0.9833	0.9857
		135.907	0.9708	0.9710	0.9716	0.9727	0.9741	0.9759	0.9780	0.9803	0.9827	0.9853
		137.906	0.9712	0.9714	0.9720	0.9731	0.9745	0.9763	0.9783	0.9806	0.9830	0.9855
		139.905	0.9716	0.9718	0.9724	0.9735	0.9749	0.9766	0.9786	0.9809	0.9832	0.9857
		141.909	0.9720	0.9722	0.9728	0.9738	0.9752	0.9769	0.9789	0.9811	0.9835	0.9859
59	Pr	140.908	0.9718	0.9720	0.9726	0.9737	0.9751	0.9768	0.9788	0.9810	0.9834	0.9858

Table A5.1. Rutherford backscattering spectrometry kinematic factors for H as a projectile (continued).

Atomic no. (Z)	El.	Isotopic mass (M2) (amu)	180°	170°	160°	150°	140°	130°	120°	110°	100°	90°
60	Nd	144.242	0.9724	0.9726	0.9733	0.9743	0.9756	0.9773	0.9793	0.9814	0.9837	0.9861
		141.908	0.9720	0.9722	0.9728	0.9738	0.9752	0.9769	0.9789	0.9811	0.9835	0.9859
		142.910	0.9722	0.9722	0.9730	0.9740	0.9754	0.9771	0.9791	0.9812	0.9836	0.9860
		143.910	0.9724	0.9726	0.9732	0.9742	0.9756	0.9773	0.9792	0.9814	0.9837	0.9861
		144.913	0.9726	0.9728	0.9734	0.9744	0.9757	0.9774	0.9794	0.9815	0.9838	0.9862
		145.913	0.9728	0.9730	0.9736	0.9746	0.9759	0.9776	0.9795	0.9816	0.9839	0.9863
		147.917	0.9731	0.9733	0.9739	0.9749	0.9762	0.9779	0.9798	0.9819	0.9841	0.9865
		149.921	0.9735	0.9737	0.9743	0.9752	0.9765	0.9782	0.9800	0.9821	0.9843	0.9866
61	Pm	145.000	0.9726	0.9726	0.9734	0.9744	0.9757	0.9774	0.9794	0.9815	0.9838	0.9862
62	Sm	150.36	0.9735	0.9737	0.9743	0.9753	0.9766	0.9782	0.9801	0.9822	0.9844	0.9867
		143.912	0.9724	0.9726	0.9732	0.9742	0.9756	0.9773	0.9792	0.9814	0.9837	0.9861
		146.915	0.9729	0.9731	0.9737	0.9747	0.9761	0.9777	0.9796	0.9818	0.9840	0.9864
		147.915	0.9731	0.9733	0.9739	0.9749	0.9762	0.9779	0.9798	0.9819	0.9841	0.9865
		148.917	0.9733	0.9735	0.9741	0.9751	0.9764	0.9780	0.9799	0.9820	0.9842	0.9866
		149.917	0.9735	0.9737	0.9743	0.9752	0.9765	0.9782	0.9800	0.9821	0.9843	0.9866
		151.920	0.9738	0.9740	0.9746	0.9755	0.9768	0.9784	0.9803	0.9824	0.9845	0.9868
		153.922	0.9741	0.9743	0.9749	0.9759	0.9771	0.9787	0.9805	0.9826	0.9847	0.9870
63	Eu	151.965	0.9738	0.9740	0.9746	0.9756	0.9768	0.9784	0.9803	0.9824	0.9846	0.9868
		150.920	0.9736	0.9738	0.9744	0.9754	0.9767	0.9783	0.9802	0.9822	0.9844	0.9867
		152.921	0.9740	0.9742	0.9748	0.9757	0.9770	0.9786	0.9804	0.9825	0.9846	0.9869
64	Gd	157.252	0.9747	0.9749	0.9754	0.9764	0.9776	0.9792	0.9810	0.9829	0.9851	0.9873
		151.921	0.9738	0.9740	0.9746	0.9755	0.9768	0.9784	0.9803	0.9824	0.9845	0.9868
		153.921	0.9741	0.9743	0.9749	0.9759	0.9771	0.9787	0.9805	0.9826	0.9847	0.9870
		154.923	0.9743	0.9745	0.9751	0.9760	0.9773	0.9789	0.9807	0.9827	0.9848	0.9871
		155.922	0.9745	0.9747	0.9752	0.9762	0.9774	0.9790	0.9808	0.9828	0.9849	0.9872
		156.924	0.9746	0.9748	0.9754	0.9763	0.9776	0.9791	0.9809	0.9829	0.9850	0.9872
		157.924	0.9748	0.9750	0.9755	0.9765	0.9777	0.9793	0.9810	0.9830	0.9851	0.9873
		159.927	0.9751	0.9753	0.9758	0.9768	0.9780	0.9795	0.9813	0.9832	0.9853	0.9875
65	Tb	158.925	0.9750	0.9751	0.9757	0.9766	0.9779	0.9794	0.9812	0.9831	0.9852	0.9874
66	Dy	162.498	0.9755	0.9757	0.9762	0.9771	0.9783	0.9798	0.9816	0.9835	0.9855	0.9877
		155.924	0.9745	0.9747	0.9752	0.9762	0.9774	0.9790	0.9808	0.9828	0.9849	0.9872
		157.924	0.9748	0.9750	0.9755	0.9765	0.9777	0.9793	0.9810	0.9830	0.9851	0.9873
		159.925	0.9751	0.9753	0.9758	0.9768	0.9780	0.9795	0.9813	0.9832	0.9853	0.9875
		160.927	0.9753	0.9754	0.9760	0.9769	0.9781	0.9796	0.9814	0.9833	0.9854	0.9876
		161.927	0.9754	0.9756	0.9761	0.9770	0.9783	0.9798	0.9815	0.9834	0.9855	0.9876
		162.929	0.9756	0.9757	0.9763	0.9772	0.9784	0.9799	0.9816	0.9835	0.9856	0.9877
		163.929	0.9757	0.9759	0.9764	0.9773	0.9785	0.9800	0.9817	0.9836	0.9857	0.9878
67	Ho	164.930	0.9759	0.9760	0.9766	0.9775	0.9786	0.9801	0.9818	0.9837	0.9857	0.9878
68	Er	167.256	0.9762	0.9764	0.9769	0.9778	0.9789	0.9804	0.9821	0.9840	0.9860	0.9880
		161.929	0.9754	0.9756	0.9761	0.9770	0.9783	0.9798	0.9815	0.9834	0.9855	0.9876
		163.929	0.9757	0.9759	0.9764	0.9773	0.9785	0.9800	0.9817	0.9836	0.9857	0.9878
		165.930	0.9760	0.9762	0.9767	0.9776	0.9788	0.9802	0.9819	0.9838	0.9858	0.9879
		166.932	0.9761	0.9763	0.9769	0.9777	0.9789	0.9804	0.9821	0.9839	0.9859	0.9880
		167.932	0.9763	0.9765	0.9770	0.9779	0.9790	0.9805	0.9822	0.9840	0.9860	0.9881
		169.936	0.9766	0.9767	0.9773	0.9781	0.9793	0.9807	0.9824	0.9842	0.9862	0.9882
69	Tm	168.934	0.9764	0.9766	0.9771	0.9780	0.9791	0.9806	0.9823	0.9841	0.9861	0.9881

Table A5.1. Rutherford backscattering spectrometry kinematic factors for H as a projectile (continued).

Atomic no. (Z)	El.	Isotopic mass (M₂) (amu)	180°	170°	160°	150°	140°	130°	120°	110°	100°	90°
70	Yb	173.034	0.9770	0.9771	0.9777	0.9785	0.9796	0.9810	0.9827	0.9845	0.9864	0.9884
		167.934	0.9763	0.9765	0.9770	0.9779	0.9790	0.9805	0.9822	0.9840	0.9860	0.9881
		169.935	0.9766	0.9767	0.9773	0.9781	0.9793	0.9807	0.9824	0.9842	0.9862	0.9882
		170.936	0.9767	0.9769	0.9774	0.9782	0.9794	0.9808	0.9825	0.9843	0.9863	0.9883
		171.936	0.9768	0.9770	0.9775	0.9784	0.9795	0.9809	0.9826	0.9844	0.9864	0.9883
		172.938	0.9770	0.9771	0.9776	0.9785	0.9796	0.9810	0.9827	0.9845	0.9864	0.9884
		173.939	0.9771	0.9773	0.9778	0.9786	0.9797	0.9811	0.9828	0.9846	0.9865	0.9885
		175.943	0.9773	0.9775	0.9780	0.9788	0.9800	0.9814	0.9830	0.9847	0.9866	0.9886
71	Lu	174.967	0.9772	0.9774	0.9779	0.9787	0.9799	0.9813	0.9829	0.9847	0.9866	0.9885
		174.941	0.9772	0.9774	0.9779	0.9787	0.9799	0.9813	0.9829	0.9847	0.9866	0.9885
		175.943	0.9773	0.9775	0.9780	0.9788	0.9800	0.9814	0.9830	0.9847	0.9866	0.9886
72	Hf	178.49	0.9777	0.9778	0.9783	0.9791	0.9803	0.9816	0.9832	0.9850	0.9868	0.9888
		173.940	0.9771	0.9773	0.9778	0.9786	0.9797	0.9811	0.9828	0.9846	0.9865	0.9885
		175.941	0.9773	0.9775	0.9780	0.9788	0.9800	0.9814	0.9830	0.9847	0.9866	0.9886
		176.943	0.9775	0.9776	0.9781	0.9790	0.9801	0.9815	0.9831	0.9848	0.9867	0.9887
		177.944	0.9776	0.9778	0.9783	0.9791	0.9802	0.9816	0.9832	0.9849	0.9868	0.9887
		178.946	0.9777	0.9779	0.9784	0.9792	0.9803	0.9817	0.9833	0.9850	0.9869	0.9888
		179.947	0.9778	0.9780	0.9785	0.9793	0.9804	0.9818	0.9834	0.9851	0.9869	0.9889
73	Ta	180.948	0.9780	0.9781	0.9786	0.9794	0.9805	0.9819	0.9834	0.9852	0.9870	0.9889
		179.947	0.9778	0.9780	0.9785	0.9793	0.9804	0.9818	0.9833	0.9851	0.9869	0.9889
		180.948	0.9780	0.9781	0.9786	0.9794	0.9805	0.9819	0.9834	0.9852	0.9870	0.9889
74	W	183.849	0.9783	0.9785	0.9790	0.9797	0.9808	0.9821	0.9837	0.9854	0.9872	0.9891
		179.947	0.9778	0.9780	0.9785	0.9793	0.9804	0.9818	0.9833	0.9851	0.9869	0.9889
		181.948	0.9781	0.9783	0.9787	0.9795	0.9806	0.9820	0.9835	0.9852	0.9871	0.9890
		182.950	0.9782	0.9784	0.9789	0.9797	0.9807	0.9821	0.9836	0.9853	0.9872	0.9890
		183.951	0.9783	0.9785	0.9790	0.9798	0.9808	0.9822	0.9837	0.9854	0.9872	0.9891
		185.954	0.9786	0.9787	0.9792	0.9800	0.9810	0.9824	0.9839	0.9856	0.9874	0.9892
75	Re	186.207	0.9786	0.9787	0.9792	0.9800	0.9811	0.9824	0.9839	0.9856	0.9874	0.9892
		184.953	0.9784	0.9786	0.9791	0.9799	0.9809	0.9823	0.9838	0.9855	0.9873	0.9892
		186.956	0.9787	0.9788	0.9793	0.9801	0.9811	0.9824	0.9840	0.9856	0.9874	0.9893
76	Os	190.24	0.9790	0.9792	0.9797	0.9804	0.9815	0.9827	0.9842	0.9859	0.9876	0.9895
		183.952	0.9783	0.9785	0.9790	0.9798	0.9808	0.9822	0.9837	0.9854	0.9872	0.9891
		185.954	0.9786	0.9787	0.9792	0.9800	0.9810	0.9824	0.9839	0.9856	0.9874	0.9892
		186.956	0.9787	0.9788	0.9793	0.9801	0.9811	0.9824	0.9840	0.9856	0.9874	0.9893
		187.956	0.9788	0.9789	0.9794	0.9802	0.9812	0.9825	0.9840	0.9857	0.9875	0.9893
		188.958	0.9789	0.9791	0.9795	0.9803	0.9813	0.9826	0.9841	0.9858	0.9876	0.9894
		189.958	0.9790	0.9792	0.9796	0.9804	0.9814	0.9827	0.9842	0.9859	0.9876	0.9894
		191.962	0.9792	0.9794	0.9798	0.9806	0.9816	0.9829	0.9844	0.9860	0.9878	0.9896
77	Ir	192.216	0.9792	0.9794	0.9799	0.9806	0.9817	0.9829	0.9844	0.9860	0.9878	0.9896
		190.961	0.9791	0.9793	0.9797	0.9805	0.9815	0.9828	0.9843	0.9859	0.9877	0.9895
		192.963	0.9793	0.9795	0.9799	0.9807	0.9817	0.9830	0.9845	0.9861	0.9878	0.9896
78	Pt	195.08	0.9795	0.9797	0.9802	0.9809	0.9819	0.9832	0.9846	0.9862	0.9879	0.9896
		189.960	0.9790	0.9792	0.9796	0.9804	0.9814	0.9827	0.9842	0.9859	0.9876	0.9894
		191.961	0.9792	0.9794	0.9798	0.9806	0.9816	0.9829	0.9844	0.9860	0.9878	0.9896
		193.963	0.9794	0.9796	0.9800	0.9808	0.9818	0.9831	0.9845	0.9862	0.9879	0.9897
		194.965	0.9795	0.9797	0.9801	0.9809	0.9819	0.9832	0.9846	0.9862	0.9879	0.9897
		195.965	0.9796	0.9798	0.9802	0.9810	0.9820	0.9832	0.9847	0.9863	0.9880	0.9898
		197.968	0.9798	0.9800	0.9804	0.9812	0.9822	0.9834	0.9848	0.9864	0.9881	0.9899

Table A5.1. Rutherford backscattering spectrometry kinematic factors for H as a projectile (continued).

Atomic no. (Z)	El.	Isotopic mass (M_2) (amu)	180°	170°	160°	150°	140°	130°	120°	110°	100°	90°
79	Au	196.967	0.9797	0.9799	0.9803	0.9811	0.9821	0.9833	0.9848	0.9864	0.9881	0.9898
80	Hg	200.588	0.9801	0.9803	0.9807	0.9814	0.9824	0.9836	0.9850	0.9866	0.9883	0.9900
		195.966	0.9796	0.9798	0.9802	0.9810	0.9820	0.9832	0.9847	0.9863	0.9880	0.9898
		197.967	0.9798	0.9800	0.9804	0.9812	0.9822	0.9834	0.9848	0.9864	0.9881	0.9899
		198.968	0.9799	0.9801	0.9805	0.9813	0.9823	0.9835	0.9849	0.9865	0.9882	0.9899
		199.968	0.9800	0.9802	0.9806	0.9814	0.9824	0.9836	0.9850	0.9866	0.9882	0.9900
		200.970	0.9801	0.9803	0.9807	0.9815	0.9824	0.9837	0.9851	0.9866	0.9883	0.9900
		201.971	0.9802	0.9804	0.9808	0.9815	0.9825	0.9837	0.9851	0.9867	0.9884	0.9901
81	Tl	203.973	0.9804	0.9806	0.9810	0.9817	0.9827	0.9839	0.9853	0.9868	0.9885	0.9902
		204.383	0.9805	0.9806	0.9811	0.9818	0.9827	0.9839	0.9853	0.9869	0.9885	0.9902
		202.972	0.9803	0.9805	0.9809	0.9816	0.9826	0.9838	0.9852	0.9868	0.9884	0.9901
		204.974	0.9805	0.9807	0.9811	0.9818	0.9828	0.9840	0.9854	0.9869	0.9885	0.9902
82	Pb	207.217	0.9807	0.9809	0.9813	0.9820	0.9830	0.9841	0.9855	0.9870	0.9886	0.9903
		203.973	0.9804	0.9806	0.9810	0.9817	0.9827	0.9839	0.9853	0.9868	0.9885	0.9902
		205.975	0.9806	0.9808	0.9812	0.9819	0.9829	0.9841	0.9854	0.9870	0.9886	0.9903
		206.976	0.9807	0.9809	0.9813	0.9820	0.9829	0.9841	0.9855	0.9870	0.9886	0.9903
		207.977	0.9808	0.9809	0.9814	0.9821	0.9830	0.9842	0.9856	0.9871	0.9887	0.9904
83	Bi	208.980	0.9809	0.9810	0.9815	0.9822	0.9831	0.9843	0.9856	0.9871	0.9887	0.9904
84	Po	208.982	0.9809	0.9810	0.9815	0.9822	0.9831	0.9843	0.9856	0.9871	0.9887	0.9904
85	At	210.000	0.9810	0.9811	0.9816	0.9822	0.9832	0.9844	0.9857	0.9872	0.9888	0.9910
86	Rn	222.018	0.9820	0.9821	0.9825	0.9832	0.9841	0.9852	0.9865	0.9879	0.9894	0.9910
87	Fr	223.000	0.9821	0.9822	0.9826	0.9833	0.9842	0.9853	0.9865	0.9879	0.9894	0.9911
88	Ra	226.025	0.9823	0.9825	0.9829	0.9835	0.9844	0.9855	0.9867	0.9881	0.9896	0.9912
89	Ac	227.000	0.9824	0.9825	0.9829	0.9836	0.9844	0.9855	0.9868	0.9882	0.9896	0.9914
90	Th	232.038	0.9828	0.9829	0.9833	0.9839	0.9848	0.9858	0.9871	0.9884	0.9899	0.9914
91	Pa	231.036	0.9827	0.9828	0.9832	0.9839	0.9847	0.9858	0.9870	0.9884	0.9898	0.9913
92	U	238.018	0.9832	0.9833	0.9837	0.9843	0.9852	0.9862	0.9874	0.9887	0.9901	0.9916
		234.041	0.9829	0.9831	0.9834	0.9841	0.9849	0.9860	0.9872	0.9885	0.9899	0.9914
		235.044	0.9830	0.9831	0.9835	0.9841	0.9850	0.9860	0.9872	0.9886	0.9900	0.9915
		238.040	0.9832	0.9833	0.9837	0.9843	0.9852	0.9862	0.9874	0.9887	0.9901	0.9916
93	Np	237.048	0.9831	0.9833	0.9836	0.9843	0.9851	0.9861	0.9873	0.9887	0.9901	0.9915
94	Pu	244.064	0.9836	0.9837	0.9841	0.9847	0.9855	0.9865	0.9877	0.9890	0.9904	0.9918
95	Am	243.061	0.9836	0.9837	0.9840	0.9846	0.9855	0.9865	0.9876	0.9889	0.9903	0.9917

Table A5.2. Rutherford backscattering spectrometry kinematic factors for He as a projectile.

This table gives the RBS kinematic factors K_{M_2}, defined by Eq. (A4.2), for He as a projectile (M_1=4.0026 amu) and for the isotopic masses (M_2) of the elements. The kinematic factors are given as a function of scattering angle from 180° to 90° measured in the laboratory frame of reference. The first row for each element gives the average atomic weight of that element and the average kine-matic factor (Average K_{M_2}) for that element. The subsequent rows give the isotopic masses for that element and the K_{M_2} for those isotopic masses. The average K_{M_2} is calculated as the weighted average of the K_{M_2}, in which the relative abundances (ϖ_{M_2}) for the M_2 are given in Appendix 1:

$$\text{Average } K_{M_2} = \Sigma\, \varpi_{M_2} K_{M_2} \text{ , summed over the masses } M_2.$$

Atomic no. (Z)	El.	Isotopic mass (M2) (amu)	180°	170°	160°	150°	140°	130°	120°	110°	100°	90°
3	Li	6.941	0.0722	0.0735	0.0774	0.0845	0.0955	0.1115	0.1343	0.1662	0.2099	0.2680
		6.015	0.0404	0.0412	0.0438	0.0485	0.0560	0.0675	0.0846	0.1102	0.1476	0.2009
		7.016	0.0748	0.0761	0.0802	0.0874	0.0987	0.1151	0.1384	0.1708	0.2150	0.2735
4	Be	9.012	0.1482	0.1502	0.1564	0.1671	0.1832	0.2056	0.2356	0.2747	0.3241	0.3849
5	B	10.811	0.2111	0.2135	0.2208	0.2333	0.2516	0.2766	0.3090	0.3499	0.3999	0.4592
		10.013	0.1839	0.1861	0.1930	0.2049	0.2225	0.2465	0.2782	0.3186	0.3686	0.4288
		11.009	0.2178	0.2203	0.2277	0.2403	0.2589	0.2841	0.3167	0.3577	0.4076	0.4667
6	C	12.011	0.2501	0.2526	0.2604	0.2736	0.2928	0.3187	0.3518	0.3929	0.4423	0.5001
		12.000	0.2498	0.2523	0.2600	0.2733	0.2925	0.3183	0.3515	0.3926	0.4420	0.4998
		13.003	0.2801	0.2828	0.2908	0.3044	0.3240	0.3502	0.3835	0.4244	0.4731	0.5293
7	N	14.007	0.3086	0.3113	0.3194	0.3333	0.3530	0.3795	0.4127	0.4532	0.5009	0.5555
		14.003	0.3085	0.3112	0.3193	0.3332	0.3530	0.3794	0.4126	0.4531	0.5008	0.5554
		15.000	0.3349	0.3377	0.3459	0.3599	0.3798	0.4062	0.4392	0.4790	0.5257	0.5787
8	O	15.999	0.3597	0.3625	0.3708	0.3848	0.4047	0.4309	0.4635	0.5027	0.5483	0.5998
		15.995	0.3596	0.3624	0.3707	0.3847	0.4046	0.4308	0.4634	0.5026	0.5482	0.5997
		16.999	0.3830	0.3857	0.3940	0.4080	0.4278	0.4538	0.4860	0.5244	0.5689	0.6188
		17.999	0.4047	0.4074	0.4157	0.4296	0.4493	0.4750	0.5067	0.5443	0.5877	0.6362
9	F	18.998	0.4251	0.4278	0.4360	0.4498	0.4693	0.4946	0.5258	0.5627	0.6050	0.6520
10	Ne	20.179	0.4473	0.4500	0.4582	0.4718	0.4910	0.5159	0.5464	0.5824	0.6234	0.6688
		19.992	0.4441	0.4468	0.4549	0.4686	0.4879	0.5128	0.5434	0.5795	0.6207	0.6664
		20.994	0.4621	0.4647	0.4728	0.4863	0.5053	0.5299	0.5599	0.5952	0.6354	0.6797
		21.991	0.4789	0.4816	0.4896	0.5029	0.5216	0.5458	0.5752	0.6098	0.6490	0.6920
11	Na	22.990	0.4948	0.4974	0.5053	0.5185	0.5370	0.5607	0.5896	0.6233	0.6615	0.7034
12	Mg	24.305	0.5142	0.5168	0.5245	0.5375	0.5555	0.5787	0.6069	0.6397	0.6766	0.7171
		23.985	0.5098	0.5123	0.5201	0.5331	0.5513	0.5746	0.6029	0.6360	0.6732	0.7140
		24.986	0.5240	0.5265	0.5342	0.5470	0.5649	0.5878	0.6155	0.6478	0.6842	0.7238
		25.983	0.5373	0.5399	0.5474	0.5600	0.5776	0.6001	0.6273	0.6589	0.6944	0.7330
13	Al	26.982	0.5500	0.5525	0.5600	0.5724	0.5897	0.6117	0.6384	0.6693	0.7040	0.7416
14	Si	28.086	0.5632	0.5657	0.5730	0.5852	0.6022	0.6238	0.6499	0.6801	0.7139	0.7505
		27.977	0.5620	0.5645	0.5718	0.5840	0.6010	0.6227	0.6489	0.6791	0.7130	0.7497
		28.976	0.5734	0.5759	0.5831	0.5951	0.6118	0.6331	0.6588	0.6884	0.7215	0.7573
		29.974	0.5843	0.5867	0.5938	0.6056	0.6220	0.6430	0.6681	0.6971	0.7294	0.7644
15	P	30.974	0.5946	0.5970	0.6040	0.6156	0.6318	0.6523	0.6770	0.7054	0.7370	0.7711
16	S	32.064	0.6053	0.6076	0.6145	0.6259	0.6418	0.6619	0.6861	0.7139	0.7448	0.7780
		31.972	0.6045	0.6068	0.6137	0.6251	0.6410	0.6612	0.6854	0.7132	0.7441	0.7775
		32.972	0.6139	0.6161	0.6229	0.6342	0.6498	0.6696	0.6934	0.7206	0.7509	0.7835
		33.968	0.6228	0.6250	0.6317	0.6428	0.6582	0.6776	0.7009	0.7277	0.7573	0.7892
		35.967	0.6395	0.6417	0.6482	0.6589	0.6738	0.6926	0.7150	0.7408	0.7692	0.7997

Table A5.2. Rutherford backscattering spectrometry kinematic factors for He as a projectile (continued).

Atomic no. (Z)	El.	Isotopic mass (M₂) (amu)	180°	170°	160°	150°	140°	130°	120°	110°	100°	90°
17	Cl	35.453	0.6353	0.6374	0.6440	0.6548	0.6698	0.6887	0.7114	0.7374	0.7662	0.7970
		34.969	0.6314	0.6336	0.6402	0.6511	0.6662	0.6853	0.7082	0.7344	0.7634	0.7946
		36.966	0.6474	0.6495	0.6559	0.6665	0.6811	0.6996	0.7216	0.7468	0.7747	0.8046
18	Ar	39.948	0.6689	0.6709	0.6770	0.6871	0.7010	0.7186	0.7395	0.7634	0.7897	0.8179
		35.967	0.6396	0.6417	0.6482	0.6589	0.6738	0.6926	0.7150	0.7408	0.7692	0.7997
		37.963	0.6549	0.6570	0.6633	0.6737	0.6880	0.7062	0.7279	0.7526	0.7800	0.8092
		39.962	0.6690	0.6710	0.6771	0.6872	0.7011	0.7187	0.7396	0.7635	0.7898	0.8179
19	K	39.098	0.6630	0.6651	0.6712	0.6815	0.6956	0.7134	0.7346	0.7589	0.7850	0.8142
		38.964	0.6621	0.6642	0.6703	0.6806	0.6947	0.7126	0.7339	0.7582	0.7850	0.8137
		40.000	0.6692	0.6713	0.6774	0.6874	0.7014	0.7189	0.7398	0.7637	0.7900	0.8181
		40.962	0.6756	0.6776	0.6836	0.6936	0.7073	0.7246	0.7451	0.7686	0.7944	0.8220
20	Ca	40.078	0.6697	0.6717	0.6778	0.6879	0.7018	0.7193	0.7402	0.7641	0.7903	0.8184
		39.963	0.6690	0.6710	0.6771	0.6872	0.7011	0.7187	0.7396	0.7635	0.7898	0.8179
		41.959	0.6820	0.6840	0.6899	0.6997	0.7132	0.7302	0.7504	0.7734	0.7988	0.8258
		42.959	0.6881	0.6901	0.6959	0.7055	0.7188	0.7356	0.7555	0.7781	0.8030	0.8295
		43.956	0.6940	0.6959	0.7017	0.7112	0.7243	0.7407	0.7603	0.7826	0.8070	0.8331
		45.954	0.7052	0.7071	0.7126	0.7218	0.7345	0.7505	0.7695	0.7910	0.8146	0.8398
		47.952	0.7156	0.7174	0.7228	0.7318	0.7441	0.7596	0.7780	0.7988	0.8217	0.8459
21	Sc	44.956	0.6997	0.7016	0.7073	0.7166	0.7295	0.7457	0.7650	0.7869	0.8109	0.8365
22	Ti	47.878	0.7152	0.7170	0.7224	0.7314	0.7437	0.7592	0.7776	0.7985	0.8214	0.8457
		45.953	0.7052	0.7071	0.7126	0.7218	0.7345	0.7505	0.7695	0.7910	0.8146	0.8398
		46.952	0.7105	0.7123	0.7178	0.7269	0.7394	0.7551	0.7738	0.7950	0.8182	0.8429
		47.948	0.7156	0.7174	0.7228	0.7318	0.7441	0.7596	0.7780	0.7988	0.8217	0.8459
		48.948	0.7205	0.7223	0.7276	0.7365	0.7486	0.7639	0.7820	0.8025	0.8250	0.8488
		49.945	0.7252	0.7270	0.7323	0.7410	0.7530	0.7680	0.7858	0.8060	0.8282	0.8516
23	V	50.942	0.7298	0.7316	0.7368	0.7454	0.7572	0.7720	0.7896	0.8095	0.8312	0.8543
		49.947	0.7253	0.7270	0.7323	0.7410	0.7530	0.7680	0.7858	0.8061	0.8282	0.8516
		50.944	0.7298	0.7316	0.7368	0.7454	0.7572	0.7720	0.7896	0.8095	0.8312	0.8543
24	Cr	51.996	0.7345	0.7362	0.7414	0.7498	0.7615	0.7761	0.7934	0.8129	0.8344	0.8570
		49.946	0.7252	0.7270	0.7323	0.7410	0.7530	0.7680	0.7858	0.8061	0.8282	0.8516
		51.940	0.7343	0.7360	0.7411	0.7496	0.7613	0.7759	0.7932	0.8128	0.8342	0.8569
		52.941	0.7386	0.7403	0.7454	0.7537	0.7652	0.7796	0.7967	0.8160	0.8371	0.8594
		53.939	0.7428	0.7444	0.7494	0.7577	0.7690	0.7832	0.8001	0.8191	0.8399	0.8618
25	Mn	54.938	0.7468	0.7485	0.7534	0.7615	0.7727	0.7867	0.8033	0.8221	0.8425	0.8642
26	Fe	55.847	0.7504	0.7520	0.7569	0.7649	0.7760	0.7898	0.8062	0.8247	0.8449	0.8662
		53.940	0.7428	0.7444	0.7494	0.7577	0.7690	0.7832	0.8001	0.8191	0.8399	0.8618
		55.935	0.7507	0.7524	0.7572	0.7653	0.7763	0.7901	0.8065	0.8250	0.8451	0.8664
		56.935	0.7545	0.7561	0.7610	0.7689	0.7798	0.7934	0.8095	0.8278	0.8476	0.8686
		57.933	0.7582	0.7598	0.7646	0.7724	0.7831	0.7966	0.8125	0.8305	0.8501	0.8708
27	Co	58.933	0.7618	0.7634	0.7681	0.7758	0.7864	0.7997	0.8154	0.8331	0.8524	0.8728
28	Ni	58.688	0.7608	0.7624	0.7671	0.7749	0.7855	0.7989	0.8146	0.8324	0.8518	0.8723
		57.935	0.7582	0.7598	0.7646	0.7724	0.7831	0.7966	0.8125	0.8305	0.8501	0.8708
		59.931	0.7653	0.7668	0.7714	0.7791	0.7896	0.8027	0.8182	0.8356	0.8547	0.8748
		60.931	0.7686	0.7702	0.7747	0.7823	0.7926	0.8056	0.8209	0.8381	0.8569	0.8767
		61.928	0.7719	0.7734	0.7779	0.7854	0.7956	0.8084	0.8235	0.8405	0.8590	0.8786
		63.928	0.7782	0.7797	0.7841	0.7914	0.8013	0.8138	0.8285	0.8451	0.8631	0.8822

Table A5.2. Rutherford backscattering spectrometry kinematic factors for He as a projectile (continued).

Atomic no. (Z)	El.	Isotopic mass (M_2) (amu)	180°	170°	160°	150°	140°	130°	120°	110°	100°	90°
29	Cu	63.546	0.7770	0.7785	0.7829	0.7902	0.8002	0.8128	0.8276	0.8442	0.8624	0.8815
		62.930	0.7751	0.7766	0.7811	0.7884	0.7985	0.8112	0.8260	0.8428	0.8611	0.8804
		64.928	0.7812	0.7827	0.7870	0.7942	0.8041	0.8164	0.8309	0.8473	0.8651	0.8839
30	Zn	65.396	0.7825	0.7840	0.7883	0.7955	0.8053	0.8175	0.8320	0.8482	0.8659	0.8846
		63.929	0.7782	0.7797	0.7841	0.7914	0.8013	0.8138	0.8285	0.8451	0.8631	0.8822
		65.926	0.7842	0.7856	0.7899	0.7970	0.8068	0.8189	0.8333	0.8494	0.8670	0.8855
		66.927	0.7870	0.7884	0.7927	0.7997	0.8094	0.8214	0.8356	0.8515	0.8689	0.8871
		67.925	0.7898	0.7912	0.7954	0.8024	0.8119	0.8238	0.8378	0.8535	0.8707	0.8887
31	Ga	69.925	0.7952	0.7965	0.8007	0.8075	0.8168	0.8284	0.8420	0.8574	0.8741	0.8917
		69.723	0.7946	0.7960	0.8001	0.8069	0.8162	0.8279	0.8416	0.8570	0.8738	0.8914
		68.926	0.7925	0.7939	0.7981	0.8049	0.8144	0.8261	0.8399	0.8555	0.8724	0.8902
		70.925	0.7977	0.7991	0.8032	0.8099	0.8191	0.8306	0.8441	0.8593	0.8758	0.8932
32	Ge	72.632	0.8019	0.8032	0.8073	0.8138	0.8229	0.8341	0.8474	0.8623	0.8785	0.8955
		69.924	0.7952	0.7965	0.8007	0.8074	0.8167	0.8284	0.8420	0.8574	0.8741	0.8917
		71.922	0.8002	0.8016	0.8056	0.8123	0.8214	0.8327	0.8461	0.8611	0.8774	0.8946
		72.924	0.8027	0.8040	0.8080	0.8146	0.8236	0.8348	0.8480	0.8629	0.8790	0.8959
		73.921	0.8051	0.8064	0.8104	0.8169	0.8258	0.8369	0.8499	0.8646	0.8805	0.8973
		75.921	0.8097	0.8110	0.8149	0.8212	0.8299	0.8408	0.8536	0.8679	0.8835	0.8998
33	As	74.922	0.8074	0.8087	0.8126	0.8191	0.8279	0.8389	0.8518	0.8663	0.8820	0.8986
34	Se	78.993	0.8163	0.8176	0.8213	0.8275	0.8359	0.8464	0.8588	0.8727	0.8877	0.9035
		73.923	0.8051	0.8064	0.8104	0.8169	0.8258	0.8369	0.8499	0.8646	0.8805	0.8973
		75.919	0.8097	0.8110	0.8149	0.8212	0.8299	0.8408	0.8536	0.8679	0.8835	0.8998
		76.920	0.8119	0.8132	0.8170	0.8233	0.8320	0.8427	0.8553	0.8695	0.8849	0.9011
		77.917	0.8141	0.8154	0.8192	0.8254	0.8339	0.8446	0.8570	0.8711	0.8863	0.9023
		79.916	0.8183	0.8196	0.8233	0.8294	0.8377	0.8481	0.8604	0.8741	0.8890	0.9046
		81.917	0.8223	0.8236	0.8272	0.8332	0.8414	0.8516	0.8635	0.8770	0.8916	0.9068
35	Br	79.904	0.8183	0.8195	0.8232	0.8293	0.8377	0.8481	0.8603	0.8741	0.8890	0.9046
		78.918	0.8162	0.8175	0.8212	0.8274	0.8358	0.8464	0.8587	0.8726	0.8877	0.9035
		80.916	0.8203	0.8216	0.8253	0.8313	0.8396	0.8499	0.8620	0.8756	0.8903	0.9057
36	Kr	83.8	0.8259	0.8271	0.8307	0.8366	0.8446	0.8546	0.8664	0.8796	0.8938	0.9088
		77.920	0.8141	0.8154	0.8192	0.8254	0.8339	0.8446	0.8571	0.8711	0.8863	0.9023
		79.916	0.8183	0.8196	0.8233	0.8294	0.8377	0.8481	0.8604	0.8741	0.8890	0.9046
		81.913	0.8223	0.8236	0.8272	0.8332	0.8414	0.8516	0.8635	0.8770	0.8915	0.9068
		82.914	0.8243	0.8255	0.8291	0.8350	0.8431	0.8532	0.8651	0.8784	0.8928	0.9079
		83.911	0.8262	0.8274	0.8309	0.8368	0.8448	0.8548	0.8666	0.8797	0.8940	0.9089
		85.911	0.8299	0.8310	0.8345	0.8403	0.8482	0.8580	0.8695	0.8824	0.8963	0.9110
37	Rb	85.468	0.8290	0.8302	0.8337	0.8395	0.8474	0.8573	0.8688	0.8818	0.8958	0.9105
		84.912	0.8280	0.8292	0.8328	0.8386	0.8465	0.8564	0.8680	0.8811	0.8952	0.9100
38	Sr	86.909	0.8316	0.8328	0.8363	0.8420	0.8498	0.8595	0.8709	0.8836	0.8975	0.9119
		87.617	0.8329	0.8340	0.8375	0.8431	0.8509	0.8605	0.8718	0.8845	0.8982	0.9126
		83.913	0.8262	0.8274	0.8309	0.8368	0.8448	0.8548	0.8666	0.8797	0.8940	0.9089
		85.909	0.8299	0.8310	0.8345	0.8403	0.8482	0.8580	0.8695	0.8824	0.8963	0.9110
		86.909	0.8316	0.8328	0.8363	0.8420	0.8498	0.8595	0.8709	0.8836	0.8975	0.9119
39	Y	87.906	0.8334	0.8345	0.8380	0.8436	0.8513	0.8610	0.8722	0.8849	0.8986	0.9129
		88.906	0.8351	0.8362	0.8396	0.8452	0.8529	0.8624	0.8736	0.8861	0.8996	0.9138

Table A5.2. Rutherford backscattering spectrometry kinematic factors for He as a projectile (continued).

Atomic no. (Z)	El.	Isotopic mass (M₂) (amu)	180°	170°	160°	150°	140°	130°	120°	110°	100°	90°
40	Zr	91.221	0.8389	0.8400	0.8433	0.8488	0.8563	0.8656	0.8765	0.8888	0.9020	0.9159
		89.905	0.8368	0.8379	0.8413	0.8468	0.8544	0.8638	0.8749	0.8873	0.9007	0.9148
		90.906	0.8384	0.8395	0.8429	0.8484	0.8559	0.8652	0.8762	0.8885	0.9017	0.9157
		91.905	0.8400	0.8411	0.8445	0.8499	0.8573	0.8666	0.8774	0.8896	0.9028	0.9165
		93.906	0.8432	0.8443	0.8475	0.8528	0.8601	0.8692	0.8799	0.8918	0.9047	0.9182
		95.908	0.8462	0.8472	0.8504	0.8557	0.8629	0.8718	0.8822	0.8940	0.9066	0.9199
41	Nb	92.906	0.8416	0.8427	0.8460	0.8514	0.8588	0.8679	0.8787	0.8907	0.9038	0.9174
42	Mo	95.931	0.8461	0.8472	0.8504	0.8556	0.8628	0.8718	0.8822	0.8939	0.9066	0.9199
		91.907	0.8400	0.8411	0.8445	0.8499	0.8573	0.8666	0.8774	0.8896	0.9028	0.9165
		93.905	0.8432	0.8443	0.8475	0.8528	0.8601	0.8692	0.8799	0.8918	0.9047	0.9182
		94.906	0.8447	0.8458	0.8490	0.8543	0.8615	0.8705	0.8811	0.8929	0.9057	0.9191
		95.905	0.8462	0.8472	0.8504	0.8557	0.8629	0.8718	0.8822	0.8940	0.9066	0.9199
		96.906	0.8476	0.8487	0.8519	0.8571	0.8642	0.8730	0.8834	0.8950	0.9075	0.9207
		97.905	0.8491	0.8501	0.8533	0.8584	0.8655	0.8742	0.8845	0.8960	0.9084	0.9214
		99.908	0.8519	0.8529	0.8560	0.8611	0.8680	0.8766	0.8867	0.8980	0.9102	0.9230
43	Tc	98.000	0.8492	0.8503	0.8534	0.8585	0.8656	0.8743	0.8846	0.8961	0.9085	0.9215
44	Ru	101.019	0.8533	0.8543	0.8574	0.8624	0.8693	0.8778	0.8878	0.8990	0.9111	0.9237
		95.908	0.8462	0.8472	0.8504	0.8557	0.8629	0.8718	0.8822	0.8940	0.9066	0.9199
		97.905	0.8491	0.8501	0.8533	0.8584	0.8655	0.8742	0.8845	0.8960	0.9084	0.9214
		98.506	0.8499	0.8510	0.8541	0.8592	0.8662	0.8750	0.8852	0.8966	0.9090	0.9219
		99.904	0.8519	0.8529	0.8560	0.8611	0.8680	0.8766	0.8867	0.8980	0.9102	0.9230
		100.906	0.8532	0.8542	0.8573	0.8623	0.8692	0.8777	0.8877	0.8989	0.9110	0.9237
		101.904	0.8545	0.8556	0.8586	0.8636	0.8704	0.8789	0.8888	0.8999	0.9119	0.9244
		103.905	0.8571	0.8581	0.8611	0.8660	0.8727	0.8811	0.8908	0.9017	0.9135	0.9258
45	Rh	102.906	0.8558	0.8569	0.8599	0.8648	0.8716	0.8800	0.8898	0.9008	0.9127	0.9251
46	Pd	106.415	0.8602	0.8612	0.8641	0.8689	0.8755	0.8837	0.8932	0.9039	0.9154	0.9275
		101.906	0.8545	0.8556	0.8586	0.8636	0.8704	0.8789	0.8888	0.8999	0.9119	0.9244
		103.904	0.8571	0.8581	0.8611	0.8660	0.8727	0.8811	0.8908	0.9017	0.9135	0.9258
		104.905	0.8584	0.8594	0.8624	0.8672	0.8739	0.8821	0.8918	0.9026	0.9143	0.9265
		105.904	0.8596	0.8606	0.8636	0.8684	0.8750	0.8832	0.8928	0.9035	0.9151	0.9272
		107.904	0.8620	0.8630	0.8659	0.8707	0.8771	0.8852	0.8946	0.9052	0.9166	0.9285
		109.905	0.8644	0.8653	0.8682	0.8729	0.8792	0.8872	0.8965	0.9068	0.9180	0.9297
47	Ag	107.868	0.8620	0.8630	0.8659	0.8706	0.8771	0.8852	0.8946	0.9052	0.9165	0.9284
		106.905	0.8609	0.8618	0.8647	0.8695	0.8761	0.8842	0.8937	0.9043	0.9158	0.9278
		108.905	0.8632	0.8642	0.8671	0.8718	0.8782	0.8862	0.8956	0.9060	0.9173	0.9291
48	Cd	112.412	0.8672	0.8681	0.8709	0.8755	0.8817	0.8895	0.8986	0.9088	0.9198	0.9312
		105.907	0.8596	0.8606	0.8636	0.8684	0.8750	0.8832	0.8928	0.9035	0.9151	0.9272
		107.904	0.8620	0.8630	0.8659	0.8707	0.8771	0.8852	0.8946	0.9052	0.9166	0.9285
		109.903	0.8644	0.8653	0.8682	0.8729	0.8792	0.8872	0.8965	0.9068	0.9180	0.9297
		110.904	0.8655	0.8665	0.8693	0.8739	0.8803	0.8881	0.8973	0.9076	0.9187	0.9303
		111.903	0.8666	0.8676	0.8704	0.8750	0.8813	0.8891	0.8982	0.9084	0.9194	0.9309
		112.904	0.8677	0.8687	0.8715	0.8760	0.8823	0.8900	0.8991	0.9092	0.9201	0.9315
		113.903	0.8688	0.8697	0.8725	0.8770	0.8832	0.8909	0.8999	0.9100	0.9208	0.9321
		115.905	0.8709	0.8718	0.8746	0.8790	0.8851	0.8927	0.9015	0.9114	0.9221	0.9332
49	In	114.818	0.8698	0.8707	0.8735	0.8780	0.8841	0.8917	0.9007	0.9106	0.9214	0.9326
		112.904	0.8677	0.8687	0.8715	0.8760	0.8823	0.8900	0.8991	0.9092	0.9201	0.9315
		114.904	0.8699	0.8708	0.8735	0.8780	0.8842	0.8918	0.9007	0.9107	0.9215	0.9327

Table A5.2. Rutherford backscattering spectrometry kinematic factors for He as a projectile (continued).

Atomic no. (Z)	El.	Isotopic mass (M_2) (amu)	180°	170°	160°	150°	140°	130°	120°	110°	100°	90°
50	Sn	118.613	0.8737	0.8745	0.8772	0.8816	0.8876	0.8950	0.9037	0.9133	0.9238	0.9347
		111.905	0.8666	0.8676	0.8704	0.8750	0.8813	0.8891	0.8982	0.9084	0.9194	0.9309
		113.903	0.8688	0.8697	0.8725	0.8770	0.8832	0.8909	0.8999	0.9100	0.9208	0.9321
		114.903	0.8699	0.8708	0.8735	0.8780	0.8842	0.8918	0.9007	0.9107	0.9215	0.9327
		115.902	0.8709	0.8718	0.8746	0.8790	0.8851	0.8927	0.9015	0.9114	0.9221	0.9332
		116.903	0.8720	0.8729	0.8756	0.8800	0.8860	0.8936	0.9023	0.9122	0.9227	0.9338
		117.602	0.8727	0.8736	0.8763	0.8807	0.8867	0.8942	0.9029	0.9127	0.9232	0.9342
		118.903	0.8740	0.8749	0.8775	0.8819	0.8879	0.8953	0.9039	0.9136	0.9240	0.9349
		119.902	0.8750	0.8758	0.8785	0.8828	0.8887	0.8961	0.9047	0.9143	0.9246	0.9354
		121.903	0.8769	0.8778	0.8804	0.8846	0.8905	0.8977	0.9062	0.9156	0.9258	0.9364
		123.905	0.8787	0.8796	0.8822	0.8864	0.8921	0.8993	0.9076	0.9169	0.9270	0.9374
51	Sb	121.758	0.8767	0.8776	0.8802	0.8845	0.8903	0.8976	0.9060	0.9155	0.9257	0.9363
		120.904	0.8759	0.8768	0.8794	0.8837	0.8896	0.8969	0.9054	0.9149	0.9252	0.9359
		122.904	0.8778	0.8787	0.8813	0.8855	0.8913	0.8985	0.9069	0.9163	0.9264	0.9369
52	Te	127.586	0.8820	0.8828	0.8853	0.8894	0.8950	0.9020	0.9101	0.9192	0.9290	0.9391
		119.904	0.8750	0.8758	0.8785	0.8828	0.8887	0.8961	0.9047	0.9143	0.9246	0.9354
		121.903	0.8769	0.8778	0.8804	0.8846	0.8905	0.8977	0.9062	0.9156	0.9258	0.9364
		122.904	0.8778	0.8787	0.8813	0.8855	0.8913	0.8985	0.9069	0.9163	0.9264	0.9369
		123.903	0.8787	0.8796	0.8822	0.8864	0.8921	0.8993	0.9076	0.9169	0.9269	0.9374
		124.904	0.8797	0.8805	0.8831	0.8872	0.8929	0.9000	0.9083	0.9176	0.9275	0.9379
		125.903	0.8806	0.8814	0.8839	0.8881	0.8937	0.9008	0.9090	0.9182	0.9281	0.9384
		127.905	0.8823	0.8831	0.8856	0.8897	0.8953	0.9023	0.9104	0.9194	0.9292	0.9393
		129.906	0.8840	0.8848	0.8873	0.8913	0.8969	0.9037	0.9117	0.9206	0.9302	0.9402
53	I	126.905	0.8814	0.8823	0.8848	0.8889	0.8945	0.9015	0.9097	0.9188	0.9286	0.9388
54	Xe	131.293	0.8851	0.8860	0.8884	0.8924	0.8979	0.9046	0.9125	0.9214	0.9309	0.9408
		123.906	0.8787	0.8796	0.8822	0.8864	0.8921	0.8993	0.9076	0.9169	0.9270	0.9374
		125.904	0.8806	0.8814	0.8839	0.8881	0.8937	0.9008	0.9090	0.9182	0.9281	0.9384
		127.904	0.8823	0.8831	0.8856	0.8897	0.8953	0.9023	0.9104	0.9194	0.9292	0.9393
		128.905	0.8832	0.8840	0.8865	0.8905	0.8961	0.9030	0.9110	0.9200	0.9297	0.9398
		129.904	0.8840	0.8848	0.8873	0.8913	0.8968	0.9037	0.9117	0.9206	0.9302	0.9402
		130.905	0.8848	0.8857	0.8881	0.8921	0.8976	0.9044	0.9123	0.9212	0.9307	0.9407
		131.904	0.8857	0.8865	0.8889	0.8929	0.8983	0.9051	0.9130	0.9218	0.9312	0.9411
		133.905	0.8873	0.8881	0.8905	0.8944	0.8998	0.9064	0.9142	0.9229	0.9322	0.9420
		135.907	0.8888	0.8896	0.8920	0.8959	0.9012	0.9077	0.9154	0.9240	0.9332	0.9428
55	Cs	132.905	0.8865	0.8873	0.8897	0.8937	0.8991	0.9058	0.9136	0.9223	0.9317	0.9415
56	Ba	137.327	0.8899	0.8907	0.8931	0.8969	0.9021	0.9086	0.9162	0.9247	0.9338	0.9434
		129.906	0.8840	0.8848	0.8873	0.8913	0.8969	0.9037	0.9117	0.9206	0.9302	0.9402
		131.905	0.8857	0.8865	0.8889	0.8929	0.8983	0.9051	0.9130	0.9218	0.9312	0.9411
		133.904	0.8873	0.8881	0.8905	0.8944	0.8998	0.9064	0.9142	0.9229	0.9322	0.9420
		134.906	0.8881	0.8889	0.8912	0.8952	0.9005	0.9071	0.9148	0.9234	0.9327	0.9424
		135.905	0.8888	0.8896	0.8920	0.8959	0.9012	0.9077	0.9154	0.9240	0.9332	0.9428
		136.906	0.8896	0.8904	0.8927	0.8966	0.9019	0.9084	0.9160	0.9245	0.9337	0.9432
		137.905	0.8904	0.8911	0.8935	0.8973	0.9025	0.9090	0.9166	0.9250	0.9341	0.9436
57	La	138.905	0.8911	0.8919	0.8942	0.8980	0.9032	0.9096	0.9172	0.9256	0.9346	0.9440
		137.907	0.8904	0.8911	0.8935	0.8973	0.9025	0.9090	0.9166	0.9250	0.9341	0.9436
		138.906	0.8911	0.8919	0.8942	0.8980	0.9032	0.9096	0.9172	0.9256	0.9346	0.9440
58	Ce	140.115	0.8920	0.8928	0.8951	0.8988	0.9040	0.9104	0.9178	0.9262	0.9351	0.9445
		135.907	0.8888	0.8896	0.8920	0.8959	0.9012	0.9077	0.9154	0.9240	0.9332	0.9428

Table A5.2. Rutherford backscattering spectrometry kinematic factors for He as a projectile (continued).

Atomic no. (Z)	El.	Isotopic mass (M₂) (amu)	180°	170°	160°	150°	140°	130°	120°	110°	100°	90°
		137.906	0.8904	0.8911	0.8935	0.8973	0.9025	0.9090	0.9166	0.9250	0.9341	0.9436
		139.905	0.8918	0.8926	0.8949	0.8987	0.9039	0.9103	0.9177	0.9261	0.9350	0.9444
		141.909	0.8933	0.8940	0.8963	0.9001	0.9052	0.9115	0.9188	0.9271	0.9359	0.9451
59	Pr	140.908	0.8926	0.8933	0.8956	0.8994	0.9045	0.9109	0.9183	0.9266	0.9355	0.9448
60	Nd	144.242	0.8949	0.8956	0.8979	0.9016	0.9066	0.9128	0.9201	0.9282	0.9369	0.9460
		141.908	0.8933	0.8940	0.8963	0.9001	0.9052	0.9115	0.9188	0.9271	0.9359	0.9451
		142.910	0.8940	0.8948	0.8970	0.9007	0.9058	0.9121	0.9194	0.9276	0.9364	0.9455
		143.910	0.8947	0.8954	0.8977	0.9014	0.9064	0.9126	0.9199	0.9280	0.9368	0.9459
		144.913	0.8954	0.8961	0.8984	0.9020	0.9070	0.9132	0.9205	0.9285	0.9372	0.9462
		145.913	0.8961	0.8968	0.8990	0.9027	0.9076	0.9138	0.9210	0.9290	0.9376	0.9466
		147.917	0.8974	0.8981	0.9003	0.9039	0.9088	0.9149	0.9220	0.9299	0.9384	0.9473
		149.921	0.8987	0.8994	0.9016	0.9051	0.9100	0.9160	0.9230	0.9308	0.9392	0.9480
61	Pm	145.000	0.8954	0.8962	0.8984	0.9021	0.9071	0.9133	0.9205	0.9286	0.9372	0.9463
62	Sm	150.36	0.8989	0.8997	0.9018	0.9054	0.9102	0.9162	0.9232	0.9310	0.9394	0.9481
		143.912	0.8947	0.8954	0.8977	0.9014	0.9064	0.9126	0.9199	0.9280	0.9368	0.9459
		146.915	0.8967	0.8975	0.8997	0.9033	0.9082	0.9144	0.9215	0.9295	0.9380	0.9470
		147.915	0.8974	0.8981	0.9003	0.9039	0.9088	0.9149	0.9220	0.9299	0.9384	0.9473
		148.917	0.8980	0.8988	0.9010	0.9045	0.9094	0.9155	0.9225	0.9304	0.9388	0.9477
		149.917	0.8987	0.8994	0.9016	0.9051	0.9100	0.9160	0.9230	0.9308	0.9392	0.9480
		151.920	0.9000	0.9007	0.9028	0.9063	0.9111	0.9171	0.9240	0.9317	0.9400	0.9487
		153.922	0.9012	0.9019	0.9040	0.9075	0.9122	0.9181	0.9249	0.9326	0.9408	0.9493
63	Eu	151.965	0.9000	0.9007	0.9028	0.9064	0.9111	0.9171	0.9240	0.9317	0.9400	0.9487
		150.920	0.8993	0.9001	0.9022	0.9057	0.9106	0.9165	0.9235	0.9313	0.9396	0.9483
		152.921	0.9006	0.9013	0.9034	0.9069	0.9117	0.9176	0.9245	0.9321	0.9404	0.9490
64	Gd	157.252	0.9032	0.9039	0.9059	0.9093	0.9140	0.9197	0.9265	0.9339	0.9420	0.9504
		151.920	0.9000	0.9007	0.9028	0.9063	0.9111	0.9171	0.9240	0.9317	0.9400	0.9487
		153.921	0.9012	0.9019	0.9040	0.9075	0.9122	0.9181	0.9249	0.9326	0.9408	0.9493
		154.923	0.9018	0.9025	0.9046	0.9081	0.9128	0.9186	0.9254	0.9330	0.9411	0.9496
		155.922	0.9024	0.9031	0.9052	0.9086	0.9133	0.9191	0.9259	0.9334	0.9415	0.9499
		156.924	0.9030	0.9037	0.9058	0.9092	0.9138	0.9196	0.9263	0.9338	0.9419	0.9503
		157.924	0.9036	0.9043	0.9063	0.9097	0.9143	0.9201	0.9268	0.9342	0.9422	0.9506
		159.927	0.9047	0.9054	0.9075	0.9108	0.9154	0.9210	0.9277	0.9350	0.9429	0.9512
65	Tb	158.925	0.9041	0.9048	0.9069	0.9103	0.9149	0.9206	0.9272	0.9346	0.9426	0.9509
66	Dy	162.498	0.9061	0.9068	0.9088	0.9121	0.9167	0.9222	0.9287	0.9360	0.9438	0.9519
		155.924	0.9024	0.9031	0.9052	0.9086	0.9133	0.9191	0.9259	0.9334	0.9415	0.9499
		157.924	0.9036	0.9043	0.9063	0.9097	0.9144	0.9201	0.9268	0.9342	0.9422	0.9506
		159.925	0.9047	0.9054	0.9075	0.9108	0.9154	0.9210	0.9276	0.9350	0.9429	0.9512
		160.927	0.9053	0.9060	0.9080	0.9113	0.9159	0.9215	0.9281	0.9354	0.9433	0.9515
		161.927	0.9058	0.9065	0.9085	0.9119	0.9164	0.9220	0.9285	0.9358	0.9436	0.9518
		162.929	0.9064	0.9071	0.9091	0.9124	0.9169	0.9224	0.9289	0.9362	0.9440	0.9520
		163.929	0.9069	0.9076	0.9096	0.9129	0.9174	0.9229	0.9294	0.9366	0.9443	0.9523
67	Ho	164.930	0.9075	0.9081	0.9101	0.9134	0.9178	0.9233	0.9298	0.9369	0.9446	0.9526
68	Er	167.256	0.9087	0.9094	0.9113	0.9145	0.9189	0.9244	0.9307	0.9378	0.9454	0.9533
		161.929	0.9058	0.9065	0.9085	0.9119	0.9164	0.9220	0.9285	0.9358	0.9436	0.9518
		163.929	0.9069	0.9076	0.9096	0.9129	0.9174	0.9229	0.9294	0.9366	0.9443	0.9523
		165.930	0.9080	0.9087	0.9106	0.9139	0.9183	0.9238	0.9302	0.9373	0.9449	0.9529
		166.932	0.9085	0.9092	0.9112	0.9144	0.9188	0.9242	0.9306	0.9377	0.9453	0.9532
		167.932	0.9090	0.9097	0.9117	0.9149	0.9192	0.9247	0.9310	0.9380	0.9456	0.9534
		169.936	0.9101	0.9107	0.9127	0.9158	0.9202	0.9255	0.9318	0.9387	0.9462	0.9540

Table A5.2. Rutherford backscattering spectrometry kinematic factors for He as a projectile (continued).

Atomic no. (Z)	El.	Isotopic mass (M₂) (amu)	180°	170°	160°	150°	140°	130°	120°	110°	100°	90°
69	Tm	168.934	0.9096	0.9102	0.9122	0.9154	0.9197	0.9251	0.9314	0.9384	0.9459	0.9537
70	Yb	173.034	0.9116	0.9122	0.9141	0.9173	0.9215	0.9268	0.9329	0.9398	0.9471	0.9548
		167.934	0.9090	0.9097	0.9117	0.9149	0.9192	0.9247	0.9310	0.9380	0.9456	0.9534
		169.935	0.9101	0.9107	0.9127	0.9158	0.9202	0.9255	0.9318	0.9387	0.9462	0.9540
		170.936	0.9106	0.9112	0.9131	0.9163	0.9206	0.9259	0.9321	0.9391	0.9465	0.9542
		171.936	0.9111	0.9117	0.9136	0.9168	0.9210	0.9264	0.9325	0.9394	0.9468	0.9545
		172.938	0.9116	0.9122	0.9141	0.9172	0.9215	0.9268	0.9329	0.9398	0.9471	0.9548
		173.939	0.9120	0.9127	0.9146	0.9177	0.9219	0.9272	0.9333	0.9401	0.9474	0.9550
		175.943	0.9130	0.9136	0.9155	0.9186	0.9228	0.9280	0.9340	0.9408	0.9480	0.9555
71	Lu	174.967	0.9125	0.9132	0.9151	0.9182	0.9224	0.9276	0.9337	0.9404	0.9477	0.9553
		174.941	0.9125	0.9132	0.9151	0.9181	0.9224	0.9276	0.9337	0.9404	0.9477	0.9553
		175.943	0.9130	0.9136	0.9155	0.9186	0.9228	0.9280	0.9340	0.9408	0.9480	0.9555
72	Hf	178.49	0.9142	0.9148	0.9167	0.9197	0.9238	0.9290	0.9349	0.9416	0.9487	0.9561
		173.940	0.9120	0.9127	0.9146	0.9177	0.9219	0.9272	0.9333	0.9401	0.9474	0.9550
		175.941	0.9130	0.9136	0.9155	0.9186	0.9228	0.9280	0.9340	0.9408	0.9480	0.9555
		176.943	0.9135	0.9141	0.9160	0.9190	0.9232	0.9284	0.9344	0.9411	0.9483	0.9558
		177.944	0.9139	0.9146	0.9164	0.9195	0.9236	0.9287	0.9347	0.9414	0.9486	0.9560
		178.946	0.9144	0.9150	0.9169	0.9199	0.9240	0.9291	0.9351	0.9417	0.9488	0.9562
		179.947	0.9149	0.9155	0.9173	0.9203	0.9244	0.9295	0.9354	0.9420	0.9491	0.9565
73	Ta	180.948	0.9153	0.9159	0.9178	0.9207	0.9248	0.9299	0.9358	0.9423	0.9494	0.9567
		179.947	0.9149	0.9155	0.9173	0.9203	0.9244	0.9295	0.9354	0.9420	0.9491	0.9565
		180.948	0.9153	0.9159	0.9178	0.9207	0.9248	0.9299	0.9358	0.9423	0.9494	0.9567
74	W	183.849	0.9166	0.9172	0.9190	0.9219	0.9260	0.9310	0.9368	0.9432	0.9502	0.9574
		179.947	0.9149	0.9155	0.9173	0.9203	0.9244	0.9295	0.9354	0.9420	0.9491	0.9565
		181.948	0.9158	0.9164	0.9182	0.9212	0.9252	0.9303	0.9361	0.9427	0.9497	0.9569
		182.950	0.9162	0.9168	0.9186	0.9216	0.9256	0.9306	0.9365	0.9430	0.9499	0.9572
		183.951	0.9166	0.9172	0.9190	0.9220	0.9260	0.9310	0.9368	0.9433	0.9502	0.9574
		185.954	0.9175	0.9181	0.9199	0.9228	0.9268	0.9317	0.9375	0.9439	0.9507	0.9579
75	Re	186.207	0.9176	0.9182	0.9200	0.9229	0.9269	0.9318	0.9375	0.9439	0.9508	0.9579
		184.953	0.9171	0.9177	0.9195	0.9224	0.9264	0.9314	0.9371	0.9436	0.9505	0.9576
		186.956	0.9179	0.9185	0.9203	0.9232	0.9272	0.9321	0.9378	0.9441	0.9510	0.9581
76	Os	190.24	0.9193	0.9199	0.9216	0.9245	0.9284	0.9332	0.9388	0.9451	0.9518	0.9588
		183.952	0.9166	0.9172	0.9190	0.9220	0.9260	0.9310	0.9368	0.9433	0.9502	0.9574
		185.954	0.9175	0.9181	0.9199	0.9228	0.9268	0.9317	0.9375	0.9439	0.9507	0.9579
		186.956	0.9179	0.9185	0.9203	0.9232	0.9272	0.9321	0.9378	0.9441	0.9510	0.9581
		187.956	0.9183	0.9189	0.9207	0.9236	0.9275	0.9324	0.9381	0.9444	0.9512	0.9583
		188.958	0.9187	0.9193	0.9211	0.9240	0.9279	0.9328	0.9384	0.9447	0.9515	0.9585
		189.958	0.9192	0.9197	0.9215	0.9244	0.9283	0.9331	0.9387	0.9450	0.9517	0.9587
		191.962	0.9200	0.9206	0.9223	0.9251	0.9290	0.9338	0.9394	0.9456	0.9522	0.9591
77	Ir	192.216	0.9201	0.9207	0.9224	0.9252	0.9291	0.9339	0.9394	0.9456	0.9523	0.9592
		190.961	0.9196	0.9202	0.9219	0.9247	0.9286	0.9334	0.9390	0.9453	0.9520	0.9589
		192.963	0.9204	0.9209	0.9227	0.9255	0.9293	0.9341	0.9397	0.9458	0.9525	0.9594
78	Pt	195.08	0.9212	0.9218	0.9235	0.9263	0.9301	0.9348	0.9403	0.9464	0.9530	0.9598
		189.960	0.9192	0.9197	0.9215	0.9244	0.9283	0.9331	0.9387	0.9450	0.9517	0.9587
		191.961	0.9200	0.9206	0.9223	0.9251	0.9290	0.9338	0.9394	0.9456	0.9522	0.9591
		193.963	0.9208	0.9213	0.9231	0.9259	0.9297	0.9344	0.9400	0.9461	0.9527	0.9596
		194.965	0.9212	0.9217	0.9234	0.9262	0.9300	0.9348	0.9403	0.9464	0.9529	0.9598
		195.965	0.9215	0.9221	0.9238	0.9266	0.9304	0.9351	0.9406	0.9466	0.9532	0.9600
		197.968	0.9223	0.9229	0.9246	0.9273	0.9311	0.9357	0.9411	0.9472	0.9536	0.9604

Table A5.2. Rutherford backscattering spectrometry kinematic factors for He as a projectile (continued).

Atomic no. (Z)	El.	Isotopic mass (M$_2$) (amu)	180°	170°	160°	150°	140°	130°	120°	110°	100°	90°
79	Au	196.967	0.9219	0.9225	0.9242	0.9270	0.9307	0.9354	0.9408	0.9469	0.9534	0.9602
80	Hg	200.588	0.9233	0.9238	0.9255	0.9282	0.9319	0.9365	0.9419	0.9478	0.9542	0.9609
		195.966	0.9215	0.9221	0.9238	0.9266	0.9304	0.9351	0.9406	0.9466	0.9532	0.9600
		197.967	0.9223	0.9229	0.9246	0.9273	0.9311	0.9357	0.9411	0.9472	0.9536	0.9604
		198.968	0.9227	0.9232	0.9249	0.9277	0.9314	0.9360	0.9414	0.9474	0.9539	0.9606
		199.968	0.9230	0.9236	0.9253	0.9280	0.9317	0.9363	0.9417	0.9477	0.9541	0.9608
		200.970	0.9234	0.9240	0.9256	0.9284	0.9321	0.9366	0.9420	0.9479	0.9543	0.9609
		201.971	0.9238	0.9243	0.9260	0.9287	0.9324	0.9370	0.9423	0.9482	0.9545	0.9611
81	Tl	203.973	0.9245	0.9251	0.9267	0.9294	0.9330	0.9376	0.9428	0.9487	0.9550	0.9615
		204.383	0.9246	0.9252	0.9268	0.9295	0.9332	0.9377	0.9429	0.9488	0.9551	0.9616
		202.972	0.9241	0.9247	0.9263	0.9290	0.9327	0.9373	0.9425	0.9484	0.9548	0.9613
		204.974	0.9249	0.9254	0.9270	0.9297	0.9333	0.9378	0.9431	0.9489	0.9552	0.9617
82	Pb	207.217	0.9256	0.9262	0.9278	0.9304	0.9340	0.9385	0.9437	0.9495	0.9557	0.9621
		203.973	0.9245	0.9251	0.9267	0.9294	0.9330	0.9376	0.9428	0.9487	0.9550	0.9615
		205.975	0.9252	0.9258	0.9274	0.9300	0.9337	0.9381	0.9434	0.9492	0.9554	0.9619
		206.976	0.9256	0.9261	0.9277	0.9304	0.9340	0.9384	0.9436	0.9494	0.9556	0.9621
		207.977	0.9259	0.9264	0.9280	0.9307	0.9343	0.9387	0.9439	0.9496	0.9558	0.9622
83	Bi	208.980	0.9262	0.9268	0.9284	0.9310	0.9346	0.9390	0.9442	0.9499	0.9560	0.9624
84	Po	208.982	0.9262	0.9268	0.9284	0.9310	0.9346	0.9390	0.9442	0.9499	0.9560	0.9624
85	At	210.000	0.9266	0.9271	0.9287	0.9313	0.9349	0.9393	0.9444	0.9501	0.9562	0.9626
86	Rn	222.018	0.9304	0.9309	0.9324	0.9349	0.9383	0.9425	0.9473	0.9528	0.9586	0.9646
87	Fr	223.000	0.9307	0.9312	0.9327	0.9352	0.9386	0.9427	0.9476	0.9530	0.9587	0.9647
88	Ra	226.025	0.9316	0.9321	0.9336	0.9360	0.9394	0.9435	0.9483	0.9536	0.9593	0.9652
89	Ac	227.000	0.9319	0.9324	0.9339	0.9363	0.9396	0.9437	0.9485	0.9538	0.9595	0.9653
90	Th	232.038	0.9333	0.9338	0.9353	0.9376	0.9409	0.9449	0.9496	0.9548	0.9603	0.9661
91	Pa	231.036	0.9330	0.9335	0.9350	0.9374	0.9406	0.9447	0.9493	0.9546	0.9601	0.9659
92	U	238.018	0.9349	0.9354	0.9368	0.9392	0.9423	0.9462	0.9508	0.9559	0.9613	0.9669
		234.041	0.9339	0.9344	0.9358	0.9382	0.9414	0.9454	0.9500	0.9551	0.9606	0.9664
		235.044	0.9341	0.9346	0.9361	0.9384	0.9416	0.9456	0.9502	0.9553	0.9608	0.9665
		238.040	0.9349	0.9354	0.9366	0.9392	0.9423	0.9462	0.9508	0.9559	0.9613	0.9669
93	Np	237.048	0.9347	0.9352	0.9366	0.9389	0.9421	0.9460	0.9506	0.9557	0.9611	0.9668
94	Pu	244.064	0.9365	0.9370	0.9384	0.9406	0.9437	0.9475	0.9520	0.9569	0.9622	0.9677
95	Am	243.061	0.9362	0.9367	0.9381	0.9404	0.9435	0.9473	0.9518	0.9568	0.9621	0.9676

Table A5.3. Rutherford backscattering spectrometry kinematic factors for Li as a projectile.

This table gives the RBS kinematic factors K_{M_2}, defined by Eq. (A4.2), for Li as a projectile (M_1=7.016 amu) and for the isotopic masses (M_2) of the elements. The kinematic factors are given as a function of scattering angle from 180° to 90° measured in the laboratory frame of reference. The first row for each element gives the average atomic weight of that element and the average kine-matic factor (Average K_{M_2}) for that element. The subsequent rows give the isotopic masses for that element and the K_{M_2} for those isotopic masses. The average K_{M_2} is calculated as the weighted average of the K_{M_2} in which the relative abundances (ϖ_{M_2}) for the M_2 are given in Appendix 1:

$$\text{Average } K_{M_2} = \sum \varpi_{M_2} K_{M_2} \text{, summed over the masses } M_2.$$

Atomic no. (Z)	El.	Isotopic mass (M_2) (amu)	180°	170°	160°	150°	140°	130°	120°	110°	100°	90°
4	Be	9.012	0.0155	0.0159	0.0171	0.0193	0.0229	0.0289	0.0386	0.0546	0.0812	0.1245
5	B	10.811	0.0455	0.0464	0.0492	0.0543	0.0625	0.0747	0.0929	0.1196	0.1582	0.2125
		10.013	0.0310	0.0316	0.0337	0.0376	0.0438	0.0535	0.0683	0.0910	0.1253	0.1760
		11.009	0.0491	0.0500	0.0530	0.0585	0.0671	0.0800	0.0990	0.1267	0.1664	0.2215
6	C	12.011	0.0689	0.0702	0.0740	0.0809	0.0916	0.1073	0.1297	0.1612	0.2046	0.2625
		12.000	0.0687	0.0699	0.0738	0.0806	0.0913	0.1070	0.1294	0.1609	0.2042	0.2621
		13.003	0.0894	0.0909	0.0955	0.1036	0.1161	0.1341	0.1592	0.1936	0.2395	0.2991
7	N	14.007	0.1106	0.1123	0.1175	0.1267	0.1407	0.1606	0.1879	0.2244	0.2721	0.3325
		14.003	0.1105	0.1122	0.1174	0.1267	0.1406	0.1605	0.1878	0.2243	0.2720	0.3324
		15.000	0.1315	0.1334	0.1392	0.1493	0.1646	0.1860	0.2149	0.2530	0.3018	0.3626
8	O	15.999	0.1523	0.1544	0.1607	0.1716	0.1879	0.2105	0.2407	0.2800	0.3295	0.3903
		15.995	0.1523	0.1543	0.1606	0.1715	0.1878	0.2104	0.2406	0.2799	0.3294	0.3902
		16.999	0.1728	0.1750	0.1817	0.1933	0.2104	0.2340	0.2652	0.3053	0.3552	0.4157
		17.999	0.1928	0.1951	0.2021	0.2142	0.2321	0.2565	0.2884	0.3290	0.3791	0.4391
9	F	18.998	0.2122	0.2145	0.2219	0.2344	0.2528	0.2778	0.3104	0.3513	0.4013	0.4606
10	Ne	20.179	0.2342	0.2367	0.2442	0.2572	0.2761	0.3017	0.3346	0.3756	0.4254	0.4838
		19.992	0.2308	0.2333	0.2409	0.2538	0.2726	0.2981	0.3310	0.3721	0.4219	0.4805
		20.994	0.2490	0.2516	0.2593	0.2725	0.2918	0.3176	0.3507	0.3918	0.4413	0.4990
		21.991	0.2665	0.2691	0.2770	0.2905	0.3099	0.3360	0.3693	0.4103	0.4593	0.5163
11	Na	22.990	0.2834	0.2860	0.2941	0.3077	0.3274	0.3536	0.3869	0.4278	0.4763	0.5324
12	Mg	24.305	0.3046	0.3073	0.3154	0.3292	0.3490	0.3754	0.4086	0.4491	0.4970	0.5518
		23.985	0.2996	0.3023	0.3104	0.3242	0.3440	0.3703	0.4036	0.4442	0.4922	0.5474
		24.986	0.3153	0.3180	0.3262	0.3401	0.3600	0.3863	0.4195	0.4598	0.5073	0.5615
		25.983	0.3304	0.3331	0.3413	0.3553	0.3752	0.4016	0.4346	0.4746	0.5215	0.5748
13	Al	26.982	0.3449	0.3476	0.3559	0.3699	0.3899	0.4161	0.4490	0.4886	0.5349	0.5873
14	Si	28.086	0.3603	0.3630	0.3713	0.3853	0.4052	0.4314	0.4640	0.5032	0.5488	0.6002
		27.977	0.3588	0.3616	0.3698	0.3839	0.4038	0.4300	0.4626	0.5018	0.5475	0.5990
		28.976	0.3723	0.3750	0.3833	0.3973	0.4172	0.4433	0.4757	0.5145	0.5595	0.6101
		29.974	0.3852	0.3880	0.3963	0.4102	0.4301	0.4560	0.4881	0.5265	0.5709	0.6207
15	P	30.974	0.3977	0.4005	0.4087	0.4227	0.4424	0.4682	0.5000	0.5380	0.5817	0.6306
16	S	32.064	0.4108	0.4135	0.4218	0.4357	0.4553	0.4809	0.5124	0.5498	0.5929	0.6409
		31.972	0.4097	0.4125	0.4207	0.4346	0.4543	0.4798	0.5114	0.5489	0.5920	0.6401
		32.972	0.4213	0.4241	0.4323	0.4461	0.4657	0.4910	0.5223	0.5593	0.6018	0.6491
		33.968	0.4325	0.4352	0.4434	0.4571	0.4766	0.5017	0.5327	0.5693	0.6111	0.6576
		35.967	0.4537	0.4564	0.4645	0.4781	0.4972	0.5219	0.5522	0.5879	0.6286	0.6735
17	Cl	35.453	0.4482	0.4509	0.4591	0.4727	0.4919	0.5168	0.5472	0.5831	0.6241	0.6695
		34.969	0.4433	0.4460	0.4541	0.4678	0.4871	0.5121	0.5427	0.5788	0.6201	0.6658
		36.966	0.4637	0.4664	0.4745	0.4880	0.5069	0.5315	0.5614	0.5967	0.6368	0.6810

Table A5.3. Rutherford backscattering spectrometry kinematic factors for Li as a projectile (continued).

Atomic no. (Z)	El.	Isotopic mass (M₂) (amu)	180°	170°	160°	150°	140°	130°	120°	110°	100°	90°
18	Ar	39.948	0.4917	0.4943	0.5022	0.5154	0.5340	0.5578	0.5868	0.6207	0.6591	0.7012
		35.967	0.4537	0.4564	0.4645	0.4781	0.4972	0.5219	0.5522	0.5879	0.6286	0.6735
		37.963	0.4734	0.4761	0.4841	0.4975	0.5163	0.5406	0.5702	0.6050	0.6445	0.6880
		39.962	0.4918	0.4945	0.5024	0.5156	0.5341	0.5579	0.5869	0.6208	0.6592	0.7013
19	K	39.098	0.4840	0.4866	0.4946	0.5079	0.5265	0.5505	0.5798	0.6141	0.6530	0.6957
		38.964	0.4828	0.4854	0.4934	0.5067	0.5254	0.5494	0.5787	0.6131	0.6520	0.6948
		40.000	0.4922	0.4948	0.5027	0.5159	0.5344	0.5582	0.5872	0.6211	0.6595	0.7015
20	Ca	40.962	0.5006	0.5032	0.5111	0.5242	0.5425	0.5661	0.5948	0.6282	0.6661	0.7075
		40.078	0.4928	0.4954	0.5033	0.5165	0.5350	0.5588	0.5878	0.6216	0.6599	0.7020
		39.963	0.4918	0.4945	0.5024	0.5156	0.5341	0.5579	0.5869	0.6208	0.6592	0.7013
		41.959	0.5091	0.5117	0.5194	0.5324	0.5506	0.5740	0.6023	0.6354	0.6727	0.7135
		42.959	0.5173	0.5199	0.5276	0.5405	0.5585	0.5816	0.6096	0.6423	0.6790	0.7192
		43.956	0.5252	0.5278	0.5354	0.5482	0.5661	0.5889	0.6166	0.6489	0.6851	0.7247
		45.954	0.5404	0.5429	0.5504	0.5630	0.5805	0.6029	0.6300	0.6614	0.6967	0.7351
		47.952	0.5546	0.5571	0.5645	0.5768	0.5940	0.6159	0.6424	0.6731	0.7074	0.7447
21	Sc	44.956	0.5329	0.5354	0.5430	0.5557	0.5734	0.5960	0.6234	0.6553	0.6910	0.7300
22	Ti	47.878	0.5540	0.5565	0.5639	0.5763	0.5935	0.6154	0.6419	0.6726	0.7070	0.7443
		45.953	0.5404	0.5429	0.5504	0.5630	0.5805	0.6029	0.6300	0.6614	0.6967	0.7351
		46.952	0.5476	0.5501	0.5576	0.5700	0.5873	0.6095	0.6363	0.6673	0.7022	0.7400
		47.948	0.5546	0.5571	0.5645	0.5768	0.5940	0.6159	0.6424	0.6731	0.7074	0.7447
		48.948	0.5614	0.5639	0.5712	0.5834	0.6004	0.6221	0.6483	0.6786	0.7125	0.7493
		49.945	0.5680	0.5704	0.5777	0.5898	0.6067	0.6282	0.6540	0.6840	0.7174	0.7537
23	V	50.942	0.5744	0.5768	0.5840	0.5960	0.6127	0.6340	0.6596	0.6891	0.7222	0.7579
		49.947	0.5680	0.5704	0.5777	0.5898	0.6067	0.6282	0.6541	0.6840	0.7174	0.7537
24	Cr	50.944	0.5744	0.5768	0.5840	0.5960	0.6127	0.6340	0.6596	0.6892	0.7222	0.7579
		51.996	0.5809	0.5833	0.5905	0.6024	0.6189	0.6399	0.6652	0.6944	0.7270	0.7622
		49.946	0.5680	0.5704	0.5777	0.5898	0.6067	0.6282	0.6540	0.6840	0.7174	0.7537
		51.940	0.5806	0.5830	0.5902	0.6021	0.6186	0.6396	0.6650	0.6942	0.7268	0.7620
		52.941	0.5867	0.5891	0.5962	0.6079	0.6243	0.6451	0.6702	0.6990	0.7312	0.7660
		53.939	0.5926	0.5949	0.6020	0.6136	0.6298	0.6505	0.6752	0.7037	0.7355	0.7698
25	Mn	54.938	0.5983	0.6006	0.6076	0.6192	0.6352	0.6556	0.6801	0.7083	0.7397	0.7735
26	Fe	55.847	0.6034	0.6057	0.6126	0.6241	0.6400	0.6602	0.6844	0.7123	0.7433	0.7768
		53.940	0.5926	0.5949	0.6020	0.6136	0.6299	0.6505	0.6752	0.7038	0.7355	0.7698
		55.935	0.6039	0.6062	0.6131	0.6245	0.6404	0.6606	0.6849	0.7127	0.7437	0.7771
		56.935	0.6093	0.6116	0.6184	0.6298	0.6455	0.6655	0.6895	0.7170	0.7476	0.7806
		57.933	0.6146	0.6168	0.6236	0.6349	0.6505	0.6703	0.6940	0.7212	0.7514	0.7840
27	Co	58.933	0.6197	0.6220	0.6287	0.6398	0.6553	0.6749	0.6983	0.7253	0.7551	0.7872
28	Ni	58.688	0.6184	0.6206	0.6274	0.6385	0.6540	0.6737	0.6972	0.7242	0.7541	0.7864
		57.935	0.6146	0.6169	0.6236	0.6349	0.6505	0.6703	0.6940	0.7212	0.7514	0.7840
		59.931	0.6247	0.6270	0.6336	0.6447	0.6600	0.6794	0.7026	0.7292	0.7587	0.7904
		60.931	0.6296	0.6318	0.6384	0.6494	0.6645	0.6837	0.7067	0.7330	0.7622	0.7935
		61.928	0.6344	0.6366	0.6431	0.6539	0.6690	0.6880	0.7107	0.7367	0.7655	0.7965
29	Cu	63.928	0.6435	0.6457	0.6521	0.6628	0.6775	0.6961	0.7184	0.7439	0.7720	0.8022
		63.546	0.6418	0.6439	0.6504	0.6611	0.6759	0.6946	0.7169	0.7425	0.7708	0.8011
		62.930	0.6390	0.6412	0.6477	0.6584	0.6733	0.6921	0.7146	0.7404	0.7688	0.7994
		64.928	0.6480	0.6501	0.6565	0.6670	0.6816	0.7001	0.7221	0.7473	0.7751	0.8050

Table A5.3. Rutherford backscattering spectrometry kinematic factors for Li as a projectile (continued).

Atomic no. (Z)	El.	Isotopic mass (M$_2$) (amu)	180°	170°	160°	150°	140°	130°	120°	110°	100°	90°
30	Zn	65.396	0.6499	0.6520	0.6583	0.6688	0.6834	0.7017	0.7237	0.7488	0.7765	0.8061
		63.929	0.6435	0.6457	0.6521	0.6628	0.6775	0.6961	0.7184	0.7439	0.7720	0.8022
		65.926	0.6523	0.6544	0.6607	0.6711	0.6856	0.7039	0.7257	0.7506	0.7782	0.8076
		66.927	0.6565	0.6586	0.6648	0.6752	0.6895	0.7076	0.7292	0.7539	0.7811	0.8102
		67.925	0.6606	0.6627	0.6689	0.6791	0.6933	0.7113	0.7326	0.7570	0.7840	0.8128
		69.925	0.6685	0.6706	0.6767	0.6867	0.7007	0.7183	0.7392	0.7631	0.7895	0.8176
31	Ga	69.723	0.6677	0.6697	0.6758	0.6859	0.6999	0.7175	0.7385	0.7625	0.7889	0.8171
		68.926	0.6646	0.6667	0.6728	0.6830	0.6971	0.7148	0.7360	0.7601	0.7868	0.8152
		70.925	0.6723	0.6744	0.6804	0.6904	0.7042	0.7217	0.7424	0.7661	0.7921	0.8200
32	Ge	72.632	0.6786	0.6806	0.6865	0.6964	0.7100	0.7271	0.7475	0.7708	0.7964	0.8237
		69.924	0.6685	0.6706	0.6767	0.6867	0.7007	0.7183	0.7392	0.7631	0.7895	0.8176
		71.922	0.6761	0.6781	0.6841	0.6940	0.7077	0.7250	0.7455	0.7689	0.7947	0.8222
		72.924	0.6797	0.6817	0.6877	0.6975	0.7111	0.7282	0.7485	0.7717	0.7973	0.8245
		73.921	0.6833	0.6853	0.6912	0.7009	0.7144	0.7313	0.7515	0.7744	0.7997	0.8266
		75.921	0.6902	0.6922	0.6980	0.7076	0.7208	0.7374	0.7572	0.7797	0.8045	0.8308
33	As	74.922	0.6868	0.6888	0.6946	0.7043	0.7176	0.7344	0.7544	0.7771	0.8021	0.8287
34	Se	78.993	0.7002	0.7021	0.7078	0.7171	0.7300	0.7462	0.7654	0.7873	0.8113	0.8368
		73.923	0.6833	0.6853	0.6912	0.7009	0.7144	0.7313	0.7515	0.7745	0.7997	0.8266
		75.919	0.6902	0.6922	0.6980	0.7076	0.7208	0.7374	0.7572	0.7797	0.8045	0.8308
		76.920	0.6936	0.6955	0.7013	0.7108	0.7239	0.7404	0.7600	0.7823	0.8068	0.8328
		77.917	0.6969	0.6988	0.7045	0.7139	0.7269	0.7432	0.7626	0.7847	0.8090	0.8348
		79.916	0.7032	0.7051	0.7107	0.7200	0.7327	0.7488	0.7679	0.7895	0.8133	0.8386
		81.917	0.7093	0.7112	0.7167	0.7258	0.7384	0.7541	0.7729	0.7941	0.8174	0.8422
35	Br	79.904	0.7032	0.7050	0.7106	0.7199	0.7327	0.7487	0.7678	0.7895	0.8133	0.8385
		78.918	0.7001	0.7020	0.7076	0.7170	0.7298	0.7461	0.7653	0.7872	0.8112	0.8367
		80.916	0.7063	0.7082	0.7137	0.7229	0.7356	0.7515	0.7704	0.7919	0.8154	0.8404
36	Kr	83.8	0.7148	0.7166	0.7221	0.7310	0.7434	0.7589	0.7773	0.7982	0.8211	0.8455
		77.920	0.6969	0.6988	0.7045	0.7139	0.7269	0.7432	0.7627	0.7847	0.8090	0.8348
		79.916	0.7032	0.7051	0.7107	0.7200	0.7327	0.7488	0.7679	0.7895	0.8133	0.8386
		81.913	0.7093	0.7112	0.7167	0.7258	0.7383	0.7541	0.7729	0.7941	0.8174	0.8422
		82.914	0.7123	0.7141	0.7196	0.7286	0.7411	0.7567	0.7753	0.7963	0.8194	0.8440
		83.911	0.7152	0.7170	0.7224	0.7314	0.7437	0.7592	0.7776	0.7985	0.8214	0.8457
		85.911	0.7208	0.7226	0.7279	0.7368	0.7489	0.7642	0.7822	0.8027	0.8252	0.8490
37	Rb	85.468	0.7196	0.7213	0.7267	0.7356	0.7477	0.7631	0.7812	0.8018	0.8243	0.8483
		84.912	0.7180	0.7198	0.7252	0.7341	0.7463	0.7617	0.7800	0.8006	0.8233	0.8474
		86.909	0.7235	0.7253	0.7306	0.7394	0.7514	0.7665	0.7844	0.8048	0.8270	0.8506
38	Sr	87.617	0.7254	0.7272	0.7325	0.7412	0.7531	0.7682	0.7860	0.8062	0.8283	0.8517
		83.913	0.7152	0.7170	0.7224	0.7314	0.7437	0.7592	0.7776	0.7985	0.8214	0.8457
		85.909	0.7208	0.7226	0.7279	0.7367	0.7489	0.7641	0.7822	0.8027	0.8252	0.8490
		86.909	0.7235	0.7253	0.7306	0.7394	0.7514	0.7665	0.7844	0.8048	0.8270	0.8506
		87.906	0.7262	0.7280	0.7332	0.7419	0.7539	0.7689	0.7866	0.8068	0.8288	0.8522
39	Y	88.906	0.7288	0.7306	0.7358	0.7444	0.7563	0.7711	0.7888	0.8087	0.8306	0.8537
40	Zr	91.221	0.7347	0.7364	0.7415	0.7500	0.7616	0.7762	0.7935	0.8131	0.8345	0.8571
		89.905	0.7314	0.7331	0.7383	0.7469	0.7586	0.7734	0.7908	0.8106	0.8323	0.8552
		90.906	0.7339	0.7357	0.7408	0.7493	0.7609	0.7756	0.7929	0.8125	0.8340	0.8567
		91.905	0.7364	0.7381	0.7432	0.7516	0.7632	0.7777	0.7949	0.8144	0.8356	0.8581
		93.906	0.7413	0.7429	0.7480	0.7562	0.7676	0.7819	0.7988	0.8180	0.8388	0.8610
		95.908	0.7459	0.7476	0.7525	0.7607	0.7719	0.7860	0.8026	0.8214	0.8419	0.8637

Table A5.3. Rutherford backscattering spectrometry kinematic factors for Li as a projectile (continued).

Atomic no. (Z)	El.	Isotopic mass (M_2) (amu)	180°	170°	160°	150°	140°	130°	120°	110°	100°	90°
41	Nb	92.906	0.7389	0.7406	0.7456	0.7540	0.7655	0.7799	0.7969	0.8162	0.8373	0.8596
42	Mo	95.931	0.7459	0.7475	0.7525	0.7606	0.7719	0.7859	0.8025	0.8214	0.8419	0.8636
		91.907	0.7364	0.7381	0.7432	0.7516	0.7632	0.7777	0.7949	0.8144	0.8356	0.8582
		93.905	0.7413	0.7429	0.7480	0.7562	0.7676	0.7819	0.7988	0.8180	0.8388	0.8610
		94.906	0.7436	0.7453	0.7503	0.7585	0.7698	0.7840	0.8007	0.8197	0.8404	0.8623
		95.905	0.7459	0.7476	0.7525	0.7607	0.7719	0.7860	0.8026	0.8214	0.8419	0.8637
		96.906	0.7482	0.7498	0.7547	0.7628	0.7740	0.7879	0.8044	0.8231	0.8434	0.8650
		97.905	0.7504	0.7520	0.7569	0.7650	0.7760	0.7899	0.8062	0.8247	0.8449	0.8663
		99.908	0.7548	0.7564	0.7612	0.7691	0.7800	0.7936	0.8097	0.8279	0.8478	0.8688
43	Tc	98.000	0.7506	0.7523	0.7571	0.7652	0.7762	0.7900	0.8064	0.8249	0.8451	0.8664
44	Ru	101.019	0.7570	0.7586	0.7634	0.7713	0.7820	0.7956	0.8115	0.8296	0.8493	0.8701
		95.908	0.7459	0.7476	0.7525	0.7607	0.7719	0.7860	0.8026	0.8214	0.8419	0.8637
		97.905	0.7504	0.7520	0.7569	0.7650	0.7760	0.7899	0.8062	0.8247	0.8449	0.8663
		98.506	0.7517	0.7534	0.7582	0.7662	0.7772	0.7910	0.8073	0.8257	0.8458	0.8670
		99.904	0.7547	0.7564	0.7612	0.7691	0.7800	0.7936	0.8097	0.8279	0.8478	0.8688
		100.906	0.7569	0.7585	0.7632	0.7711	0.7819	0.7954	0.8114	0.8295	0.8492	0.8700
		101.904	0.7589	0.7605	0.7653	0.7731	0.7838	0.7972	0.8131	0.8310	0.8505	0.8712
		103.905	0.7630	0.7646	0.7692	0.7769	0.7875	0.8007	0.8163	0.8340	0.8532	0.8735
45	Rh	102.906	0.7610	0.7626	0.7673	0.7750	0.7857	0.7990	0.8147	0.8325	0.8519	0.8723
46	Pd	106.415	0.7678	0.7694	0.7740	0.7815	0.7919	0.8049	0.8202	0.8375	0.8564	0.8763
		101.906	0.7589	0.7605	0.7653	0.7731	0.7838	0.7972	0.8131	0.8310	0.8505	0.8712
		103.904	0.7630	0.7646	0.7692	0.7769	0.7875	0.8007	0.8163	0.8340	0.8532	0.8735
		104.905	0.7650	0.7665	0.7712	0.7788	0.7893	0.8024	0.8179	0.8354	0.8545	0.8746
		105.904	0.7669	0.7685	0.7731	0.7807	0.7911	0.8041	0.8195	0.8369	0.8558	0.8757
		107.904	0.7707	0.7722	0.7768	0.7843	0.7945	0.8074	0.8225	0.8396	0.8583	0.8779
		109.905	0.7744	0.7759	0.7804	0.7877	0.7979	0.8105	0.8255	0.8423	0.8607	0.8800
47	Ag	107.868	0.7706	0.7721	0.7767	0.7842	0.7944	0.8073	0.8225	0.8396	0.8582	0.8779
		106.905	0.7688	0.7704	0.7749	0.7825	0.7928	0.8058	0.8210	0.8383	0.8570	0.8768
		108.905	0.7726	0.7741	0.7786	0.7860	0.7962	0.8090	0.8240	0.8410	0.8595	0.8790
48	Cd	112.412	0.7788	0.7802	0.7847	0.7919	0.8019	0.8143	0.8290	0.8455	0.8635	0.8825
		105.907	0.7669	0.7685	0.7731	0.7807	0.7911	0.8041	0.8195	0.8369	0.8558	0.8757
		107.904	0.7707	0.7722	0.7768	0.7843	0.7945	0.8074	0.8225	0.8396	0.8583	0.8779
		109.903	0.7744	0.7759	0.7804	0.7877	0.7979	0.8105	0.8255	0.8423	0.8607	0.8800
		110.904	0.7762	0.7777	0.7821	0.7894	0.7995	0.8121	0.8269	0.8436	0.8618	0.8810
		111.903	0.7779	0.7794	0.7838	0.7911	0.8011	0.8136	0.8283	0.8449	0.8630	0.8820
		112.904	0.7797	0.7811	0.7855	0.7928	0.8027	0.8151	0.8297	0.8462	0.8641	0.8830
		113.903	0.7814	0.7828	0.7872	0.7944	0.8042	0.8166	0.8311	0.8474	0.8652	0.8840
		115.905	0.7847	0.7862	0.7905	0.7976	0.8073	0.8194	0.8337	0.8499	0.8674	0.8858
49	In	114.818	0.7829	0.7844	0.7887	0.7958	0.8056	0.8179	0.8323	0.8485	0.8662	0.8848
		112.904	0.7797	0.7811	0.7855	0.7928	0.8027	0.8151	0.8297	0.8462	0.8641	0.8830
		114.904	0.7831	0.7845	0.7889	0.7960	0.8058	0.8180	0.8324	0.8486	0.8663	0.8849
50	Sn	118.613	0.7888	0.7902	0.7945	0.8014	0.8110	0.8229	0.8370	0.8528	0.8700	0.8882
		111.905	0.7779	0.7794	0.7838	0.7911	0.8011	0.8136	0.8283	0.8449	0.8630	0.8820
		113.903	0.7814	0.7828	0.7872	0.7944	0.8042	0.8166	0.8311	0.8474	0.8652	0.8840
		114.903	0.7831	0.7845	0.7888	0.7960	0.8058	0.8180	0.8324	0.8486	0.8663	0.8849
		115.902	0.7847	0.7862	0.7905	0.7976	0.8073	0.8194	0.8337	0.8498	0.8674	0.8858
		116.903	0.7864	0.7878	0.7921	0.7991	0.8088	0.8208	0.8350	0.8510	0.8684	0.8868

Table A5.3. Rutherford backscattering spectrometry kinematic factors for Li as a projectile (continued).

Atomic no. (Z)	El.	Isotopic mass (M_2) (amu)	180°	170°	160°	150°	140°	130°	120°	110°	100°	90°
		117.602	0.7875	0.7889	0.7932	0.8002	0.8098	0.8218	0.8359	0.8519	0.8692	0.8874
		118.903	0.7895	0.7910	0.7952	0.8021	0.8117	0.8236	0.8376	0.8534	0.8705	0.8886
		119.902	0.7911	0.7925	0.7967	0.8036	0.8131	0.8249	0.8388	0.8545	0.8715	0.8894
		121.903	0.7942	0.7955	0.7997	0.8065	0.8158	0.8275	0.8412	0.8567	0.8735	0.8912
		123.905	0.7971	0.7985	0.8026	0.8093	0.8185	0.8301	0.8436	0.8588	0.8754	0.8928
51	Sb	121.758	0.7939	0.7953	0.7995	0.8063	0.8156	0.8273	0.8411	0.8565	0.8733	0.8910
		120.904	0.7926	0.7940	0.7982	0.8051	0.8145	0.8262	0.8400	0.8556	0.8725	0.8903
		122.904	0.7957	0.7970	0.8012	0.8079	0.8172	0.8288	0.8424	0.8578	0.8745	0.8920
52	Te	127.586	0.8023	0.8037	0.8077	0.8142	0.8232	0.8345	0.8477	0.8626	0.8788	0.8957
		119.904	0.7911	0.7925	0.7967	0.8036	0.8131	0.8249	0.8388	0.8545	0.8715	0.8894
		121.903	0.7942	0.7955	0.7997	0.8065	0.8158	0.8275	0.8412	0.8567	0.8735	0.8912
		122.904	0.7957	0.7970	0.8012	0.8079	0.8172	0.8288	0.8424	0.8578	0.8745	0.8920
		123.903	0.7971	0.7985	0.8026	0.8093	0.8185	0.8301	0.8436	0.8588	0.8754	0.8928
		124.904	0.7986	0.7999	0.8040	0.8107	0.8199	0.8313	0.8448	0.8599	0.8763	0.8936
		125.903	0.8000	0.8014	0.8054	0.8120	0.8211	0.8325	0.8459	0.8609	0.8773	0.8944
		127.905	0.8028	0.8042	0.8081	0.8147	0.8237	0.8349	0.8481	0.8630	0.8791	0.8960
		129.906	0.8055	0.8069	0.8108	0.8173	0.8262	0.8372	0.8503	0.8649	0.8808	0.8975
53	I	126.905	0.8014	0.8028	0.8068	0.8134	0.8224	0.8337	0.8470	0.8620	0.8782	0.8952
54	Xe	131.293	0.8073	0.8087	0.8126	0.8190	0.8278	0.8388	0.8517	0.8662	0.8820	0.8985
		123.906	0.7971	0.7985	0.8026	0.8093	0.8185	0.8301	0.8436	0.8588	0.8754	0.8928
		125.904	0.8000	0.8014	0.8054	0.8120	0.8212	0.8325	0.8459	0.8609	0.8773	0.8944
		127.904	0.8028	0.8042	0.8081	0.8147	0.8237	0.8349	0.8481	0.8630	0.8791	0.8960
		128.905	0.8042	0.8055	0.8095	0.8160	0.8249	0.8361	0.8492	0.8639	0.8799	0.8968
		129.904	0.8055	0.8069	0.8108	0.8173	0.8262	0.8372	0.8503	0.8649	0.8808	0.8975
		130.905	0.8069	0.8082	0.8121	0.8185	0.8274	0.8384	0.8513	0.8659	0.8817	0.8983
		131.904	0.8082	0.8095	0.8134	0.8198	0.8286	0.8395	0.8524	0.8668	0.8825	0.8990
		133.905	0.8108	0.8121	0.8159	0.8222	0.8309	0.8417	0.8544	0.8687	0.8842	0.9004
		135.907	0.8133	0.8146	0.8184	0.8246	0.8332	0.8438	0.8564	0.8705	0.8858	0.9018
55	Cs	132.905	0.8095	0.8108	0.8147	0.8210	0.8297	0.8406	0.8534	0.8678	0.8833	0.8997
56	Ba	137.327	0.8150	0.8163	0.8201	0.8263	0.8347	0.8453	0.8578	0.8717	0.8869	0.9028
		129.906	0.8055	0.8069	0.8108	0.8173	0.8262	0.8372	0.8503	0.8649	0.8808	0.8975
		131.905	0.8082	0.8095	0.8134	0.8198	0.8286	0.8395	0.8524	0.8668	0.8825	0.8990
		133.904	0.8108	0.8121	0.8159	0.8222	0.8309	0.8417	0.8544	0.8687	0.8842	0.9004
		134.906	0.8120	0.8133	0.8171	0.8234	0.8320	0.8428	0.8554	0.8696	0.8850	0.9011
		135.905	0.8133	0.8146	0.8184	0.8246	0.8332	0.8438	0.8564	0.8705	0.8858	0.9018
		136.906	0.8145	0.8158	0.8196	0.8258	0.8343	0.8449	0.8574	0.8714	0.8866	0.9025
		137.905	0.8157	0.8170	0.8207	0.8269	0.8354	0.8459	0.8583	0.8722	0.8873	0.9032
57	La	138.905	0.8169	0.8182	0.8219	0.8281	0.8365	0.8470	0.8593	0.8731	0.8881	0.9038
		137.907	0.8157	0.8170	0.8207	0.8269	0.8354	0.8459	0.8583	0.8722	0.8873	0.9032
		138.906	0.8169	0.8182	0.8219	0.8281	0.8365	0.8470	0.8604	0.8741	0.8890	0.9046
58	Ce	140.115	0.8184	0.8196	0.8233	0.8294	0.8378	0.8482	0.8564	0.8705	0.8858	0.9018
		135.907	0.8133	0.8146	0.8184	0.8246	0.8332	0.8438	0.8583	0.8722	0.8873	0.9032
		137.906	0.8157	0.8170	0.8207	0.8269	0.8354	0.8459	0.8602	0.8740	0.8889	0.9045
		139.905	0.8181	0.8194	0.8231	0.8292	0.8375	0.8480	0.8620	0.8756	0.8903	0.9058
		141.909	0.8204	0.8217	0.8253	0.8314	0.8396	0.8499	0.8611	0.8748	0.8896	0.9051
59	Pr	140.908	0.8193	0.8205	0.8242	0.8303	0.8386	0.8490				

Table A5.3. Rutherford backscattering spectrometry kinematic factors for Li as a projectile (continued).

Atomic no. (Z)	El.	Isotopic mass (M₂) (amu)	180°	170°	160°	150°	140°	130°	120°	110°	100°	90°
60	Nd	144.242	0.8230	0.8243	0.8279	0.8338	0.8420	0.8522	0.8641	0.8775	0.8920	0.9072
		141.908	0.8204	0.8217	0.8253	0.8314	0.8396	0.8499	0.8620	0.8756	0.8903	0.9058
		142.910	0.8216	0.8228	0.8265	0.8325	0.8407	0.8509	0.8629	0.8764	0.8911	0.9064
		143.910	0.8227	0.8239	0.8276	0.8335	0.8417	0.8519	0.8638	0.8772	0.8918	0.9070
		144.913	0.8238	0.8250	0.8286	0.8346	0.8427	0.8528	0.8647	0.8780	0.8925	0.9076
		145.913	0.8249	0.8261	0.8297	0.8356	0.8437	0.8537	0.8656	0.8788	0.8932	0.9082
		147.917	0.8271	0.8283	0.8318	0.8376	0.8456	0.8556	0.8673	0.8804	0.8946	0.9094
		149.921	0.8292	0.8304	0.8339	0.8396	0.8475	0.8574	0.8689	0.8819	0.8959	0.9106
61	Pm	145.000	0.8239	0.8251	0.8287	0.8347	0.8428	0.8529	0.8648	0.8781	0.8926	0.9077
62	Sm	150.36	0.8296	0.8308	0.8343	0.8400	0.8479	0.8577	0.8692	0.8822	0.8961	0.9108
		143.912	0.8227	0.8239	0.8276	0.8335	0.8417	0.8519	0.8638	0.8772	0.8918	0.9070
		146.915	0.8260	0.8272	0.8308	0.8366	0.8447	0.8547	0.8664	0.8796	0.8939	0.9088
		147.915	0.8271	0.8283	0.8318	0.8376	0.8456	0.8556	0.8673	0.8804	0.8946	0.9094
		148.917	0.8281	0.8293	0.8328	0.8386	0.8466	0.8565	0.8681	0.8811	0.8952	0.9100
		149.917	0.8292	0.8303	0.8339	0.8396	0.8475	0.8574	0.8689	0.8819	0.8959	0.9106
		151.920	0.8312	0.8324	0.8359	0.8416	0.8494	0.8591	0.8705	0.8833	0.8972	0.9117
		153.922	0.8332	0.8344	0.8378	0.8435	0.8512	0.8608	0.8721	0.8848	0.8985	0.9128
63	Eu	151.965	0.8313	0.8324	0.8359	0.8416	0.8494	0.8591	0.8706	0.8834	0.8972	0.9117
		150.920	0.8302	0.8314	0.8349	0.8406	0.8485	0.8582	0.8697	0.8826	0.8965	0.9112
		152.921	0.8322	0.8334	0.8368	0.8425	0.8503	0.8600	0.8713	0.8841	0.8978	0.9123
64	Gd	157.252	0.8364	0.8376	0.8410	0.8465	0.8541	0.8635	0.8746	0.8870	0.9005	0.9146
		151.920	0.8312	0.8324	0.8359	0.8416	0.8494	0.8591	0.8705	0.8833	0.8972	0.9117
		153.921	0.8332	0.8344	0.8378	0.8435	0.8512	0.8608	0.8721	0.8848	0.8985	0.9128
		154.923	0.8342	0.8354	0.8388	0.8444	0.8521	0.8616	0.8729	0.8855	0.8991	0.9134
		155.922	0.8352	0.8363	0.8397	0.8453	0.8530	0.8625	0.8736	0.8862	0.8997	0.9139
		156.924	0.8361	0.8373	0.8407	0.8462	0.8538	0.8633	0.8744	0.8868	0.9003	0.9144
		157.924	0.8371	0.8382	0.8416	0.8471	0.8547	0.8641	0.8751	0.8875	0.9009	0.9149
		159.927	0.8390	0.8401	0.8434	0.8489	0.8564	0.8657	0.8766	0.8888	0.9021	0.9159
65	Tb	158.925	0.8380	0.8392	0.8425	0.8480	0.8555	0.8649	0.8759	0.8882	0.9015	0.9154
66	Dy	162.498	0.8413	0.8424	0.8457	0.8511	0.8585	0.8677	0.8784	0.8905	0.9036	0.9172
		155.924	0.8352	0.8363	0.8397	0.8453	0.8530	0.8625	0.8736	0.8862	0.8997	0.9139
		157.924	0.8371	0.8382	0.8416	0.8471	0.8547	0.8641	0.8751	0.8875	0.9009	0.9149
		159.925	0.8390	0.8401	0.8434	0.8489	0.8564	0.8657	0.8766	0.8888	0.9021	0.9159
		160.927	0.8399	0.8410	0.8443	0.8497	0.8572	0.8665	0.8773	0.8895	0.9027	0.9164
		161.927	0.8408	0.8419	0.8452	0.8506	0.8580	0.8672	0.8780	0.8901	0.9032	0.9169
		162.929	0.8417	0.8428	0.8461	0.8515	0.8588	0.8680	0.8787	0.8908	0.9038	0.9174
		163.929	0.8426	0.8437	0.8469	0.8523	0.8596	0.8687	0.8794	0.8914	0.9044	0.9179
67	Ho	164.930	0.8434	0.8445	0.8478	0.8531	0.8604	0.8695	0.8801	0.8920	0.9049	0.9184
68	Er	167.256	0.8454	0.8465	0.8497	0.8550	0.8622	0.8712	0.8817	0.8934	0.9062	0.9195
		161.929	0.8408	0.8419	0.8452	0.8506	0.8580	0.8672	0.8780	0.8901	0.9032	0.9169
		163.929	0.8426	0.8437	0.8469	0.8523	0.8596	0.8687	0.8794	0.8914	0.9044	0.9179
		165.930	0.8443	0.8454	0.8486	0.8539	0.8612	0.8702	0.8808	0.8926	0.9055	0.9189
		166.932	0.8452	0.8463	0.8495	0.8547	0.8620	0.8709	0.8815	0.8933	0.9060	0.9193
		167.932	0.8460	0.8471	0.8503	0.8555	0.8627	0.8717	0.8821	0.8939	0.9065	0.9198
69	Tm	169.936	0.8477	0.8488	0.8519	0.8571	0.8642	0.8731	0.8834	0.8950	0.9076	0.9207
		168.934	0.8469	0.8479	0.8511	0.8563	0.8635	0.8724	0.8828	0.8945	0.9071	0.9203

Table A5.3. Rutherford backscattering spectrometry kinematic factors for Li as a projectile (continued).

Atomic no. (Z)	El.	Isotopic mass (M_2) (amu)	180°	170°	160°	150°	140°	130°	120°	110°	100°	90°
70	Yb	173.034	0.8502	0.8512	0.8544	0.8595	0.8665	0.8752	0.8854	0.8968	0.9092	0.9221
		167.934	0.8460	0.8471	0.8503	0.8555	0.8627	0.8717	0.8821	0.8939	0.9065	0.9198
		169.935	0.8477	0.8488	0.8519	0.8571	0.8642	0.8731	0.8834	0.8950	0.9076	0.9207
		170.936	0.8485	0.8496	0.8527	0.8579	0.8650	0.8738	0.8841	0.8956	0.9081	0.9211
		171.936	0.8493	0.8504	0.8535	0.8587	0.8657	0.8745	0.8847	0.8962	0.9086	0.9216
		172.938	0.8501	0.8512	0.8543	0.8594	0.8664	0.8751	0.8853	0.8968	0.9091	0.9220
		173.939	0.8509	0.8520	0.8551	0.8602	0.8671	0.8758	0.8860	0.8973	0.9096	0.9225
		175.943	0.8525	0.8535	0.8566	0.8617	0.8686	0.8771	0.8872	0.8984	0.9106	0.9233
71	Lu	174.967	0.8517	0.8528	0.8559	0.8609	0.8679	0.8765	0.8866	0.8979	0.9101	0.9229
		174.941	0.8517	0.8528	0.8558	0.8609	0.8678	0.8765	0.8866	0.8979	0.9101	0.9229
		175.943	0.8525	0.8535	0.8566	0.8617	0.8686	0.8771	0.8872	0.8984	0.9106	0.9233
72	Hf	178.49	0.8544	0.8555	0.8585	0.8635	0.8703	0.8788	0.8887	0.8998	0.9118	0.9244
		173.940	0.8509	0.8520	0.8551	0.8602	0.8671	0.8758	0.8860	0.8973	0.9096	0.9225
		175.941	0.8525	0.8535	0.8566	0.8616	0.8685	0.8771	0.8872	0.8984	0.9106	0.9233
		176.943	0.8533	0.8543	0.8574	0.8624	0.8692	0.8778	0.8878	0.8990	0.9111	0.9237
		177.944	0.8540	0.8550	0.8581	0.8631	0.8699	0.8784	0.8884	0.8995	0.9116	0.9241
		178.946	0.8548	0.8558	0.8588	0.8638	0.8706	0.8791	0.8890	0.9001	0.9120	0.9245
		179.947	0.8555	0.8565	0.8596	0.8645	0.8713	0.8797	0.8896	0.9006	0.9125	0.9249
73	Ta	180.948	0.8563	0.8573	0.8603	0.8652	0.8719	0.8803	0.8901	0.9011	0.9130	0.9253
		179.947	0.8555	0.8565	0.8596	0.8645	0.8713	0.8797	0.8896	0.9006	0.9125	0.9249
74	W	180.948	0.8563	0.8573	0.8603	0.8652	0.8719	0.8803	0.8901	0.9011	0.9130	0.9253
		183.849	0.8584	0.8594	0.8623	0.8672	0.8738	0.8821	0.8918	0.9026	0.9143	0.9265
		179.947	0.8555	0.8565	0.8596	0.8645	0.8713	0.8797	0.8896	0.9006	0.9125	0.9249
		181.948	0.8570	0.8580	0.8610	0.8659	0.8726	0.8809	0.8907	0.9016	0.9134	0.9257
		182.950	0.8577	0.8587	0.8617	0.8666	0.8733	0.8816	0.8913	0.9021	0.9139	0.9261
		183.951	0.8584	0.8594	0.8624	0.8673	0.8739	0.8822	0.8918	0.9026	0.9143	0.9265
		185.954	0.8599	0.8608	0.8638	0.8686	0.8752	0.8834	0.8929	0.9036	0.9152	0.9273
75	Re	186.207	0.8600	0.8610	0.8639	0.8688	0.8753	0.8835	0.8931	0.9038	0.9153	0.9274
		184.953	0.8592	0.8601	0.8631	0.8679	0.8745	0.8828	0.8924	0.9031	0.9148	0.9269
		186.956	0.8606	0.8615	0.8645	0.8693	0.8758	0.8839	0.8935	0.9041	0.9156	0.9277
76	Os	190.24	0.8628	0.8637	0.8666	0.8713	0.8778	0.8858	0.8952	0.9057	0.9170	0.9289
		183.952	0.8584	0.8594	0.8624	0.8673	0.8739	0.8822	0.8918	0.9026	0.9143	0.9265
		185.954	0.8599	0.8608	0.8638	0.8686	0.8752	0.8834	0.8929	0.9036	0.9152	0.9273
		186.956	0.8606	0.8615	0.8645	0.8693	0.8758	0.8839	0.8935	0.9041	0.9156	0.9277
		187.956	0.8612	0.8622	0.8651	0.8699	0.8764	0.8845	0.8940	0.9046	0.9161	0.9280
		188.958	0.8619	0.8629	0.8658	0.8705	0.8770	0.8851	0.8945	0.9051	0.9165	0.9284
		189.958	0.8626	0.8636	0.8665	0.8712	0.8776	0.8857	0.8951	0.9056	0.9169	0.9288
		191.962	0.8639	0.8649	0.8677	0.8724	0.8788	0.8868	0.8961	0.9065	0.9177	0.9295
77	Ir	192.216	0.8641	0.8651	0.8679	0.8726	0.8790	0.8869	0.8962	0.9066	0.9179	0.9296
		190.961	0.8633	0.8642	0.8671	0.8718	0.8782	0.8862	0.8956	0.9060	0.9173	0.9291
		192.963	0.8646	0.8655	0.8684	0.8731	0.8794	0.8873	0.8966	0.9070	0.9182	0.9298
78	Pt	195.08	0.8660	0.8669	0.8697	0.8743	0.8807	0.8885	0.8977	0.9079	0.9190	0.9306
		189.960	0.8626	0.8636	0.8665	0.8712	0.8776	0.8857	0.8951	0.9056	0.9169	0.9288
		191.961	0.8639	0.8649	0.8677	0.8724	0.8788	0.8868	0.8961	0.9065	0.9177	0.9295
		193.963	0.8652	0.8662	0.8690	0.8737	0.8800	0.8879	0.8971	0.9074	0.9186	0.9302
		194.965	0.8659	0.8668	0.8696	0.8743	0.8806	0.8884	0.8976	0.9079	0.9190	0.9305
		195.965	0.8665	0.8675	0.8703	0.8749	0.8812	0.8890	0.8981	0.9083	0.9194	0.9309
		197.968	0.8678	0.8687	0.8715	0.8761	0.8823	0.8900	0.8991	0.9092	0.9201	0.9315

Table A5.3. Rutherford backscattering spectrometry kinematic factors for Li as a projectile (continued).

Atomic no. (Z)	El.	Isotopic mass (M₂) (amu)	180°	170°	160°	150°	140°	130°	120°	110°	100°	90°
79	Au	196.967	0.8672	0.8681	0.8709	0.8755	0.8817	0.8895	0.8986	0.9088	0.9198	0.9312
80	Hg	200.588	0.8694	0.8703	0.8731	0.8776	0.8837	0.8914	0.9003	0.9103	0.9211	0.9324
		195.966	0.8665	0.8675	0.8703	0.8749	0.8812	0.8890	0.8981	0.9083	0.9194	0.9309
		197.967	0.8678	0.8687	0.8715	0.8761	0.8823	0.8900	0.8991	0.9092	0.9201	0.9315
		198.968	0.8684	0.8693	0.8721	0.8766	0.8828	0.8906	0.8996	0.9097	0.9205	0.9319
	Hg	199.968	0.8690	0.8699	0.8727	0.8772	0.8834	0.8911	0.9000	0.9101	0.9209	0.9322
		200.970	0.8696	0.8705	0.8733	0.8778	0.8839	0.8916	0.9005	0.9105	0.9213	0.9325
		201.971	0.8702	0.8711	0.8739	0.8784	0.8845	0.8921	0.9010	0.9109	0.9217	0.9329
		203.973	0.8714	0.8723	0.8750	0.8795	0.8856	0.8931	0.9019	0.9118	0.9224	0.9335
81	Tl	204.383	0.8716	0.8726	0.8753	0.8797	0.8858	0.8933	0.9021	0.9119	0.9226	0.9336
		202.972	0.8708	0.8717	0.8745	0.8789	0.8850	0.8926	0.9015	0.9114	0.9220	0.9332
		204.974	0.8720	0.8729	0.8756	0.8800	0.8861	0.8936	0.9024	0.9122	0.9228	0.9338
82	Pb	207.217	0.8733	0.8742	0.8769	0.8813	0.8872	0.8947	0.9034	0.9131	0.9236	0.9345
		203.973	0.8714	0.8723	0.8750	0.8795	0.8856	0.8931	0.9019	0.9118	0.9224	0.9335
		205.975	0.8726	0.8735	0.8762	0.8806	0.8866	0.8941	0.9028	0.9126	0.9231	0.9341
		206.976	0.8732	0.8741	0.8767	0.8811	0.8871	0.8946	0.9033	0.9130	0.9235	0.9344
		207.977	0.8737	0.8746	0.8773	0.8817	0.8876	0.8950	0.9037	0.9134	0.9238	0.9347
83	Bi	208.980	0.8743	0.8752	0.8778	0.8822	0.8881	0.8955	0.9041	0.9138	0.9242	0.9350
84	Po	208.982	0.8743	0.8752	0.8778	0.8822	0.8881	0.8955	0.9042	0.9138	0.9242	0.9350
85	At	210.000	0.8749	0.8758	0.8784	0.8827	0.8886	0.8960	0.9046	0.9142	0.9245	0.9353
86	Rn	222.018	0.8812	0.8821	0.8846	0.8887	0.8943	0.9013	0.9095	0.9187	0.9285	0.9387
87	Fr	223.000	0.8817	0.8826	0.8851	0.8892	0.8948	0.9018	0.9099	0.9190	0.9288	0.9390
88	Ra	226.025	0.8832	0.8840	0.8865	0.8906	0.8961	0.9030	0.9111	0.9200	0.9297	0.9398
89	Ac	227.000	0.8837	0.8845	0.8870	0.8910	0.8965	0.9034	0.9114	0.9204	0.9300	0.9400
90	Th	232.038	0.8860	0.8869	0.8893	0.8933	0.8987	0.9054	0.9133	0.9220	0.9315	0.9413
91	Pa	231.036	0.8856	0.8864	0.8888	0.8928	0.8983	0.9050	0.9129	0.9217	0.9312	0.9411
92	U	238.018	0.8888	0.8895	0.8919	0.8958	0.9011	0.9077	0.9153	0.9239	0.9331	0.9427
		234.041	0.8870	0.8878	0.8902	0.8941	0.8995	0.9062	0.9140	0.9227	0.9320	0.9418
		235.044	0.8874	0.8882	0.8906	0.8945	0.8999	0.9066	0.9143	0.9230	0.9323	0.9420
		238.040	0.8888	0.8896	0.8919	0.8958	0.9011	0.9077	0.9153	0.9239	0.9331	0.9427
93	Np	237.048	0.8883	0.8891	0.8915	0.8954	0.9007	0.9073	0.9150	0.9236	0.9329	0.9425
94	Pu	244.064	0.8914	0.8921	0.8944	0.8982	0.9034	0.9098	0.9173	0.9257	0.9347	0.9441
95	Am	243.061	0.8909	0.8917	0.8940	0.8978	0.9030	0.9095	0.9170	0.9254	0.9345	0.9439

Table A5.4. Rutherford backscattering spectrometry kinematic factors for C as a projectile.

This table gives the RBS kinematic factors K_{M_2} defined by Eq. (A4.2), for C as a projectile ($M_1=12$ amu) and for the isotopic masses (M_2) of the elements. The kinematic factors are given as a function of scattering angle from 180° to 90° measured in the laboratory frame of reference. The first row for each element gives the average atomic weight of that element and the average kine- matic factor (Average K_{M_2}) for that element. The subsequent rows give the isotopic masses for that element and the K_{M_2} for those isotopic masses. The average K_{M_2} is calculated as the weighted average of the K_{M_2} in which the relative abundances (ϖ_{M_2}) for the M_2 are given in Appendix 1:

$$\text{Average } K_{M_2} = \sum_{M_2} \varpi_{M_2} K_{M_2}\text{ , summed over the masses } M_2.$$

Atomic no. (Z)	El.	Isotopic mass (M_2) (amu)	180°	170°	160°	150°	140°	130°	120°	110°	100°	90°
7	N	14.007	0.0060	0.0061	0.0066	0.0076	0.0092	0.0121	0.0170	0.0261	0.0437	0.0772
		14.003	0.0059	0.0061	0.0066	0.0076	0.0092	0.0120	0.0169	0.0261	0.0436	0.0770
		15.000	0.0123	0.0127	0.0136	0.0154	0.0185	0.0235	0.0318	0.0459	0.0702	0.1111
8	O	15.999	0.0204	0.0209	0.0224	0.0251	0.0297	0.0369	0.0485	0.0670	0.0966	0.1428
		15.995	0.0204	0.0208	0.0223	0.0251	0.0296	0.0369	0.0484	0.0669	0.0965	0.1427
		16.999	0.0297	0.0304	0.0324	0.0361	0.0422	0.0516	0.0660	0.0883	0.1222	0.1724
		17.999	0.0400	0.0408	0.0434	0.0481	0.0555	0.0669	0.0840	0.1094	0.1468	0.2000
9	F	18.998	0.0510	0.0520	0.0551	0.0606	0.0695	0.0827	0.1021	0.1302	0.1703	0.2258
10	Ne	20.179	0.0647	0.0658	0.0695	0.0761	0.0863	0.1014	0.1232	0.1539	0.1966	0.2539
		19.992	0.0624	0.0636	0.0671	0.0736	0.0836	0.0985	0.1199	0.1503	0.1926	0.2498
		20.994	0.0743	0.0756	0.0797	0.0869	0.0981	0.1144	0.1377	0.1700	0.2141	0.2726
		21.991	0.0864	0.0878	0.0923	0.1003	0.1125	0.1302	0.1550	0.1890	0.2346	0.2939
11	Na	22.990	0.0987	0.1002	0.1051	0.1137	0.1269	0.1458	0.1719	0.2073	0.2541	0.3141
12	Mg	24.305	0.1149	0.1166	0.1220	0.1314	0.1456	0.1658	0.1934	0.2303	0.2782	0.3387
		23.985	0.1109	0.1126	0.1179	0.1271	0.1411	0.1610	0.1883	0.2249	0.2726	0.3331
		24.986	0.1233	0.1251	0.1307	0.1405	0.1552	0.1761	0.2044	0.2419	0.2904	0.3511
		25.983	0.1355	0.1374	0.1433	0.1536	0.1691	0.1908	0.2200	0.2583	0.3073	0.3681
13	Al	26.982	0.1477	0.1497	0.1559	0.1667	0.1827	0.2051	0.2351	0.2741	0.3235	0.3843
14	Si	28.086	0.1610	0.1631	0.1696	0.1808	0.1974	0.2205	0.2512	0.2908	0.3406	0.4012
		27.977	0.1597	0.1618	0.1683	0.1794	0.1960	0.2190	0.2497	0.2892	0.3390	0.3997
		28.976	0.1716	0.1738	0.1805	0.1920	0.2091	0.2327	0.2639	0.3039	0.3538	0.4143
		29.974	0.1834	0.1856	0.1925	0.2044	0.2219	0.2459	0.2776	0.3179	0.3680	0.4282
15	P	30.974	0.1949	0.1973	0.2043	0.2165	0.2344	0.2589	0.2909	0.3315	0.3816	0.4415
16	S	32.064	0.2073	0.2097	0.2169	0.2294	0.2477	0.2725	0.3049	0.3458	0.3958	0.4553
		31.972	0.2063	0.2087	0.2159	0.2283	0.2466	0.2714	0.3038	0.3446	0.3947	0.4542
		32.972	0.2175	0.2199	0.2273	0.2399	0.2585	0.2836	0.3163	0.3572	0.4072	0.4663
		33.968	0.2284	0.2309	0.2384	0.2513	0.2701	0.2955	0.3283	0.3694	0.4192	0.4779
17	Cl	35.453	0.2442	0.2467	0.2544	0.2675	0.2867	0.3124	0.3454	0.3865	0.4361	0.4941
		34.969	0.2391	0.2417	0.2493	0.2624	0.2814	0.3070	0.3400	0.3811	0.4308	0.4890
		36.966	0.2600	0.2625	0.2704	0.2838	0.3031	0.3291	0.3623	0.4034	0.4526	0.5099
18	Ar	39.948	0.2894	0.2921	0.3001	0.3138	0.3336	0.3598	0.3931	0.4339	0.4823	0.5380
		35.967	0.2497	0.2522	0.2599	0.2732	0.2924	0.3182	0.3514	0.3925	0.4419	0.4997
		37.963	0.2700	0.2726	0.2806	0.2941	0.3136	0.3397	0.3730	0.4139	0.4629	0.5196
		39.962	0.2896	0.2922	0.3003	0.3140	0.3337	0.3600	0.3933	0.4340	0.4824	0.5381
19	K	39.098	0.2812	0.2838	0.2918	0.3055	0.3251	0.3513	0.3846	0.4255	0.4741	0.5303
		38.964	0.2799	0.2826	0.2905	0.3042	0.3238	0.3500	0.3833	0.4242	0.4728	0.5291
		39.964	0.2899	0.2926	0.3007	0.3144	0.3341	0.3604	0.3937	0.4344	0.4828	0.5385
		40.962	0.2990	0.3017	0.3098	0.3236	0.3434	0.3697	0.4030	0.4436	0.4917	0.5468

Table A5.4. Rutherford backscattering spectrometry kinematic factors for C as a projectile (continued).

Atomic no. (Z)	El.	Isotopic mass (M₂) (amu)	180°	170°	160°	150°	140°	130°	120°	110°	100°	90°
20	Ca	40.078	0.2906	0.2933	0.3014	0.3151	0.3348	0.3611	0.3944	0.4351	0.4834	0.5391
		39.963	0.2896	0.2922	0.3003	0.3140	0.3337	0.3600	0.3933	0.4341	0.4824	0.5381
		41.959	0.3083	0.3110	0.3191	0.3330	0.3528	0.3792	0.4124	0.4528	0.5006	0.5552
		42.959	0.3173	0.3200	0.3282	0.3421	0.3620	0.3884	0.4215	0.4618	0.5092	0.5633
		43.956	0.3261	0.3289	0.3371	0.3510	0.3710	0.3973	0.4304	0.4705	0.5175	0.5711
		45.954	0.3433	0.3460	0.3543	0.3682	0.3882	0.4145	0.4474	0.4870	0.5334	0.5859
		47.952	0.3596	0.3624	0.3707	0.3847	0.4046	0.4308	0.4634	0.5026	0.5482	0.5997
21	Sc	44.956	0.3348	0.3375	0.3458	0.3597	0.3797	0.4060	0.4390	0.4789	0.5256	0.5786
22	Ti	47.878	0.3590	0.3617	0.3700	0.3840	0.4040	0.4301	0.4628	0.5020	0.5476	0.5991
		45.953	0.3432	0.3460	0.3543	0.3682	0.3882	0.4145	0.4474	0.4870	0.5334	0.5859
		46.952	0.3515	0.3543	0.3625	0.3765	0.3965	0.4227	0.4555	0.4949	0.5409	0.5929
		47.948	0.3596	0.3623	0.3706	0.3846	0.4046	0.4308	0.4634	0.5026	0.5482	0.5997
		48.948	0.3675	0.3703	0.3786	0.3926	0.4125	0.4386	0.4711	0.5100	0.5553	0.6062
		49.945	0.3752	0.3780	0.3863	0.4003	0.4202	0.4462	0.4785	0.5173	0.5621	0.6126
23	V	50.942	0.3828	0.3855	0.3938	0.4078	0.4277	0.4536	0.4858	0.5243	0.5688	0.6187
		49.947	0.3752	0.3780	0.3863	0.4003	0.4202	0.4462	0.4786	0.5173	0.5621	0.6126
		50.944	0.3828	0.3856	0.3939	0.4078	0.4277	0.4536	0.4858	0.5243	0.5688	0.6187
24	Cr	51.996	0.3906	0.3933	0.4016	0.4156	0.4354	0.4612	0.4932	0.5314	0.5755	0.6249
		49.946	0.3752	0.3780	0.3863	0.4003	0.4202	0.4462	0.4786	0.5173	0.5621	0.6126
		51.940	0.3902	0.3929	0.4012	0.4152	0.4350	0.4609	0.4929	0.5311	0.5752	0.6246
		52.941	0.3974	0.4002	0.4085	0.4224	0.4422	0.4679	0.4998	0.5377	0.5815	0.6304
		53.939	0.4045	0.4073	0.4155	0.4295	0.4492	0.4748	0.5065	0.5442	0.5876	0.6360
25	Mn	54.938	0.4115	0.4142	0.4225	0.4363	0.4560	0.4815	0.5131	0.5505	0.5935	0.6415
26	Fe	55.847	0.4176	0.4204	0.4286	0.4425	0.4620	0.4875	0.5188	0.5560	0.5987	0.6462
		53.940	0.4045	0.4073	0.4156	0.4295	0.4492	0.4748	0.5065	0.5442	0.5876	0.6360
		55.935	0.4182	0.4210	0.4292	0.4431	0.4626	0.4881	0.5194	0.5566	0.5992	0.6467
		56.935	0.4249	0.4276	0.4359	0.4497	0.4692	0.4945	0.5256	0.5625	0.6048	0.6518
		57.933	0.4314	0.4341	0.4423	0.4561	0.4755	0.5007	0.5317	0.5683	0.6103	0.6568
27	Co	58.933	0.4378	0.4405	0.4487	0.4624	0.4818	0.5068	0.5376	0.5740	0.6156	0.6617
28	Ni	58.688	0.4361	0.4389	0.4470	0.4608	0.4801	0.5052	0.5361	0.5725	0.6142	0.6604
		57.935	0.4314	0.4341	0.4423	0.4561	0.4755	0.5007	0.5317	0.5683	0.6103	0.6568
		59.931	0.4440	0.4467	0.4549	0.4686	0.4878	0.5128	0.5434	0.5795	0.6207	0.6663
		60.931	0.4501	0.4528	0.4610	0.4746	0.4938	0.5186	0.5490	0.5848	0.6257	0.6709
		61.928	0.4561	0.4588	0.4669	0.4805	0.4996	0.5243	0.5545	0.5901	0.6306	0.6754
		63.928	0.4677	0.4704	0.4785	0.4919	0.5108	0.5353	0.5651	0.6002	0.6400	0.6839
29	Cu	63.546	0.4655	0.4682	0.4762	0.4897	0.5087	0.5331	0.5631	0.5982	0.6382	0.6823
		62.930	0.4620	0.4647	0.4728	0.4863	0.5053	0.5298	0.5599	0.5952	0.6354	0.6797
		64.928	0.4734	0.4760	0.4841	0.4975	0.5163	0.5406	0.5702	0.6050	0.6445	0.6880
30	Zn	65.396	0.4758	0.4785	0.4865	0.4999	0.5187	0.5429	0.5724	0.6071	0.6465	0.6898
		63.929	0.4677	0.4704	0.4785	0.4919	0.5109	0.5353	0.5651	0.6002	0.6400	0.6839
		65.926	0.4789	0.4815	0.4895	0.5029	0.5216	0.5458	0.5752	0.6098	0.6489	0.6920
		66.927	0.4843	0.4870	0.4949	0.5082	0.5268	0.5509	0.5801	0.6144	0.6532	0.6959
		67.925	0.4896	0.4922	0.5002	0.5134	0.5319	0.5558	0.5849	0.6189	0.6574	0.6997
		69.925	0.4999	0.5025	0.5104	0.5235	0.5419	0.5655	0.5941	0.6277	0.6655	0.7070
31	Ga	69.723	0.4989	0.5015	0.5093	0.5224	0.5408	0.5645	0.5932	0.6268	0.6647	0.7063
		68.926	0.4948	0.4974	0.5053	0.5185	0.5370	0.5607	0.5896	0.6233	0.6615	0.7034
		70.925	0.5049	0.5075	0.5153	0.5284	0.5467	0.5701	0.5986	0.6319	0.6695	0.7106

Table A5.4. Rutherford backscattering spectrometry kinematic factors for C as a projectile (continued).

Atomic no. (Z)	El.	Isotopic mass (M₂) (amu)	180°	170°	160°	150°	140°	130°	120°	110°	100°	90°
32	Ge	72.632	0.5131	0.5157	0.5235	0.5364	0.5545	0.5777	0.6059	0.6388	0.6758	0.7163
		69.924	0.4999	0.5025	0.5104	0.5235	0.5419	0.5654	0.5941	0.6277	0.6655	0.7070
		71.922	0.5098	0.5124	0.5202	0.5332	0.5514	0.5747	0.6030	0.6360	0.6733	0.7140
		72.924	0.5147	0.5172	0.5250	0.5379	0.5560	0.5792	0.6073	0.6401	0.6770	0.7174
		73.921	0.5194	0.5219	0.5296	0.5425	0.5605	0.5835	0.6115	0.6440	0.6807	0.7207
		75.921	0.5286	0.5311	0.5388	0.5515	0.5693	0.5920	0.6196	0.6517	0.6877	0.7270
33	As	74.922	0.5240	0.5266	0.5342	0.5470	0.5649	0.5878	0.6156	0.6479	0.6842	0.7239
34	Se	78.993	0.5419	0.5444	0.5520	0.5645	0.5820	0.6043	0.6313	0.6627	0.6979	0.7361
		73.923	0.5194	0.5219	0.5297	0.5425	0.5605	0.5835	0.6115	0.6440	0.6807	0.7207
		75.919	0.5286	0.5311	0.5387	0.5515	0.5692	0.5920	0.6196	0.6517	0.6877	0.7270
		76.920	0.5330	0.5356	0.5432	0.5558	0.5735	0.5961	0.6235	0.6554	0.6911	0.7301
		77.917	0.5374	0.5399	0.5475	0.5601	0.5777	0.6002	0.6274	0.6590	0.6945	0.7331
		79.916	0.5460	0.5485	0.5559	0.5684	0.5858	0.6080	0.6349	0.6660	0.7009	0.7389
		81.917	0.5542	0.5567	0.5641	0.5764	0.5936	0.6156	0.6421	0.6728	0.7071	0.7445
35	Br	79.904	0.5459	0.5484	0.5559	0.5683	0.5857	0.6079	0.6348	0.6659	0.7009	0.7388
		78.918	0.5417	0.5442	0.5518	0.5643	0.5818	0.6041	0.6312	0.6625	0.6977	0.7360
		80.916	0.5501	0.5526	0.5601	0.5725	0.5897	0.6118	0.6385	0.6694	0.7041	0.7417
36	Kr	83.8	0.5617	0.5641	0.5714	0.5837	0.6007	0.6224	0.6485	0.6788	0.7127	0.7494
		77.920	0.5374	0.5400	0.5475	0.5601	0.5777	0.6002	0.6274	0.6590	0.6945	0.7331
		79.916	0.5460	0.5485	0.5559	0.5684	0.5858	0.6080	0.6349	0.6660	0.7009	0.7389
		81.913	0.5542	0.5567	0.5641	0.5764	0.5936	0.6156	0.6421	0.6728	0.7071	0.7444
		82.914	0.5582	0.5607	0.5681	0.5803	0.5974	0.6192	0.6456	0.6760	0.7101	0.7471
		83.911	0.5622	0.5646	0.5719	0.5841	0.6011	0.6228	0.6490	0.6792	0.7131	0.7498
		85.911	0.5698	0.5723	0.5795	0.5916	0.6084	0.6298	0.6556	0.6855	0.7188	0.7549
37	Rb	85.468	0.5681	0.5706	0.5778	0.5899	0.6068	0.6283	0.6542	0.6841	0.7175	0.7537
		84.912	0.5660	0.5685	0.5758	0.5879	0.6048	0.6264	0.6523	0.6824	0.7160	0.7524
		86.909	0.5736	0.5760	0.5832	0.5952	0.6119	0.6332	0.6589	0.6885	0.7216	0.7574
38	Sr	87.617	0.5762	0.5786	0.5858	0.5977	0.6144	0.6356	0.6611	0.6906	0.7235	0.7591
		83.913	0.5622	0.5646	0.5719	0.5841	0.6012	0.6228	0.6490	0.6792	0.7131	0.7498
		85.909	0.5698	0.5723	0.5795	0.5916	0.6084	0.6298	0.6556	0.6855	0.7188	0.7549
		86.909	0.5736	0.5760	0.5832	0.5952	0.6119	0.6332	0.6589	0.6885	0.7216	0.7574
		87.906	0.5773	0.5797	0.5868	0.5988	0.6154	0.6366	0.6620	0.6915	0.7243	0.7598
39	Y	88.906	0.5809	0.5833	0.5904	0.6023	0.6188	0.6399	0.6652	0.6944	0.7269	0.7622
40	Zr	91.221	0.5890	0.5913	0.5984	0.6101	0.6264	0.6472	0.6721	0.7008	0.7329	0.7674
		89.905	0.5844	0.5868	0.5939	0.6057	0.6222	0.6431	0.6682	0.6972	0.7296	0.7645
		90.906	0.5879	0.5903	0.5974	0.6091	0.6255	0.6463	0.6712	0.7000	0.7321	0.7668
		91.905	0.5914	0.5937	0.6008	0.6125	0.6287	0.6494	0.6742	0.7028	0.7346	0.7690
		93.906	0.5981	0.6004	0.6074	0.6190	0.6350	0.6555	0.6800	0.7082	0.7395	0.7734
		95.908	0.6046	0.6069	0.6138	0.6253	0.6412	0.6613	0.6855	0.7133	0.7443	0.7776
41	Nb	92.906	0.5948	0.5971	0.6041	0.6158	0.6319	0.6524	0.6771	0.7055	0.7371	0.7712
42	Mo	95.931	0.6046	0.6069	0.6138	0.6252	0.6411	0.6613	0.6854	0.7133	0.7442	0.7775
		91.907	0.5914	0.5937	0.6008	0.6125	0.6287	0.6494	0.6742	0.7028	0.7346	0.7690
		93.905	0.5981	0.6004	0.6074	0.6190	0.6350	0.6554	0.6799	0.7082	0.7395	0.7734
		94.906	0.6014	0.6037	0.6107	0.6222	0.6381	0.6584	0.6828	0.7108	0.7419	0.7755
		95.905	0.6046	0.6069	0.6138	0.6253	0.6412	0.6613	0.6855	0.7133	0.7443	0.7776
		96.906	0.6078	0.6101	0.6170	0.6283	0.6441	0.6642	0.6882	0.7159	0.7466	0.7796
		97.905	0.6109	0.6132	0.6200	0.6314	0.6471	0.6670	0.6909	0.7183	0.7488	0.7816
		99.908	0.6171	0.6193	0.6261	0.6373	0.6528	0.6725	0.6961	0.7232	0.7532	0.7855

Table A5.4. Rutherford backscattering spectrometry kinematic factors for C as a projectile (continued).

Atomic no. (Z)	El.	Isotopic mass (M_2) (amu)	180°	170°	160°	150°	140°	130°	120°	110°	100°	90°
43	Tc	98.000	0.6112	0.6135	0.6203	0.6316	0.6473	0.6673	0.6911	0.7186	0.7490	0.7818
44	Ru	101.019	0.6203	0.6225	0.6292	0.6404	0.6558	0.6754	0.6988	0.7257	0.7555	0.7876
		95.908	0.6046	0.6069	0.6138	0.6253	0.6412	0.6613	0.6855	0.7133	0.7443	0.7776
		97.905	0.6109	0.6132	0.6200	0.6314	0.6471	0.6670	0.6909	0.7183	0.7488	0.7816
		98.506	0.6128	0.6151	0.6219	0.6332	0.6488	0.6687	0.6925	0.7198	0.7502	0.7828
		99.904	0.6171	0.6193	0.6261	0.6373	0.6528	0.6725	0.6961	0.7232	0.7532	0.7855
		100.906	0.6201	0.6223	0.6290	0.6401	0.6556	0.6752	0.6986	0.7255	0.7554	0.7874
		101.904	0.6230	0.6252	0.6319	0.6430	0.6583	0.6778	0.7011	0.7278	0.7575	0.7893
		103.905	0.6287	0.6310	0.6376	0.6485	0.6637	0.6829	0.7059	0.7323	0.7616	0.7929
45	Rh	102.906	0.6259	0.6281	0.6348	0.6458	0.6610	0.6804	0.7035	0.7301	0.7595	0.7911
46	Pd	106.415	0.6357	0.6378	0.6444	0.6552	0.6702	0.6891	0.7118	0.7377	0.7665	0.7973
		101.906	0.6230	0.6252	0.6319	0.6430	0.6583	0.6778	0.7011	0.7278	0.7575	0.7893
		103.904	0.6287	0.6310	0.6376	0.6485	0.6637	0.6829	0.7059	0.7323	0.7616	0.7929
		104.905	0.6316	0.6338	0.6403	0.6512	0.6663	0.6855	0.7083	0.7345	0.7636	0.7947
		105.904	0.6343	0.6365	0.6431	0.6539	0.6689	0.6879	0.7106	0.7367	0.7655	0.7964
		107.904	0.6397	0.6419	0.6484	0.6591	0.6740	0.6928	0.7152	0.7409	0.7694	0.7998
		109.905	0.6450	0.6472	0.6536	0.6642	0.6789	0.6974	0.7196	0.7450	0.7731	0.8031
47	Ag	107.868	0.6396	0.6418	0.6483	0.6590	0.6739	0.6927	0.7151	0.7408	0.7693	0.7998
		106.905	0.6371	0.6392	0.6457	0.6565	0.6715	0.6904	0.7129	0.7388	0.7675	0.7982
		108.905	0.6424	0.6446	0.6510	0.6617	0.6764	0.6951	0.7174	0.7430	0.7712	0.8015
48	Cd	112.412	0.6513	0.6535	0.6598	0.6703	0.6847	0.7031	0.7249	0.7499	0.7775	0.8070
		105.907	0.6343	0.6365	0.6431	0.6539	0.6689	0.6879	0.7107	0.7367	0.7655	0.7964
		107.904	0.6397	0.6419	0.6484	0.6591	0.6740	0.6928	0.7152	0.7409	0.7694	0.7998
		109.903	0.6450	0.6471	0.6536	0.6642	0.6789	0.6974	0.7196	0.7450	0.7731	0.8031
		110.904	0.6476	0.6497	0.6561	0.6666	0.6813	0.6997	0.7218	0.7470	0.7749	0.8047
		111.903	0.6501	0.6522	0.6586	0.6691	0.6836	0.7020	0.7239	0.7490	0.7767	0.8063
		112.904	0.6526	0.6547	0.6610	0.6715	0.6859	0.7042	0.7260	0.7509	0.7784	0.8079
		113.903	0.6551	0.6572	0.6635	0.6739	0.6882	0.7064	0.7281	0.7528	0.7801	0.8094
		115.905	0.6599	0.6620	0.6682	0.6785	0.6927	0.7107	0.7321	0.7565	0.7835	0.8124
49	In	114.818	0.6573	0.6594	0.6657	0.6760	0.6903	0.7084	0.7299	0.7545	0.7817	0.8108
		112.904	0.6526	0.6547	0.6610	0.6715	0.6859	0.7042	0.7260	0.7509	0.7784	0.8079
		114.904	0.6575	0.6596	0.6659	0.6762	0.6905	0.7086	0.7301	0.7547	0.7818	0.8109
50	Sn	118.613	0.6659	0.6679	0.6741	0.6842	0.6983	0.7160	0.7370	0.7611	0.7877	0.8160
		111.905	0.6501	0.6522	0.6586	0.6691	0.6836	0.7020	0.7239	0.7490	0.7767	0.8063
		113.903	0.6551	0.6572	0.6635	0.6739	0.6882	0.7064	0.7281	0.7528	0.7801	0.8094
		114.903	0.6575	0.6596	0.6659	0.6762	0.6905	0.7086	0.7301	0.7547	0.7818	0.8109
		115.902	0.6599	0.6620	0.6682	0.6785	0.6927	0.7107	0.7321	0.7565	0.7835	0.8124
		116.903	0.6623	0.6644	0.6705	0.6808	0.6949	0.7128	0.7341	0.7584	0.7852	0.8138
		117.602	0.6639	0.6660	0.6722	0.6823	0.6964	0.7142	0.7354	0.7596	0.7863	0.8148
		118.903	0.6669	0.6690	0.6751	0.6852	0.6992	0.7169	0.7379	0.7619	0.7884	0.8167
		119.902	0.6692	0.6712	0.6773	0.6874	0.7013	0.7189	0.7398	0.7637	0.7900	0.8180
		121.903	0.6737	0.6757	0.6817	0.6917	0.7054	0.7228	0.7435	0.7671	0.7930	0.8208
		123.905	0.6780	0.6800	0.6860	0.6958	0.7095	0.7266	0.7471	0.7704	0.7960	0.8234
51	Sb	121.758	0.6733	0.6753	0.6814	0.6914	0.7051	0.7225	0.7432	0.7668	0.7928	0.8206
		120.904	0.6714	0.6735	0.6795	0.6896	0.7034	0.7209	0.7417	0.7654	0.7915	0.8194
		122.904	0.6758	0.6778	0.6838	0.6938	0.7075	0.7247	0.7453	0.7687	0.7946	0.8221

Table A5.4. Rutherford backscattering spectrometry kinematic factors for C as a projectile (continued).

Atomic no. (Z)	El.	Isotopic mass (M₂) (amu)	180°	170°	160°	150°	140°	130°	120°	110°	100°	90°
52	Te	127.586	0.6856	0.6876	0.6935	0.7031	0.7165	0.7334	0.7534	0.7762	0.8013	0.8280
		119.904	0.6692	0.6712	0.6773	0.6874	0.7013	0.7189	0.7398	0.7637	0.7900	0.8180
		121.903	0.6737	0.6757	0.6817	0.6917	0.7054	0.7228	0.7435	0.7671	0.7930	0.8208
		122.904	0.6758	0.6778	0.6838	0.6938	0.7075	0.7247	0.7453	0.7687	0.7946	0.8221
		123.903	0.6780	0.6800	0.6860	0.6958	0.7095	0.7266	0.7471	0.7704	0.7960	0.8234
		124.904	0.6801	0.6821	0.6881	0.6979	0.7114	0.7285	0.7488	0.7720	0.7975	0.8247
		125.903	0.6822	0.6842	0.6901	0.6999	0.7134	0.7304	0.7506	0.7736	0.7990	0.8260
		127.905	0.6863	0.6883	0.6942	0.7038	0.7172	0.7340	0.7540	0.7767	0.8018	0.8285
		129.906	0.6904	0.6923	0.6981	0.7077	0.7209	0.7375	0.7573	0.7798	0.8045	0.8309
53	I	126.905	0.6843	0.6863	0.6921	0.7019	0.7153	0.7322	0.7523	0.7752	0.8004	0.8272
54	Xe	131.293	0.6930	0.6949	0.7007	0.7102	0.7233	0.7399	0.7595	0.7818	0.8064	0.8325
		123.906	0.6780	0.6800	0.6860	0.6958	0.7095	0.7266	0.7471	0.7704	0.7960	0.8234
		125.904	0.6822	0.6842	0.6901	0.6999	0.7134	0.7304	0.7506	0.7736	0.7990	0.8260
		127.904	0.6863	0.6883	0.6941	0.7038	0.7172	0.7340	0.7540	0.7767	0.8018	0.8285
		128.905	0.6884	0.6903	0.6961	0.7058	0.7190	0.7358	0.7556	0.7783	0.8032	0.8297
		129.904	0.6903	0.6923	0.6981	0.7077	0.7209	0.7375	0.7573	0.7798	0.8045	0.8309
		130.905	0.6923	0.6942	0.7000	0.7095	0.7227	0.7392	0.7589	0.7813	0.8059	0.8321
		131.905	0.6943	0.6962	0.7019	0.7114	0.7245	0.7409	0.7605	0.7828	0.8072	0.8332
		133.905	0.6981	0.7000	0.7057	0.7151	0.7280	0.7443	0.7636	0.7856	0.8098	0.8355
		135.907	0.7018	0.7037	0.7093	0.7186	0.7314	0.7476	0.7667	0.7885	0.8123	0.8377
55	Cs	132.905	0.6962	0.6981	0.7038	0.7132	0.7262	0.7426	0.7621	0.7842	0.8085	0.8344
56	Ba	137.327	0.7044	0.7062	0.7118	0.7211	0.7338	0.7498	0.7688	0.7904	0.8141	0.8393
		129.906	0.6904	0.6923	0.6981	0.7077	0.7209	0.7375	0.7573	0.7798	0.8045	0.8309
		131.905	0.6943	0.6962	0.7019	0.7114	0.7245	0.7409	0.7605	0.7828	0.8072	0.8332
		133.904	0.6981	0.7000	0.7057	0.7150	0.7280	0.7443	0.7636	0.7856	0.8098	0.8355
		134.906	0.7000	0.7018	0.7075	0.7168	0.7297	0.7459	0.7652	0.7871	0.8111	0.8366
		135.905	0.7018	0.7037	0.7093	0.7186	0.7314	0.7476	0.7667	0.7885	0.8123	0.8377
		136.906	0.7036	0.7055	0.7111	0.7204	0.7331	0.7492	0.7682	0.7898	0.8136	0.8388
		137.905	0.7054	0.7073	0.7129	0.7221	0.7348	0.7507	0.7697	0.7912	0.8148	0.8399
57	La	138.905	0.7072	0.7091	0.7146	0.7238	0.7364	0.7523	0.7711	0.7925	0.8160	0.8410
		137.907	0.7054	0.7073	0.7129	0.7221	0.7348	0.7507	0.7697	0.7912	0.8148	0.8399
		138.906	0.7072	0.7091	0.7146	0.7238	0.7364	0.7523	0.7711	0.7925	0.8160	0.8410
58	Ce	140.115	0.7093	0.7112	0.7167	0.7258	0.7384	0.7541	0.7729	0.7941	0.8174	0.8422
		135.907	0.7018	0.7037	0.7093	0.7186	0.7314	0.7476	0.7667	0.7885	0.8123	0.8377
		137.906	0.7054	0.7073	0.7129	0.7221	0.7348	0.7507	0.7697	0.7912	0.8148	0.8399
		139.905	0.7090	0.7108	0.7163	0.7255	0.7380	0.7538	0.7726	0.7939	0.8172	0.8420
		141.909	0.7124	0.7143	0.7197	0.7288	0.7412	0.7569	0.7754	0.7965	0.8196	0.8441
59	Pr	140.908	0.7107	0.7126	0.7181	0.7271	0.7396	0.7554	0.7740	0.7952	0.8184	0.8430
60	Nd	144.242	0.7163	0.7181	0.7236	0.7325	0.7448	0.7603	0.7786	0.7994	0.8222	0.8464
		141.908	0.7124	0.7143	0.7197	0.7288	0.7412	0.7569	0.7754	0.7965	0.8195	0.8441
		142.910	0.7141	0.7160	0.7214	0.7304	0.7428	0.7583	0.7768	0.7977	0.8207	0.8451
		143.910	0.7158	0.7176	0.7231	0.7320	0.7443	0.7598	0.7782	0.7990	0.8218	0.8461
		144.913	0.7175	0.7193	0.7247	0.7336	0.7459	0.7613	0.7795	0.8003	0.8230	0.8470
		145.913	0.7191	0.7209	0.7263	0.7352	0.7474	0.7627	0.7809	0.8015	0.8241	0.8480
		147.917	0.7224	0.7242	0.7295	0.7382	0.7503	0.7655	0.7835	0.8039	0.8262	0.8499
		149.921	0.7255	0.7273	0.7326	0.7413	0.7532	0.7683	0.7861	0.8063	0.8284	0.8518

Table A5.4. Rutherford backscattering spectrometry kinematic factors for C as a projectile (continued).

Atomic no. (Z)	El.	Isotopic mass (M₂) (amu)	180°	170°	160°	150°	140°	130°	120°	110°	100°	90°
61	Pm	145.000	0.7176	0.7194	0.7248	0.7337	0.7460	0.7614	0.7796	0.8004	0.8231	0.8471
62	Sm	150.36	0.7261	0.7279	0.7332	0.7419	0.7538	0.7688	0.7866	0.8067	0.8288	0.8521
		143.912	0.7158	0.7176	0.7231	0.7320	0.7443	0.7598	0.7782	0.7990	0.8218	0.8461
		146.915	0.7208	0.7226	0.7279	0.7367	0.7489	0.7641	0.7822	0.8027	0.8252	0.8490
		147.915	0.7224	0.7241	0.7295	0.7382	0.7503	0.7655	0.7835	0.8039	0.8262	0.8499
		148.917	0.7240	0.7257	0.7310	0.7398	0.7518	0.7669	0.7848	0.8051	0.8273	0.8509
		149.917	0.7255	0.7273	0.7326	0.7413	0.7532	0.7683	0.7861	0.8063	0.8284	0.8518
		151.920	0.7286	0.7304	0.7356	0.7442	0.7561	0.7710	0.7886	0.8086	0.8304	0.8536
		153.922	0.7316	0.7334	0.7385	0.7471	0.7588	0.7736	0.7910	0.8108	0.8324	0.8554
63	Eu	151.965	0.7287	0.7304	0.7356	0.7443	0.7561	0.7710	0.7886	0.8086	0.8305	0.8536
		150.920	0.7271	0.7288	0.7341	0.7427	0.7547	0.7696	0.7873	0.8074	0.8294	0.8527
		152.921	0.7301	0.7319	0.7371	0.7456	0.7575	0.7723	0.7898	0.8097	0.8314	0.8545
64	Gd	157.252	0.7365	0.7382	0.7433	0.7517	0.7633	0.7778	0.7950	0.8144	0.8357	0.8582
		151.920	0.7286	0.7304	0.7356	0.7442	0.7561	0.7710	0.7886	0.8086	0.8304	0.8536
		153.921	0.7316	0.7334	0.7385	0.7471	0.7588	0.7736	0.7910	0.8108	0.8324	0.8554
		154.923	0.7331	0.7348	0.7400	0.7485	0.7602	0.7749	0.7922	0.8119	0.8334	0.8562
		155.922	0.7346	0.7363	0.7414	0.7499	0.7615	0.7761	0.7934	0.8130	0.8344	0.8571
		156.924	0.7360	0.7377	0.7429	0.7513	0.7629	0.7774	0.7946	0.8141	0.8354	0.8579
		157.924	0.7375	0.7392	0.7443	0.7526	0.7642	0.7786	0.7958	0.8151	0.8363	0.8588
		159.927	0.7403	0.7420	0.7470	0.7553	0.7668	0.7811	0.7981	0.8172	0.8382	0.8604
65	Tb	158.925	0.7389	0.7406	0.7457	0.7540	0.7655	0.7799	0.7969	0.8162	0.8373	0.8596
66	Dy	162.498	0.7438	0.7455	0.7505	0.7587	0.7700	0.7842	0.8009	0.8199	0.8406	0.8625
		155.924	0.7346	0.7363	0.7414	0.7499	0.7615	0.7761	0.7934	0.8130	0.8344	0.8571
		157.924	0.7375	0.7392	0.7443	0.7526	0.7642	0.7786	0.7958	0.8151	0.8363	0.8588
		159.925	0.7403	0.7420	0.7470	0.7553	0.7668	0.7811	0.7981	0.8172	0.8382	0.8604
		160.927	0.7417	0.7434	0.7484	0.7567	0.7680	0.7823	0.7992	0.8183	0.8391	0.8612
		161.927	0.7431	0.7447	0.7497	0.7580	0.7693	0.7835	0.8003	0.8193	0.8400	0.8620
		162.929	0.7444	0.7461	0.7511	0.7593	0.7705	0.7847	0.8014	0.8203	0.8410	0.8628
		163.929	0.7458	0.7474	0.7524	0.7606	0.7718	0.7858	0.8025	0.8213	0.8418	0.8636
67	Ho	164.930	0.7471	0.7488	0.7537	0.7618	0.7730	0.7870	0.8036	0.8223	0.8427	0.8644
68	Er	167.256	0.7501	0.7518	0.7567	0.7647	0.7758	0.7896	0.8060	0.8245	0.8447	0.8661
		161.929	0.7431	0.7447	0.7497	0.7580	0.7693	0.7835	0.8003	0.8193	0.8401	0.8620
		163.929	0.7458	0.7474	0.7524	0.7606	0.7718	0.7858	0.8025	0.8213	0.8418	0.8636
		165.930	0.7484	0.7501	0.7550	0.7631	0.7742	0.7881	0.8046	0.8233	0.8436	0.8651
		166.932	0.7497	0.7514	0.7563	0.7643	0.7754	0.7893	0.8057	0.8242	0.8445	0.8659
		167.932	0.7510	0.7527	0.7575	0.7655	0.7766	0.7904	0.8067	0.8252	0.8453	0.8666
		169.936	0.7536	0.7552	0.7600	0.7680	0.7789	0.7926	0.8088	0.8271	0.8470	0.8681
69	Tm	168.934	0.7523	0.7539	0.7588	0.7668	0.7777	0.7915	0.8077	0.8261	0.8462	0.8674
70	Yb	173.034	0.7574	0.7590	0.7638	0.7716	0.7824	0.7959	0.8118	0.8299	0.8495	0.8703
		167.934	0.7510	0.7527	0.7575	0.7655	0.7766	0.7904	0.8067	0.8252	0.8453	0.8666
		169.935	0.7536	0.7552	0.7600	0.7680	0.7789	0.7926	0.8088	0.8271	0.8470	0.8681
		170.936	0.7548	0.7564	0.7612	0.7692	0.7800	0.7937	0.8098	0.8280	0.8478	0.8688
		171.936	0.7561	0.7577	0.7625	0.7703	0.7812	0.7947	0.8108	0.8289	0.8486	0.8695
		172.938	0.7573	0.7589	0.7637	0.7715	0.7823	0.7958	0.8118	0.8298	0.8495	0.8702
		173.939	0.7585	0.7601	0.7649	0.7727	0.7834	0.7969	0.8127	0.8307	0.8503	0.8709
		175.943	0.7609	0.7625	0.7672	0.7749	0.7856	0.7989	0.8147	0.8325	0.8518	0.8723

Table A5.4. Rutherford backscattering spectrometry kinematic factors for C as a projectile (continued).

Atomic no. (Z)	El.	Isotopic mass (M₂) (amu)	180°	170°	160°	150°	140°	130°	120°	110°	100°	90°
71	Lu	174.967	0.7597	0.7613	0.7661	0.7738	0.7845	0.7979	0.8137	0.8316	0.8511	0.8716
		174.941	0.7597	0.7613	0.7660	0.7738	0.7845	0.7979	0.8137	0.8316	0.8511	0.8716
		175.943	0.7609	0.7625	0.7672	0.7749	0.7856	0.7989	0.8147	0.8325	0.8518	0.8723
72	Hf	178.49	0.7639	0.7654	0.7701	0.7778	0.7883	0.8015	0.8171	0.8346	0.8538	0.8740
		173.940	0.7585	0.7601	0.7649	0.7727	0.7834	0.7969	0.8127	0.8307	0.8503	0.8709
		175.941	0.7609	0.7625	0.7672	0.7749	0.7856	0.7989	0.8147	0.8325	0.8518	0.8723
		176.943	0.7621	0.7637	0.7684	0.7761	0.7867	0.8000	0.8156	0.8333	0.8526	0.8730
		177.944	0.7633	0.7648	0.7695	0.7772	0.7877	0.8010	0.8166	0.8342	0.8534	0.8736
		178.946	0.7644	0.7660	0.7706	0.7783	0.7888	0.8020	0.8175	0.8350	0.8541	0.8743
		179.947	0.7656	0.7671	0.7717	0.7794	0.7898	0.8029	0.8184	0.8359	0.8549	0.8750
73	Ta	180.948	0.7667	0.7682	0.7729	0.7804	0.7909	0.8039	0.8193	0.8367	0.8556	0.8756
		179.947	0.7656	0.7671	0.7717	0.7794	0.7898	0.8029	0.8184	0.8359	0.8549	0.8750
		180.948	0.7667	0.7682	0.7729	0.7804	0.7909	0.8039	0.8193	0.8367	0.8556	0.8756
74	W	183.849	0.7699	0.7714	0.7760	0.7835	0.7938	0.8067	0.8219	0.8391	0.8577	0.8774
		179.947	0.7656	0.7671	0.7717	0.7794	0.7898	0.8029	0.8184	0.8359	0.8549	0.8750
		181.948	0.7678	0.7694	0.7740	0.7815	0.7919	0.8049	0.8202	0.8375	0.8564	0.8763
		182.950	0.7689	0.7705	0.7750	0.7826	0.7929	0.8059	0.8211	0.8383	0.8571	0.8769
		183.951	0.7700	0.7716	0.7761	0.7836	0.7939	0.8068	0.8220	0.8391	0.8578	0.8775
		185.954	0.7722	0.7737	0.7783	0.7857	0.7959	0.8087	0.8237	0.8407	0.8592	0.8788
75	Re	186.207	0.7725	0.7740	0.7785	0.7859	0.7961	0.8089	0.8240	0.8409	0.8594	0.8789
		184.953	0.7711	0.7727	0.7772	0.7847	0.7949	0.8077	0.8229	0.8399	0.8585	0.8781
		186.956	0.7733	0.7748	0.7793	0.7867	0.7969	0.8096	0.8246	0.8415	0.8599	0.8794
76	Os	190.24	0.7767	0.7782	0.7827	0.7900	0.8000	0.8126	0.8273	0.8440	0.8622	0.8813
		183.952	0.7700	0.7716	0.7761	0.7836	0.7939	0.8068	0.8220	0.8391	0.8578	0.8775
		185.954	0.7722	0.7737	0.7783	0.7857	0.7959	0.8087	0.8237	0.8407	0.8592	0.8788
		186.956	0.7733	0.7748	0.7793	0.7867	0.7969	0.8096	0.8246	0.8415	0.8599	0.8794
		187.956	0.7744	0.7759	0.7803	0.7877	0.7978	0.8105	0.8254	0.8423	0.8606	0.8800
		188.958	0.7754	0.7769	0.7814	0.7887	0.7988	0.8114	0.8263	0.8431	0.8613	0.8806
		189.958	0.7764	0.7779	0.7824	0.7897	0.7998	0.8123	0.8271	0.8438	0.8620	0.8812
		191.962	0.7785	0.7800	0.7844	0.7917	0.8016	0.8141	0.8288	0.8453	0.8633	0.8823
77	Ir	192.216	0.7788	0.7802	0.7846	0.7919	0.8019	0.8143	0.8290	0.8455	0.8635	0.8825
		190.961	0.7775	0.7790	0.7834	0.7907	0.8007	0.8132	0.8280	0.8446	0.8627	0.8818
		192.963	0.7795	0.7810	0.7854	0.7926	0.8025	0.8150	0.8296	0.8461	0.8640	0.8829
78	Pt	195.08	0.7816	0.7831	0.7875	0.7946	0.8045	0.8168	0.8313	0.8476	0.8654	0.8841
		189.960	0.7765	0.7779	0.7824	0.7897	0.7998	0.8123	0.8271	0.8438	0.8620	0.8812
		191.961	0.7785	0.7800	0.7844	0.7917	0.8016	0.8141	0.8288	0.8453	0.8633	0.8823
		193.963	0.7805	0.7820	0.7864	0.7936	0.8035	0.8158	0.8304	0.8468	0.8647	0.8835
		194.965	0.7815	0.7830	0.7873	0.7945	0.8044	0.8167	0.8312	0.8475	0.8653	0.8840
		195.965	0.7825	0.7840	0.7883	0.7955	0.8053	0.8175	0.8320	0.8482	0.8659	0.8846
		197.968	0.7845	0.7859	0.7902	0.7973	0.8070	0.8192	0.8335	0.8497	0.8672	0.8857
79	Au	196.967	0.7835	0.7849	0.7893	0.7964	0.8062	0.8184	0.8327	0.8490	0.8666	0.8851
80	Hg	200.588	0.7869	0.7884	0.7926	0.7997	0.8093	0.8213	0.8355	0.8515	0.8688	0.8871
		195.966	0.7825	0.7840	0.7883	0.7955	0.8053	0.8175	0.8320	0.8482	0.8659	0.8846
		197.967	0.7845	0.7859	0.7902	0.7973	0.8070	0.8192	0.8335	0.8497	0.8672	0.8857
		198.968	0.7854	0.7869	0.7912	0.7982	0.8079	0.8200	0.8343	0.8504	0.8678	0.8862
		199.968	0.7864	0.7878	0.7921	0.7991	0.8088	0.8208	0.8350	0.8511	0.8685	0.8868

Table A5.4. Rutherford backscattering spectrometry kinematic factors for C as a projectile (continued).

Atomic no. (Z)	El.	Isotopic mass (M$_2$) (amu)	180°	170°	160°	150°	140°	130°	120°	110°	100°	90°
		200.970	0.7873	0.7887	0.7930	0.8000	0.8096	0.8216	0.8358	0.8517	0.8691	0.8873
		201.971	0.7883	0.7897	0.7939	0.8009	0.8105	0.8225	0.8365	0.8524	0.8697	0.8878
		203.973	0.7901	0.7915	0.7957	0.8027	0.8122	0.8240	0.8380	0.8538	0.8709	0.8889
81	Tl	204.383	0.7905	0.7919	0.7961	0.8030	0.8125	0.8244	0.8383	0.8540	0.8711	0.8891
		202.972	0.7892	0.7906	0.7948	0.8018	0.8113	0.8232	0.8373	0.8531	0.8703	0.8884
		204.974	0.7910	0.7924	0.7966	0.8035	0.8130	0.8248	0.8387	0.8544	0.8715	0.8894
82	Pb	207.217	0.7930	0.7944	0.7986	0.8054	0.8148	0.8265	0.8403	0.8559	0.8728	0.8905
		203.973	0.7901	0.7915	0.7957	0.8027	0.8122	0.8240	0.8380	0.8538	0.8709	0.8889
		205.975	0.7919	0.7933	0.7975	0.8044	0.8138	0.8256	0.8395	0.8551	0.8720	0.8899
		206.976	0.7928	0.7942	0.7984	0.8052	0.8146	0.8264	0.8402	0.8557	0.8726	0.8904
		207.977	0.7937	0.7951	0.7992	0.8061	0.8154	0.8271	0.8409	0.8564	0.8732	0.8909
83	Bi	208.980	0.7946	0.7960	0.8001	0.8069	0.8162	0.8279	0.8416	0.8570	0.8738	0.8914
84	Po	208.982	0.7946	0.7960	0.8001	0.8069	0.8162	0.8279	0.8416	0.8570	0.8738	0.8914
85	At	210.000	0.7955	0.7969	0.8010	0.8077	0.8170	0.8286	0.8423	0.8576	0.8743	0.8919
86	Rn	222.018	0.8054	0.8067	0.8107	0.8172	0.8260	0.8371	0.8502	0.8648	0.8807	0.8974
87	Fr	223.000	0.8062	0.8075	0.8114	0.8179	0.8267	0.8378	0.8508	0.8654	0.8812	0.8979
88	Ra	226.025	0.8085	0.8098	0.8137	0.8201	0.8288	0.8398	0.8526	0.8671	0.8827	0.8992
89	Ac	227.000	0.8092	0.8105	0.8144	0.8208	0.8295	0.8404	0.8532	0.8676	0.8832	0.8996
90	Th	232.038	0.8130	0.8143	0.8181	0.8243	0.8329	0.8436	0.8562	0.8703	0.8856	0.9017
91	Pa	231.036	0.8123	0.8135	0.8174	0.8236	0.8322	0.8430	0.8556	0.8698	0.8851	0.9012
92	U	238.018	0.8172	0.8185	0.8222	0.8283	0.8367	0.8472	0.8595	0.8733	0.8883	0.9040
		234.041	0.8144	0.8157	0.8195	0.8257	0.8342	0.8448	0.8573	0.8713	0.8865	0.9025
		235.044	0.8151	0.8164	0.8202	0.8264	0.8349	0.8454	0.8579	0.8718	0.8870	0.9029
		238.040	0.8172	0.8185	0.8222	0.8284	0.8368	0.8472	0.8595	0.8733	0.8883	0.9040
93	Np	237.048	0.8166	0.8178	0.8216	0.8277	0.8361	0.8466	0.8590	0.8728	0.8879	0.9036
94	Pu	244.064	0.8213	0.8226	0.8262	0.8322	0.8405	0.8507	0.8627	0.8763	0.8909	0.9063
95	Am	243.061	0.8207	0.8219	0.8256	0.8316	0.8398	0.8501	0.8622	0.8758	0.8905	0.9059

APPENDIX

6

RUTHERFORD CROSS SECTIONS

Compiled by

A. Dick and J. R. Tesmer

Los Alamos National Laboratory, Los Alamos, New Mexico, USA

CONTENTS

A6.1 RUTHERFORD CROSS SECTIONS

The equation used to calculate the Rutherford cross sections is:

$$\sigma_R(E,\theta) = \left(\frac{Z_1 Z_2 e^2}{4E}\right)^2$$

$$\times \frac{4\left[\left(M_2^2 - M_1^2 \sin^2\theta\right)^{1/2} + M_2 \cos\theta\right]^2}{M_2 \sin^4\theta \left(M_2^2 - M_1^2 \sin^2\theta\right)^{1/2}}, \qquad (A6.1)$$

where M_1 and M_2 are the incident and target masses, respectively, and Z_1 and Z_2 are the corresponding atomic numbers. θ is the scattering angle (see Appendix 4). The cross sections calculated in this appendix use the masses from *Nuclides and Isotopes*, 14th Ed. (Walker *et al.*, 1989). Tables A6.1 through A6.9 give the cross sections for 1 MeV protons, deuterons, and ^{3}He, ^{4}He, ^{7}Li, ^{12}C, ^{15}N, ^{16}O, and ^{28}Si ions scattered from a range of target nuclei at various angles. The mass for the scattering element is an average over the isotopic masses. For Rutherford cross sections at other energies, denoted by E, the values in the Tables should be divided by a value of E^2 (where E takes units of MeV).

In some cases, Tables A6.1 through A6.9 must be corrected for electron screening to obtain accurate scattering cross sections. The corrections, called F-factors, can be found in Table A6.10 and were calculated from the expression given by L'Ecuyer *et al.* (1979)

$$\sigma_{sc}/\sigma_R = 1 - \frac{0.049 Z_1 Z_2^{4/3}}{E_{cm}} = F, \qquad (A6.2)$$

where E_{cm}, the center-of-mass kinetic energy in keV, is replaced by the laboratory energy, E_{lab}, with negligible error. See Chapter 4, Section 4.2.2.3, and Chapter 15, Section 15.4.1, for more detailed information. Although the F-factor is relatively insensitive to angles normally used for backscattering, it is valid only for scattering angles greater than ~90° (see Hautala and Luomajarvi, 1980). The corrections for forward angles can be much worse! See references in Chapters 4 and 15 for more information.

This L'Ecuyer *et al.* correction is incorporated in the main IBA program codes. The uncertainty in the screening correction is cited as 0.5% by Wätjen and Bax (1994) for the case of 1.5 MeV ^{4}He on Bi, and this uncertainty can be scaled for other cases from Eq. (A6.2). However, for small scattering angles or for heavy-ion beams, L'Ecuyer *et al.*'s correction is not as accurate as that proposed by Andersen *et al.* (1980), which takes into account the dependence of screening on the scattering angle. Andersen *et al.*'s correction is also incorporated in the main IBA codes and should be used.

Table A6.1. Rutherford scattering cross sections (barns) of the elements for 1 MeV ^{1}H (protons).

Element	Atomic no. (Z_2)	Mass (M_2)	Scattering angle (degrees)										
			30	60	90	120	135	140	150	160	165	170	175
He	2	4.003	1.154E+00	8.223E-02	2.005E-02	8.550E-03	6.459E-03	5.995E-03	5.307E-03	4.868E-03	4.723E-03	4.623E-03	4.564E-03
Li	3	6.941	2.597E+00	1.860E-01	4.613E-02	2.023E-02	1.551E-02	1.446E-02	1.290E-02	1.190E-02	1.158E-02	1.135E-02	1.121E-02
Be	4	9.012	4.617E+00	3.310E-01	8.237E-02	3.632E-02	2.792E-02	2.606E-02	2.329E-02	2.151E-02	2.093E-02	2.052E-02	2.029E-02
B	5	10.811	7.215E+00	5.175E-01	1.289E-01	5.700E-02	4.388E-02	4.096E-02	3.663E-02	3.386E-02	3.295E-02	3.231E-02	3.194E-02
C	6	12.011	1.039E+01	7.453E-01	1.858E-01	8.223E-02	6.334E-02	5.914E-02	5.290E-02	4.891E-02	4.760E-02	4.669E-02	4.615E-02
N	7	14.007	1.414E+01	1.015E+00	2.532E-01	1.122E-01	8.645E-02	8.073E-02	7.224E-02	6.681E-02	6.502E-02	6.378E-02	6.305E-02
O	8	15.999	1.847E+01	1.326E+00	3.309E-01	1.467E-01	1.131E-01	1.056E-01	9.456E-02	8.746E-02	8.513E-02	8.350E-02	8.255E-02
F	9	18.998	2.338E+01	1.678E+00	4.190E-01	1.859E-01	1.434E-01	1.339E-01	1.199E-01	1.109E-01	1.080E-01	1.059E-01	1.047E-01
Ne	10	20.180	2.886E+01	2.072E+00	5.174E-01	2.296E-01	1.771E-01	1.655E-01	1.481E-01	1.370E-01	1.334E-01	1.309E-01	1.294E-01
Na	11	22.990	3.492E+01	2.507E+00	6.262E-01	2.780E-01	2.145E-01	2.004E-01	1.794E-01	1.660E-01	1.616E-01	1.585E-01	1.567E-01
Mg	12	24.305	4.156E+01	2.983E+00	7.453E-01	3.309E-01	2.553E-01	2.385E-01	2.136E-01	1.976E-01	1.924E-01	1.887E-01	1.866E-01
Al	13	26.982	4.878E+01	3.501E+00	8.749E-01	3.885E-01	2.998E-01	2.801E-01	2.508E-01	2.321E-01	2.259E-01	2.216E-01	2.191E-01
Si	14	28.086	5.657E+01	4.061E+00	1.015E+00	4.506E-01	3.478E-01	3.249E-01	2.909E-01	2.692E-01	2.621E-01	2.571E-01	2.542E-01
P	15	30.974	6.494E+01	4.662E+00	1.165E+00	5.174E-01	3.994E-01	3.731E-01	3.341E-01	3.092E-01	3.010E-01	2.953E-01	2.919E-01
S	16	32.066	7.388E+01	5.304E+00	1.326E+00	5.888E-01	4.544E-01	4.246E-01	3.802E-01	3.518E-01	3.425E-01	3.360E-01	3.322E-01
Cl	17	35.453	8.341E+01	5.988E+00	1.497E+00	6.648E-01	5.131E-01	4.794E-01	4.294E-01	3.973E-01	3.868E-01	3.794E-01	3.751E-01
Ar	18	39.948	9.351E+01	6.713E+00	1.678E+00	7.454E-01	5.754E-01	5.376E-01	4.815E-01	4.456E-01	4.338E-01	4.255E-01	4.207E-01
K	19	39.098	1.042E+02	7.480E+00	1.870E+00	8.305E-01	6.411E-01	5.990E-01	5.365E-01	4.964E-01	4.833E-01	4.741E-01	4.687E-01
Ca	20	40.078	1.154E+02	8.288E+00	2.072E+00	9.203E-01	7.104E-01	6.637E-01	5.944E-01	5.501E-01	5.355E-01	5.253E-01	5.194E-01
Sc	21	44.959	1.273E+02	9.138E+00	2.284E+00	1.015E+00	7.834E-01	7.319E-01	6.555E-01	6.066E-01	5.905E-01	5.793E-01	5.727E-01
Ti	22	47.880	1.397E+02	1.003E+01	2.507E+00	1.114E+00	8.598E-01	8.034E-01	7.195E-01	6.659E-01	6.482E-01	6.359E-01	6.287E-01
V	23	50.942	1.527E+02	1.096E+01	2.740E+00	1.217E+00	9.398E-01	8.781E-01	7.865E-01	7.278E-01	7.085E-01	6.951E-01	6.872E-01
Cr	24	51.996	1.662E+02	1.194E+01	2.983E+00	1.326E+00	1.023E+00	9.562E-01	8.564E-01	7.925E-01	7.715E-01	7.569E-01	7.483E-01
Mn	25	54.938	1.804E+02	1.295E+01	3.237E+00	1.438E+00	1.110E+00	1.038E+00	9.293E-01	8.600E-01	8.372E-01	8.213E-01	8.120E-01
Fe	26	55.847	1.951E+02	1.401E+01	3.501E+00	1.556E+00	1.201E+00	1.122E+00	1.005E+00	9.302E-01	9.055E-01	8.884E-01	8.783E-01
Co	27	58.933	2.104E+02	1.511E+01	3.776E+00	1.678E+00	1.295E+00	1.210E+00	1.084E+00	1.003E+00	9.766E-01	9.581E-01	9.472E-01
Ni	28	58.690	2.263E+02	1.625E+01	4.061E+00	1.804E+00	1.393E+00	1.302E+00	1.166E+00	1.079E+00	1.050E+00	1.030E+00	1.019E+00
Cu	29	63.546	2.427E+02	1.743E+01	4.356E+00	1.936E+00	1.494E+00	1.396E+00	1.251E+00	1.157E+00	1.126E+00	1.105E+00	1.093E+00
Zn	30	65.390	2.598E+02	1.865E+01	4.662E+00	2.072E+00	1.599E+00	1.494E+00	1.338E+00	1.239E+00	1.206E+00	1.183E+00	1.169E+00
Ga	31	69.723	2.774E+02	1.991E+01	4.978E+00	2.212E+00	1.708E+00	1.596E+00	1.429E+00	1.323E+00	1.288E+00	1.263E+00	1.249E+00
Ge	32	72.610	2.955E+02	2.122E+01	5.304E+00	2.357E+00	1.820E+00	1.700E+00	1.523E+00	1.409E+00	1.372E+00	1.346E+00	1.331E+00
As	33	74.922	3.143E+02	2.257E+01	5.641E+00	2.507E+00	1.935E+00	1.808E+00	1.620E+00	1.499E+00	1.459E+00	1.432E+00	1.415E+00
Se	34	78.960	3.336E+02	2.395E+01	5.988E+00	2.661E+00	2.054E+00	1.920E+00	1.719E+00	1.591E+00	1.549E+00	1.520E+00	1.502E+00
Br	35	79.904	3.536E+02	2.538E+01	6.345E+00	2.820E+00	2.177E+00	2.034E+00	1.822E+00	1.686E+00	1.641E+00	1.610E+00	1.592E+00
Kr	36	83.800	3.740E+02	2.685E+01	6.713E+00	2.983E+00	2.303E+00	2.152E+00	1.928E+00	1.784E+00	1.737E+00	1.704E+00	1.684E+00
Rb	37	85.468	3.951E+02	2.837E+01	7.091E+00	3.151E+00	2.433E+00	2.273E+00	2.036E+00	1.884E+00	1.834E+00	1.800E+00	1.779E+00
Sr	38	87.620	4.168E+02	2.992E+01	7.480E+00	3.324E+00	2.566E+00	2.398E+00	2.148E+00	1.988E+00	1.935E+00	1.898E+00	1.877E+00
Y	39	88.906	4.390E+02	3.152E+01	7.879E+00	3.501E+00	2.703E+00	2.526E+00	2.262E+00	2.094E+00	2.038E+00	2.000E+00	1.977E+00
Zr	40	91.224	4.618E+02	3.315E+01	8.288E+00	3.683E+00	2.844E+00	2.657E+00	2.380E+00	2.203E+00	2.144E+00	2.103E+00	2.080E+00
Nb	41	92.906	4.852E+02	3.483E+01	8.708E+00	3.870E+00	2.988E+00	2.792E+00	2.500E+00	2.314E+00	2.253E+00	2.210E+00	2.185E+00
Mo	42	95.940	5.091E+02	3.655E+01	9.138E+00	4.061E+00	3.135E+00	2.929E+00	2.624E+00	2.428E+00	2.364E+00	2.319E+00	2.293E+00
Tc	43	98.000	5.336E+02	3.831E+01	9.578E+00	4.257E+00	3.286E+00	3.071E+00	2.750E+00	2.545E+00	2.478E+00	2.431E+00	2.403E+00
Ru	44	101.070	5.588E+02	4.012E+01	1.003E+01	4.457E+00	3.441E+00	3.215E+00	2.880E+00	2.665E+00	2.594E+00	2.545E+00	2.516E+00
Rh	45	102.906	5.844E+02	4.196E+01	1.049E+01	4.662E+00	3.599E+00	3.363E+00	3.012E+00	2.788E+00	2.714E+00	2.662E+00	2.632E+00
Pd	46	106.420	6.107E+02	4.385E+01	1.096E+01	4.871E+00	3.761E+00	3.514E+00	3.148E+00	2.913E+00	2.836E+00	2.782E+00	2.750E+00

Table A6.1. Rutherford scattering cross sections (barns) of the elements for 1 MeV ^{1}H (protons) (continued).

Element	Atomic no. (Z_2)	Mass (M_2)	175	170	165	160	150	140	135	120	90	60	30
													Scattering angle (degrees)
Ag	47	107.868	2.871E+00	2.904E+00	2.960E+00	3.041E+00	3.286E+00	3.669E+00	3.926E+00	5.086E+00	1.144E+01	4.577E+01	6.375E+02
Cd	48	122.411	2.995E+00	3.029E+00	3.088E+00	3.172E+00	3.427E+00	3.826E+00	4.095E+00	5.304E+00	1.194E+01	4.774E+01	6.650E+02
In	49	114.820	3.121E+00	3.157E+00	3.218E+00	3.305E+00	3.572E+00	3.987E+00	4.268E+00	5.528E+00	1.244E+01	4.975E+01	6.930E+02
Sn	50	118.710	3.250E+00	3.287E+00	3.350E+00	3.442E+00	3.719E+00	4.152E+00	4.444E+00	5.756E+00	1.295E+01	5.180E+01	7.215E+02
Sb	51	121.750	3.381E+00	3.420E+00	3.486E+00	3.581E+00	3.869E+00	4.320E+00	4.623E+00	5.988E+00	1.347E+01	5.390E+01	7.507E+02
Te	52	127.600	3.515E+00	3.555E+00	3.624E+00	3.723E+00	4.022E+00	4.491E+00	4.806E+00	6.225E+00	1.401E+01	5.603E+01	7.804E+02
I	53	126.904	3.651E+00	3.693E+00	3.765E+00	3.867E+00	4.179E+00	4.665E+00	4.993E+00	6.467E+00	1.455E+01	5.821E+01	8.107E+02
Xe	54	131.290	3.790E+00	3.834E+00	3.908E+00	4.015E+00	4.338E+00	4.843E+00	5.183E+00	6.713E+00	1.511E+01	6.042E+01	8.416E+02
Cs	55	132.905	3.932E+00	3.977E+00	4.054E+00	4.165E+00	4.500E+00	5.024E+00	5.377E+00	6.964E+00	1.567E+01	6.268E+01	8.731E+02
Ba	56	137.327	4.076E+00	4.123E+00	4.203E+00	4.317E+00	4.665E+00	5.208E+00	5.574E+00	7.220E+00	1.625E+01	6.498E+01	9.051E+02
La	57	138.906	4.223E+00	4.272E+00	4.354E+00	4.473E+00	4.833E+00	5.396E+00	5.775E+00	7.480E+00	1.683E+01	6.732E+01	9.377E+02
Ce	58	140.115	4.373E+00	4.423E+00	4.509E+00	4.631E+00	5.004E+00	5.587E+00	5.980E+00	7.745E+00	1.743E+01	6.971E+01	9.709E+02
Pr	59	140.908	4.525E+00	4.577E+00	4.665E+00	4.792E+00	5.178E+00	5.781E+00	6.187E+00	8.014E+00	1.803E+01	7.213E+01	1.005E+03
Nd	60	144.240	4.680E+00	4.734E+00	4.825E+00	4.956E+00	5.355E+00	5.979E+00	6.399E+00	8.288E+00	1.865E+01	7.460E+01	1.039E+03
Pm	61	145.000	4.837E+00	4.893E+00	4.987E+00	5.123E+00	5.535E+00	6.180E+00	6.614E+00	8.567E+00	1.928E+01	7.710E+01	1.074E+03
Sm	62	150.360	4.997E+00	5.054E+00	5.152E+00	5.292E+00	5.718E+00	6.384E+00	6.833E+00	8.850E+00	1.991E+01	7.965E+01	1.109E+03
Eu	63	151.965	5.159E+00	5.219E+00	5.320E+00	5.464E+00	5.904E+00	6.592E+00	7.055E+00	9.138E+00	2.056E+01	8.224E+01	1.146E+03
Gd	64	157.250	5.325E+00	5.386E+00	5.490E+00	5.639E+00	6.093E+00	6.803E+00	7.281E+00	9.430E+00	2.122E+01	8.488E+01	1.182E+03
Tb	65	158.925	5.492E+00	5.555E+00	5.663E+00	5.817E+00	6.285E+00	7.017E+00	7.510E+00	9.727E+00	2.189E+01	8.755E+01	1.219E+03
Dy	66	162.500	5.663E+00	5.728E+00	5.838E+00	5.997E+00	6.480E+00	7.235E+00	7.743E+00	1.003E+01	2.257E+01	9.026E+01	1.257E+03
Ho	67	164.930	5.835E+00	5.903E+00	6.017E+00	6.180E+00	6.678E+00	7.456E+00	7.979E+00	1.034E+01	2.325E+01	9.302E+01	1.296E+03
Er	68	167.260	6.011E+00	6.080E+00	6.197E+00	6.366E+00	6.879E+00	7.680E+00	8.219E+00	1.065E+01	2.395E+01	9.582E+01	1.335E+03
Tm	69	168.934	6.189E+00	6.260E+00	6.381E+00	6.555E+00	7.083E+00	7.907E+00	8.463E+00	1.096E+01	2.466E+01	9.866E+01	1.374E+03
Yb	70	173.040	6.370E+00	6.443E+00	6.567E+00	6.746E+00	7.290E+00	8.138E+00	8.710E+00	1.128E+01	2.538E+01	1.015E+02	1.414E+03
Lu	71	174.967	6.553E+00	6.628E+00	6.756E+00	6.940E+00	7.499E+00	8.372E+00	8.961E+00	1.161E+01	2.611E+01	1.045E+02	1.455E+03
Hf	72	178.490	6.739E+00	6.817E+00	6.948E+00	7.137E+00	7.712E+00	8.610E+00	9.215E+00	1.194E+01	2.685E+01	1.074E+02	1.496E+03
Ta	73	180.948	6.927E+00	7.007E+00	7.142E+00	7.337E+00	7.928E+00	8.851E+00	9.473E+00	1.227E+01	2.761E+01	1.104E+02	1.538E+03
W	74	183.850	7.119E+00	7.201E+00	7.340E+00	7.539E+00	8.146E+00	9.095E+00	9.734E+00	1.261E+01	2.837E+01	1.135E+02	1.580E+03
Re	75	186.207	7.312E+00	7.396E+00	7.539E+00	7.745E+00	8.368E+00	9.342E+00	9.999E+00	1.295E+01	2.914E+01	1.166E+02	1.623E+03
Os	76	190.200	7.509E+00	7.595E+00	7.742E+00	7.952E+00	8.593E+00	9.593E+00	1.027E+01	1.330E+01	2.992E+01	1.197E+02	1.667E+03
Ir	77	192.220	7.708E+00	7.796E+00	7.947E+00	8.163E+00	8.820E+00	9.847E+00	1.054E+01	1.365E+01	3.071E+01	1.229E+02	1.711E+03
Pt	78	195.080	7.909E+00	8.000E+00	8.154E+00	8.377E+00	9.051E+00	1.010E+01	1.081E+01	1.401E+01	3.152E+01	1.261E+02	1.756E+03
Au	79	196.967	8.113E+00	8.207E+00	8.365E+00	8.593E+00	9.285E+00	1.037E+01	1.109E+01	1.437E+01	3.233E+01	1.293E+02	1.801E+03
Hg	80	200.590	8.320E+00	8.416E+00	8.578E+00	8.812E+00	9.521E+00	1.063E+01	1.138E+01	1.473E+01	3.315E+01	1.326E+02	1.847E+03
Tl	81	204.383	8.529E+00	8.627E+00	8.794E+00	9.033E+00	9.761E+00	1.090E+01	1.166E+01	1.511E+01	3.399E+01	1.360E+02	1.894E+03
Pb	82	207.200	8.741E+00	8.842E+00	9.012E+00	9.258E+00	1.000E+01	1.117E+01	1.195E+01	1.548E+01	3.483E+01	1.393E+02	1.941E+03
Bi	83	208.980	8.956E+00	9.059E+00	9.233E+00	9.485E+00	1.025E+01	1.144E+01	1.225E+01	1.586E+01	3.569E+01	1.428E+02	1.988E+03
Po	84	209.000	9.173E+00	9.278E+00	9.457E+00	9.715E+00	1.050E+01	1.172E+01	1.254E+01	1.625E+01	3.655E+01	1.462E+02	2.036E+03
At	85	210.000	9.392E+00	9.500E+00	9.684E+00	9.948E+00	1.075E+01	1.200E+01	1.284E+01	1.663E+01	3.743E+01	1.497E+02	2.085E+03
Rn	86	222.000	9.615E+00	9.725E+00	9.913E+00	1.018E+01	1.100E+01	1.228E+01	1.315E+01	1.703E+01	3.831E+01	1.533E+02	2.135E+03
Fr	87	223.000	9.840E+00	9.953E+00	1.014E+01	1.042E+01	1.126E+01	1.257E+01	1.345E+01	1.743E+01	3.921E+01	1.568E+02	2.185E+03
Ra	88	226.025	1.007E+01	1.018E+01	1.038E+01	1.066E+01	1.152E+01	1.286E+01	1.377E+01	1.783E+01	4.012E+01	1.605E+02	2.235E+03
Ac	89	227.028	1.030E+01	1.042E+01	1.062E+01	1.091E+01	1.178E+01	1.316E+01	1.408E+01	1.824E+01	4.103E+01	1.641E+02	2.286E+03
Th	90	232.038	1.053E+01	1.065E+01	1.086E+01	1.115E+01	1.205E+01	1.345E+01	1.440E+01	1.865E+01	4.196E+01	1.678E+02	2.338E+03
Pa	91	231.036	1.077E+01	1.089E+01	1.110E+01	1.140E+01	1.232E+01	1.375E+01	1.472E+01	1.907E+01	4.290E+01	1.716E+02	2.390E+03
U	92	238.029	1.100E+01	1.113E+01	1.134E+01	1.165E+01	1.259E+01	1.406E+01	1.505E+01	1.949E+01	4.385E+01	1.754E+02	2.443E+03
Np	93	237.048	1.124E+01	1.137E+01	1.159E+01	1.191E+01	1.287E+01	1.437E+01	1.537E+01	1.991E+01	4.480E+01	1.792E+02	2.496E+03
Pu	94	244.000	1.149E+01	1.162E+01	1.184E+01	1.217E+01	1.315E+01	1.468E+01	1.571E+01	2.034E+01	4.577E+01	1.831E+02	2.550E+03

Table A6.2. Rutherford scattering cross sections (barns) of the elements for 1 MeV ^{2}H (deuterons).

Element	Atomic no. (Z_2)	Mass (M_2)	30	60	90	120	135	140	150	160	165	170	175
He	2	4.003	1.152E+00	8.023E-02	1.791E-02	6.550E-03	4.588E-03	4.165E-03	3.551E-03	3.167E-03	3.042E-03	2.956E-03	2.906E-03
Li	3	6.941	2.596E+00	1.845E-01	4.462E-02	1.875E-02	1.406E-02	1.302E-02	1.148E-02	1.050E-02	1.018E-02	9.951E-03	9.819E-03
Be	4	9.012	4.616E+00	3.295E-01	8.079E-02	3.476E-02	2.639E-02	2.453E-02	2.177E-02	2.000E-02	1.942E-02	1.902E-02	1.878E-02
B	5	10.811	7.213E+00	5.158E-01	1.272E-01	5.531E-02	4.220E-02	3.929E-02	3.497E-02	3.221E-02	3.130E-02	3.066E-02	3.029E-02
C	6	12.011	1.039E+01	7.434E-01	1.839E-01	8.026E-02	6.138E-02	5.719E-02	5.096E-02	4.698E-02	4.566E-02	4.475E-02	4.422E-02
N	7	14.007	1.414E+01	1.013E+00	2.512E-01	1.102E-01	8.449E-02	7.877E-02	7.029E-02	6.487E-02	6.308E-02	6.184E-02	6.110E-02
O	8	15.999	1.847E+01	1.324E+00	3.289E-01	1.447E-01	1.111E-01	1.037E-01	9.260E-02	8.551E-02	8.318E-02	8.155E-02	8.060E-02
F	9	18.998	2.338E+01	1.676E+00	4.172E-01	1.841E-01	1.416E-01	1.322E-01	1.182E-01	1.092E-01	1.062E-01	1.042E-01	1.030E-01
Ne	10	20.180	2.886E+01	2.070E+00	5.155E-01	2.277E-01	1.752E-01	1.635E-01	1.462E-01	1.351E-01	1.315E-01	1.289E-01	1.274E-01
Na	11	22.990	3.492E+01	2.505E+00	6.244E-01	2.762E-01	2.127E-01	1.986E-01	1.776E-01	1.642E-01	1.598E-01	1.567E-01	1.549E-01
Mg	12	24.305	4.156E+01	2.981E+00	7.434E-01	3.290E-01	2.534E-01	2.366E-01	2.117E-01	1.957E-01	1.905E-01	1.868E-01	1.847E-01
Al	13	26.982	4.877E+01	3.500E+00	8.730E-01	3.867E-01	2.980E-01	2.783E-01	2.490E-01	2.303E-01	2.241E-01	2.198E-01	2.173E-01
Si	14	28.086	5.657E+01	4.059E+00	1.013E+00	4.487E-01	3.458E-01	3.229E-01	2.890E-01	2.673E-01	2.601E-01	2.551E-01	2.522E-01
P	15	30.974	6.494E+01	4.660E+00	1.163E+00	5.156E-01	3.975E-01	3.713E-01	3.323E-01	3.073E-01	2.991E-01	2.934E-01	2.901E-01
S	16	32.066	7.388E+01	5.302E+00	1.324E+00	5.868E-01	4.525E-01	4.226E-01	3.782E-01	3.499E-01	3.405E-01	3.340E-01	3.302E-01
Cl	17	35.453	8.341E+01	5.986E+00	1.495E+00	6.630E-01	5.113E-01	4.776E-01	4.275E-01	3.955E-01	3.850E-01	3.776E-01	3.733E-01
Ar	18	39.948	9.351E+01	6.712E+00	1.676E+00	7.438E-01	5.738E-01	5.360E-01	4.799E-01	4.440E-01	4.322E-01	4.239E-01	4.191E-01
K	19	39.098	1.042E+02	7.478E+00	1.868E+00	8.287E-01	6.392E-01	5.971E-01	5.346E-01	4.946E-01	4.814E-01	4.722E-01	4.668E-01
Ca	20	40.078	1.154E+02	8.286E+00	2.070E+00	9.183E-01	7.084E-01	6.618E-01	5.925E-01	5.481E-01	5.335E-01	5.234E-01	5.174E-01
Sc	21	44.959	1.273E+02	9.136E+00	2.282E+00	1.013E+00	7.816E-01	7.302E-01	6.538E-01	6.049E-01	5.888E-01	5.776E-01	5.710E-01
Ti	22	47.880	1.397E+02	1.003E+01	2.505E+00	1.112E+00	8.582E-01	8.017E-01	7.179E-01	6.642E-01	6.465E-01	6.342E-01	6.270E-01
V	23	50.942	1.527E+02	1.096E+01	2.738E+00	1.216E+00	9.382E-01	8.765E-01	7.849E-01	7.262E-01	7.069E-01	6.935E-01	6.856E-01
Cr	24	51.996	1.662E+02	1.193E+01	2.982E+00	1.324E+00	1.022E+00	9.545E-01	8.547E-01	7.908E-01	7.698E-01	7.552E-01	7.466E-01
Mn	25	54.938	1.804E+02	1.295E+01	3.236E+00	1.437E+00	1.109E+00	1.036E+00	9.277E-01	8.584E-01	8.356E-01	8.197E-01	8.104E-01
Fe	26	55.847	1.951E+02	1.401E+01	3.500E+00	1.554E+00	1.199E+00	1.121E+00	1.003E+00	9.285E-01	9.038E-01	8.867E-01	8.766E-01
Co	27	58.933	2.104E+02	1.510E+01	3.774E+00	1.676E+00	1.294E+00	1.209E+00	1.082E+00	1.002E+00	9.749E-01	9.564E-01	9.455E-01
Ni	28	58.690	2.263E+02	1.624E+01	4.059E+00	1.803E+00	1.391E+00	1.300E+00	1.164E+00	1.077E+00	1.048E+00	1.029E+00	1.017E+00
Cu	29	63.546	2.427E+02	1.742E+01	4.355E+00	1.934E+00	1.493E+00	1.395E+00	1.249E+00	1.156E+00	1.125E+00	1.104E+00	1.091E+00
Zn	30	65.390	2.598E+02	1.865E+01	4.660E+00	2.070E+00	1.598E+00	1.493E+00	1.337E+00	1.237E+00	1.204E+00	1.181E+00	1.168E+00
Ga	31	69.723	2.774E+02	1.991E+01	4.976E+00	2.211E+00	1.706E+00	1.594E+00	1.428E+00	1.321E+00	1.286E+00	1.262E+00	1.247E+00
Ge	32	72.610	2.955E+02	2.122E+01	5.303E+00	2.356E+00	1.818E+00	1.699E+00	1.521E+00	1.408E+00	1.371E+00	1.345E+00	1.329E+00
As	33	74.922	3.143E+02	2.256E+01	5.639E+00	2.505E+00	1.934E+00	1.807E+00	1.618E+00	1.497E+00	1.458E+00	1.430E+00	1.414E+00
Se	34	78.960	3.336E+02	2.395E+01	5.987E+00	2.660E+00	2.053E+00	1.918E+00	1.718E+00	1.590E+00	1.548E+00	1.518E+00	1.501E+00
Br	35	79.904	3.536E+02	2.538E+01	6.344E+00	2.818E+00	2.176E+00	2.033E+00	1.820E+00	1.685E+00	1.640E+00	1.609E+00	1.591E+00
Kr	36	83.800	3.740E+02	2.685E+01	6.712E+00	2.982E+00	2.302E+00	2.151E+00	1.926E+00	1.783E+00	1.735E+00	1.702E+00	1.683E+00
Rb	37	85.468	3.951E+02	2.837E+01	7.090E+00	3.150E+00	2.432E+00	2.272E+00	2.035E+00	1.883E+00	1.833E+00	1.798E+00	1.778E+00
Sr	38	87.620	4.168E+02	2.992E+01	7.479E+00	3.323E+00	2.565E+00	2.396E+00	2.146E+00	1.986E+00	1.934E+00	1.897E+00	1.875E+00
Y	39	88.906	4.390E+02	3.152E+01	7.877E+00	3.500E+00	2.702E+00	2.524E+00	2.261E+00	2.092E+00	2.037E+00	1.998E+00	1.975E+00
Zr	40	91.224	4.618E+02	3.315E+01	8.287E+00	3.682E+00	2.842E+00	2.656E+00	2.378E+00	2.201E+00	2.143E+00	2.102E+00	2.078E+00
Nb	41	92.906	4.852E+02	3.483E+01	8.706E+00	3.868E+00	2.986E+00	2.790E+00	2.499E+00	2.312E+00	2.251E+00	2.208E+00	2.183E+00
Mo	42	95.940	5.091E+02	3.655E+01	9.136E+00	4.059E+00	3.134E+00	2.928E+00	2.622E+00	2.427E+00	2.362E+00	2.318E+00	2.291E+00
Tc	43	98.000	5.336E+02	3.831E+01	9.577E+00	4.255E+00	3.285E+00	3.069E+00	2.749E+00	2.544E+00	2.476E+00	2.429E+00	2.402E+00
Ru	44	101.070	5.588E+02	4.012E+01	1.003E+01	4.455E+00	3.440E+00	3.214E+00	2.878E+00	2.664E+00	2.593E+00	2.544E+00	2.515E+00
Rh	45	102.906	5.844E+02	4.196E+01	1.049E+01	4.660E+00	3.598E+00	3.361E+00	3.011E+00	2.786E+00	2.712E+00	2.661E+00	2.631E+00
Pd	46	106.420	6.107E+02	4.384E+01	1.096E+01	4.870E+00	3.760E+00	3.513E+00	3.146E+00	2.912E+00	2.834E+00	2.781E+00	2.749E+00

Scattering angle (degrees)

Table A6.2. Rutherford scattering cross sections (barns) of the elements for 1 MeV ^{2}H (deuterons) (continued).

| Element | Atomic no. (Z_2) | Mass (M_2) | \multicolumn{11}{c}{Scattering angle (degrees)} | | | | | | | | | | |
			175	170	165	160	150	140	135	120	90	60	30
Ag	47	107.868	2.870E+00	2.903E+00	2.959E+00	3.040E+00	3.284E+00	3.667E+00	3.925E+00	5.084E+00	1.144E+01	4.577E+01	6.375E+02
Cd	48	122.411	2.994E+00	3.028E+00	3.087E+00	3.171E+00	3.426E+00	3.825E+00	4.094E+00	5.303E+00	1.193E+01	4.774E+01	6.650E+02
In	49	114.820	3.119E+00	3.155E+00	3.216E+00	3.304E+00	3.570E+00	3.986E+00	4.266E+00	5.526E+00	1.244E+01	4.975E+01	6.930E+02
Sn	50	118.710	3.248E+00	3.286E+00	3.349E+00	3.440E+00	3.717E+00	4.151E+00	4.442E+00	5.754E+00	1.295E+01	5.180E+01	7.215E+02
Sb	51	121.750	3.380E+00	3.418E+00	3.484E+00	3.579E+00	3.868E+00	4.318E+00	4.622E+00	5.987E+00	1.347E+01	5.390E+01	7.507E+02
Te	52	127.600	3.514E+00	3.554E+00	3.623E+00	3.721E+00	4.021E+00	4.490E+00	4.805E+00	6.224E+00	1.401E+01	5.603E+01	7.804E+02
I	53	126.904	3.650E+00	3.692E+00	3.763E+00	3.866E+00	4.177E+00	4.664E+00	4.992E+00	6.466E+00	1.455E+01	5.821E+01	8.107E+02
Xe	54	131.290	3.789E+00	3.833E+00	3.907E+00	4.013E+00	4.336E+00	4.842E+00	5.182E+00	6.712E+00	1.510E+01	6.042E+01	8.416E+02
Cs	55	132.905	3.931E+00	3.976E+00	4.053E+00	4.163E+00	4.499E+00	5.023E+00	5.376E+00	6.963E+00	1.567E+01	6.268E+01	8.731E+02
Ba	56	137.327	4.075E+00	4.122E+00	4.202E+00	4.316E+00	4.664E+00	5.207E+00	5.573E+00	7.219E+00	1.624E+01	6.498E+01	9.051E+02
La	57	138.906	4.222E+00	4.271E+00	4.353E+00	4.472E+00	4.832E+00	5.395E+00	5.774E+00	7.479E+00	1.683E+01	6.732E+01	9.377E+02
Ce	58	140.115	4.372E+00	4.422E+00	4.507E+00	4.630E+00	5.003E+00	5.586E+00	5.978E+00	7.743E+00	1.743E+01	6.971E+01	9.709E+02
Pr	59	140.908	4.524E+00	4.576E+00	4.664E+00	4.791E+00	5.177E+00	5.780E+00	6.186E+00	8.013E+00	1.803E+01	7.213E+01	1.005E+03
Nd	60	144.240	4.678E+00	4.732E+00	4.824E+00	4.955E+00	5.354E+00	5.978E+00	6.398E+00	8.287E+00	1.865E+01	7.460E+01	1.039E+03
Pm	61	145.000	4.836E+00	4.891E+00	4.986E+00	5.122E+00	5.534E+00	6.179E+00	6.613E+00	8.565E+00	1.927E+01	7.710E+01	1.074E+03
Sm	62	150.360	4.996E+00	5.053E+00	5.151E+00	5.291E+00	5.717E+00	6.383E+00	6.831E+00	8.849E+00	1.991E+01	7.965E+01	1.109E+03
Eu	63	151.965	5.158E+00	5.217E+00	5.318E+00	5.463E+00	5.903E+00	6.591E+00	7.054E+00	9.136E+00	2.056E+01	8.224E+01	1.146E+03
Gd	64	157.250	5.323E+00	5.385E+00	5.488E+00	5.638E+00	6.092E+00	6.802E+00	7.279E+00	9.429E+00	2.122E+01	8.487E+01	1.182E+03
Tb	65	158.925	5.491E+00	5.554E+00	5.661E+00	5.816E+00	6.284E+00	7.016E+00	7.509E+00	9.726E+00	2.189E+01	8.755E+01	1.219E+03
Dy	66	162.500	5.661E+00	5.726E+00	5.837E+00	5.996E+00	6.479E+00	7.233E+00	7.742E+00	1.003E+01	2.256E+01	9.026E+01	1.257E+03
Ho	67	164.930	5.834E+00	5.901E+00	6.015E+00	6.179E+00	6.677E+00	7.454E+00	7.978E+00	1.033E+01	2.325E+01	9.302E+01	1.296E+03
Er	68	167.260	6.010E+00	6.079E+00	6.196E+00	6.365E+00	6.878E+00	7.679E+00	8.218E+00	1.064E+01	2.395E+01	9.581E+01	1.335E+03
Tm	69	168.934	6.188E+00	6.259E+00	6.380E+00	6.554E+00	7.081E+00	7.906E+00	8.462E+00	1.096E+01	2.466E+01	9.865E+01	1.374E+03
Yb	70	173.040	6.368E+00	6.442E+00	6.566E+00	6.745E+00	7.288E+00	8.137E+00	8.709E+00	1.128E+01	2.538E+01	1.015E+02	1.414E+03
Lu	71	174.967	6.552E+00	6.627E+00	6.755E+00	6.939E+00	7.498E+00	8.371E+00	8.959E+00	1.160E+01	2.611E+01	1.045E+02	1.455E+03
Hf	72	178.490	6.738E+00	6.815E+00	6.947E+00	7.136E+00	7.711E+00	8.609E+00	9.214E+00	1.193E+01	2.685E+01	1.074E+02	1.496E+03
Ta	73	180.948	6.926E+00	7.006E+00	7.141E+00	7.336E+00	7.926E+00	8.850E+00	9.471E+00	1.227E+01	2.760E+01	1.104E+02	1.538E+03
W	74	183.850	7.117E+00	7.199E+00	7.338E+00	7.538E+00	8.145E+00	9.094E+00	9.733E+00	1.261E+01	2.837E+01	1.135E+02	1.580E+03
Re	75	186.207	7.311E+00	7.395E+00	7.538E+00	7.743E+00	8.367E+00	9.341E+00	9.997E+00	1.295E+01	2.914E+01	1.166E+02	1.623E+03
Os	76	190.200	7.507E+00	7.594E+00	7.740E+00	7.951E+00	8.592E+00	9.592E+00	1.027E+01	1.330E+01	2.992E+01	1.197E+02	1.667E+03
Ir	77	192.220	7.706E+00	7.795E+00	7.945E+00	8.162E+00	8.819E+00	9.846E+00	1.054E+01	1.365E+01	3.071E+01	1.229E+02	1.711E+03
Pt	78	195.080	7.908E+00	7.999E+00	8.153E+00	8.375E+00	9.050E+00	1.010E+01	1.081E+01	1.401E+01	3.152E+01	1.261E+02	1.756E+03
Au	79	196.967	8.112E+00	8.205E+00	8.364E+00	8.591E+00	9.283E+00	1.036E+01	1.109E+01	1.437E+01	3.233E+01	1.293E+02	1.801E+03
Hg	80	200.590	8.319E+00	8.414E+00	8.577E+00	8.810E+00	9.520E+00	1.063E+01	1.138E+01	1.473E+01	3.315E+01	1.326E+02	1.847E+03
Tl	81	204.383	8.528E+00	8.626E+00	8.793E+00	9.032E+00	9.759E+00	1.090E+01	1.166E+01	1.510E+01	3.399E+01	1.360E+02	1.894E+03
Pb	82	207.200	8.740E+00	8.840E+00	9.011E+00	9.257E+00	1.000E+01	1.117E+01	1.195E+01	1.548E+01	3.483E+01	1.393E+02	1.941E+03
Bi	83	208.980	8.954E+00	9.057E+00	9.232E+00	9.484E+00	1.025E+01	1.144E+01	1.224E+01	1.586E+01	3.569E+01	1.427E+02	1.988E+03
Po	84	209.000	9.171E+00	9.277E+00	9.456E+00	9.714E+00	1.050E+01	1.172E+01	1.254E+01	1.624E+01	3.655E+01	1.462E+02	2.036E+03
At	85	210.000	9.391E+00	9.499E+00	9.683E+00	9.946E+00	1.075E+01	1.200E+01	1.284E+01	1.663E+01	3.743E+01	1.497E+02	2.085E+03
Rn	86	222.000	9.614E+00	9.724E+00	9.912E+00	1.018E+01	1.100E+01	1.228E+01	1.315E+01	1.703E+01	3.831E+01	1.533E+02	2.135E+03
Fr	87	223.000	9.838E+00	9.952E+00	1.014E+01	1.042E+01	1.126E+01	1.257E+01	1.345E+01	1.743E+01	3.921E+01	1.568E+02	2.185E+03
Ra	88	226.025	1.007E+01	1.018E+01	1.038E+01	1.066E+01	1.152E+01	1.286E+01	1.376E+01	1.783E+01	4.012E+01	1.605E+02	2.235E+03
Ac	89	227.028	1.030E+01	1.041E+01	1.062E+01	1.090E+01	1.178E+01	1.315E+01	1.408E+01	1.824E+01	4.103E+01	1.641E+02	2.286E+03
Th	90	232.038	1.053E+01	1.065E+01	1.086E+01	1.115E+01	1.205E+01	1.345E+01	1.440E+01	1.865E+01	4.196E+01	1.678E+02	2.338E+03
Pa	91	231.036	1.076E+01	1.089E+01	1.110E+01	1.140E+01	1.232E+01	1.375E+01	1.472E+01	1.906E+01	4.290E+01	1.716E+02	2.390E+03
U	92	238.029	1.100E+01	1.113E+01	1.134E+01	1.165E+01	1.259E+01	1.406E+01	1.504E+01	1.949E+01	4.385E+01	1.754E+02	2.443E+03
Np	93	237.048	1.124E+01	1.137E+01	1.159E+01	1.191E+01	1.287E+01	1.436E+01	1.537E+01	1.991E+01	4.480E+01	1.792E+02	2.496E+03
Pu	94	244.000	1.149E+01	1.162E+01	1.184E+01	1.216E+01	1.314E+01	1.467E+01	1.571E+01	2.034E+01	4.577E+01	1.831E+02	2.550E+03

Table A6.3. Rutherford scattering cross sections (barns) of the elements for 1 MeV ^{3}He.

Element	Atomic no. (Z$_2$)	Mass (M$_2$)	Scattering angle (degrees)										
			175	170	165	160	150	140	135	120	90	60	30
Li	3	6.941	3.084E-02	3.133E-02	3.216E-02	3.336E-02	3.705E-02	4.289E-02	4.687E-02	6.509E-02	1.680E-01	7.282E-01	1.037E+01
Be	4	9.012	6.567E-02	6.659E-02	6.816E-02	7.042E-02	7.732E-02	8.816E-02	9.549E-02	1.287E-01	3.124E-01	1.307E+00	1.845E+01
B	5	10.811	1.106E-01	1.121E-01	1.146E-01	1.182E-01	1.291E-01	1.462E-01	1.578E-01	2.100E-01	4.975E-01	2.052E+00	2.884E+01
C	6	12.011	1.644E-01	1.666E-01	1.702E-01	1.754E-01	1.912E-01	2.160E-01	2.327E-01	3.079E-01	7.221E-01	2.960E+00	4.154E+01
N	7	14.007	2.318E-01	2.347E-01	2.397E-01	2.468E-01	2.684E-01	3.022E-01	3.250E-01	4.277E-01	9.915E-01	4.038E+00	5.655E+01
O	8	15.999	3.097E-01	3.135E-01	3.200E-01	3.293E-01	3.576E-01	4.018E-01	4.316E-01	5.658E-01	1.302E+00	5.281E+00	7.386E+01
F	9	18.998	4.003E-01	4.052E-01	4.134E-01	4.252E-01	4.610E-01	5.171E-01	5.549E-01	7.248E-01	1.657E+00	6.693E+00	9.349E+01
Ne	10	20.180	4.971E-01	5.031E-01	5.132E-01	5.278E-01	5.721E-01	6.413E-01	6.880E-01	8.978E-01	2.049E+00	8.265E+00	1.154E+02
Na	11	22.990	6.078E-01	6.151E-01	6.273E-01	6.450E-01	6.986E-01	7.824E-01	8.388E-01	1.093E+00	2.486E+00	1.001E+01	1.397E+02
Mg	12	24.305	7.260E-01	7.346E-01	7.492E-01	7.703E-01	8.341E-01	9.338E-01	1.001E+00	1.303E+00	2.961E+00	1.191E+01	1.662E+02
Al	13	26.982	8.571E-01	8.672E-01	8.843E-01	9.090E-01	9.839E-01	1.101E+00	1.180E+00	1.535E+00	3.480E+00	1.399E+01	1.951E+02
Si	14	28.086	9.959E-01	1.008E+00	1.028E+00	1.056E+00	1.143E+00	1.279E+00	1.370E+00	1.782E+00	4.038E+00	1.622E+01	2.263E+02
P	15	30.974	1.148E+00	1.161E+00	1.184E+00	1.217E+00	1.317E+00	1.473E+00	1.578E+00	2.050E+00	4.640E+00	1.863E+01	2.597E+02
S	16	32.066	1.308E+00	1.323E+00	1.349E+00	1.387E+00	1.500E+00	1.677E+00	1.797E+00	2.334E+00	5.281E+00	2.120E+01	2.955E+02
Cl	17	35.453	1.481E+00	1.499E+00	1.528E+00	1.570E+00	1.698E+00	1.898E+00	2.033E+00	2.640E+00	5.967E+00	2.393E+01	3.336E+02
Ar	18	39.948	1.666E+00	1.685E+00	1.718E+00	1.765E+00	1.909E+00	2.133E+00	2.285E+00	2.965E+00	6.695E+00	2.684E+01	3.740E+02
K	19	39.098	1.855E+00	1.877E+00	1.913E+00	1.966E+00	2.126E+00	2.376E+00	2.545E+00	3.302E+00	7.458E+00	2.990E+01	4.167E+02
Ca	20	40.078	2.057E+00	2.081E+00	2.121E+00	2.180E+00	2.357E+00	2.634E+00	2.821E+00	3.660E+00	8.265E+00	3.313E+01	4.618E+02
Sc	21	44.959	2.273E+00	2.299E+00	2.344E+00	2.408E+00	2.604E+00	2.909E+00	3.115E+00	4.041E+00	9.118E+00	3.653E+01	5.091E+02
Ti	22	47.880	2.497E+00	2.526E+00	2.575E+00	2.646E+00	2.860E+00	3.196E+00	3.422E+00	4.438E+00	1.001E+01	4.010E+01	5.587E+02
V	23	50.942	2.732E+00	2.763E+00	2.817E+00	2.894E+00	3.129E+00	3.495E+00	3.742E+00	4.853E+00	1.094E+01	4.383E+01	6.107E+02
Cr	24	51.996	2.975E+00	3.010E+00	3.068E+00	3.152E+00	3.408E+00	3.807E+00	4.076E+00	5.285E+00	1.192E+01	4.772E+01	6.649E+02
Mn	25	54.938	3.231E+00	3.268E+00	3.331E+00	3.423E+00	3.700E+00	4.133E+00	4.425E+00	5.736E+00	1.293E+01	5.178E+01	7.215E+02
Fe	26	55.847	3.495E+00	3.535E+00	3.604E+00	3.703E+00	4.002E+00	4.471E+00	4.786E+00	6.205E+00	1.399E+01	5.601E+01	7.804E+02
Co	27	58.933	3.771E+00	3.815E+00	3.889E+00	3.995E+00	4.318E+00	4.824E+00	5.164E+00	6.694E+00	1.509E+01	6.040E+01	8.416E+02
Ni	28	58.690	4.056E+00	4.102E+00	4.182E+00	4.296E+00	4.644E+00	5.187E+00	5.553E+00	7.199E+00	1.622E+01	6.496E+01	9.051E+02
Cu	29	63.546	4.354E+00	4.404E+00	4.489E+00	4.612E+00	4.985E+00	5.568E+00	5.960E+00	7.726E+00	1.741E+01	6.969E+01	9.709E+02
Zn	30	65.390	4.660E+00	4.714E+00	4.806E+00	4.937E+00	5.336E+00	5.960E+00	6.380E+00	8.269E+00	1.863E+01	7.458E+01	1.039E+03
Ga	31	69.723	4.979E+00	5.036E+00	5.134E+00	5.274E+00	5.700E+00	6.366E+00	6.815E+00	8.832E+00	1.989E+01	7.964E+01	1.109E+03
Ge	32	72.610	5.307E+00	5.368E+00	5.472E+00	5.621E+00	6.076E+00	6.785E+00	7.263E+00	9.412E+00	2.120E+01	8.486E+01	1.182E+03
As	33	74.922	5.645E+00	5.710E+00	5.820E+00	5.979E+00	6.462E+00	7.217E+00	7.725E+00	1.001E+01	2.255E+01	9.025E+01	1.257E+03
Se	34	78.960	5.994E+00	6.063E+00	6.180E+00	6.349E+00	6.862E+00	7.663E+00	8.202E+00	1.063E+01	2.394E+01	9.580E+01	1.335E+03
Br	35	79.904	6.352E+00	6.425E+00	6.550E+00	6.729E+00	7.272E+00	8.121E+00	8.692E+00	1.126E+01	2.537E+01	1.015E+02	1.414E+03
Kr	36	83.800	6.722E+00	6.800E+00	6.931E+00	7.120E+00	7.695E+00	8.593E+00	9.198E+00	1.192E+01	2.684E+01	1.074E+02	1.496E+03
Rb	37	85.468	7.101E+00	7.183E+00	7.322E+00	7.522E+00	8.129E+00	9.078E+00	9.717E+00	1.259E+01	2.835E+01	1.135E+02	1.580E+03
Sr	38	87.620	7.491E+00	7.578E+00	7.724E+00	7.935E+00	8.575E+00	9.576E+00	1.025E+01	1.328E+01	2.990E+01	1.197E+02	1.667E+03
Y	39	88.906	7.891E+00	7.982E+00	8.137E+00	8.359E+00	9.033E+00	1.009E+01	1.080E+01	1.399E+01	3.150E+01	1.261E+02	1.756E+03
Zr	40	91.224	8.302E+00	8.398E+00	8.560E+00	8.794E+00	9.503E+00	1.061E+01	1.136E+01	1.472E+01	3.314E+01	1.326E+02	1.847E+03
Nb	41	92.906	8.723E+00	8.824E+00	8.994E+00	9.240E+00	9.985E+00	1.115E+01	1.193E+01	1.546E+01	3.481E+01	1.393E+02	1.941E+03
Mo	42	95.940	9.155E+00	9.261E+00	9.440E+00	9.697E+00	1.048E+01	1.170E+01	1.252E+01	1.623E+01	3.653E+01	1.462E+02	2.036E+03
Tc	43	98.000	9.597E+00	9.708E+00	9.895E+00	1.017E+01	1.099E+01	1.227E+01	1.313E+01	1.701E+01	3.830E+01	1.532E+02	2.135E+03
Ru	44	101.070	1.005E+01	1.017E+01	1.036E+01	1.064E+01	1.150E+01	1.284E+01	1.375E+01	1.781E+01	4.010E+01	1.605E+02	2.235E+03
Rh	45	102.906	1.051E+01	1.063E+01	1.084E+01	1.113E+01	1.203E+01	1.344E+01	1.438E+01	1.863E+01	4.194E+01	1.678E+02	2.338E+03
Pd	46	106.420	1.099E+01	1.111E+01	1.133E+01	1.164E+01	1.257E+01	1.404E+01	1.503E+01	1.947E+01	4.383E+01	1.754E+02	2.443E+03
Ag	47	107.868	1.147E+01	1.160E+01	1.183E+01	1.215E+01	1.313E+01	1.466E+01	1.569E+01	2.033E+01	4.576E+01	1.831E+02	2.550E+03
Cd	48	122.411	1.197E+01	1.210E+01	1.234E+01	1.267E+01	1.370E+01	1.529E+01	1.637E+01	2.120E+01	4.773E+01	1.910E+02	2.660E+03

Table A6.3. Rutherford scattering cross sections (barns) of the elements for 1 MeV ^{3}He (continued).

Element	Atomic no. (Z_2)	Mass (M_2)	Scattering angle (degrees)										
			175	170	165	160	150	140	135	120	90	60	30
In	49	114.820	1.247E+01	1.261E+01	1.286E+01	1.321E+01	1.427E+01	1.593E+01	1.706E+01	2.210E+01	4.974E+01	1.990E+02	2.772E+03
Sn	50	118.710	1.298E+01	1.313E+01	1.339E+01	1.375E+01	1.486E+01	1.659E+01	1.776E+01	2.301E+01	5.179E+01	2.072E+02	2.886E+03
Sb	51	121.750	1.351E+01	1.366E+01	1.393E+01	1.431E+01	1.546E+01	1.726E+01	1.848E+01	2.394E+01	5.388E+01	2.156E+02	3.003E+03
Te	52	127.600	1.405E+01	1.421E+01	1.448E+01	1.488E+01	1.608E+01	1.795E+01	1.921E+01	2.489E+01	5.602E+01	2.241E+02	3.122E+03
I	53	126.904	1.459E+01	1.476E+01	1.504E+01	1.545E+01	1.670E+01	1.865E+01	1.996E+01	2.585E+01	5.819E+01	2.328E+02	3.243E+03
Xe	54	131.290	1.515E+01	1.532E+01	1.562E+01	1.604E+01	1.734E+01	1.936E+01	2.072E+01	2.684E+01	6.041E+01	2.417E+02	3.366E+03
Cs	55	132.905	1.571E+01	1.590E+01	1.620E+01	1.664E+01	1.799E+01	2.008E+01	2.149E+01	2.784E+01	6.267E+01	2.507E+02	3.492E+03
Ba	56	137.327	1.629E+01	1.648E+01	1.680E+01	1.726E+01	1.865E+01	2.082E+01	2.228E+01	2.887E+01	6.497E+01	2.599E+02	3.620E+03
La	57	138.906	1.688E+01	1.707E+01	1.740E+01	1.788E+01	1.932E+01	2.157E+01	2.309E+01	2.991E+01	6.731E+01	2.693E+02	3.751E+03
Ce	58	140.115	1.748E+01	1.768E+01	1.802E+01	1.851E+01	2.000E+01	2.233E+01	2.390E+01	3.096E+01	6.969E+01	2.788E+02	3.884E+03
Pr	59	140.908	1.809E+01	1.829E+01	1.865E+01	1.916E+01	2.070E+01	2.311E+01	2.474E+01	3.204E+01	7.212E+01	2.885E+02	4.019E+03
Nd	60	144.240	1.870E+01	1.892E+01	1.929E+01	1.981E+01	2.141E+01	2.390E+01	2.558E+01	3.314E+01	7.458E+01	2.984E+02	4.156E+03
Pm	61	145.000	1.933E+01	1.956E+01	1.993E+01	2.048E+01	2.213E+01	2.471E+01	2.644E+01	3.425E+01	7.709E+01	3.084E+02	4.296E+03
Sm	62	150.360	1.997E+01	2.020E+01	2.059E+01	2.115E+01	2.286E+01	2.552E+01	2.732E+01	3.539E+01	7.964E+01	3.186E+02	4.438E+03
Eu	63	151.965	2.062E+01	2.086E+01	2.126E+01	2.184E+01	2.360E+01	2.635E+01	2.821E+01	3.654E+01	8.223E+01	3.290E+02	4.582E+03
Gd	64	157.250	2.128E+01	2.153E+01	2.195E+01	2.254E+01	2.436E+01	2.720E+01	2.911E+01	3.771E+01	8.486E+01	3.395E+02	4.729E+03
Tb	65	158.925	2.195E+01	2.221E+01	2.264E+01	2.325E+01	2.513E+01	2.805E+01	3.003E+01	3.889E+01	8.753E+01	3.502E+02	4.878E+03
Dy	66	162.500	2.264E+01	2.290E+01	2.334E+01	2.398E+01	2.591E+01	2.893E+01	3.096E+01	4.010E+01	9.025E+01	3.610E+02	5.029E+03
Ho	67	164.930	2.333E+01	2.360E+01	2.405E+01	2.471E+01	2.670E+01	2.981E+01	3.190E+01	4.133E+01	9.300E+01	3.721E+02	5.182E+03
Er	68	167.260	2.403E+01	2.431E+01	2.478E+01	2.545E+01	2.750E+01	3.071E+01	3.286E+01	4.257E+01	9.580E+01	3.833E+02	5.338E+03
Tm	69	168.934	2.474E+01	2.503E+01	2.551E+01	2.621E+01	2.832E+01	3.162E+01	3.384E+01	4.383E+01	9.864E+01	3.946E+02	5.496E+03
Yb	70	173.040	2.547E+01	2.576E+01	2.626E+01	2.697E+01	2.914E+01	3.254E+01	3.483E+01	4.511E+01	1.015E+02	4.061E+02	5.657E+03
Lu	71	174.967	2.620E+01	2.650E+01	2.701E+01	2.775E+01	2.998E+01	3.348E+01	3.583E+01	4.641E+01	1.044E+02	4.178E+02	5.820E+03
Hf	72	178.490	2.694E+01	2.725E+01	2.778E+01	2.854E+01	3.083E+01	3.443E+01	3.685E+01	4.773E+01	1.074E+02	4.297E+02	5.985E+03
Ta	73	180.948	2.770E+01	2.802E+01	2.856E+01	2.933E+01	3.170E+01	3.539E+01	3.788E+01	4.906E+01	1.104E+02	4.417E+02	6.152E+03
W	74	183.850	2.846E+01	2.879E+01	2.934E+01	3.014E+01	3.257E+01	3.637E+01	3.892E+01	5.042E+01	1.135E+02	4.539E+02	6.322E+03
Re	75	186.207	2.924E+01	2.957E+01	3.014E+01	3.096E+01	3.346E+01	3.736E+01	3.998E+01	5.179E+01	1.165E+02	4.662E+02	6.494E+03
Os	76	190.200	3.002E+01	3.037E+01	3.095E+01	3.180E+01	3.436E+01	3.836E+01	4.106E+01	5.318E+01	1.197E+02	4.787E+02	6.668E+03
Ir	77	192.220	3.082E+01	3.117E+01	3.177E+01	3.264E+01	3.527E+01	3.938E+01	4.214E+01	5.459E+01	1.228E+02	4.914E+02	6.845E+03
Pt	78	195.080	3.162E+01	3.199E+01	3.260E+01	3.349E+01	3.619E+01	4.041E+01	4.325E+01	5.602E+01	1.261E+02	5.043E+02	7.024E+03
Au	79	196.967	3.244E+01	3.281E+01	3.345E+01	3.436E+01	3.712E+01	4.145E+01	4.436E+01	5.746E+01	1.293E+02	5.173E+02	7.205E+03
Hg	80	200.590	3.327E+01	3.365E+01	3.430E+01	3.523E+01	3.807E+01	4.251E+01	4.549E+01	5.893E+01	1.326E+02	5.305E+02	7.389E+03
Tl	81	204.383	3.410E+01	3.450E+01	3.516E+01	3.612E+01	3.903E+01	4.358E+01	4.664E+01	6.041E+01	1.359E+02	5.438E+02	7.574E+03
Pb	82	207.200	3.495E+01	3.535E+01	3.604E+01	3.702E+01	4.000E+01	4.466E+01	4.780E+01	6.191E+01	1.393E+02	5.573E+02	7.763E+03
Bi	83	208.980	3.581E+01	3.622E+01	3.692E+01	3.793E+01	4.098E+01	4.575E+01	4.897E+01	6.343E+01	1.427E+02	5.710E+02	7.953E+03
Po	84	209.000	3.668E+01	3.710E+01	3.782E+01	3.885E+01	4.197E+01	4.686E+01	5.016E+01	6.497E+01	1.462E+02	5.848E+02	8.146E+03
At	85	210.000	3.756E+01	3.799E+01	3.872E+01	3.978E+01	4.298E+01	4.799E+01	5.136E+01	6.652E+01	1.497E+02	5.988E+02	8.341E+03
Rn	86	222.000	3.845E+01	3.889E+01	3.964E+01	4.072E+01	4.400E+01	4.912E+01	5.258E+01	6.810E+01	1.532E+02	6.130E+02	8.538E+03
Fr	87	223.000	3.935E+01	3.980E+01	4.057E+01	4.167E+01	4.503E+01	5.027E+01	5.381E+01	6.969E+01	1.568E+02	6.274E+02	8.738E+03
Ra	88	226.025	4.026E+01	4.072E+01	4.151E+01	4.264E+01	4.607E+01	5.144E+01	5.505E+01	7.130E+01	1.605E+02	6.419E+02	8.940E+03
Ac	89	227.028	4.118E+01	4.165E+01	4.245E+01	4.361E+01	4.712E+01	5.261E+01	5.631E+01	7.293E+01	1.641E+02	6.565E+02	9.144E+03
Th	90	232.038	4.211E+01	4.259E+01	4.341E+01	4.460E+01	4.819E+01	5.380E+01	5.758E+01	7.458E+01	1.678E+02	6.714E+02	9.351E+03
Pa	91	231.036	4.305E+01	4.354E+01	4.438E+01	4.559E+01	4.927E+01	5.500E+01	5.887E+01	7.625E+01	1.716E+02	6.864E+02	9.560E+03
U	92	238.029	4.400E+01	4.451E+01	4.537E+01	4.660E+01	5.035E+01	5.622E+01	6.017E+01	7.794E+01	1.754E+02	7.015E+02	9.771E+03
Np	93	237.048	4.496E+01	4.548E+01	4.636E+01	4.762E+01	5.146E+01	5.745E+01	6.148E+01	7.964E+01	1.792E+02	7.169E+02	9.985E+03
Pu	94	244.000	4.593E+01	4.646E+01	4.736E+01	4.865E+01	5.257E+01	5.869E+01	6.281E+01	8.136E+01	1.831E+02	7.324E+02	1.020E+04

Table A6.4. Rutherford scattering cross sections (barns) of the elements for 1 MeV ^{4}He.

Element	Atomic no. (Z_2)	Mass (M_2)	Scattering angle (degrees)										
			30	60	90	120	135	140	150	160	165	170	175
Li	3	6.941	1.036E+01	7.145E-01	1.524E-01	5.137E-02	3.466E-02	3.114E-02	2.609E-02	2.299E-02	2.199E-02	2.130E-02	2.090E-02
Be	4	9.012	1.844E+01	1.293E+00	2.971E-01	1.143E-01	8.201E-02	7.497E-02	6.465E-02	5.815E-02	5.603E-02	5.456E-02	5.370E-02
B	5	10.811	2.883E+01	2.036E+00	4.812E-01	1.945E-01	1.429E-01	1.316E-01	1.148E-01	1.042E-01	1.007E-01	9.833E-02	9.691E-02
C	6	12.011	4.152E+01	2.942E+00	7.033E-01	2.899E-01	2.152E-01	1.987E-01	1.743E-01	1.588E-01	1.537E-01	1.501E-01	1.481E-01
N	7	14.007	5.653E+01	4.020E+00	9.730E-01	4.096E-01	3.074E-01	2.848E-01	2.512E-01	2.298E-01	2.228E-01	2.179E-01	2.150E-01
O	8	15.999	7.384E+01	5.263E+00	1.284E+00	5.478E-01	4.139E-01	3.842E-01	3.402E-01	3.121E-01	3.028E-01	2.964E-01	2.926E-01
F	9	18.998	9.347E+01	6.676E+00	1.641E+00	7.086E-01	5.389E-01	5.012E-01	4.453E-01	4.095E-01	3.978E-01	3.896E-01	3.848E-01
Ne	10	20.180	1.154E+02	8.248E+00	2.031E+00	8.801E-01	6.705E-01	6.239E-01	5.548E-01	5.107E-01	4.961E-01	4.860E-01	4.800E-01
Na	11	22.990	1.397E+02	9.991E+00	2.469E+00	1.076E+00	8.225E-01	7.661E-01	6.824E-01	6.289E-01	6.113E-01	5.990E-01	5.918E-01
Mg	12	24.305	1.662E+02	1.190E+01	2.943E+00	1.286E+00	9.836E-01	9.165E-01	8.168E-01	7.531E-01	7.321E-01	7.175E-01	7.089E-01
Al	13	26.982	1.951E+02	1.397E+01	3.463E+00	1.518E+00	1.163E+00	1.084E+00	9.674E-01	8.926E-01	8.679E-01	8.508E-01	8.407E-01
Si	14	28.086	2.262E+02	1.620E+01	4.020E+00	1.764E+00	1.353E+00	1.261E+00	1.125E+00	1.039E+00	1.010E+00	9.901E-01	9.784E-01
P	15	30.974	2.597E+02	1.861E+01	4.623E+00	2.033E+00	1.561E+00	1.456E+00	1.300E+00	1.201E+00	1.168E+00	1.145E+00	1.131E+00
S	16	32.066	2.955E+02	2.118E+01	5.263E+00	2.316E+00	1.779E+00	1.660E+00	1.482E+00	1.369E+00	1.332E+00	1.306E+00	1.290E+00
Cl	17	35.453	3.336E+02	2.392E+01	5.950E+00	2.623E+00	2.017E+00	1.882E+00	1.682E+00	1.554E+00	1.512E+00	1.482E+00	1.465E+00
Ar	18	39.948	3.740E+02	2.682E+01	6.680E+00	2.950E+00	2.270E+00	2.119E+00	1.895E+00	1.751E+00	1.704E+00	1.671E+00	1.651E+00
K	19	39.098	4.167E+02	2.988E+01	7.441E+00	3.285E+00	2.528E+00	2.359E+00	2.109E+00	1.949E+00	1.897E+00	1.860E+00	1.838E+00
Ca	20	40.078	4.617E+02	3.311E+01	8.247E+00	3.642E+00	2.803E+00	2.616E+00	2.339E+00	2.162E+00	2.103E+00	2.063E+00	2.039E+00
Sc	21	44.959	5.091E+02	3.652E+01	9.102E+00	4.025E+00	3.100E+00	2.894E+00	2.588E+00	2.393E+00	2.328E+00	2.284E+00	2.257E+00
Ti	22	47.880	5.587E+02	4.008E+01	9.994E+00	4.422E+00	3.406E+00	3.181E+00	2.845E+00	2.631E+00	2.560E+00	2.511E+00	2.482E+00
V	23	50.942	6.107E+02	4.381E+01	1.093E+01	4.838E+00	3.728E+00	3.481E+00	3.114E+00	2.880E+00	2.803E+00	2.749E+00	2.717E+00
Cr	24	51.996	6.649E+02	4.771E+01	1.190E+01	5.269E+00	4.060E+00	3.792E+00	3.392E+00	3.137E+00	3.053E+00	2.995E+00	2.960E+00
Mn	25	54.938	7.215E+02	5.177E+01	1.292E+01	5.722E+00	4.410E+00	4.118E+00	3.685E+00	3.408E+00	3.317E+00	3.253E+00	3.216E+00
Fe	26	55.847	7.804E+02	5.600E+01	1.397E+01	6.190E+00	4.771E+00	4.455E+00	3.987E+00	3.687E+00	3.589E+00	3.520E+00	3.479E+00
Co	27	58.933	8.416E+02	6.039E+01	1.507E+01	6.679E+00	5.149E+00	4.809E+00	4.303E+00	3.980E+00	3.874E+00	3.800E+00	3.756E+00
Ni	28	58.690	9.051E+02	6.495E+01	1.621E+01	7.183E+00	5.537E+00	5.171E+00	4.628E+00	4.280E+00	4.166E+00	4.086E+00	4.039E+00
Cu	29	63.546	9.709E+02	6.967E+01	1.739E+01	7.711E+00	5.945E+00	5.553E+00	4.970E+00	4.597E+00	4.475E+00	4.389E+00	4.339E+00
Zn	30	65.390	1.039E+03	7.456E+01	1.861E+01	8.254E+00	6.365E+00	5.945E+00	5.321E+00	4.922E+00	4.791E+00	4.699E+00	4.645E+00
Ga	31	69.723	1.109E+03	7.962E+01	1.988E+01	8.818E+00	6.800E+00	6.352E+00	5.686E+00	5.260E+00	5.120E+00	5.022E+00	4.965E+00
Ge	32	72.610	1.182E+03	8.484E+01	2.119E+01	9.398E+00	7.249E+00	6.771E+00	6.062E+00	5.608E+00	5.458E+00	5.354E+00	5.293E+00
As	33	74.922	1.257E+03	9.023E+01	2.253E+01	9.997E+00	7.711E+00	7.203E+00	6.448E+00	5.966E+00	5.807E+00	5.696E+00	5.631E+00
Se	34	78.960	1.335E+03	9.579E+01	2.392E+01	1.062E+01	8.189E+00	7.650E+00	6.849E+00	6.336E+00	6.167E+00	6.050E+00	5.981E+00
Br	35	79.904	1.414E+03	1.015E+02	2.535E+01	1.125E+01	8.679E+00	8.107E+00	7.258E+00	6.715E+00	6.536E+00	6.412E+00	6.338E+00
Kr	36	83.800	1.496E+03	1.074E+02	2.682E+01	1.191E+01	9.185E+00	8.580E+00	7.682E+00	7.107E+00	6.918E+00	6.786E+00	6.709E+00
Rb	37	85.468	1.580E+03	1.134E+02	2.834E+01	1.258E+01	9.703E+00	9.064E+00	8.116E+00	7.509E+00	7.309E+00	7.170E+00	7.088E+00
Sr	38	87.620	1.667E+03	1.197E+02	2.989E+01	1.327E+01	1.024E+01	9.563E+00	8.562E+00	7.922E+00	7.711E+00	7.564E+00	7.478E+00
Y	39	88.906	1.756E+03	1.260E+02	3.149E+01	1.398E+01	1.078E+01	1.007E+01	9.020E+00	8.345E+00	8.123E+00	7.969E+00	7.878E+00
Zr	40	91.224	1.847E+03	1.326E+02	3.312E+01	1.470E+01	1.134E+01	1.060E+01	9.490E+00	8.780E+00	8.547E+00	8.384E+00	8.288E+00
Nb	41	92.906	1.941E+03	1.393E+02	3.480E+01	1.545E+01	1.192E+01	1.114E+01	9.971E+00	9.226E+00	8.980E+00	8.810E+00	8.709E+00
Mo	42	95.940	2.036E+03	1.462E+02	3.652E+01	1.621E+01	1.251E+01	1.169E+01	1.047E+01	9.684E+00	9.426E+00	9.247E+00	9.141E+00
Tc	43	98.000	2.135E+03	1.532E+02	3.828E+01	1.700E+01	1.312E+01	1.225E+01	1.097E+01	1.015E+01	9.882E+00	9.694E+00	9.583E+00
Ru	44	101.070	2.235E+03	1.604E+02	4.009E+01	1.780E+01	1.373E+01	1.283E+01	1.149E+01	1.063E+01	1.035E+01	1.015E+01	1.004E+01
Rh	45	102.906	2.338E+03	1.678E+02	4.193E+01	1.862E+01	1.437E+01	1.342E+01	1.202E+01	1.112E+01	1.083E+01	1.062E+01	1.050E+01
Pd	46	106.420	2.443E+03	1.754E+02	4.382E+01	1.946E+01	1.501E+01	1.403E+01	1.256E+01	1.162E+01	1.131E+01	1.110E+01	1.097E+01
Ag	47	107.868	2.550E+03	1.831E+02	4.574E+01	2.031E+01	1.568E+01	1.464E+01	1.311E+01	1.213E+01	1.181E+01	1.159E+01	1.146E+01
Cd	48	112.411	2.660E+03	1.909E+02	4.772E+01	2.119E+01	1.636E+01	1.528E+01	1.369E+01	1.266E+01	1.233E+01	1.209E+01	1.196E+01

Table A6.4. Rutherford scattering cross sections (barns) of the elements for 1 MeV ^{4}He (continued).

Element	Atomic no. (Z₂)	Mass (M₂)	Scattering angle (degrees)										
			175	170	165	160	150	140	135	120	90	60	30
In	49	114.820	1.246E+01	1.260E+01	1.284E+01	1.319E+01	1.426E+01	1.592E+01	1.704E+01	2.208E+01	4.972E+01	1.990E+02	2.772E+03
Sn	50	118.710	1.297E+01	1.312E+01	1.337E+01	1.374E+01	1.485E+01	1.658E+01	1.775E+01	2.299E+01	5.177E+01	2.072E+02	2.886E+03
Sb	51	121.750	1.350E+01	1.365E+01	1.392E+01	1.430E+01	1.545E+01	1.725E+01	1.847E+01	2.393E+01	5.387E+01	2.156E+02	3.003E+03
Te	52	127.600	1.403E+01	1.420E+01	1.447E+01	1.486E+01	1.606E+01	1.794E+01	1.920E+01	2.488E+01	5.600E+01	2.241E+02	3.122E+03
I	53	126.904	1.458E+01	1.475E+01	1.503E+01	1.544E+01	1.669E+01	1.863E+01	1.994E+01	2.584E+01	5.818E+01	2.328E+02	3.243E+03
Xe	54	131.290	1.514E+01	1.531E+01	1.561E+01	1.603E+01	1.732E+01	1.935E+01	2.071E+01	2.683E+01	6.040E+01	2.417E+02	3.366E+03
Cs	55	132.905	1.570E+01	1.588E+01	1.619E+01	1.663E+01	1.797E+01	2.007E+01	2.148E+01	2.783E+01	6.265E+01	2.507E+02	3.492E+03
Ba	56	137.327	1.628E+01	1.647E+01	1.679E+01	1.724E+01	1.863E+01	2.081E+01	2.227E+01	2.885E+01	6.496E+01	2.599E+02	3.620E+03
La	57	138.906	1.687E+01	1.706E+01	1.739E+01	1.787E+01	1.931E+01	2.156E+01	2.307E+01	2.989E+01	6.730E+01	2.693E+02	3.751E+03
Ce	58	140.115	1.746E+01	1.767E+01	1.801E+01	1.850E+01	1.999E+01	2.232E+01	2.389E+01	3.095E+01	6.968E+01	2.788E+02	3.884E+03
Pr	59	140.908	1.807E+01	1.828E+01	1.863E+01	1.914E+01	2.069E+01	2.310E+01	2.472E+01	3.203E+01	7.210E+01	2.885E+02	4.019E+03
Nd	60	144.240	1.869E+01	1.891E+01	1.927E+01	1.980E+01	2.139E+01	2.389E+01	2.557E+01	3.313E+01	7.457E+01	2.984E+02	4.156E+03
Pm	61	145.000	1.932E+01	1.954E+01	1.992E+01	2.046E+01	2.211E+01	2.469E+01	2.643E+01	3.424E+01	7.708E+01	3.084E+02	4.296E+03
Sm	62	150.360	1.996E+01	2.019E+01	2.058E+01	2.114E+01	2.285E+01	2.551E+01	2.730E+01	3.537E+01	7.963E+01	3.186E+02	4.438E+03
Eu	63	151.965	2.061E+01	2.085E+01	2.125E+01	2.183E+01	2.359E+01	2.634E+01	2.819E+01	3.652E+01	8.222E+01	3.289E+02	4.582E+03
Gd	64	157.250	2.127E+01	2.152E+01	2.193E+01	2.253E+01	2.435E+01	2.719E+01	2.910E+01	3.770E+01	8.485E+01	3.395E+02	4.729E+03
Tb	65	158.925	2.194E+01	2.220E+01	2.262E+01	2.324E+01	2.512E+01	2.804E+01	3.001E+01	3.888E+01	8.752E+01	3.502E+02	4.878E+03
Dy	66	162.500	2.262E+01	2.289E+01	2.333E+01	2.396E+01	2.590E+01	2.891E+01	3.095E+01	4.009E+01	9.024E+01	3.610E+02	5.029E+03
Ho	67	164.930	2.332E+01	2.358E+01	2.404E+01	2.470E+01	2.669E+01	2.980E+01	3.189E+01	4.131E+01	9.299E+01	3.720E+02	5.182E+03
Er	68	167.260	2.402E+01	2.429E+01	2.476E+01	2.544E+01	2.749E+01	3.069E+01	3.285E+01	4.256E+01	9.579E+01	3.832E+02	5.338E+03
Tm	69	168.934	2.473E+01	2.502E+01	2.550E+01	2.619E+01	2.830E+01	3.160E+01	3.383E+01	4.382E+01	9.863E+01	3.946E+02	5.496E+03
Yb	70	173.040	2.545E+01	2.575E+01	2.624E+01	2.696E+01	2.913E+01	3.253E+01	3.481E+01	4.510E+01	1.015E+02	4.061E+02	5.657E+03
Lu	71	174.967	2.619E+01	2.649E+01	2.700E+01	2.774E+01	2.997E+01	3.346E+01	3.582E+01	4.640E+01	1.044E+02	4.178E+02	5.820E+03
Hf	72	178.490	2.693E+01	2.724E+01	2.777E+01	2.852E+01	3.082E+01	3.441E+01	3.683E+01	4.772E+01	1.074E+02	4.297E+02	5.985E+03
Ta	73	180.948	2.768E+01	2.800E+01	2.854E+01	2.932E+01	3.169E+01	3.538E+01	3.786E+01	4.905E+01	1.104E+02	4.417E+02	6.152E+03
W	74	183.850	2.845E+01	2.878E+01	2.933E+01	3.013E+01	3.256E+01	3.635E+01	3.891E+01	5.040E+01	1.134E+02	4.539E+02	6.322E+03
Re	75	186.207	2.922E+01	2.956E+01	3.013E+01	3.095E+01	3.345E+01	3.734E+01	3.997E+01	5.178E+01	1.165E+02	4.662E+02	6.494E+03
Os	76	190.200	3.001E+01	3.036E+01	3.094E+01	3.179E+01	3.435E+01	3.835E+01	4.104E+01	5.317E+01	1.197E+02	4.787E+02	6.668E+03
Ir	77	192.220	3.081E+01	3.116E+01	3.176E+01	3.263E+01	3.526E+01	3.936E+01	4.213E+01	5.458E+01	1.228E+02	4.914E+02	6.845E+03
Pt	78	195.080	3.161E+01	3.198E+01	3.259E+01	3.348E+01	3.618E+01	4.039E+01	4.323E+01	5.600E+01	1.260E+02	5.043E+02	7.024E+03
Au	79	196.967	3.243E+01	3.280E+01	3.343E+01	3.435E+01	3.711E+01	4.144E+01	4.435E+01	5.745E+01	1.293E+02	5.173E+02	7.205E+03
Hg	80	200.590	3.325E+01	3.364E+01	3.429E+01	3.522E+01	3.806E+01	4.249E+01	4.548E+01	5.892E+01	1.326E+02	5.304E+02	7.389E+03
Tl	81	204.383	3.409E+01	3.448E+01	3.515E+01	3.611E+01	3.902E+01	4.356E+01	4.663E+01	6.040E+01	1.359E+02	5.438E+02	7.574E+03
Pb	82	207.200	3.494E+01	3.534E+01	3.602E+01	3.701E+01	3.999E+01	4.465E+01	4.779E+01	6.190E+01	1.393E+02	5.573E+02	7.763E+03
Bi	83	208.980	3.580E+01	3.621E+01	3.691E+01	3.792E+01	4.097E+01	4.574E+01	4.896E+01	6.342E+01	1.427E+02	5.710E+02	7.953E+03
Po	84	209.000	3.667E+01	3.709E+01	3.780E+01	3.883E+01	4.196E+01	4.685E+01	5.015E+01	6.496E+01	1.462E+02	5.848E+02	8.146E+03
At	85	210.000	3.754E+01	3.798E+01	3.871E+01	3.976E+01	4.297E+01	4.797E+01	5.135E+01	6.651E+01	1.497E+02	5.988E+02	8.341E+03
Rn	86	222.000	3.844E+01	3.888E+01	3.963E+01	4.071E+01	4.399E+01	4.911E+01	5.256E+01	6.809E+01	1.532E+02	6.130E+02	8.538E+03
Fr	87	223.000	3.933E+01	3.979E+01	4.056E+01	4.166E+01	4.502E+01	5.026E+01	5.379E+01	6.968E+01	1.568E+02	6.273E+02	8.738E+03
Ra	88	226.025	4.024E+01	4.071E+01	4.149E+01	4.263E+01	4.606E+01	5.142E+01	5.504E+01	7.129E+01	1.604E+02	6.418E+02	8.940E+03
Ac	89	227.028	4.117E+01	4.164E+01	4.244E+01	4.360E+01	4.711E+01	5.260E+01	5.630E+01	7.292E+01	1.641E+02	6.565E+02	9.144E+03
Th	90	232.038	4.210E+01	4.258E+01	4.340E+01	4.459E+01	4.818E+01	5.379E+01	5.757E+01	7.457E+01	1.678E+02	6.714E+02	9.351E+03
Pa	91	231.036	4.304E+01	4.353E+01	4.437E+01	4.558E+01	4.925E+01	5.499E+01	5.886E+01	7.624E+01	1.716E+02	6.864E+02	9.560E+03
U	92	238.029	4.399E+01	4.450E+01	4.536E+01	4.659E+01	5.034E+01	5.621E+01	6.016E+01	7.793E+01	1.754E+02	7.015E+02	9.771E+03
Np	93	237.048	4.495E+01	4.547E+01	4.635E+01	4.761E+01	5.144E+01	5.744E+01	6.147E+01	7.963E+01	1.792E+02	7.169E+02	9.985E+03
Pu	94	244.000	4.592E+01	4.645E+01	4.735E+01	4.864E+01	5.256E+01	5.868E+01	6.280E+01	8.135E+01	1.831E+02	7.324E+02	1.020E+04

Table A6.5. Rutherford scattering cross sections (barns) of the elements for 1 MeV ^{7}Li.

Element	Atomic no. (Z_2)	Mass (M_2)	Scattering angle (degrees)										
			30	60	90	120	135	140	150	160	165	170	175
Be	4	9.012	4.139E+01	2.755E+00	4.682E-01	1.022E-01	5.832E-02	5.021E-02	3.935E-02	3.314E-02	3.123E-02	2.993E-02	2.919E-02
B	5	10.811	6.473E+01	4.412E+00	8.868E-01	2.681E-01	1.726E-01	1.532E-01	1.258E-01	1.092E-01	1.040E-01	1.004E-01	9.832E-02
C	6	12.011	9.326E+01	6.423E+00	1.362E+00	4.548E-01	3.056E-01	2.743E-01	2.293E-01	2.018E-01	1.929E-01	1.868E-01	1.833E-01
N	7	14.007	1.270E+02	8.848E+00	1.977E+00	7.249E-01	5.083E-01	4.616E-01	3.936E-01	3.512E-01	3.374E-01	3.279E-01	3.223E-01
O	8	15.999	1.660E+02	1.165E+01	2.682E+00	1.036E+00	7.450E-01	6.815E-01	5.882E-01	5.295E-01	5.103E-01	4.970E-01	4.892E-01
F	9	18.998	2.102E+02	1.485E+01	3.510E+00	1.419E+00	1.043E+00	9.605E-01	8.384E-01	7.610E-01	7.356E-01	7.180E-01	7.076E-01
Ne	10	20.180	2.595E+02	1.837E+01	4.371E+00	1.788E+00	1.323E+00	1.220E+00	1.068E+00	9.717E-01	9.400E-01	9.180E-01	9.051E-01
Na	11	22.990	3.140E+02	2.230E+01	5.372E+00	2.243E+00	1.676E+00	1.551E+00	1.365E+00	1.247E+00	1.208E+00	1.181E+00	1.165E+00
Mg	12	24.305	3.738E+02	2.657E+01	6.428E+00	2.703E+00	2.027E+00	1.878E+00	1.656E+00	1.514E+00	1.468E+00	1.436E+00	1.417E+00
Al	13	26.982	4.387E+02	3.125E+01	7.608E+00	3.234E+00	2.440E+00	2.263E+00	2.002E+00	1.835E+00	1.780E+00	1.742E+00	1.720E+00
Si	14	28.086	5.088E+02	3.627E+01	8.849E+00	3.775E+00	2.853E+00	2.648E+00	2.345E+00	2.151E+00	2.087E+00	2.043E+00	2.017E+00
P	15	30.974	5.842E+02	4.169E+01	1.022E+01	4.392E+00	3.332E+00	3.097E+00	2.748E+00	2.525E+00	2.451E+00	2.400E+00	2.370E+00
S	16	32.066	6.647E+02	4.746E+01	1.165E+01	5.018E+00	3.812E+00	3.544E+00	3.146E+00	2.892E+00	2.809E+00	2.751E+00	2.716E+00
Cl	17	35.453	7.504E+02	5.363E+01	1.321E+01	5.724E+00	4.361E+00	4.058E+00	3.609E+00	3.322E+00	3.227E+00	3.161E+00	3.123E+00
Ar	18	39.948	8.414E+02	6.019E+01	1.487E+01	6.480E+00	4.952E+00	4.612E+00	4.108E+00	3.785E+00	3.679E+00	3.605E+00	3.562E+00
K	19	39.098	9.374E+02	6.705E+01	1.656E+01	7.209E+00	5.506E+00	5.127E+00	4.565E+00	4.206E+00	4.088E+00	4.006E+00	3.957E+00
Ca	20	40.078	1.039E+03	7.431E+01	1.836E+01	8.002E+00	6.115E+00	5.695E+00	5.073E+00	4.675E+00	4.544E+00	4.452E+00	4.399E+00
Sc	21	44.959	1.145E+03	8.199E+01	2.031E+01	8.887E+00	6.806E+00	6.343E+00	5.656E+00	5.217E+00	5.072E+00	4.972E+00	4.913E+00
Ti	22	47.880	1.257E+03	9.002E+01	2.232E+01	9.787E+00	7.502E+00	6.994E+00	6.240E+00	5.758E+00	5.599E+00	5.488E+00	5.423E+00
V	23	50.942	1.374E+03	9.842E+01	2.443E+01	1.073E+01	8.230E+00	7.675E+00	6.851E+00	6.323E+00	6.150E+00	6.029E+00	5.958E+00
Cr	24	51.996	1.496E+03	1.072E+02	2.661E+01	1.169E+01	8.971E+00	8.367E+00	7.469E+00	6.895E+00	6.706E+00	6.575E+00	6.497E+00
Mn	25	54.938	1.623E+03	1.163E+02	2.890E+01	1.271E+01	9.762E+00	9.106E+00	8.132E+00	7.509E+00	7.304E+00	7.161E+00	7.077E+00
Fe	26	55.847	1.756E+03	1.258E+02	3.127E+01	1.376E+01	1.057E+01	9.857E+00	8.804E+00	8.130E+00	7.908E+00	7.754E+00	7.663E+00
Co	27	58.933	1.893E+03	1.357E+02	3.375E+01	1.486E+01	1.142E+01	1.066E+01	9.521E+00	8.794E+00	8.555E+00	8.388E+00	8.290E+00
Ni	28	58.680	2.036E+03	1.460E+02	3.629E+01	1.598E+01	1.228E+01	1.146E+01	1.024E+01	9.456E+00	9.198E+00	9.019E+00	8.914E+00
Cu	29	63.546	2.184E+03	1.566E+02	3.897E+01	1.719E+01	1.322E+01	1.233E+01	1.102E+01	1.018E+01	9.908E+00	9.716E+00	9.602E+00
Zn	30	65.390	2.338E+03	1.676E+02	4.172E+01	1.841E+01	1.416E+01	1.321E+01	1.181E+01	1.091E+01	1.062E+01	1.041E+01	1.029E+01
Ga	31	69.723	2.496E+03	1.790E+02	4.458E+01	1.969E+01	1.515E+01	1.414E+01	1.264E+01	1.168E+01	1.137E+01	1.115E+01	1.102E+01
Ge	32	72.610	2.660E+03	1.907E+02	4.752E+01	2.100E+01	1.616E+01	1.508E+01	1.349E+01	1.247E+01	1.213E+01	1.190E+01	1.176E+01
As	33	74.922	2.828E+03	2.029E+02	5.055E+01	2.234E+01	1.720E+01	1.606E+01	1.436E+01	1.327E+01	1.292E+01	1.267E+01	1.252E+01
Se	34	78.960	3.003E+03	2.154E+02	5.368E+01	2.374E+01	1.828E+01	1.707E+01	1.527E+01	1.411E+01	1.373E+01	1.347E+01	1.331E+01
Br	35	79.904	3.182E+03	2.282E+02	5.689E+01	2.516E+01	1.938E+01	1.809E+01	1.618E+01	1.496E+01	1.456E+01	1.428E+01	1.411E+01
Kr	36	83.800	3.366E+03	2.415E+02	6.021E+01	2.664E+01	2.052E+01	1.916E+01	1.714E+01	1.585E+01	1.542E+01	1.513E+01	1.495E+01
Rb	37	85.468	3.556E+03	2.551E+02	6.361E+01	2.815E+01	2.169E+01	2.025E+01	1.812E+01	1.675E+01	1.630E+01	1.599E+01	1.580E+01
Sr	38	87.620	3.751E+03	2.691E+02	6.711E+01	2.971E+01	2.289E+01	2.137E+01	1.912E+01	1.768E+01	1.720E+01	1.687E+01	1.668E+01
Y	39	88.906	3.951E+03	2.834E+02	7.069E+01	3.130E+01	2.411E+01	2.252E+01	2.015E+01	1.863E+01	1.813E+01	1.778E+01	1.758E+01
Zr	40	91.224	4.156E+03	2.982E+02	7.438E+01	3.293E+01	2.538E+01	2.370E+01	2.120E+01	1.961E+01	1.908E+01	1.872E+01	1.850E+01
Nb	41	92.906	4.366E+03	3.133E+02	7.815E+01	3.461E+01	2.667E+01	2.491E+01	2.229E+01	2.061E+01	2.006E+01	1.967E+01	1.945E+01
Mo	42	95.940	4.582E+03	3.288E+02	8.202E+01	3.633E+01	2.800E+01	2.615E+01	2.340E+01	2.164E+01	2.106E+01	2.066E+01	2.042E+01
Tc	43	98.000	4.803E+03	3.446E+02	8.599E+01	3.809E+01	2.936E+01	2.742E+01	2.454E+01	2.269E+01	2.208E+01	2.166E+01	2.141E+01
Ru	44	101.070	5.029E+03	3.608E+02	9.005E+01	3.990E+01	3.076E+01	2.872E+01	2.571E+01	2.377E+01	2.314E+01	2.270E+01	2.243E+01
Rh	45	102.906	5.260E+03	3.774E+02	9.419E+01	4.174E+01	3.218E+01	3.005E+01	2.690E+01	2.487E+01	2.421E+01	2.375E+01	2.347E+01
Pd	46	106.420	5.496E+03	3.944E+02	9.844E+01	4.363E+01	3.364E+01	3.142E+01	2.812E+01	2.601E+01	2.531E+01	2.483E+01	2.454E+01
Ag	47	107.868	5.738E+03	4.117E+02	1.028E+02	4.556E+01	3.512E+01	3.280E+01	2.936E+01	2.716E+01	2.643E+01	2.593E+01	2.563E+01
Cd	48	122.411	5.985E+03	4.295E+02	1.072E+02	4.757E+01	3.668E+01	3.427E+01	3.067E+01	2.837E+01	2.762E+01	2.709E+01	2.678E+01

Table A6.5. Rutherford scattering cross sections (barns) of the elements for 1 MeV ^{7}Li (continued).

Element	Atomic no. (Z$_2$)	Mass (M$_2$)	175	170	165	160	150	140	135	120	90	60	30
In	49	114.820	2.788E+01	2.821E+01	2.876E+01	2.954E+01	3.194E+01	3.568E+01	3.820E+01	4.954E+01	1.117E+02	4.476E+02	6.236E+03
Sn	50	118.710	2.905E+01	2.938E+01	2.996E+01	3.078E+01	3.327E+01	3.717E+01	3.979E+01	5.160E+01	1.164E+02	4.660E+02	6.494E+03
Sb	51	121.750	3.023E+01	3.058E+01	3.118E+01	3.203E+01	3.463E+01	3.868E+01	4.141E+01	5.370E+01	1.211E+02	4.849E+02	6.756E+03
Te	52	127.600	3.145E+01	3.181E+01	3.243E+01	3.332E+01	3.602E+01	4.023E+01	4.307E+01	5.584E+01	1.259E+02	5.041E+02	7.024E+03
I	53	126.904	3.267E+01	3.304E+01	3.369E+01	3.461E+01	3.741E+01	4.179E+01	4.474E+01	5.801E+01	1.308E+02	5.237E+02	7.296E+03
Xe	54	131.290	3.392E+01	3.432E+01	3.498E+01	3.594E+01	3.885E+01	4.340E+01	4.646E+01	6.023E+01	1.358E+02	5.436E+02	7.574E+03
Cs	55	132.905	3.520E+01	3.560E+01	3.630E+01	3.729E+01	4.031E+01	4.502E+01	4.820E+01	6.249E+01	1.408E+02	5.639E+02	7.857E+03
Ba	56	137.327	3.650E+01	3.692E+01	3.764E+01	3.867E+01	4.180E+01	4.669E+01	4.998E+01	6.479E+01	1.460E+02	5.847E+02	8.146E+03
La	57	138.906	3.782E+01	3.826E+01	3.900E+01	4.007E+01	4.331E+01	4.838E+01	5.179E+01	6.713E+01	1.513E+02	6.057E+02	8.439E+03
Ce	58	140.115	3.916E+01	3.962E+01	4.039E+01	4.149E+01	4.485E+01	5.009E+01	5.362E+01	6.951E+01	1.566E+02	6.272E+02	8.738E+03
Pr	59	140.908	4.053E+01	4.100E+01	4.179E+01	4.294E+01	4.641E+01	5.184E+01	5.549E+01	7.193E+01	1.621E+02	6.490E+02	9.042E+03
Nd	60	144.240	4.192E+01	4.241E+01	4.323E+01	4.441E+01	4.800E+01	5.362E+01	5.740E+01	7.440E+01	1.676E+02	6.712E+02	9.351E+03
Pm	61	145.000	4.333E+01	4.384E+01	4.469E+01	4.591E+01	4.962E+01	5.542E+01	5.933E+01	7.690E+01	1.733E+02	6.937E+02	9.665E+03
Sm	62	150.360	4.478E+01	4.530E+01	4.618E+01	4.744E+01	5.128E+01	5.727E+01	6.130E+01	7.946E+01	1.790E+02	7.167E+02	9.985E+03
Eu	63	151.965	4.624E+01	4.678E+01	4.768E+01	4.899E+01	5.295E+01	5.913E+01	6.330E+01	8.205E+01	1.849E+02	7.400E+02	1.031E+04
Gd	64	157.250	4.773E+01	4.829E+01	4.922E+01	5.057E+01	5.465E+01	6.104E+01	6.534E+01	8.469E+01	1.908E+02	7.637E+02	1.064E+04
Tb	65	158.925	4.924E+01	4.981E+01	5.078E+01	5.216E+01	5.638E+01	6.297E+01	6.740E+01	8.736E+01	1.968E+02	7.877E+02	1.097E+04
Dy	66	162.500	5.078E+01	5.136E+01	5.236E+01	5.379E+01	5.814E+01	6.493E+01	6.950E+01	9.007E+01	2.029E+02	8.122E+02	1.131E+04
Ho	67	164.930	5.233E+01	5.294E+01	5.396E+01	5.544E+01	5.992E+01	6.692E+01	7.163E+01	9.283E+01	2.091E+02	8.370E+02	1.166E+04
Er	68	167.260	5.391E+01	5.454E+01	5.559E+01	5.711E+01	6.172E+01	6.893E+01	7.379E+01	9.563E+01	2.154E+02	8.622E+02	1.201E+04
Tm	69	168.934	5.551E+01	5.616E+01	5.724E+01	5.881E+01	6.356E+01	7.098E+01	7.598E+01	9.846E+01	2.218E+02	8.877E+02	1.237E+04
Yb	70	173.040	5.714E+01	5.780E+01	5.892E+01	6.053E+01	6.542E+01	7.306E+01	7.821E+01	1.013E+02	2.283E+02	9.136E+02	1.273E+04
Lu	71	174.967	5.879E+01	5.947E+01	6.062E+01	6.228E+01	6.731E+01	7.517E+01	8.046E+01	1.043E+02	2.348E+02	9.399E+02	1.309E+04
Hf	72	178.490	6.047E+01	6.117E+01	6.235E+01	6.405E+01	6.923E+01	7.731E+01	8.275E+01	1.072E+02	2.415E+02	9.666E+02	1.347E+04
Ta	73	180.948	6.216E+01	6.288E+01	6.410E+01	6.585E+01	7.117E+01	7.947E+01	8.507E+01	1.102E+02	2.483E+02	9.936E+02	1.384E+04
W	74	183.850	6.389E+01	6.462E+01	6.587E+01	6.767E+01	7.314E+01	8.167E+01	8.742E+01	1.133E+02	2.551E+02	1.021E+03	1.422E+04
Re	75	186.207	6.563E+01	6.639E+01	6.767E+01	6.952E+01	7.513E+01	8.390E+01	8.981E+01	1.164E+02	2.621E+02	1.049E+03	1.461E+04
Os	76	190.200	6.740E+01	6.818E+01	6.950E+01	7.139E+01	7.716E+01	8.616E+01	9.223E+01	1.195E+02	2.691E+02	1.077E+03	1.500E+04
Ir	77	192.220	6.919E+01	6.999E+01	7.134E+01	7.329E+01	7.920E+01	8.845E+01	9.467E+01	1.227E+02	2.762E+02	1.106E+03	1.540E+04
Pt	78	195.080	7.100E+01	7.182E+01	7.321E+01	7.521E+01	8.128E+01	9.076E+01	9.715E+01	1.259E+02	2.835E+02	1.134E+03	1.580E+04
Au	79	196.967	7.284E+01	7.368E+01	7.510E+01	7.715E+01	8.338E+01	9.311E+01	9.966E+01	1.291E+02	2.908E+02	1.164E+03	1.621E+04
Hg	80	200.590	7.470E+01	7.556E+01	7.702E+01	7.913E+01	8.551E+01	9.549E+01	1.022E+02	1.324E+02	2.982E+02	1.193E+03	1.662E+04
Tl	81	204.383	7.659E+01	7.747E+01	7.897E+01	8.112E+01	8.767E+01	9.790E+01	1.048E+02	1.358E+02	3.057E+02	1.223E+03	1.704E+04
Pb	82	207.200	7.849E+01	7.940E+01	8.094E+01	8.314E+01	8.985E+01	1.003E+02	1.074E+02	1.392E+02	3.133E+02	1.254E+03	1.747E+04
Bi	83	208.980	8.042E+01	8.135E+01	8.292E+01	8.519E+01	9.206E+01	1.028E+02	1.100E+02	1.426E+02	3.210E+02	1.285E+03	1.789E+04
Po	84	209.000	8.237E+01	8.332E+01	8.493E+01	8.725E+01	9.429E+01	1.053E+02	1.127E+02	1.460E+02	3.288E+02	1.316E+03	1.833E+04
At	85	210.000	8.435E+01	8.532E+01	8.697E+01	8.934E+01	9.655E+01	1.078E+02	1.154E+02	1.495E+02	3.367E+02	1.347E+03	1.877E+04
Rn	86	222.000	8.636E+01	8.736E+01	8.905E+01	9.148E+01	9.886E+01	1.104E+02	1.182E+02	1.531E+02	3.447E+02	1.379E+03	1.921E+04
Fr	87	223.000	8.839E+01	8.940E+01	9.113E+01	9.362E+01	1.012E+02	1.130E+02	1.209E+02	1.567E+02	3.527E+02	1.411E+03	1.966E+04
Ra	88	226.025	9.043E+01	9.148E+01	9.325E+01	9.579E+01	1.035E+02	1.156E+02	1.237E+02	1.603E+02	3.609E+02	1.444E+03	2.012E+04
Ac	89	227.028	9.250E+01	9.357E+01	9.538E+01	9.798E+01	1.059E+02	1.182E+02	1.265E+02	1.640E+02	3.691E+02	1.477E+03	2.057E+04
Th	90	232.038	9.460E+01	9.569E+01	9.754E+01	1.002E+02	1.083E+02	1.209E+02	1.294E+02	1.677E+02	3.775E+02	1.510E+03	2.104E+04
Pa	91	231.036	9.671E+01	9.783E+01	9.972E+01	1.024E+02	1.107E+02	1.236E+02	1.323E+02	1.714E+02	3.859E+02	1.544E+03	2.151E+04
U	92	238.029	9.886E+01	1.000E+02	1.019E+02	1.047E+02	1.132E+02	1.264E+02	1.352E+02	1.752E+02	3.945E+02	1.578E+03	2.199E+04
Np	93	237.048	1.010E+02	1.022E+02	1.042E+02	1.070E+02	1.156E+02	1.291E+02	1.382E+02	1.790E+02	4.031E+02	1.613E+03	2.247E+04
Pu	94	244.000	1.032E+02	1.044E+02	1.064E+02	1.093E+02	1.181E+02	1.319E+02	1.412E+02	1.829E+02	4.118E+02	1.648E+03	2.295E+04

Scattering angle (degrees)

74

Table A6.6. Rutherford scattering cross sections (barns) of the elements for 1 MeV ^{12}C.

Element	Atomic no. (Z_2)	Mass (M_2)	30	60	90	120	135	140	150	160	165	170	175
N	7	14.007	5.067E+02	3.320E+01	4.713E+00	7.040E-01	3.599E-01	3.023E-01	2.285E-01	1.882E-01	1.760E-01	1.679E-01	1.632E-01
O	8	15.999	6.623E+02	4.433E+01	7.894E+00	1.891E+00	1.114E+00	9.668E-01	7.664E-01	6.504E-01	6.143E-01	5.898E-01	5.757E-01
F	9	18.998	8.390E+02	5.736E+01	1.171E+01	3.646E+00	2.377E+00	2.116E+00	1.746E+00	1.522E+00	1.451E+00	1.402E+00	1.374E+00
Ne	10	20.180	1.036E+03	7.124E+01	1.499E+01	4.935E+00	3.295E+00	2.952E+00	2.462E+00	2.162E+00	2.065E+00	1.999E+00	1.961E+00
Na	11	22.990	1.254E+03	8.714E+01	1.925E+01	6.911E+00	4.799E+00	4.346E+00	3.689E+00	3.281E+00	3.148E+00	3.057E+00	3.004E+00
Mg	12	24.305	1.493E+03	1.041E+02	2.335E+01	8.619E+00	6.063E+00	5.511E+00	4.706E+00	4.204E+00	4.040E+00	3.928E+00	3.861E+00
Al	13	26.982	1.753E+03	1.229E+02	2.823E+01	1.086E+01	7.788E+00	7.119E+00	6.138E+00	5.520E+00	5.319E+00	5.179E+00	5.098E+00
Si	14	28.086	2.033E+03	1.428E+02	3.305E+01	1.287E+01	9.294E+00	8.511E+00	7.361E+00	6.636E+00	6.399E+00	6.235E+00	6.138E+00
P	15	30.974	2.335E+03	1.647E+02	3.868E+01	1.547E+01	1.132E+01	1.040E+01	9.056E+00	8.202E+00	7.923E+00	7.729E+00	7.615E+00
S	16	32.066	2.657E+03	1.876E+02	4.427E+01	1.785E+01	1.310E+01	1.206E+01	1.052E+01	9.543E+00	9.222E+00	9.000E+00	8.870E+00
Cl	17	35.453	3.000E+03	2.125E+02	5.072E+01	2.085E+01	1.546E+01	1.426E+01	1.251E+01	1.139E+01	1.102E+01	1.076E+01	1.061E+01
Ar	18	39.948	3.364E+03	2.390E+02	5.763E+01	2.411E+01	1.804E+01	1.670E+01	1.471E+01	1.344E+01	1.302E+01	1.273E+01	1.256E+01
K	19	39.098	3.748E+03	2.661E+02	6.408E+01	2.673E+01	1.997E+01	1.847E+01	1.626E+01	1.485E+01	1.438E+01	1.406E+01	1.387E+01
Ca	20	40.078	4.153E+03	2.950E+02	7.118E+01	2.979E+01	2.229E+01	2.063E+01	1.818E+01	1.661E+01	1.609E+01	1.574E+01	1.553E+01
Sc	21	44.959	4.579E+03	3.260E+02	7.926E+01	3.361E+01	2.532E+01	2.348E+01	2.076E+01	1.902E+01	1.844E+01	1.805E+01	1.781E+01
Ti	22	47.880	5.026E+03	3.582E+02	8.738E+01	3.727E+01	2.816E+01	2.614E+01	2.314E+01	2.123E+01	2.060E+01	2.016E+01	1.991E+01
V	23	50.942	5.494E+03	3.919E+02	9.588E+01	4.110E+01	3.114E+01	2.892E+01	2.564E+01	2.355E+01	2.286E+01	2.238E+01	2.210E+01
Cr	24	51.996	5.982E+03	4.268E+02	1.045E+02	4.487E+01	3.402E+01	3.161E+01	2.804E+01	2.575E+01	2.500E+01	2.448E+01	2.417E+01
Mn	25	54.938	6.491E+03	4.634E+02	1.137E+02	4.902E+01	3.723E+01	3.462E+01	3.074E+01	2.825E+01	2.744E+01	2.687E+01	2.654E+01
Fe	26	55.847	7.021E+03	5.014E+02	1.231E+02	5.311E+01	4.037E+01	3.754E+01	3.334E+01	3.065E+01	2.977E+01	2.916E+01	2.879E+01
Co	27	58.933	7.572E+03	5.410E+02	1.331E+02	5.760E+01	4.385E+01	4.079E+01	3.626E+01	3.336E+01	3.241E+01	3.175E+01	3.136E+01
Ni	28	58.690	8.143E+03	5.818E+02	1.431E+02	6.192E+01	4.713E+01	4.385E+01	3.897E+01	3.586E+01	3.483E+01	3.412E+01	3.370E+01
Cu	29	63.546	8.735E+03	6.246E+02	1.540E+02	6.690E+01	5.104E+01	4.751E+01	4.228E+01	3.893E+01	3.783E+01	3.706E+01	3.661E+01
Zn	30	65.390	9.348E+03	6.685E+02	1.650E+02	7.177E+01	5.478E+01	5.101E+01	4.541E+01	4.182E+01	4.065E+01	3.983E+01	3.934E+01
Ga	31	69.723	9.982E+03	7.142E+02	1.765E+02	7.699E+01	5.885E+01	5.482E+01	4.884E+01	4.501E+01	4.375E+01	4.288E+01	4.236E+01
Ge	32	72.610	1.064E+04	7.613E+02	1.883E+02	8.226E+01	6.293E+01	5.864E+01	5.226E+01	4.818E+01	4.684E+01	4.590E+01	4.535E+01
As	33	74.922	1.131E+04	8.098E+02	2.005E+02	8.765E+01	6.709E+01	6.252E+01	5.574E+01	5.140E+01	4.997E+01	4.898E+01	4.839E+01
Se	34	78.960	1.201E+04	8.599E+02	2.131E+02	9.332E+01	7.150E+01	6.664E+01	5.944E+01	5.483E+01	5.332E+01	5.226E+01	5.164E+01
Br	35	79.904	1.273E+04	9.112E+02	2.259E+02	9.896E+01	7.582E+01	7.068E+01	6.305E+01	5.817E+01	5.656E+01	5.544E+01	5.478E+01
Kr	36	83.800	1.346E+04	9.643E+02	2.392E+02	1.049E+02	8.047E+01	7.503E+01	6.695E+01	6.178E+01	6.008E+01	5.890E+01	5.820E+01
Rb	37	85.468	1.422E+04	1.019E+03	2.528E+02	1.110E+02	8.510E+01	7.935E+01	7.082E+01	6.536E+01	6.357E+01	6.232E+01	6.158E+01
Sr	38	87.620	1.500E+04	1.075E+03	2.668E+02	1.172E+02	8.989E+01	8.383E+01	7.483E+01	6.907E+01	6.717E+01	6.586E+01	6.508E+01
Y	39	88.906	1.580E+04	1.132E+03	2.811E+02	1.235E+02	9.476E+01	8.837E+01	7.889E+01	7.283E+01	7.083E+01	6.944E+01	6.862E+01
Zr	40	91.224	1.662E+04	1.191E+03	2.958E+02	1.300E+02	9.982E+01	9.310E+01	8.313E+01	7.675E+01	7.464E+01	7.318E+01	7.232E+01
Nb	41	92.906	1.746E+04	1.251E+03	3.109E+02	1.367E+02	1.050E+02	9.791E+01	8.743E+01	8.073E+01	7.852E+01	7.698E+01	7.608E+01
Mo	42	95.940	1.833E+04	1.313E+03	3.264E+02	1.436E+02	1.103E+02	1.029E+02	9.192E+01	8.488E+01	8.256E+01	8.095E+01	8.000E+01
Tc	43	98.000	1.921E+04	1.377E+03	3.422E+02	1.507E+02	1.157E+02	1.080E+02	9.646E+01	8.908E+01	8.665E+01	8.497E+01	8.397E+01
Ru	44	101.070	2.011E+04	1.442E+03	3.585E+02	1.579E+02	1.214E+02	1.132E+02	1.012E+02	9.343E+01	9.089E+01	8.912E+01	8.808E+01
Rh	45	102.906	2.104E+04	1.508E+03	3.751E+02	1.653E+02	1.270E+02	1.185E+02	1.059E+02	9.782E+01	9.516E+01	9.331E+01	9.222E+01
Pd	46	106.420	2.198E+04	1.576E+03	3.921E+02	1.729E+02	1.329E+02	1.240E+02	1.108E+02	1.024E+02	9.961E+01	9.768E+01	9.654E+01
Ag	47	107.868	2.295E+04	1.645E+03	4.094E+02	1.805E+02	1.388E+02	1.295E+02	1.158E+02	1.070E+02	1.041E+02	1.020E+02	1.009E+02
Cd	48	112.411	2.394E+04	1.717E+03	4.276E+02	1.889E+02	1.454E+02	1.357E+02	1.213E+02	1.121E+02	1.091E+02	1.070E+02	1.058E+02
In	49	114.820	2.494E+04	1.789E+03	4.453E+02	1.966E+02	1.512E+02	1.411E+02	1.262E+02	1.166E+02	1.134E+02	1.112E+02	1.099E+02
Sn	50	118.710	2.597E+04	1.863E+03	4.638E+02	2.048E+02	1.576E+02	1.471E+02	1.315E+02	1.215E+02	1.183E+02	1.160E+02	1.146E+02
Sb	51	121.750	2.702E+04	1.938E+03	4.827E+02	2.132E+02	1.641E+02	1.532E+02	1.370E+02	1.266E+02	1.232E+02	1.208E+02	1.194E+02

Scattering angle (degrees)

Table A6.6. Rutherford scattering cross sections (barns) of the elements for 1 MeV ^{12}C (continued).

Element	Atomic no. (Z_2)	Mass (M_2)	30	60	90	120	135	140	150	160	165	170	175
Te	52	127.600	2.809E+04	2.015E+03	5.020E+02	2.219E+02	1.708E+02	1.595E+02	1.426E+02	1.318E+02	1.283E+02	1.258E+02	1.243E+02
I	53	126.904	2.918E+04	2.093E+03	5.215E+02	2.305E+02	1.774E+02	1.656E+02	1.481E+02	1.369E+02	1.332E+02	1.306E+02	1.291E+02
Xe	54	131.290	3.030E+04	2.173E+03	5.415E+02	2.394E+02	1.843E+02	1.721E+02	1.539E+02	1.423E+02	1.384E+02	1.358E+02	1.342E+02
Cs	55	132.905	3.143E+04	2.254E+03	5.618E+02	2.484E+02	1.913E+02	1.786E+02	1.597E+02	1.477E+02	1.437E+02	1.409E+02	1.393E+02
Ba	56	137.327	3.258E+04	2.337E+03	5.826E+02	2.577E+02	1.985E+02	1.853E+02	1.657E+02	1.532E+02	1.491E+02	1.462E+02	1.445E+02
La	57	138.906	3.376E+04	2.421E+03	6.037E+02	2.670E+02	2.057E+02	1.920E+02	1.718E+02	1.588E+02	1.545E+02	1.516E+02	1.498E+02
Ce	58	140.115	3.495E+04	2.507E+03	6.251E+02	2.765E+02	2.130E+02	1.989E+02	1.779E+02	1.645E+02	1.600E+02	1.570E+02	1.551E+02
Pr	59	140.908	3.617E+04	2.594E+03	6.468E+02	2.862E+02	2.204E+02	2.058E+02	1.841E+02	1.702E+02	1.656E+02	1.624E+02	1.606E+02
Nd	60	144.240	3.740E+04	2.683E+03	6.691E+02	2.961E+02	2.281E+02	2.129E+02	1.905E+02	1.761E+02	1.714E+02	1.681E+02	1.662E+02
Pm	61	145.000	3.866E+04	2.773E+03	6.916E+02	3.060E+02	2.358E+02	2.201E+02	1.969E+02	1.821E+02	1.772E+02	1.738E+02	1.718E+02
Sm	62	150.360	3.994E+04	2.865E+03	7.146E+02	3.163E+02	2.437E+02	2.276E+02	2.036E+02	1.883E+02	1.832E+02	1.797E+02	1.776E+02
Eu	63	151.965	4.124E+04	2.958E+03	7.379E+02	3.267E+02	2.517E+02	2.350E+02	2.103E+02	1.944E+02	1.892E+02	1.856E+02	1.835E+02
Gd	64	157.250	4.256E+04	3.053E+03	7.617E+02	3.373E+02	2.599E+02	2.427E+02	2.172E+02	2.008E+02	1.954E+02	1.917E+02	1.895E+02
Tb	65	158.925	4.390E+04	3.150E+03	7.857E+02	3.479E+02	2.681E+02	2.504E+02	2.240E+02	2.072E+02	2.016E+02	1.978E+02	1.955E+02
Dy	66	162.500	4.526E+04	3.247E+03	8.102E+02	3.588E+02	2.765E+02	2.583E+02	2.311E+02	2.137E+02	2.080E+02	2.040E+02	2.017E+02
Ho	67	164.930	4.664E+04	3.346E+03	8.350E+02	3.699E+02	2.851E+02	2.662E+02	2.382E+02	2.203E+02	2.144E+02	2.103E+02	2.079E+02
Er	68	167.260	4.804E+04	3.447E+03	8.601E+02	3.810E+02	2.937E+02	2.743E+02	2.454E+02	2.270E+02	2.209E+02	2.167E+02	2.142E+02
Tm	69	168.934	4.947E+04	3.549E+03	8.857E+02	3.924E+02	3.024E+02	2.824E+02	2.528E+02	2.338E+02	2.275E+02	2.232E+02	2.206E+02
Yb	70	173.040	5.091E+04	3.653E+03	9.116E+02	4.039E+02	3.114E+02	2.908E+02	2.602E+02	2.407E+02	2.343E+02	2.298E+02	2.271E+02
Lu	71	174.967	5.237E+04	3.758E+03	9.379E+02	4.156E+02	3.204E+02	2.992E+02	2.678E+02	2.477E+02	2.410E+02	2.364E+02	2.337E+02
Hf	72	178.490	5.386E+04	3.865E+03	9.646E+02	4.275E+02	3.296E+02	3.078E+02	2.755E+02	2.548E+02	2.480E+02	2.432E+02	2.404E+02
Ta	73	180.948	5.537E+04	3.973E+03	9.916E+02	4.395E+02	3.388E+02	3.165E+02	2.832E+02	2.620E+02	2.550E+02	2.501E+02	2.472E+02
W	74	183.850	5.689E+04	4.083E+03	1.019E+03	4.517E+02	3.483E+02	3.253E+02	2.911E+02	2.693E+02	2.621E+02	2.571E+02	2.541E+02
Re	75	186.207	5.844E+04	4.194E+03	1.047E+03	4.641E+02	3.578E+02	3.342E+02	2.991E+02	2.766E+02	2.693E+02	2.641E+02	2.611E+02
Os	76	190.200	6.001E+04	4.307E+03	1.075E+03	4.766E+02	3.675E+02	3.432E+02	3.072E+02	2.842E+02	2.766E+02	2.713E+02	2.682E+02
Ir	77	192.220	6.160E+04	4.421E+03	1.104E+03	4.893E+02	3.773E+02	3.524E+02	3.154E+02	2.917E+02	2.839E+02	2.785E+02	2.753E+02
Pt	78	195.080	6.321E+04	4.536E+03	1.132E+03	5.021E+02	3.872E+02	3.616E+02	3.237E+02	2.994E+02	2.914E+02	2.859E+02	2.826E+02
Au	79	196.967	6.484E+04	4.653E+03	1.162E+03	5.151E+02	3.972E+02	3.710E+02	3.321E+02	3.072E+02	2.990E+02	2.933E+02	2.899E+02
Hg	80	200.590	6.649E+04	4.772E+03	1.191E+03	5.283E+02	4.074E+02	3.806E+02	3.406E+02	3.151E+02	3.067E+02	3.008E+02	2.974E+02
Tl	81	204.383	6.817E+04	4.892E+03	1.221E+03	5.417E+02	4.178E+02	3.902E+02	3.493E+02	3.231E+02	3.145E+02	3.085E+02	3.050E+02
Pb	82	207.200	6.986E+04	5.014E+03	1.252E+03	5.552E+02	4.282E+02	4.000E+02	3.580E+02	3.312E+02	3.224E+02	3.162E+02	3.126E+02
Bi	83	208.980	7.158E+04	5.137E+03	1.283E+03	5.689E+02	4.387E+02	4.098E+02	3.669E+02	3.394E+02	3.303E+02	3.240E+02	3.203E+02
Po	84	209.000	7.331E+04	5.261E+03	1.314E+03	5.827E+02	4.494E+02	4.197E+02	3.757E+02	3.476E+02	3.383E+02	3.319E+02	3.281E+02
At	85	210.000	7.507E+04	5.387E+03	1.345E+03	5.967E+02	4.602E+02	4.298E+02	3.848E+02	3.559E+02	3.464E+02	3.398E+02	3.359E+02
Rn	86	222.000	7.684E+04	5.515E+03	1.377E+03	6.110E+02	4.713E+02	4.402E+02	3.941E+02	3.646E+02	3.549E+02	3.481E+02	3.441E+02
Fr	87	223.000	7.864E+04	5.644E+03	1.410E+03	6.253E+02	4.823E+02	4.505E+02	4.033E+02	3.731E+02	3.632E+02	3.563E+02	3.522E+02
Ra	88	226.025	8.046E+04	5.775E+03	1.442E+03	6.398E+02	4.935E+02	4.610E+02	4.127E+02	3.818E+02	3.716E+02	3.646E+02	3.604E+02
Ac	89	227.028	8.230E+04	5.907E+03	1.475E+03	6.545E+02	5.048E+02	4.716E+02	4.222E+02	3.906E+02	3.802E+02	3.729E+02	3.687E+02
Th	90	232.038	8.416E+04	6.040E+03	1.509E+03	6.694E+02	5.163E+02	4.823E+02	4.318E+02	3.995E+02	3.888E+02	3.814E+02	3.771E+02
Pa	91	231.036	8.604E+04	6.175E+03	1.542E+03	6.843E+02	5.279E+02	4.931E+02	4.414E+02	4.084E+02	3.975E+02	3.899E+02	3.855E+02
U	92	238.029	8.794E+04	6.312E+03	1.576E+03	6.995E+02	5.396E+02	5.041E+02	4.513E+02	4.175E+02	4.064E+02	3.987E+02	3.941E+02
Np	93	237.048	8.986E+04	6.450E+03	1.611E+03	7.148E+02	5.514E+02	5.151E+02	4.612E+02	4.266E+02	4.153E+02	4.074E+02	4.027E+02
Pu	94	244.000	9.181E+04	6.589E+03	1.646E+03	7.304E+02	5.635E+02	5.264E+02	4.713E+02	4.360E+02	4.244E+02	4.163E+02	4.115E+02

Scattering angle (degrees)

76

Table A6.7. Rutherford scattering cross sections (barns) of the elements for 1 MeV ^{15}N.

Element	Atomic no. (Z_2)	Mass (M_2)	\multicolumn{11}{c}{Scattering angle (degrees)}										
			30	60	90	120	135	140	150	160	165	170	175
O	8	15.999	9.005E+02	5.811E+01	5.651E+00	3.470E-01	1.499E-01	1.219E-01	8.801E-02	7.055E-02	6.543E-02	6.204E-02	6.011E-02
F	9	18.998	1.141E+03	7.575E+01	1.262E+01	2.643E+00	1.488E+00	1.278E+00	9.964E-01	8.368E-01	7.877E-01	7.545E-01	7.354E-01
Ne	10	20.180	1.409E+03	9.440E+01	1.698E+01	4.149E+00	2.462E+00	2.139E+00	1.700E+00	1.445E+00	1.366E+00	1.312E+00	1.281E+00
Na	11	22.990	1.706E+03	1.162E+02	2.328E+01	6.993E+00	4.492E+00	3.983E+00	3.266E+00	2.835E+00	2.698E+00	2.605E+00	2.550E+00
Mg	12	24.305	2.030E+03	1.391E+02	2.876E+01	9.161E+00	6.031E+00	5.382E+00	4.460E+00	3.900E+00	3.721E+00	3.598E+00	3.526E+00
Al	13	26.982	2.384E+03	1.649E+02	3.566E+01	1.233E+01	8.418E+00	7.587E+00	6.390E+00	5.650E+00	5.412E+00	5.248E+00	5.151E+00
Si	14	28.086	2.765E+03	1.918E+02	4.206E+01	1.491E+01	1.029E+01	9.306E+00	7.878E+00	6.992E+00	6.706E+00	6.508E+00	6.392E+00
P	15	30.974	3.176E+03	2.217E+02	4.997E+01	1.860E+01	1.314E+01	1.195E+01	1.023E+01	9.149E+00	8.798E+00	8.556E+00	8.413E+00
S	16	32.066	3.614E+03	2.527E+02	5.743E+01	2.169E+01	1.542E+01	1.406E+01	1.207E+01	1.082E+01	1.041E+01	1.013E+01	9.968E+00
Cl	17	35.453	4.081E+03	2.868E+02	6.647E+01	2.597E+01	1.878E+01	1.720E+01	1.489E+01	1.343E+01	1.295E+01	1.262E+01	1.243E+01
Ar	18	39.948	4.577E+03	3.231E+02	7.623E+01	3.071E+01	2.254E+01	2.074E+01	1.809E+01	1.640E+01	1.585E+01	1.547E+01	1.525E+01
K	19	39.098	5.099E+03	3.597E+02	8.462E+01	3.393E+01	2.484E+01	2.285E+01	1.990E+01	1.803E+01	1.742E+01	1.699E+01	1.674E+01
Ca	20	40.078	5.650E+03	3.990E+02	9.416E+01	3.796E+01	2.787E+01	2.565E+01	2.237E+01	2.029E+01	1.961E+01	1.914E+01	1.886E+01
Sc	21	44.959	6.231E+03	4.415E+02	1.055E+02	4.348E+01	3.228E+01	2.980E+01	2.614E+01	2.381E+01	2.305E+01	2.252E+01	2.221E+01
Ti	22	47.880	6.839E+03	4.854E+02	1.167E+02	4.854E+01	3.621E+01	3.348E+01	2.944E+01	2.687E+01	2.603E+01	2.544E+01	2.510E+01
V	23	50.942	7.476E+03	5.313E+02	1.283E+02	5.383E+01	4.032E+01	3.733E+01	3.290E+01	3.008E+01	2.915E+01	2.851E+01	2.813E+01
Cr	24	51.996	8.140E+03	5.787E+02	1.400E+02	5.887E+01	4.416E+01	4.090E+01	3.607E+01	3.299E+01	3.198E+01	3.127E+01	3.086E+01
Mn	25	54.938	8.833E+03	6.287E+02	1.526E+02	6.457E+01	4.858E+01	4.504E+01	3.979E+01	3.644E+01	3.533E+01	3.457E+01	3.412E+01
Fe	26	55.847	9.554E+03	6.802E+02	1.653E+02	7.005E+01	5.275E+01	4.892E+01	4.323E+01	3.960E+01	3.841E+01	3.758E+01	3.709E+01
Co	27	58.933	1.030E+04	7.342E+02	1.790E+02	7.623E+01	5.756E+01	5.341E+01	4.727E+01	4.335E+01	4.206E+01	4.116E+01	4.064E+01
Ni	28	58.690	1.108E+04	7.895E+02	1.924E+02	8.192E+01	6.185E+01	5.739E+01	5.079E+01	4.657E+01	4.518E+01	4.422E+01	4.365E+01
Cu	29	63.546	1.189E+04	8.479E+02	2.074E+02	8.891E+01	6.735E+01	6.256E+01	5.547E+01	5.093E+01	4.944E+01	4.840E+01	4.779E+01
Zn	30	65.390	1.272E+04	9.078E+02	2.224E+02	9.551E+01	7.242E+01	6.730E+01	5.970E+01	5.484E+01	5.324E+01	5.213E+01	5.148E+01
Ga	31	69.723	1.359E+04	9.701E+02	2.382E+02	1.028E+02	7.810E+01	7.262E+01	6.449E+01	5.930E+01	5.759E+01	5.640E+01	5.570E+01
Ge	32	72.610	1.448E+04	1.034E+03	2.543E+02	1.100E+02	8.368E+01	7.784E+01	6.918E+01	6.364E+01	6.182E+01	6.055E+01	5.980E+01
As	33	74.922	1.540E+04	1.100E+03	2.708E+02	1.173E+02	8.934E+01	8.313E+01	7.392E+01	6.802E+01	6.608E+01	6.474E+01	6.394E+01
Se	34	78.960	1.634E+04	1.168E+03	2.881E+02	1.251E+02	9.542E+01	8.883E+01	7.904E+01	7.278E+01	7.072E+01	6.928E+01	6.844E+01
Br	35	79.904	1.732E+04	1.238E+03	3.054E+02	1.327E+02	1.012E+02	9.426E+01	8.388E+01	7.725E+01	7.506E+01	7.355E+01	7.265E+01
Kr	36	83.800	1.832E+04	1.311E+03	3.237E+02	1.409E+02	1.076E+02	1.002E+02	8.926E+01	8.224E+01	7.993E+01	7.832E+01	7.737E+01
Rb	37	85.468	1.936E+04	1.385E+03	3.421E+02	1.491E+02	1.139E+02	1.061E+02	9.450E+01	8.708E+01	8.464E+01	8.294E+01	8.194E+01
Sr	38	87.620	2.042E+04	1.461E+03	3.611E+02	1.575E+02	1.204E+02	1.122E+02	9.995E+01	9.212E+01	8.954E+01	8.775E+01	8.669E+01
Y	39	88.906	2.150E+04	1.539E+03	3.806E+02	1.661E+02	1.270E+02	1.183E+02	1.054E+02	9.719E+01	9.447E+01	9.259E+01	9.147E+01
Zr	40	91.224	2.262E+04	1.619E+03	4.006E+02	1.750E+02	1.339E+02	1.248E+02	1.112E+02	1.025E+02	9.966E+01	9.768E+01	9.651E+01
Nb	41	92.906	2.377E+04	1.701E+03	4.211E+02	1.841E+02	1.409E+02	1.313E+02	1.170E+02	1.079E+02	1.049E+02	1.028E+02	1.016E+02
Mo	42	95.940	2.494E+04	1.786E+03	4.423E+02	1.935E+02	1.482E+02	1.381E+02	1.232E+02	1.136E+02	1.104E+02	1.083E+02	1.070E+02
Tc	43	98.000	2.614E+04	1.872E+03	4.638E+02	2.031E+02	1.556E+02	1.450E+02	1.293E+02	1.193E+02	1.160E+02	1.137E+02	1.124E+02
Ru	44	101.070	2.737E+04	1.960E+03	4.860E+02	2.130E+02	1.632E+02	1.522E+02	1.358E+02	1.253E+02	1.218E+02	1.194E+02	1.180E+02
Rh	45	102.906	2.863E+04	2.051E+03	5.085E+02	2.230E+02	1.709E+02	1.594E+02	1.422E+02	1.312E+02	1.276E+02	1.251E+02	1.236E+02
Pd	46	106.420	2.992E+04	2.143E+03	5.318E+02	2.334E+02	1.790E+02	1.669E+02	1.490E+02	1.375E+02	1.337E+02	1.311E+02	1.295E+02
Ag	47	107.868	3.123E+04	2.237E+03	5.553E+02	2.438E+02	1.870E+02	1.744E+02	1.556E+02	1.437E+02	1.397E+02	1.370E+02	1.353E+02
Cd	48	112.411	3.258E+04	2.335E+03	5.804E+02	2.555E+02	1.963E+02	1.831E+02	1.636E+02	1.511E+02	1.470E+02	1.441E+02	1.424E+02
In	49	114.820	3.395E+04	2.433E+03	6.042E+02	2.657E+02	2.039E+02	1.902E+02	1.699E+02	1.568E+02	1.525E+02	1.496E+02	1.478E+02
Sn	50	118.710	3.535E+04	2.533E+03	6.295E+02	2.770E+02	2.127E+02	1.984E+02	1.772E+02	1.636E+02	1.592E+02	1.561E+02	1.542E+02
Sb	51	121.750	3.678E+04	2.636E+03	6.552E+02	2.884E+02	2.216E+02	2.067E+02	1.846E+02	1.705E+02	1.659E+02	1.626E+02	1.607E+02
Te	52	127.600	3.824E+04	2.741E+03	6.816E+02	3.003E+02	2.308E+02	2.153E+02	1.924E+02	1.777E+02	1.729E+02	1.695E+02	1.675E+02

Table A6.7. Rutherford scattering cross sections (barns) of the elements for 1 MeV ^{15}N (continued).

Element	Atomic no. (Z_2)	Mass (M_2)	Scattering angle (degrees)										
			175	170	165	160	150	140	135	120	90	60	30
I	53	126.904	1.740E+02	1.761E+02	1.795E+02	1.846E+02	1.998E+02	2.236E+02	2.397E+02	3.119E+02	7.080E+02	2.847E+03	3.972E+04
Xe	54	131.290	1.810E+02	1.831E+02	1.867E+02	1.919E+02	2.078E+02	2.325E+02	2.492E+02	3.241E+02	7.353E+02	2.956E+03	4.123E+04
Cs	55	132.905	1.878E+02	1.901E+02	1.938E+02	1.992E+02	2.156E+02	2.413E+02	2.586E+02	3.364E+02	7.630E+02	3.067E+03	4.278E+04
Ba	56	137.327	1.950E+02	1.973E+02	2.012E+02	2.069E+02	2.239E+02	2.505E+02	2.684E+02	3.490E+02	7.913E+02	3.179E+03	4.434E+04
La	57	138.906	2.022E+02	2.046E+02	2.086E+02	2.144E+02	2.321E+02	2.596E+02	2.782E+02	3.617E+02	8.199E+02	3.294E+03	4.594E+04
Ce	58	140.115	2.094E+02	2.119E+02	2.161E+02	2.221E+02	2.404E+02	2.689E+02	2.881E+02	3.746E+02	8.490E+02	3.411E+03	4.757E+04
Pr	59	140.908	2.168E+02	2.193E+02	2.236E+02	2.299E+02	2.488E+02	2.783E+02	2.982E+02	3.877E+02	8.786E+02	3.529E+03	4.922E+04
Nd	60	144.240	2.244E+02	2.271E+02	2.315E+02	2.380E+02	2.575E+02	2.881E+02	3.086E+02	4.012E+02	9.089E+02	3.650E+03	5.091E+04
Pm	61	145.000	2.320E+02	2.347E+02	2.394E+02	2.460E+02	2.662E+02	2.978E+02	3.191E+02	4.147E+02	9.395E+02	3.773E+03	5.262E+04
Sm	62	150.360	2.400E+02	2.429E+02	2.476E+02	2.545E+02	2.754E+02	3.080E+02	3.300E+02	4.288E+02	9.709E+02	3.898E+03	5.436E+04
Eu	63	151.965	2.479E+02	2.509E+02	2.558E+02	2.629E+02	2.844E+02	3.181E+02	3.408E+02	4.429E+02	1.003E+03	4.025E+03	5.613E+04
Gd	64	157.250	2.562E+02	2.592E+02	2.643E+02	2.716E+02	2.939E+02	3.286E+02	3.521E+02	4.574E+02	1.035E+03	4.154E+03	5.792E+04
Tb	65	158.925	2.644E+02	2.675E+02	2.727E+02	2.803E+02	3.032E+02	3.391E+02	3.632E+02	4.719E+02	1.068E+03	4.285E+03	5.975E+04
Dy	66	162.500	2.728E+02	2.760E+02	2.814E+02	2.892E+02	3.129E+02	3.498E+02	3.747E+02	4.867E+02	1.101E+03	4.418E+03	6.160E+04
Ho	67	164.930	2.813E+02	2.846E+02	2.901E+02	2.982E+02	3.225E+02	3.606E+02	3.863E+02	5.017E+02	1.135E+03	4.553E+03	6.348E+04
Er	68	167.260	2.899E+02	2.932E+02	2.990E+02	3.073E+02	3.324E+02	3.716E+02	3.981E+02	5.169E+02	1.169E+03	4.690E+03	6.539E+04
Tm	69	168.934	2.985E+02	3.020E+02	3.079E+02	3.165E+02	3.423E+02	3.827E+02	4.099E+02	5.324E+02	1.204E+03	4.829E+03	6.733E+04
Yb	70	173.040	3.075E+02	3.111E+02	3.172E+02	3.259E+02	3.525E+02	3.941E+02	4.221E+02	5.481E+02	1.239E+03	4.971E+03	6.929E+04
Lu	71	174.967	3.164E+02	3.201E+02	3.264E+02	3.354E+02	3.628E+02	4.056E+02	4.344E+02	5.640E+02	1.275E+03	5.114E+03	7.129E+04
Hf	72	178.490	3.256E+02	3.294E+02	3.358E+02	3.451E+02	3.733E+02	4.173E+02	4.469E+02	5.802E+02	1.311E+03	5.259E+03	7.331E+04
Ta	73	180.948	3.348E+02	3.387E+02	3.454E+02	3.549E+02	3.838E+02	4.291E+02	4.595E+02	5.966E+02	1.348E+03	5.406E+03	7.536E+04
W	74	183.850	3.442E+02	3.482E+02	3.550E+02	3.648E+02	3.946E+02	4.411E+02	4.724E+02	6.132E+02	1.385E+03	5.555E+03	7.744E+04
Re	75	186.207	3.537E+02	3.578E+02	3.648E+02	3.749E+02	4.054E+02	4.532E+02	4.853E+02	6.300E+02	1.423E+03	5.707E+03	7.954E+04
Os	76	190.200	3.634E+02	3.676E+02	3.748E+02	3.851E+02	4.165E+02	4.655E+02	4.986E+02	6.471E+02	1.462E+03	5.860E+03	8.168E+04
Ir	77	192.220	3.731E+02	3.775E+02	3.848E+02	3.954E+02	4.276E+02	4.780E+02	5.119E+02	6.643E+02	1.500E+03	6.015E+03	8.384E+04
Pt	78	195.080	3.830E+02	3.875E+02	3.950E+02	4.059E+02	4.390E+02	4.906E+02	5.254E+02	6.818E+02	1.540E+03	6.173E+03	8.604E+04
Au	79	196.967	3.930E+02	3.976E+02	4.053E+02	4.165E+02	4.504E+02	5.033E+02	5.390E+02	6.995E+02	1.580E+03	6.332E+03	8.826E+04
Hg	80	200.590	4.032E+02	4.079E+02	4.158E+02	4.273E+02	4.620E+02	5.163E+02	5.529E+02	7.175E+02	1.620E+03	6.494E+03	9.051E+04
Tl	81	204.383	4.135E+02	4.183E+02	4.264E+02	4.382E+02	4.738E+02	5.295E+02	5.670E+02	7.357E+02	1.661E+03	6.657E+03	9.278E+04
Pb	82	207.200	4.239E+02	4.288E+02	4.372E+02	4.492E+02	4.857E+02	5.428E+02	5.812E+02	7.541E+02	1.702E+03	6.823E+03	9.509E+04
Bi	83	208.980	4.344E+02	4.394E+02	4.480E+02	4.603E+02	4.977E+02	5.562E+02	5.956E+02	7.727E+02	1.744E+03	6.990E+03	9.742E+04
Po	84	209.000	4.449E+02	4.501E+02	4.588E+02	4.714E+02	5.098E+02	5.697E+02	6.100E+02	7.914E+02	1.786E+03	7.160E+03	9.978E+04
At	85	210.000	4.556E+02	4.609E+02	4.699E+02	4.828E+02	5.220E+02	5.833E+02	6.247E+02	8.104E+02	1.829E+03	7.331E+03	1.022E+05
Rn	86	222.000	4.669E+02	4.723E+02	4.815E+02	4.947E+02	5.349E+02	5.977E+02	6.399E+02	8.301E+02	1.873E+03	7.505E+03	1.046E+05
Fr	87	223.000	4.778E+02	4.834E+02	4.928E+02	5.063E+02	5.474E+02	6.117E+02	6.549E+02	8.496E+02	1.917E+03	7.681E+03	1.070E+05
Ra	88	226.025	4.890E+02	4.947E+02	5.043E+02	5.181E+02	5.602E+02	6.259E+02	6.702E+02	8.693E+02	1.961E+03	7.859E+03	1.095E+05
Ac	89	227.028	5.002E+02	5.060E+02	5.159E+02	5.300E+02	5.731E+02	6.403E+02	6.856E+02	8.892E+02	2.006E+03	8.038E+03	1.120E+05
Th	90	232.038	5.117E+02	5.176E+02	5.277E+02	5.422E+02	5.862E+02	6.549E+02	7.012E+02	9.095E+02	2.052E+03	8.220E+03	1.145E+05
Pa	91	231.036	5.231E+02	5.292E+02	5.396E+02	5.543E+02	5.993E+02	6.695E+02	7.169E+02	9.298E+02	2.098E+03	8.404E+03	1.171E+05
U	92	238.029	5.349E+02	5.411E+02	5.516E+02	5.668E+02	6.128E+02	6.846E+02	7.330E+02	9.506E+02	2.144E+03	8.590E+03	1.197E+05
Np	93	237.048	5.466E+02	5.529E+02	5.637E+02	5.791E+02	6.261E+02	6.995E+02	7.490E+02	9.714E+02	2.191E+03	8.777E+03	1.223E+05
Pu	94	244.000	5.586E+02	5.651E+02	5.761E+02	5.919E+02	6.399E+02	7.149E+02	7.654E+02	9.926E+02	2.239E+03	8.967E+03	1.250E+05

Table A6.8. Rutherford scattering cross sections (barns) of the elements for 1 MeV ^{16}O.

Element	Atomic no. (Z_2)	Mass (M_2)	175	170	165	160	150	140	135	120	90	60	30
F	9	18.998	5.745E-01	5.907E-01	6.187E-01	6.605E-01	7.985E-01	1.049E+00	1.242E+00	2.372E+00	1.449E+01	9.786E+01	1.489E+03
Ne	10	20.180	1.155E+00	1.186E+00	1.238E+00	1.315E+00	1.568E+00	2.013E+00	2.347E+00	4.184E+00	2.021E+01	1.221E+02	1.839E+03
Na	11	22.990	2.690E+00	2.751E+00	2.856E+00	3.011E+00	3.502E+00	4.330E+00	4.928E+00	7.941E+00	2.882E+01	1.506E+02	2.227E+03
Mg	12	24.305	3.863E+00	3.947E+00	4.090E+00	4.299E+00	4.958E+00	6.056E+00	6.838E+00	1.069E+01	3.595E+01	1.804E+02	2.651E+03
Al	13	26.982	5.931E+00	6.047E+00	6.247E+00	6.537E+00	7.441E+00	8.919E+00	9.953E+00	1.489E+01	4.512E+01	2.141E+02	3.113E+03
Si	14	28.086	7.463E+00	7.605E+00	7.848E+00	8.200E+00	9.295E+00	1.107E+01	1.231E+01	1.817E+01	5.342E+01	2.492E+02	3.611E+03
P	15	30.974	1.009E+01	1.027E+01	1.057E+01	1.101E+01	1.237E+01	1.456E+01	1.606E+01	2.307E+01	6.388E+01	2.883E+02	4.147E+03
S	16	32.066	1.204E+01	1.225E+01	1.260E+01	1.312E+01	1.470E+01	1.723E+01	1.896E+01	2.702E+01	7.356E+01	3.288E+02	4.719E+03
Cl	17	35.453	1.528E+01	1.552E+01	1.595E+01	1.655E+01	1.842E+01	2.139E+01	2.342E+01	3.272E+01	8.551E+01	3.734E+02	5.329E+03
Ar	18	39.948	1.903E+01	1.932E+01	1.981E+01	2.052E+01	2.269E+01	2.612E+01	2.845E+01	3.905E+01	9.843E+01	4.210E+02	5.977E+03
K	19	39.098	2.085E+01	2.117E+01	2.172E+01	2.250E+01	2.491E+01	2.871E+01	3.129E+01	4.308E+01	1.092E+02	4.686E+02	6.659E+03
Ca	20	40.078	2.355E+01	2.391E+01	2.452E+01	2.539E+01	2.807E+01	3.231E+01	3.518E+01	4.828E+01	1.216E+02	5.198E+02	7.379E+03
Sc	21	44.959	2.802E+01	2.842E+01	2.911E+01	3.010E+01	3.312E+01	3.786E+01	4.108E+01	5.566E+01	1.366E+02	5.755E+02	8.137E+03
Ti	22	47.880	3.181E+01	3.226E+01	3.302E+01	3.411E+01	3.745E+01	4.270E+01	4.625E+01	6.231E+01	1.512E+02	6.329E+02	8.931E+03
V	23	50.942	3.579E+01	3.628E+01	3.712E+01	3.833E+01	4.200E+01	4.776E+01	5.165E+01	6.925E+01	1.665E+02	6.929E+02	9.763E+03
Cr	24	51.996	3.931E+01	3.985E+01	4.076E+01	4.208E+01	4.608E+01	5.237E+01	5.661E+01	7.579E+01	1.817E+02	7.548E+02	1.063E+04
Mn	25	54.938	4.359E+01	4.417E+01	4.517E+01	4.660E+01	5.097E+01	5.780E+01	6.242E+01	8.327E+01	1.982E+02	8.200E+02	1.154E+04
Fe	26	55.847	4.743E+01	4.806E+01	4.914E+01	5.069E+01	5.541E+01	6.282E+01	6.782E+01	9.037E+01	2.147E+02	8.873E+02	1.248E+04
Co	27	58.933	5.208E+01	5.276E+01	5.393E+01	5.561E+01	6.072E+01	6.872E+01	7.412E+01	9.848E+01	2.326E+02	9.578E+02	1.346E+04
Ni	28	58.690	5.593E+01	5.667E+01	5.793E+01	5.973E+01	6.522E+01	7.383E+01	7.964E+01	1.058E+02	2.501E+02	1.030E+03	1.447E+04
Cu	29	63.546	6.142E+01	6.221E+01	6.356E+01	6.551E+01	7.142E+01	8.068E+01	8.692E+01	1.151E+02	2.699E+02	1.106E+03	1.553E+04
Zn	30	65.390	6.622E+01	6.707E+01	6.852E+01	7.060E+01	7.694E+01	8.685E+01	9.353E+01	1.237E+02	2.893E+02	1.185E+03	1.662E+04
Ga	31	69.723	7.179E+01	7.270E+01	7.425E+01	7.648E+01	8.326E+01	9.386E+01	1.010E+02	1.332E+02	3.101E+02	1.266E+03	1.774E+04
Ge	32	72.610	7.716E+01	7.813E+01	7.979E+01	8.216E+01	8.939E+01	1.007E+02	1.083E+02	1.426E+02	3.312E+02	1.350E+03	1.891E+04
As	33	74.922	8.257E+01	8.360E+01	8.536E+01	8.789E+01	9.558E+01	1.076E+02	1.157E+02	1.522E+02	3.527E+02	1.436E+03	2.011E+04
Se	34	78.960	8.846E+01	8.958E+01	9.145E+01	9.414E+01	1.023E+02	1.151E+02	1.237E+02	1.625E+02	3.753E+02	1.525E+03	2.135E+04
Br	35	79.904	9.395E+01	9.511E+01	9.710E+01	9.995E+01	1.086E+02	1.221E+02	1.313E+02	1.724E+02	3.979E+02	1.616E+03	2.262E+04
Kr	36	83.800	1.001E+02	1.014E+02	1.035E+02	1.065E+02	1.157E+02	1.300E+02	1.397E+02	1.831E+02	4.218E+02	1.711E+03	2.393E+04
Rb	37	85.468	1.061E+02	1.074E+02	1.096E+02	1.128E+02	1.225E+02	1.376E+02	1.478E+02	1.938E+02	4.459E+02	1.808E+03	2.528E+04
Sr	38	87.620	1.123E+02	1.137E+02	1.160E+02	1.194E+02	1.296E+02	1.456E+02	1.563E+02	2.048E+02	4.707E+02	1.907E+03	2.666E+04
Y	39	88.906	1.185E+02	1.200E+02	1.224E+02	1.260E+02	1.367E+02	1.536E+02	1.649E+02	2.159E+02	4.961E+02	2.009E+03	2.809E+04
Zr	40	91.224	1.251E+02	1.266E+02	1.292E+02	1.329E+02	1.443E+02	1.620E+02	1.739E+02	2.276E+02	5.223E+02	2.114E+03	2.955E+04
Nb	41	92.906	1.317E+02	1.333E+02	1.361E+02	1.400E+02	1.519E+02	1.705E+02	1.830E+02	2.394E+02	5.490E+02	2.221E+03	3.104E+04
Mo	42	95.940	1.388E+02	1.404E+02	1.433E+02	1.474E+02	1.599E+02	1.794E+02	1.926E+02	2.518E+02	5.767E+02	2.331E+03	3.258E+04
Tc	43	98.000	1.458E+02	1.476E+02	1.505E+02	1.549E+02	1.680E+02	1.884E+02	2.022E+02	2.643E+02	6.048E+02	2.444E+03	3.415E+04
Ru	44	101.070	1.531E+02	1.550E+02	1.581E+02	1.626E+02	1.764E+02	1.978E+02	2.122E+02	2.772E+02	6.338E+02	2.559E+03	3.575E+04
Rh	45	102.906	1.605E+02	1.624E+02	1.657E+02	1.704E+02	1.848E+02	2.072E+02	2.223E+02	2.903E+02	6.632E+02	2.677E+03	3.740E+04
Pd	46	106.420	1.682E+02	1.702E+02	1.737E+02	1.786E+02	1.936E+02	2.170E+02	2.328E+02	3.039E+02	6.936E+02	2.798E+03	3.908E+04
Ag	47	107.868	1.758E+02	1.779E+02	1.815E+02	1.867E+02	2.023E+02	2.268E+02	2.433E+02	3.174E+02	7.243E+02	2.921E+03	4.080E+04
Cd	48	112.411	1.852E+02	1.874E+02	1.912E+02	1.966E+02	2.129E+02	2.384E+02	2.556E+02	3.330E+02	7.573E+02	3.049E+03	4.255E+04
In	49	114.820	1.921E+02	1.944E+02	1.983E+02	2.039E+02	2.209E+02	2.475E+02	2.655E+02	3.461E+02	7.883E+02	3.176E+03	4.434E+04
Sn	50	118.710	2.006E+02	2.029E+02	2.070E+02	2.128E+02	2.306E+02	2.583E+02	2.769E+02	3.609E+02	8.213E+02	3.308E+03	4.617E+04
Sb	51	121.750	2.090E+02	2.115E+02	2.157E+02	2.218E+02	2.403E+02	2.691E+02	2.885E+02	3.758E+02	8.549E+02	3.442E+03	4.804E+04
Te	52	127.600	2.180E+02	2.206E+02	2.250E+02	2.313E+02	2.505E+02	2.804E+02	3.006E+02	3.914E+02	8.894E+02	3.579E+03	4.994E+04
I	53	126.904	2.264E+02	2.291E+02	2.336E+02	2.402E+02	2.601E+02	2.912E+02	3.122E+02	4.065E+02	9.239E+02	3.718E+03	5.188E+04

Scattering angle (degrees)

Table A6.8. Rutherford scattering cross sections (barns) of the elements for 1 MeV ^{16}O (continued).

Element	Atomic no. (Z_2)	Mass (M_2)	Scattering angle (degrees)										
			175	170	165	160	150	140	135	120	90	60	30
Xe	54	131.290	2.355E+02	2.383E+02	2.430E+02	2.498E+02	2.705E+02	3.028E+02	3.246E+02	4.225E+02	9.596E+02	3.860E+03	5.386E+04
Cs	55	132.905	2.445E+02	2.474E+02	2.523E+02	2.593E+02	2.808E+02	3.143E+02	3.369E+02	4.385E+02	9.956E+02	4.004E+03	5.587E+04
Ba	56	137.327	2.539E+02	2.569E+02	2.620E+02	2.693E+02	2.916E+02	3.263E+02	3.497E+02	4.550E+02	1.033E+03	4.152E+03	5.792E+04
La	57	138.906	2.632E+02	2.663E+02	2.716E+02	2.792E+02	3.022E+02	3.383E+02	3.625E+02	4.716E+02	1.070E+03	4.302E+03	6.001E+04
Ce	58	140.115	2.727E+02	2.759E+02	2.814E+02	2.892E+02	3.131E+02	3.503E+02	3.755E+02	4.884E+02	1.108E+03	4.454E+03	6.213E+04
Pr	59	140.908	2.822E+02	2.856E+02	2.912E+02	2.994E+02	3.240E+02	3.626E+02	3.886E+02	5.055E+02	1.147E+03	4.609E+03	6.429E+04
Nd	60	144.240	2.922E+02	2.957E+02	3.015E+02	3.099E+02	3.355E+02	3.754E+02	4.022E+02	5.231E+02	1.186E+03	4.767E+03	6.649E+04
Pm	61	145.000	3.021E+02	3.057E+02	3.117E+02	3.204E+02	3.468E+02	3.881E+02	4.158E+02	5.408E+02	1.226E+03	4.927E+03	6.872E+04
Sm	62	150.360	3.127E+02	3.163E+02	3.226E+02	3.316E+02	3.588E+02	4.014E+02	4.301E+02	5.592E+02	1.267E+03	5.091E+03	7.100E+04
Eu	63	151.965	3.230E+02	3.268E+02	3.332E+02	3.425E+02	3.706E+02	4.146E+02	4.443E+02	5.776E+02	1.309E+03	5.256E+03	7.331E+04
Gd	64	157.250	3.338E+02	3.377E+02	3.444E+02	3.539E+02	3.830E+02	4.284E+02	4.590E+02	5.965E+02	1.351E+03	5.425E+03	7.565E+04
Tb	65	158.925	3.445E+02	3.485E+02	3.554E+02	3.652E+02	3.952E+02	4.420E+02	4.736E+02	6.155E+02	1.394E+03	5.596E+03	7.803E+04
Dy	66	162.500	3.555E+02	3.596E+02	3.667E+02	3.769E+02	4.078E+02	4.561E+02	4.886E+02	6.349E+02	1.437E+03	5.770E+03	8.045E+04
Ho	67	164.930	3.665E+02	3.708E+02	3.781E+02	3.886E+02	4.204E+02	4.702E+02	5.037E+02	6.545E+02	1.481E+03	5.946E+03	8.291E+04
Er	68	167.260	3.778E+02	3.822E+02	3.897E+02	4.005E+02	4.333E+02	4.845E+02	5.191E+02	6.743E+02	1.526E+03	6.125E+03	8.540E+04
Tm	69	168.934	3.891E+02	3.936E+02	4.014E+02	4.125E+02	4.463E+02	4.990E+02	5.346E+02	6.945E+02	1.571E+03	6.307E+03	8.793E+04
Yb	70	173.040	4.008E+02	4.055E+02	4.134E+02	4.249E+02	4.596E+02	5.139E+02	5.505E+02	7.151E+02	1.618E+03	6.491E+03	9.050E+04
Lu	71	174.967	4.125E+02	4.173E+02	4.255E+02	4.373E+02	4.730E+02	5.289E+02	5.665E+02	7.358E+02	1.664E+03	6.678E+03	9.311E+04
Hf	72	178.490	4.244E+02	4.294E+02	4.378E+02	4.499E+02	4.867E+02	5.442E+02	5.829E+02	7.570E+02	1.712E+03	6.868E+03	9.575E+04
Ta	73	180.948	4.365E+02	4.416E+02	4.503E+02	4.627E+02	5.005E+02	5.596E+02	5.994E+02	7.783E+02	1.760E+03	7.060E+03	9.843E+04
W	74	183.850	4.488E+02	4.540E+02	4.629E+02	4.757E+02	5.145E+02	5.752E+02	6.161E+02	8.000E+02	1.809E+03	7.255E+03	1.011E+05
Re	75	186.207	4.612E+02	4.665E+02	4.757E+02	4.888E+02	5.287E+02	5.911E+02	6.331E+02	8.220E+02	1.858E+03	7.453E+03	1.039E+05
Os	76	190.200	4.738E+02	4.794E+02	4.887E+02	5.022E+02	5.432E+02	6.072E+02	6.504E+02	8.443E+02	1.908E+03	7.653E+03	1.067E+05
Ir	77	192.220	4.865E+02	4.922E+02	5.018E+02	5.157E+02	5.577E+02	6.235E+02	6.677E+02	8.669E+02	1.959E+03	7.856E+03	1.095E+05
Pt	78	195.080	4.994E+02	5.053E+02	5.152E+02	5.294E+02	5.725E+02	6.400E+02	6.854E+02	8.897E+02	2.010E+03	8.062E+03	1.124E+05
Au	79	196.967	5.125E+02	5.184E+02	5.286E+02	5.432E+02	5.874E+02	6.566E+02	7.032E+02	9.128E+02	2.062E+03	8.270E+03	1.153E+05
Hg	80	200.590	5.258E+02	5.319E+02	5.423E+02	5.572E+02	6.026E+02	6.736E+02	7.214E+02	9.363E+02	2.115E+03	8.481E+03	1.182E+05
Tl	81	204.383	5.393E+02	5.455E+02	5.562E+02	5.715E+02	6.181E+02	6.908E+02	7.398E+02	9.601E+02	2.169E+03	8.694E+03	1.212E+05
Pb	82	207.200	5.528E+02	5.593E+02	5.702E+02	5.859E+02	6.336E+02	7.081E+02	7.583E+02	9.842E+02	2.223E+03	8.911E+03	1.242E+05
Bi	83	208.980	5.665E+02	5.731E+02	5.843E+02	6.004E+02	6.493E+02	7.256E+02	7.771E+02	1.008E+03	2.277E+03	9.129E+03	1.272E+05
Po	84	209.000	5.802E+02	5.870E+02	5.985E+02	6.149E+02	6.650E+02	7.432E+02	7.959E+02	1.033E+03	2.333E+03	9.351E+03	1.303E+05
At	85	210.000	5.942E+02	6.011E+02	6.129E+02	6.297E+02	6.810E+02	7.611E+02	8.150E+02	1.058E+03	2.388E+03	9.575E+03	1.334E+05
Rn	86	222.000	6.090E+02	6.161E+02	6.281E+02	6.454E+02	6.979E+02	7.798E+02	8.351E+02	1.083E+03	2.446E+03	9.802E+03	1.366E+05
Fr	87	223.000	6.233E+02	6.306E+02	6.429E+02	6.605E+02	7.142E+02	7.981E+02	8.547E+02	1.109E+03	2.503E+03	1.003E+04	1.398E+05
Ra	88	226.025	6.379E+02	6.453E+02	6.579E+02	6.760E+02	7.309E+02	8.168E+02	8.746E+02	1.135E+03	2.561E+03	1.026E+04	1.430E+05
Ac	89	227.028	6.525E+02	6.601E+02	6.730E+02	6.915E+02	7.477E+02	8.355E+02	8.946E+02	1.161E+03	2.620E+03	1.050E+04	1.463E+05
Th	90	232.038	6.676E+02	6.753E+02	6.885E+02	7.074E+02	7.649E+02	8.547E+02	9.151E+02	1.187E+03	2.679E+03	1.074E+04	1.496E+05
Pa	91	231.036	6.824E+02	6.904E+02	7.038E+02	7.232E+02	7.819E+02	8.737E+02	9.355E+02	1.214E+03	2.739E+03	1.098E+04	1.530E+05
U	92	238.029	6.979E+02	7.060E+02	7.198E+02	7.395E+02	7.996E+02	8.934E+02	9.566E+02	1.241E+03	2.800E+03	1.122E+04	1.563E+05
Np	93	237.048	7.131E+02	7.214E+02	7.354E+02	7.556E+02	8.170E+02	9.129E+02	9.775E+02	1.268E+03	2.861E+03	1.146E+04	1.598E+05
Pu	94	244.000	7.289E+02	7.374E+02	7.517E+02	7.724E+02	8.350E+02	9.330E+02	9.990E+02	1.296E+03	2.923E+03	1.171E+04	1.632E+05

Table A6.9. Rutherford scattering cross sections (barns) of the elements for 1 MeV ^{28}Si.

Element	Atomic no. (Z_2)	Mass (M_2)	Scattering angle (degrees)										
			175	170	165	160	150	140	135	120	90	60	30
P	15	30.974	1.956E+00	2.016E+00	2.121E+00	2.278E+00	2.809E+00	3.810E+00	4.618E+00	9.861E+00	9.804E+01	8.221E+02	1.267E+04
S	16	32.066	3.741E+00	3.851E+00	4.042E+00	4.328E+00	5.282E+00	7.045E+00	8.435E+00	1.696E+01	1.270E+02	9.412E+02	1.441E+04
Cl	17	35.453	1.053E+01	1.081E+01	1.128E+01	1.198E+01	1.427E+01	1.829E+01	2.131E+01	3.781E+01	1.802E+02	1.081E+03	1.628E+04
Ar	18	39.948	2.151E+01	2.201E+01	2.285E+01	2.410E+01	2.806E+01	3.475E+01	3.959E+01	6.405E+01	2.348E+02	1.234E+03	1.826E+04
K	19	39.098	2.199E+01	2.250E+01	2.339E+01	2.470E+01	2.885E+01	3.593E+01	4.107E+01	6.737E+01	2.561E+02	1.371E+03	2.034E+04
Ca	20	40.078	2.689E+01	2.751E+01	2.856E+01	3.011E+01	3.504E+01	4.337E+01	4.938E+01	7.973E+01	2.908E+02	1.524E+03	2.255E+04
Sc	21	44.959	4.232E+01	4.319E+01	4.467E+01	4.684E+01	5.362E+01	6.480E+01	7.267E+01	1.108E+02	3.505E+02	1.703E+03	2.487E+04
Ti	22	47.880	5.363E+01	5.467E+01	5.645E+01	5.904E+01	6.711E+01	8.026E+01	8.943E+01	1.331E+02	3.988E+02	1.880E+03	2.731E+04
V	23	50.942	6.590E+01	6.711E+01	6.919E+01	7.221E+01	8.157E+01	9.670E+01	1.072E+02	1.565E+02	4.489E+02	2.066E+03	2.985E+04
Cr	24	51.996	7.425E+01	7.560E+01	7.791E+01	8.126E+01	9.168E+01	1.083E+02	1.199E+02	1.740E+02	4.930E+02	2.253E+03	3.251E+04
Mn	25	54.938	8.754E+01	8.907E+01	9.168E+01	9.546E+01	1.071E+02	1.259E+02	1.387E+02	1.986E+02	5.461E+02	2.455E+03	3.528E+04
Fe	26	55.847	9.683E+01	9.850E+01	1.014E+02	1.055E+02	1.182E+02	1.387E+02	1.527E+02	2.178E+02	5.940E+02	2.658E+03	3.816E+04
Co	27	58.933	1.117E+02	1.135E+02	1.167E+02	1.213E+02	1.354E+02	1.580E+02	1.734E+02	2.445E+02	6.515E+02	2.876E+03	4.116E+04
Ni	28	58.690	1.195E+02	1.215E+02	1.249E+02	1.298E+02	1.450E+02	1.692E+02	1.858E+02	2.622E+02	6.998E+02	3.093E+03	4.427E+04
Cu	29	63.546	1.395E+02	1.417E+02	1.455E+02	1.510E+02	1.678E+02	1.944E+02	2.126E+02	2.959E+02	7.667E+02	3.332E+03	4.750E+04
Zn	30	65.390	1.533E+02	1.557E+02	1.598E+02	1.657E+02	1.838E+02	2.126E+02	2.321E+02	3.216E+02	8.260E+02	3.571E+03	5.083E+04
Ga	31	69.723	1.726E+02	1.752E+02	1.797E+02	1.861E+02	2.058E+02	2.369E+02	2.581E+02	3.544E+02	8.938E+02	3.824E+03	5.429E+04
Ge	32	72.610	1.894E+02	1.922E+02	1.970E+02	2.040E+02	2.251E+02	2.585E+02	2.812E+02	3.843E+02	9.594E+02	4.081E+03	5.785E+04
As	33	74.922	2.057E+02	2.087E+02	2.139E+02	2.213E+02	2.440E+02	2.796E+02	3.038E+02	4.137E+02	1.026E+03	4.345E+03	6.153E+04
Se	34	78.960	2.255E+02	2.287E+02	2.342E+02	2.422E+02	2.664E+02	3.045E+02	3.304E+02	4.474E+02	1.098E+03	4.621E+03	6.532E+04
Br	35	79.904	2.405E+02	2.440E+02	2.498E+02	2.583E+02	2.840E+02	3.245E+02	3.519E+02	4.760E+02	1.165E+03	4.898E+03	6.922E+04
Kr	36	83.800	2.609E+02	2.646E+02	2.708E+02	2.798E+02	3.072E+02	3.502E+02	3.793E+02	5.110E+02	1.240E+03	5.190E+03	7.324E+04
Rb	37	85.468	2.783E+02	2.822E+02	2.888E+02	2.983E+02	3.273E+02	3.728E+02	4.036E+02	5.428E+02	1.313E+03	5.485E+03	7.737E+04
Sr	38	87.620	2.970E+02	3.011E+02	3.081E+02	3.181E+02	3.488E+02	3.969E+02	4.294E+02	5.764E+02	1.389E+03	5.790E+03	8.161E+04
Y	39	88.906	3.148E+02	3.192E+02	3.266E+02	3.372E+02	3.695E+02	4.202E+02	4.545E+02	6.095E+02	1.466E+03	6.101E+03	8.597E+04
Zr	40	91.224	3.348E+02	3.394E+02	3.472E+02	3.584E+02	3.925E+02	4.459E+02	4.820E+02	6.452E+02	1.546E+03	6.421E+03	9.044E+04
Nb	41	92.906	3.544E+02	3.592E+02	3.674E+02	3.792E+02	4.150E+02	4.713E+02	5.092E+02	6.808E+02	1.628E+03	6.749E+03	9.502E+04
Mo	42	95.940	3.765E+02	3.816E+02	3.902E+02	4.026E+02	4.403E+02	4.994E+02	5.393E+02	7.195E+02	1.713E+03	7.088E+03	9.971E+04
Tc	43	98.000	3.977E+02	4.030E+02	4.121E+02	4.251E+02	4.647E+02	5.267E+02	5.685E+02	7.575E+02	1.799E+03	7.433E+03	1.045E+05
Ru	44	101.070	4.209E+02	4.264E+02	4.359E+02	4.496E+02	4.911E+02	5.561E+02	6.000E+02	7.980E+02	1.889E+03	7.787E+03	1.094E+05
Rh	45	102.906	4.428E+02	4.486E+02	4.586E+02	4.728E+02	5.163E+02	5.844E+02	6.303E+02	8.375E+02	1.979E+03	8.148E+03	1.145E+05
Pd	46	106.420	4.675E+02	4.736E+02	4.840E+02	4.989E+02	5.444E+02	6.156E+02	6.637E+02	8.803E+02	2.073E+03	8.519E+03	1.196E+05
Ag	47	107.868	4.899E+02	4.963E+02	5.072E+02	5.228E+02	5.703E+02	6.447E+02	6.949E+02	9.211E+02	2.166E+03	8.896E+03	1.249E+05
Cd	48	112.411	5.276E+02	5.343E+02	5.456E+02	5.620E+02	6.118E+02	6.896E+02	7.421E+02	9.784E+02	2.277E+03	9.296E+03	1.303E+05
In	49	114.820	5.416E+02	5.485E+02	5.604E+02	5.774E+02	6.291E+02	7.101E+02	7.647E+02	1.011E+03	2.364E+03	9.679E+03	1.357E+05
Sn	50	118.710	5.685E+02	5.757E+02	5.881E+02	6.058E+02	6.597E+02	7.441E+02	8.011E+02	1.057E+03	2.467E+03	1.008E+04	1.414E+05
Sb	51	121.750	5.949E+02	6.024E+02	6.153E+02	6.337E+02	6.899E+02	7.777E+02	8.370E+02	1.104E+03	2.570E+03	1.049E+04	1.471E+05
Te	52	127.600	6.246E+02	6.325E+02	6.458E+02	6.650E+02	7.235E+02	8.149E+02	8.765E+02	1.154E+03	2.679E+03	1.092E+04	1.529E+05
I	53	126.904	6.481E+02	6.563E+02	6.702E+02	6.902E+02	7.509E+02	8.458E+02	9.098E+02	1.198E+03	2.782E+03	1.134E+04	1.588E+05
Xe	54	131.290	6.773E+02	6.858E+02	7.000E+02	7.210E+02	7.840E+02	8.827E+02	9.492E+02	1.248E+03	2.893E+03	1.178E+04	1.649E+05
Cs	55	132.905	7.042E+02	7.131E+02	7.280E+02	7.496E+02	8.150E+02	9.173E+02	9.863E+02	1.297E+03	3.003E+03	1.222E+04	1.711E+05
Ba	56	137.327	7.344E+02	7.435E+02	7.590E+02	7.814E+02	8.492E+02	9.554E+02	1.027E+03	1.349E+03	3.117E+03	1.267E+04	1.773E+05
La	57	138.906	7.623E+02	7.718E+02	7.879E+02	8.110E+02	8.813E+02	9.913E+02	1.065E+03	1.399E+03	3.231E+03	1.313E+04	1.837E+05
Ce	58	140.115	7.904E+02	8.002E+02	8.169E+02	8.409E+02	9.137E+02	1.028E+03	1.104E+03	1.450E+03	3.347E+03	1.359E+04	1.902E+05
Pr	59	140.908	8.187E+02	8.288E+02	8.461E+02	8.709E+02	9.463E+02	1.064E+03	1.144E+03	1.501E+03	3.464E+03	1.407E+04	1.968E+05

Table A6.9. Rutherford scattering cross sections (barns) of the elements for 1 MeV ^{28}Si (continued).

Element	Atomic no. (Z_2)	Mass (M_2)	175	170	165	160	150	140	135	120	90	60	30
Nd	60	144.240	8.498E+02	8.603E+02	8.782E+02	9.038E+02	9.818E+02	1.104E+03	1.186E+03	1.556E+03	3.586E+03	1.455E+04	2.036E+05
Pm	61	145.000	8.791E+02	8.900E+02	9.084E+02	9.349E+02	1.016E+03	1.142E+03	1.226E+03	1.609E+03	3.707E+03	1.504E+04	2.104E+05
Sm	62	150.360	9.131E+02	9.243E+02	9.434E+02	9.708E+02	1.054E+03	1.184E+03	1.272E+03	1.667E+03	3.835E+03	1.554E+04	2.174E+05
Eu	63	151.965	9.442E+02	9.558E+02	9.755E+02	1.004E+03	1.090E+03	1.224E+03	1.315E+03	1.723E+03	3.961E+03	1.605E+04	2.245E+05
Gd	64	157.250	9.789E+02	9.909E+02	1.011E+03	1.040E+03	1.129E+03	1.268E+03	1.362E+03	1.782E+03	4.093E+03	1.657E+04	2.316E+05
Tb	65	158.925	1.011E+03	1.023E+03	1.044E+03	1.075E+03	1.166E+03	1.309E+03	1.406E+03	1.840E+03	4.223E+03	1.709E+04	2.389E+05
Dy	66	162.500	1.045E+03	1.058E+03	1.080E+03	1.111E+03	1.205E+03	1.353E+03	1.452E+03	1.900E+03	4.357E+03	1.763E+04	2.463E+05
Ho	67	164.930	1.079E+03	1.092E+03	1.115E+03	1.147E+03	1.244E+03	1.396E+03	1.499E+03	1.960E+03	4.492E+03	1.817E+04	2.539E+05
Er	68	167.260	1.113E+03	1.127E+03	1.150E+03	1.183E+03	1.283E+03	1.440E+03	1.546E+03	2.021E+03	4.629E+03	1.871E+04	2.615E+05
Tm	69	168.934	1.148E+03	1.162E+03	1.185E+03	1.219E+03	1.323E+03	1.484E+03	1.593E+03	2.082E+03	4.767E+03	1.927E+04	2.693E+05
Yb	70	173.040	1.184E+03	1.199E+03	1.223E+03	1.258E+03	1.364E+03	1.530E+03	1.642E+03	2.146E+03	4.910E+03	1.984E+04	2.771E+05
Lu	71	174.967	1.220E+03	1.235E+03	1.260E+03	1.296E+03	1.405E+03	1.576E+03	1.691E+03	2.209E+03	5.053E+03	2.041E+04	2.851E+05
Hf	72	178.490	1.257E+03	1.272E+03	1.298E+03	1.335E+03	1.447E+03	1.623E+03	1.742E+03	2.275E+03	5.199E+03	2.099E+04	2.932E+05
Ta	73	180.948	1.294E+03	1.310E+03	1.336E+03	1.374E+03	1.490E+03	1.670E+03	1.792E+03	2.340E+03	5.346E+03	2.158E+04	3.014E+05
W	74	183.850	1.332E+03	1.348E+03	1.375E+03	1.414E+03	1.533E+03	1.719E+03	1.844E+03	2.407E+03	5.495E+03	2.218E+04	3.097E+05
Re	75	186.207	1.370E+03	1.386E+03	1.414E+03	1.454E+03	1.576E+03	1.767E+03	1.896E+03	2.474E+03	5.647E+03	2.278E+04	3.181E+05
Os	76	190.200	1.409E+03	1.426E+03	1.455E+03	1.496E+03	1.621E+03	1.817E+03	1.949E+03	2.543E+03	5.801E+03	2.340E+04	3.267E+05
Ir	77	192.220	1.448E+03	1.465E+03	1.494E+03	1.537E+03	1.666E+03	1.867E+03	2.002E+03	2.612E+03	5.956E+03	2.402E+04	3.353E+05
Pt	78	195.080	1.487E+03	1.505E+03	1.535E+03	1.579E+03	1.711E+03	1.917E+03	2.056E+03	2.682E+03	6.114E+03	2.465E+04	3.441E+05
Au	79	196.967	1.527E+03	1.545E+03	1.576E+03	1.621E+03	1.756E+03	1.968E+03	2.111E+03	2.752E+03	6.273E+03	2.528E+04	3.530E+05
Hg	80	200.590	1.568E+03	1.587E+03	1.619E+03	1.664E+03	1.803E+03	2.021E+03	2.167E+03	2.825E+03	6.435E+03	2.593E+04	3.620E+05
Tl	81	204.383	1.610E+03	1.629E+03	1.662E+03	1.709E+03	1.851E+03	2.074E+03	2.224E+03	2.898E+03	6.599E+03	2.658E+04	3.711E+05
Pb	82	207.200	1.652E+03	1.671E+03	1.705E+03	1.753E+03	1.899E+03	2.127E+03	2.281E+03	2.972E+03	6.765E+03	2.725E+04	3.803E+05
Bi	83	208.980	1.693E+03	1.713E+03	1.748E+03	1.797E+03	1.946E+03	2.180E+03	2.338E+03	3.046E+03	6.932E+03	2.792E+04	3.896E+05
Po	84	209.000	1.734E+03	1.755E+03	1.790E+03	1.840E+03	1.994E+03	2.233E+03	2.394E+03	3.120E+03	7.100E+03	2.859E+04	3.991E+05
At	85	210.000	1.776E+03	1.798E+03	1.834E+03	1.885E+03	2.042E+03	2.287E+03	2.452E+03	3.195E+03	7.271E+03	2.928E+04	4.086E+05
Rn	86	222.000	1.825E+03	1.847E+03	1.884E+03	1.937E+03	2.097E+03	2.348E+03	2.517E+03	3.278E+03	7.450E+03	2.998E+04	4.183E+05
Fr	87	223.000	1.869E+03	1.891E+03	1.928E+03	1.983E+03	2.147E+03	2.404E+03	2.577E+03	3.355E+03	7.625E+03	3.068E+04	4.281E+05
Ra	88	226.025	1.913E+03	1.936E+03	1.975E+03	2.030E+03	2.198E+03	2.461E+03	2.638E+03	3.434E+03	7.802E+03	3.139E+04	4.380E+05
Ac	89	227.028	1.958E+03	1.981E+03	2.020E+03	2.077E+03	2.249E+03	2.518E+03	2.699E+03	3.513E+03	7.981E+03	3.211E+04	4.480E+05
Th	90	232.038	2.005E+03	2.028E+03	2.069E+03	2.127E+03	2.302E+03	2.577E+03	2.762E+03	3.595E+03	8.164E+03	3.284E+04	4.581E+05
Pa	91	231.036	2.049E+03	2.073E+03	2.114E+03	2.174E+03	2.353E+03	2.634E+03	2.824E+03	3.675E+03	8.346E+03	3.357E+04	4.684E+05
U	92	238.029	2.098E+03	2.123E+03	2.165E+03	2.225E+03	2.409E+03	2.696E+03	2.890E+03	3.760E+03	8.534E+03	3.432E+04	4.787E+05
Np	93	237.048	2.143E+03	2.168E+03	2.211E+03	2.273E+03	2.461E+03	2.755E+03	2.952E+03	3.842E+03	8.720E+03	3.507E+04	4.892E+05
Pu	94	244.000	2.193E+03	2.219E+03	2.263E+03	2.326E+03	2.518E+03	2.818E+03	3.020E+03	3.928E+03	8.913E+03	3.583E+04	4.998E+05

Scattering angle (degrees)

Table A6.10. *F*-Factors. E is the lab energy of the incident ion divided by its atomic number. Z_1 and Z_2 are the incident and target atom's atomic numbers. The factors are only valid for scattering angles greater than 90°.

E/Z_1 (MeV)	Z_2										
	8	14	20	29	32	39	47	56	73	79	92
0.100	0.992	0.983	0.973	0.956	0.950	0.935	0.917	0.895	0.851	0.834	0.796
0.200	0.996	0.992	0.987	0.978	0.975	0.968	0.958	0.948	0.925	0.917	0.898
0.300	0.997	0.994	0.991	0.985	0.983	0.978	0.972	0.965	0.950	0.945	0.932
0.400	0.998	0.996	0.993	0.989	0.988	0.984	0.979	0.974	0.963	0.958	0.949
0.500	0.998	0.997	0.995	0.991	0.990	0.987	0.983	0.979	0.970	0.967	0.959
0.600	0.999	0.997	0.996	0.993	0.992	0.989	0.986	0.983	0.975	0.972	0.966
0.700	0.999	0.998	0.996	0.994	0.993	0.991	0.988	0.985	0.979	0.976	0.971
0.800	0.999	0.998	0.997	0.995	0.994	0.992	0.990	0.987	0.981	0.979	0.975
0.900	0.999	0.998	0.997	0.995	0.994	0.993	0.991	0.988	0.983	0.982	0.977
1.000	0.999	0.998	0.997	0.996	0.995	0.994	0.992	0.990	0.985	0.983	0.980
1.100	0.999	0.998	0.998	0.996	0.995	0.994	0.992	0.990	0.986	0.985	0.981
1.200	0.999	0.999	0.998	0.996	0.996	0.995	0.993	0.991	0.988	0.986	0.983
1.300	0.999	0.999	0.998	0.997	0.996	0.995	0.994	0.992	0.989	0.987	0.984
1.400	0.999	0.999	0.998	0.997	0.996	0.995	0.994	0.993	0.989	0.988	0.985
1.500	0.999	0.999	0.998	0.997	0.997	0.996	0.994	0.993	0.990	0.989	0.986
1.600	1.000	0.999	0.998	0.997	0.997	0.996	0.995	0.993	0.991	0.990	0.987
1.700	1.000	0.999	0.998	0.997	0.997	0.996	0.995	0.994	0.991	0.990	0.988
1.800	1.000	0.999	0.999	0.998	0.997	0.996	0.995	0.994	0.992	0.991	0.989
1.900	1.000	0.999	0.999	0.998	0.997	0.997	0.996	0.994	0.992	0.991	0.989
2.000	1.000	0.999	0.999	0.998	0.998	0.997	0.996	0.995	0.993	0.992	0.990
2.100	1.000	0.999	0.999	0.998	0.998	0.997	0.996	0.995	0.993	0.992	0.990
2.200	1.000	0.999	0.999	0.998	0.998	0.997	0.996	0.995	0.993	0.992	0.991
2.300	1.000	0.999	0.999	0.998	0.998	0.997	0.996	0.995	0.994	0.993	0.991
2.400	1.000	0.999	0.999	0.998	0.998	0.997	0.997	0.996	0.994	0.993	0.992
2.500	1.000	0.999	0.999	0.998	0.998	0.997	0.997	0.996	0.994	0.993	0.992
2.600	1.000	0.999	0.999	0.998	0.998	0.998	0.997	0.996	0.994	0.994	0.992
2.700	1.000	0.999	0.999	0.998	0.998	0.998	0.997	0.996	0.994	0.994	0.992
2.800	1.000	0.999	0.999	0.998	0.998	0.998	0.997	0.996	0.995	0.994	0.993
2.900	1.000	0.999	0.999	0.998	0.998	0.998	0.997	0.997	0.995	0.994	0.993
3.000	1.000	0.999	0.999	0.999	0.998	0.998	0.997	0.997	0.995	0.994	0.993

REFERENCES

Andersen, H.H., Besenbacher, F., Loftager, P., and Möller, W. (1980), *Phys. Rev.* **A21**, 1891.

Hautala, M., and Luomajärvi, M. (1980), *Radiat. Eff.* **45**, 159.

L'Ecuyer, J., Davies, J.A., and Matsunami, N. (1979), *Nucl. Instrum. Methods* **160**, 337.

Walker, F.W., Parrington, J.R., and Feiner, F. (1989), *Nuclides and Isotopes, Chart of the Nuclides*, 14th Ed., General Electric, San Jose, CA.

Wätjen, U., and Bax, H. (1994), *Nucl. Instrum. Methods* **B85**, 627.

NON-RUTHERFORD ELASTIC BACKSCATTERING CROSS SECTIONS

Compiled by

R. P. Cox, J. A. Leavitt, and L. C. McIntyre, Jr.

University of Arizona, Tucson, Arizona, USA

Contributor: A. Gurbich

This appendix contains non-Rutherford cross-section information for elastic backscattering of ^{1}H and ^{4}He analysis beam ions by target elements with $Z \leq 20$. For ^{1}H projectiles, laboratory energies of 0.5–3.0 MeV are emphasized; for ^{4}He projectiles, laboratory energies of 1–10 MeV are emphasized. No data for non-Rutherford backscattering of heavier ions are included; for beam energies below 10 MeV, heavier-ion scattering is usually Rutherford (except for electron-shell corrections). Further, no useful cross section data were found in the literature for the following beam target pairs: ^{1}H–K, ^{4}He–P, and ^{4}He–S.

The non-Rutherford cross-section information in this appendix is presented in graphical form only, as plots of the ratio of the measured cross section to the Rutherford cross section, σ/σ_R, versus the incident projectile laboratory energy, E_{Lab}, at a specific, large laboratory backscattering angle, θ_{Lab}. In the few instances where the original references give the measured cross-section data in tabular form, the data are plotted as points on the graphs. In the many instances where only graphical data are presented in the original references, the data were digitized by the procedure described below and used to produce continuous curves of σ/σ_R versus E_{Lab}.

The procedure used to digitize the cross-section plots appearing in the original references was as follows: The appropriate figure in the original work was xerographically enlarged. A smooth curve was drawn by hand through the relevant experimental points. This smooth curve was digitized at irregular intervals with a SUMMA SKETCH II (Suma Graphics) digitizing tablet; the result was stored in a PC ASCII disk file. These data were converted to a σ/σ_R versus E_{Lab} file that was also stored on the disk. A plotting program, GRAFTOOL (3-D Visions), was used to produce the final plots of these data (with the digitized points connected with straight lines), showing appropriate regions of interest with appropriate vertical scales.

The absolute accuracy of the σ/σ_R ratios produced by this procedure is difficult to estimate and varies from plot to plot. The accuracy of the digitizing procedure itself was tested by digitizing plots of reported tabular data for scattering of ^{4}He by O (Leavitt *et al.*, 1990) and comparing the result of the digitizing procedure with the original tabular data. The digitized data agreed with the tabular data to within less than 5%. It should be noted that the plots digitized in this instance were larger and clearer than those normally encountered in the literature, so this test might indicate the lower limit of the uncertainty associated with the use of the digitized curves.

No systematic effort was made to judge the reliability of the data contained in the original reports themselves. However, the occurrence of several well-known instances of discrepancies between results produced by different investigators does not inspire confidence in the reliability of the quoted uncertainties. An example is the ^{1}H–^{16}O cross section for 0.6 MeV < E_{Lab} < 2.5 MeV with θ_{Lab} = 164–170°. The values reported by

Laubenstein *et al.* (1951) are a factor 2.9 larger than the values reported by Luomajarvi *et al.* (1985) and Chow *et al.* (1975). Another example is the ^{4}He–^{11}B cross section for 2 MeV < E_{Lab} < 2.4 MeV with θ_{Lab} = 151–170°. The values reported by Ramirez *et al.* (1972) are a factor of 1.5 larger than the values reported by McIntyre *et al.* (1992).

It should be noted that all available data on non-Rutherford cross sections for ^{1}H and ^{4}He projectiles are not included in the plots in this appendix. The original works frequently displayed σ_{CM} versus E_{Lab} plots for several values of θ_{CM}. The σ/σ_R versus E_{Lab} curves in this appendix usually display information for one large backscattering angle (150° < θ_{Lab} < 170°) only. Further, several original sources frequently contained essentially the same data; in such cases, data from only one of the sources is usually included here. For more complete lists of references on non-Rutherford cross sections, see Leavitt and McIntyre (1991) for ^{4}He projectiles and Rauhala (1991) for ^{1}H projectiles. Further, online data bases, such as NNDC and IBANDL (see References for details) should be checked for the most recent cross-section information. The evaluated (recommended) cross sections are available at http://www-nds.iaea.org/sigmacalc/.

The curves contained in this appendix are intended to serve as a general indication of the energy dependence of the cross section. They can be used to locate flat regions in σ/σ_R for general backscattering work or to locate resonances that can be used for depth profiling. Once the colliding pair and energy region of interest have been chosen, all relevant information in the more complete reference lists should be gathered. This additional information might also be of use in judging the accuracy of the σ/σ_R ratio.

As has been indicated, values of σ/σ_R obtained from the curves should be used with caution in actual calculations as these values might be uncertain by at least 5%. If higher accuracy is desired, or if the analyst needs σ/σ_R versus E_{Lab} data for θ_{Lab} values not given here (or in the more extensive references), the analyst might wish to actually measure σ/σ_R versus E_{Lab} at the θ_{Lab} value(s) of interest. Results of attempts to obtain σ/σ_R values (for θ_{Lab} at values not reported) by interpolation are questionable, at best. For example, result (John *et al.*, 1969) from ^{4}He–^{16}O cross section measurements fo E_{Lab} near 8.8 MeV—specifically, at θ_{Lab} = 165.6°, σ/σ_R = 28 whereas at θ_{Lab} = 145.7°, σ/σ_R = 7—provide an indication of the strong dependence of σ/σ_R on θ_{Lab}. Accurate measurement of the cross sections might require construction of a sample for tha purpose. The usual sample consists of a stable thin film containing the target element (x) of interest deposited on a ligh substrate (such as C) and covered by a thin film of heavy meta (m). Except for small corrections due to electron-shell and film-thickness effects, the cross-section ratio at energy E can be calculated from

$$\left(\frac{\sigma}{\sigma_R}\right)_E \approx \frac{(A_x/A_m)_E}{(A_x/A_m)_{E°}} ,$$

where $(A_x/A_m)_E$ and $(A_x/A_m)_{E°}$ are measured integrated peak count ratios at analysis beam energies E and E°, respectively. Here, E° is a sufficiently low energy that both cross sections (for x and m) are Rutherford; it is assumed that the cross section for m is also Rutherford at energy E. This comparison procedure effectively eliminates the dependence of the result on the accuracy of charge and solid-angle measurements. It is capable of producing results with uncertainties of ±2%. For a discussion of the corrections and application of this procedure to the measurement of ^{4}He–C cross sections, see Leavitt *et al.* (1989).

The non-Rutherford cross-section plots for ^{1}H backscattering are followed by those for ^{4}He backscattering; the references conclude this appendix. A target-element superscript on the plots indicates that the target was isotopically pure; lack of the superscript (e.g., Fig. A7.15) indicates that the target contained isotopes in the naturally occurring abundances.

We acknowledge the assistance of M.D. Ashbaugh, J.D. Frank, and Z. Lin during the initial stages of this project.

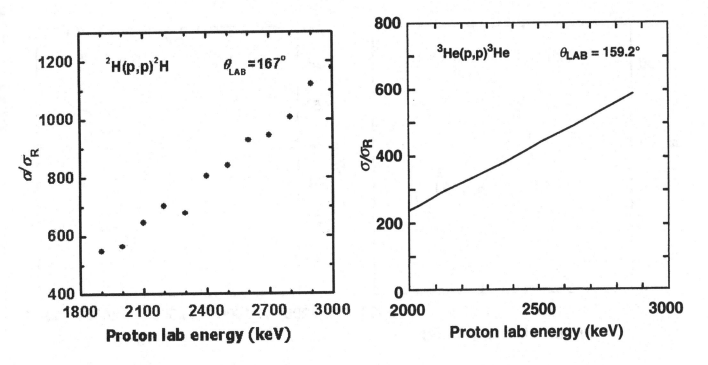

FIG. A7.1. DeSimone *et al.* (2007).

FIG. A7.3. Tombrello *et al.* (1962).

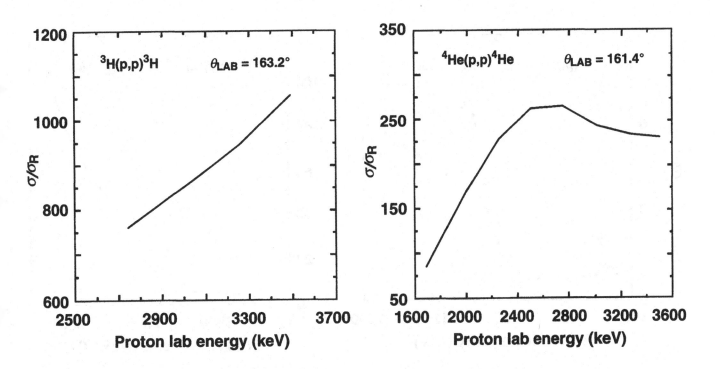

FIG. A7.2. Claassen *et al.* (1951).

FIG. A7.4. Freier *et al.* (1949).

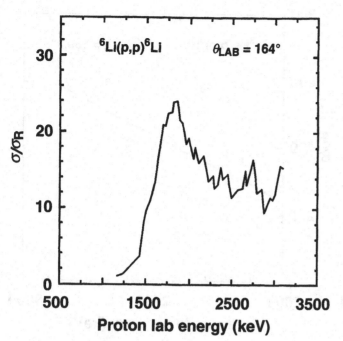

FIG. A7.5. Bashkin and Richards (1951).

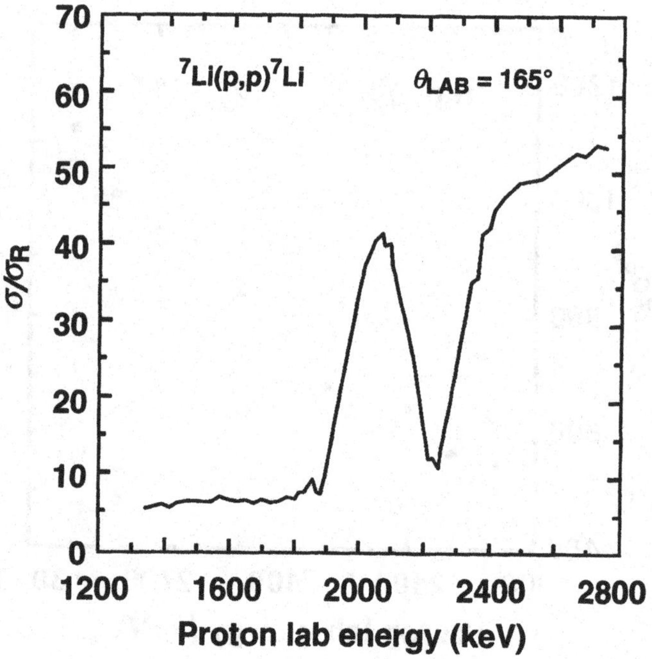

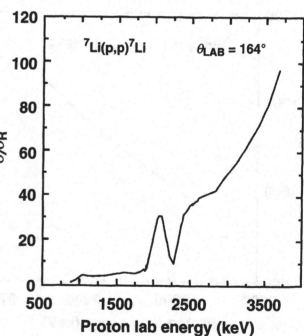

FIG. A7.7. Malmberg (1956).

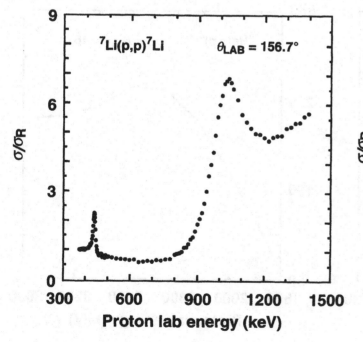

FIG. A7.6. Warters *et al.* (1953).

FIG. A7.8. Bashkin and Richards (1951).

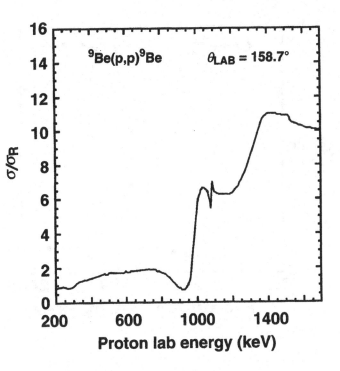

FIG. A7.9. Mozer (1956).

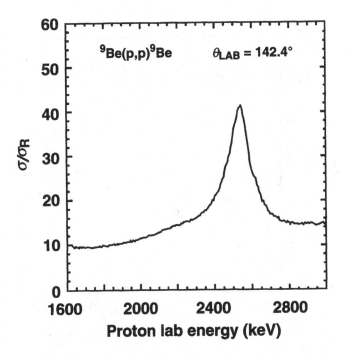

FIG. A7.10. Mozer (1956).

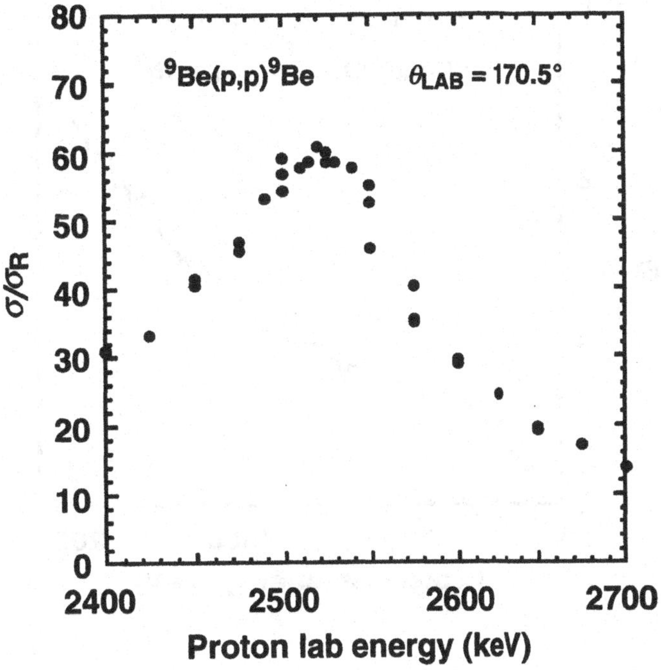

FIG. A7.11. Leavitt *et al.* (1994).

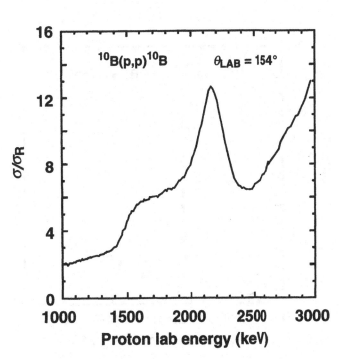

FIG. A7.12. Overley and Whaling (1962).

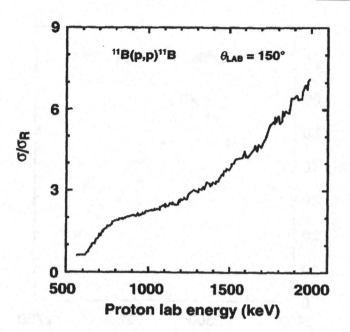

FIG. A7.13. Tautfest and Rubin (1956).

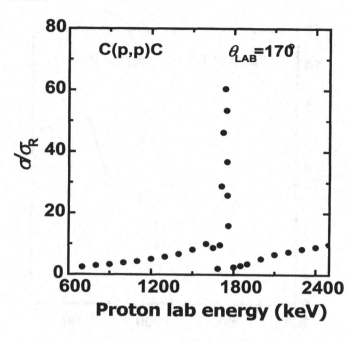

FIG. A7.15. Rauhala (1985).

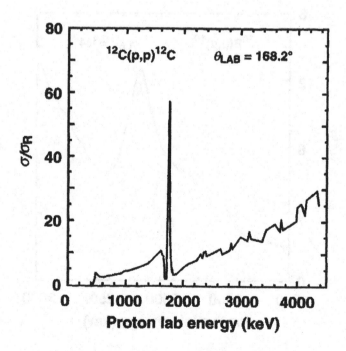

FIG. A7.14. Liu *et al.* (1993).

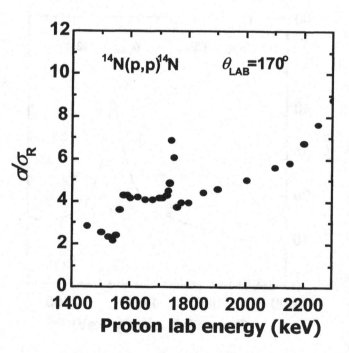

FIG. A7.16. Rauhala (1985).

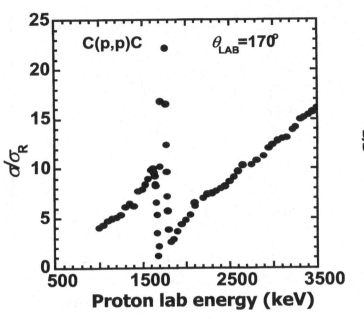

FIG. A7.17. Amirikas *et al.* (1993).

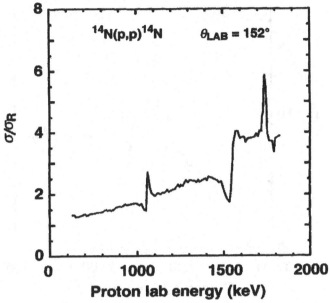

FIG. A7.19. Hagedorn *et al.* (1957).

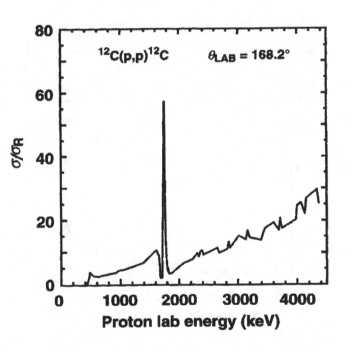

FIG. A7.18. Jackson *et al.* (1953).

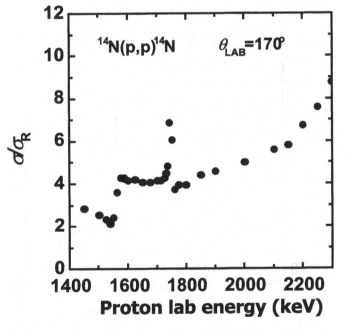

FIG. A7.20. Rauhala (1985).

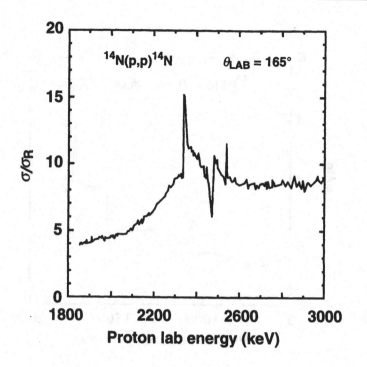

FIG. A7.21. Lambert and Durand (1967).

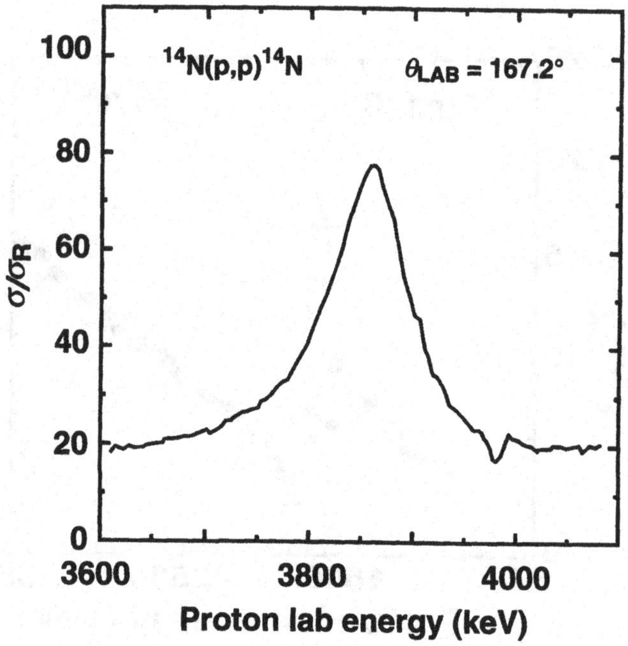

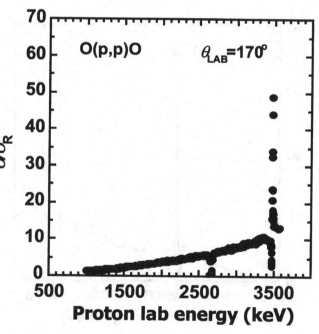

FIG. A7.23. Olness *et al.* (1958a).

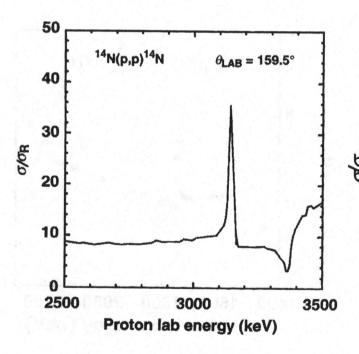

FIG. A7.22. Bashkin *et al.* (1959).

FIG. A7.24. Amirikas *et al.* (1993).

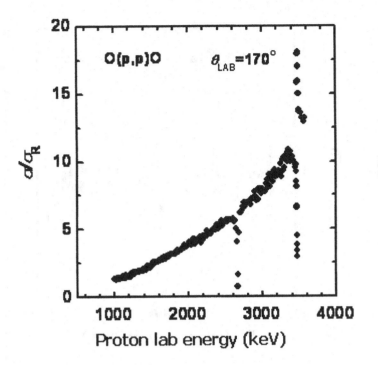

FIG. A7.25. Amirikas *et al.* (1993).

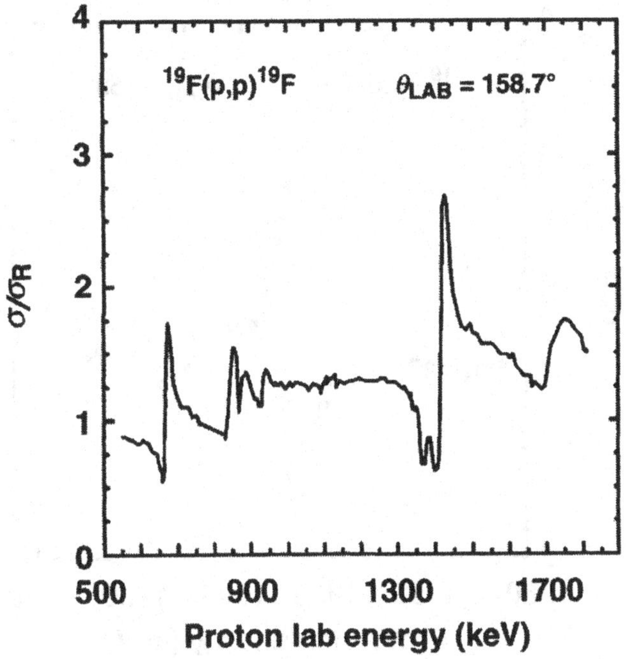

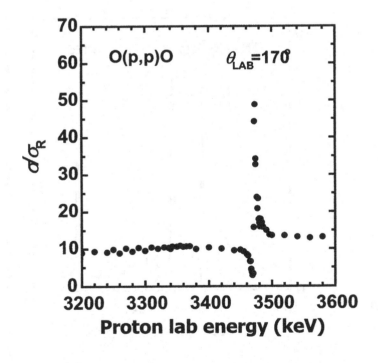

FIG. A7.26. Amirikas *et al.* (1993).

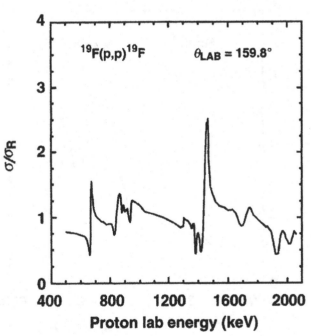

FIG. A7.27. Webb *et al.* (1955).

FIG. A7.28. Dearnaly (1956).

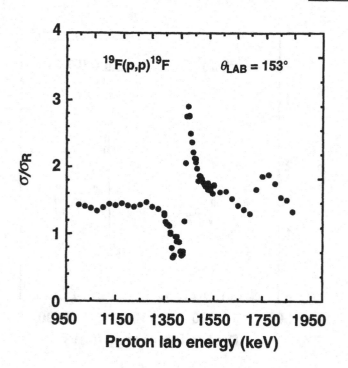

FIG. A7.29. Knox and Harmon (1989).

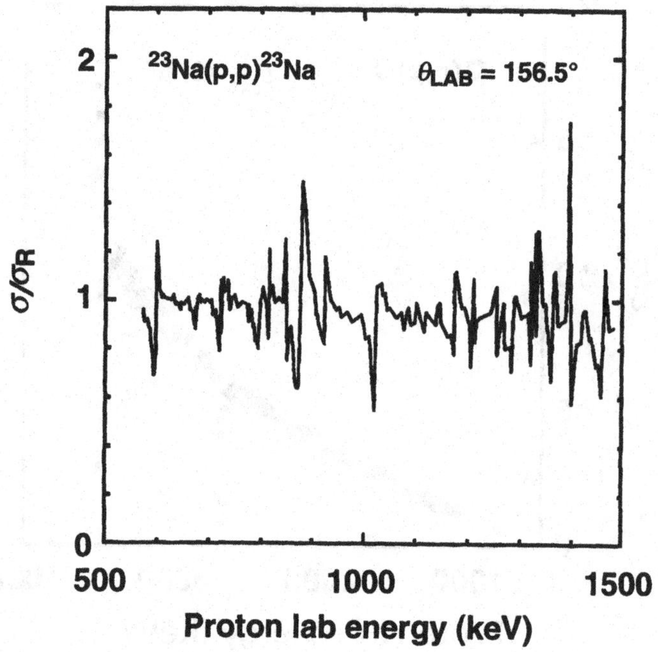

FIG. A7.31. Baumann *et al.* (1956).

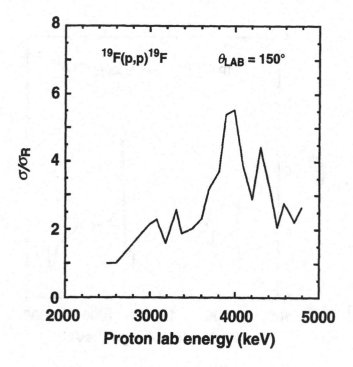

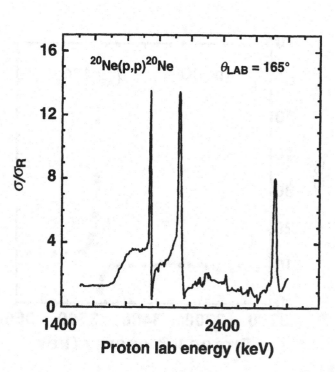

FIG. A7.30. Bogdanović *et al.* (1993).

FIG. A7.32. Lambert *et al.* (1971).

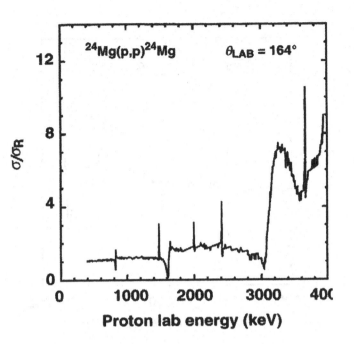

FIG. A7.33. Mooring *et al.* (1951).

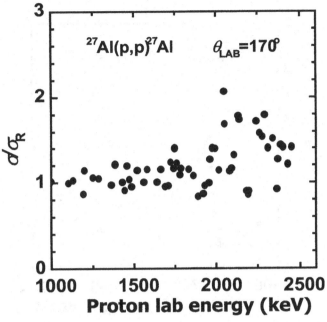

FIG. A7.35. Rauhala (1989).

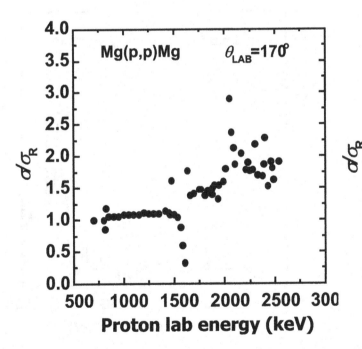

FIG. A7.34. Rauhala and Luomajärvi (1988).

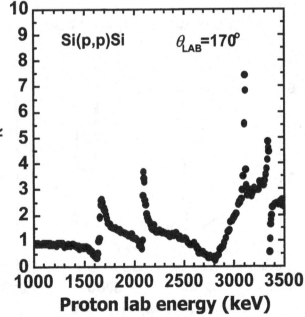

FIG. A7.36. Amirikas *et al.* (1993).

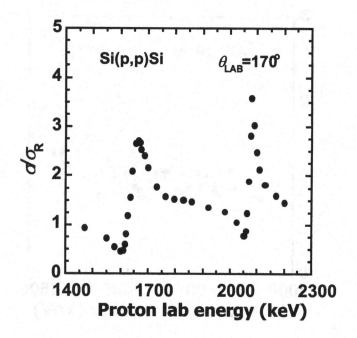

FIG. A7.37. Rauhala (1985).

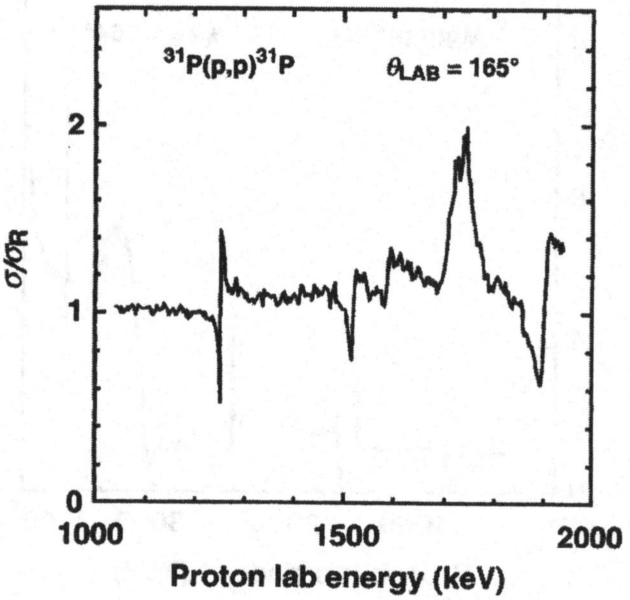

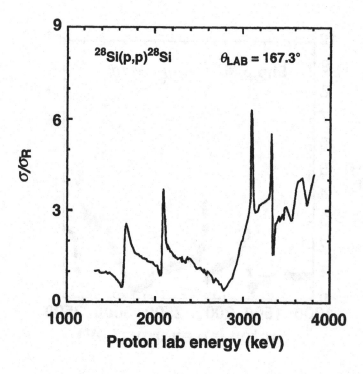

FIG. A7.38. Vorona *et al.* (1959).

FIG. A7.39. Cohen-Ganouna *et al.* (1963).

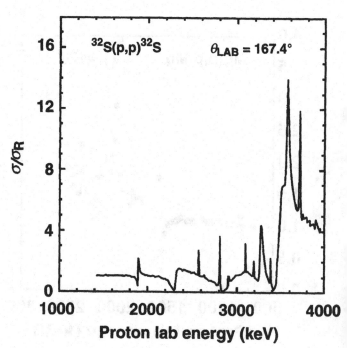

FIG. A7.40. Olness *et al.* (1958b).

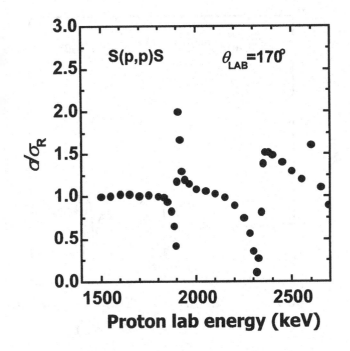

FIG. A7.41. Rauhala and Luomajärvi (1988).

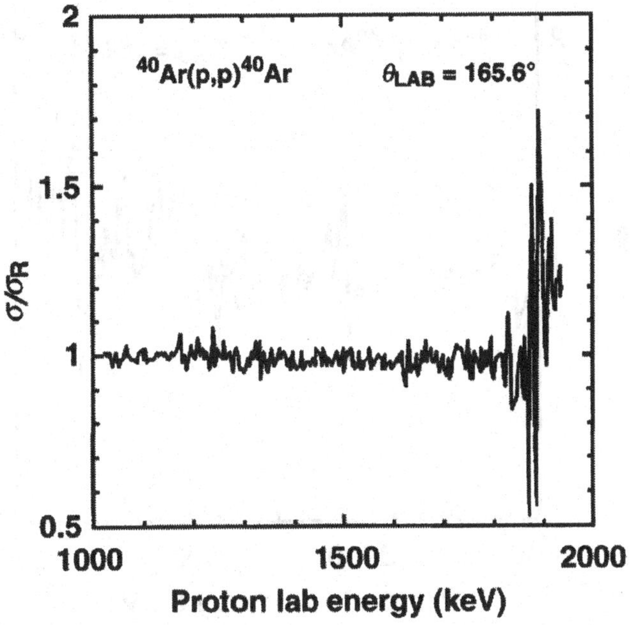

FIG. A7.43. Cohen-Ganouna *et al.* (1963).

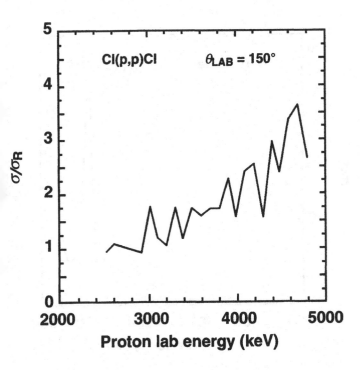

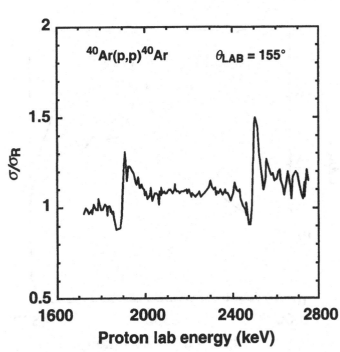

FIG. A7.42. Bogdanović *et al.* (1993).

FIG. A7.44. Freier *et al.* (1958).

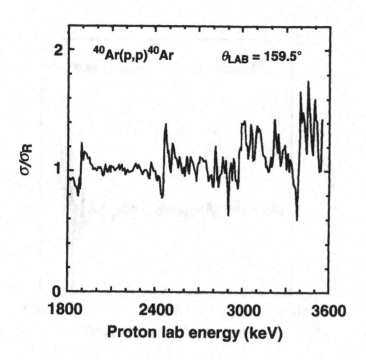

FIG. A7.45. Barnhard and Kim (1961).

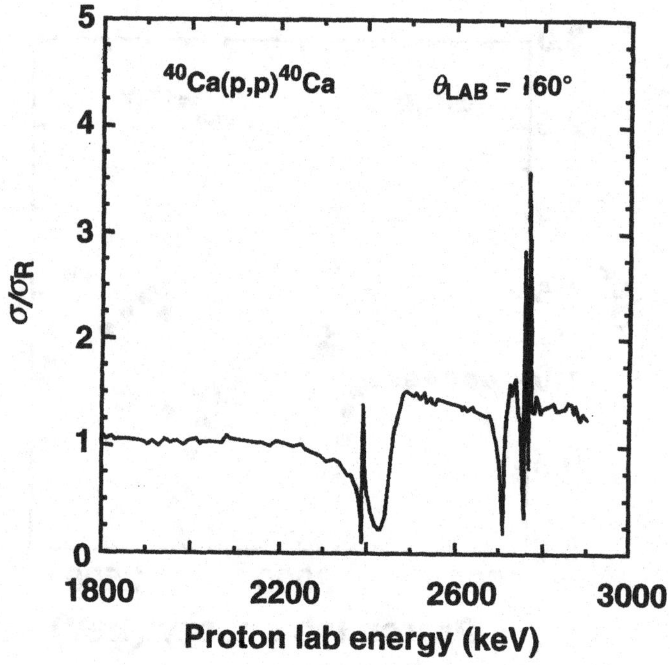

FIG. A7.46. Wilson *et al.* (1974).

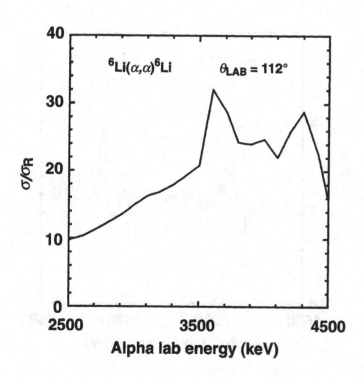

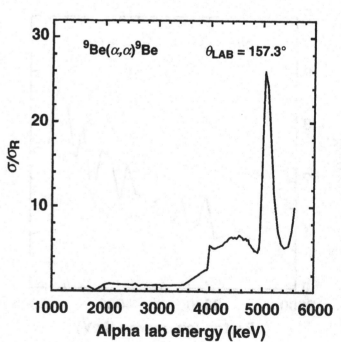

FIG. A7.47. Bohlen *et al.* (1972).

FIG. A7.49. Goss *et al.* (1973).

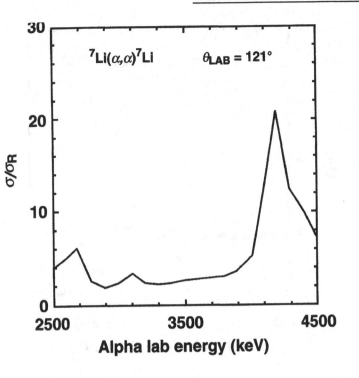

FIG. A7.48 Bohlen *et al.* (1972).

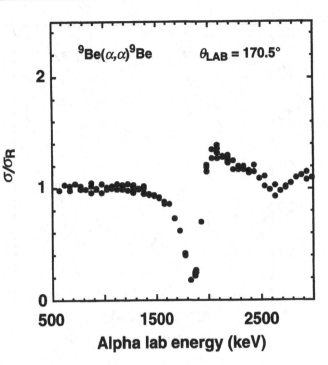

FIG. A7.50. Leavitt *et al.* (1994).

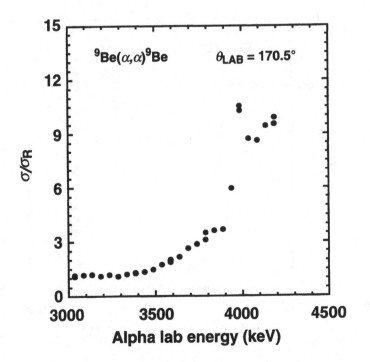

FIG. A7.51. Leavitt *et al.* (1994).

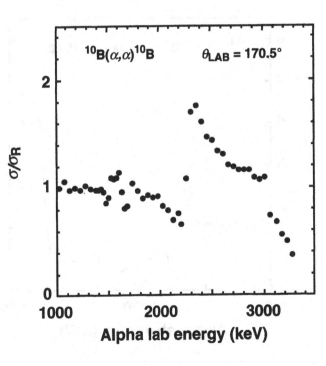

FIG. A7.53. McIntyre *et al.* (1992).

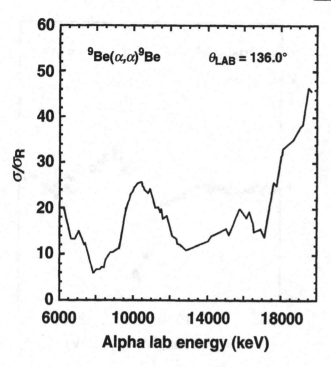

FIG. A7.52. Taylor *et al.* (1965).

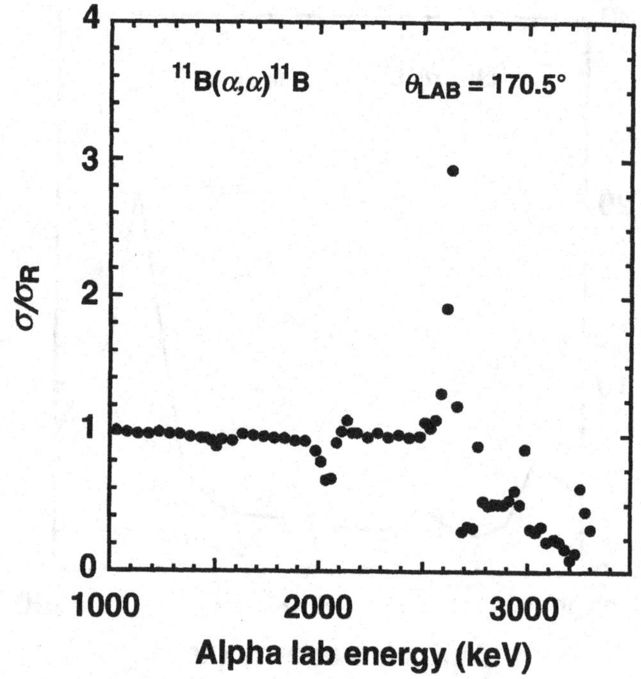

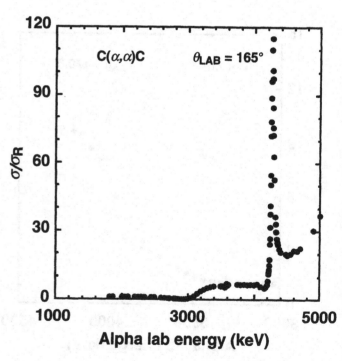

FIG. A7.54. McIntyre *et al.* (1992).

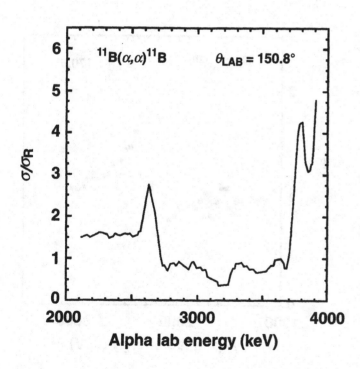

FIG. A7.55. Ramirez *et al.* (1972).

FIG. A7.57. Feng *et al.* (1994a).

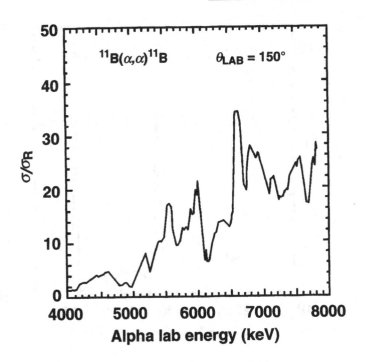

FIG. A7.56. Ott and Weller (1972).

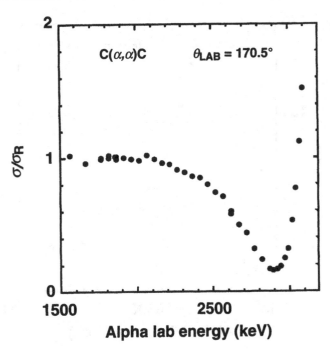

FIG. A7.58. Leavitt *et al.* (1989).

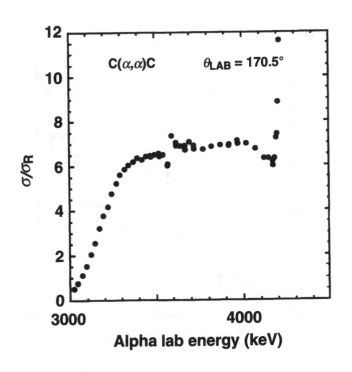

FIG. A7.59. Leavitt *et al.* (1989).

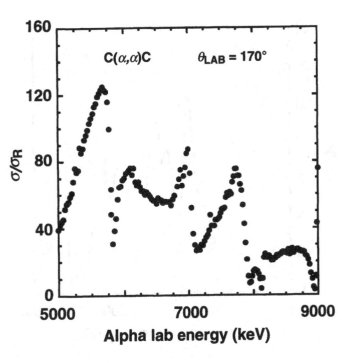

FIG. A7.61. Cheng *et al.* (1994b).

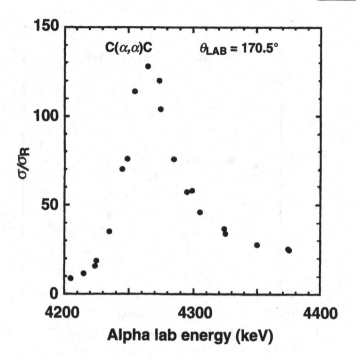

FIG. A7.60. Leavitt *et al.* (1989).

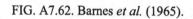

FIG. A7.62. Barnes *et al.* (1965).

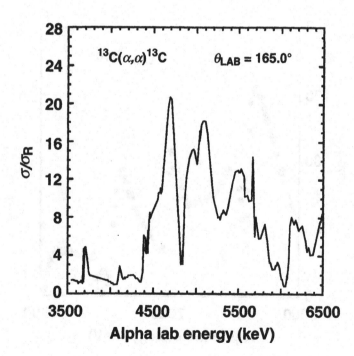

FIG. A7.63. Kerr *et al.* (1968).

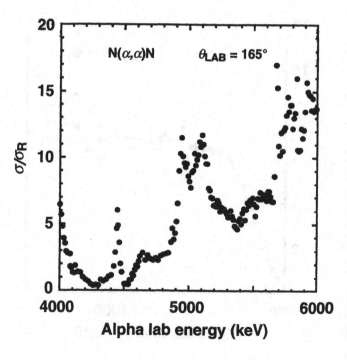

FIG. A7.65. Feng *et al.* (1994b).

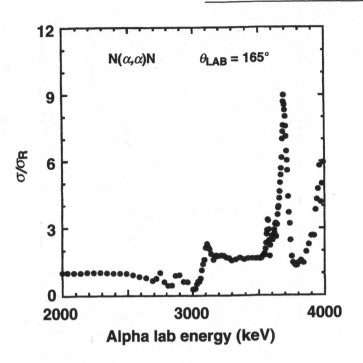

FIG. A7.64. Feng *et al.* (1994b).

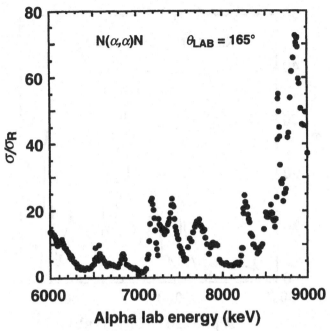

FIG. A7.66. Feng *et al.* (1994b).

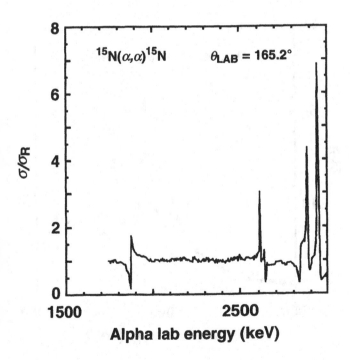

FIG. A7.67. Smotrich *et al.* (1961).

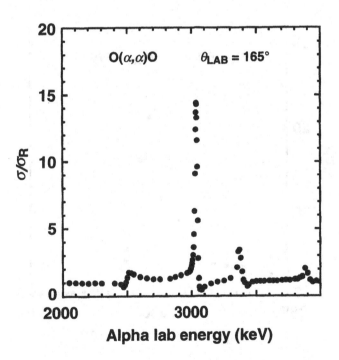

FIG. A7.69. Zhou *et al.* (1995).

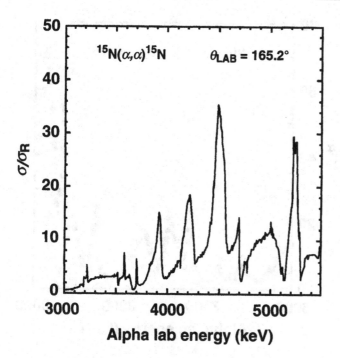

FIG. A7.68. Smotrich *et al.* (1961).

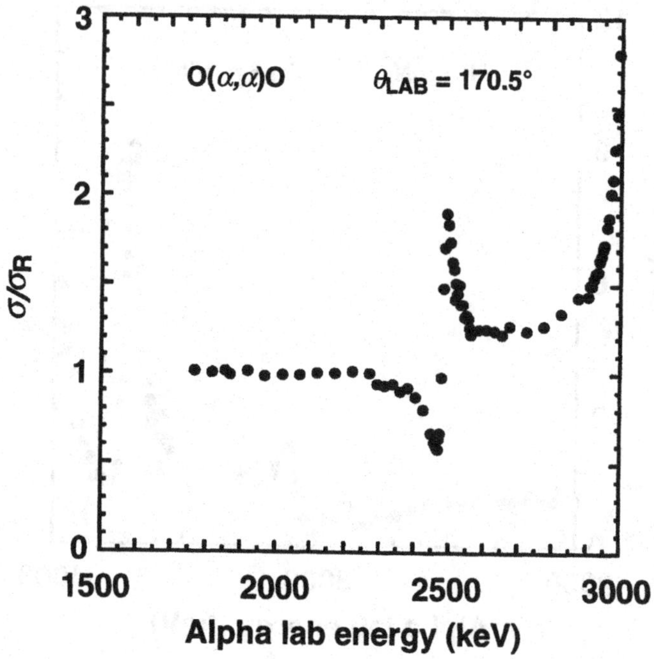

FIG. A7.70. Leavitt *et al.* (1990).

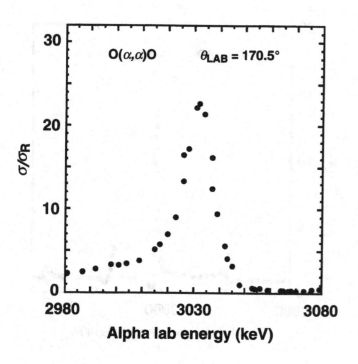

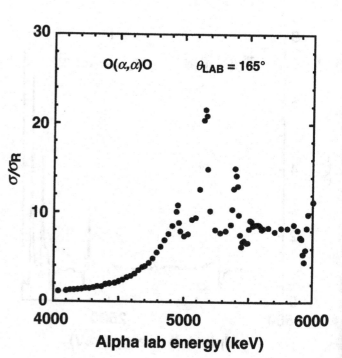

FIG. A7.71. Leavitt *et al.* (1990).

FIG. A7.73. Zhou *et al.* (1995).

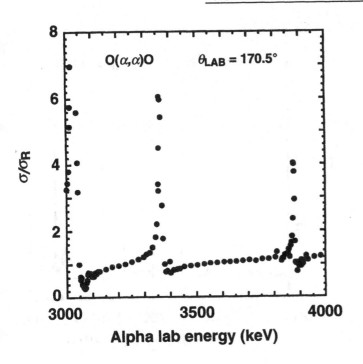

FIG. A7.72. Leavitt *et al.* (1990).

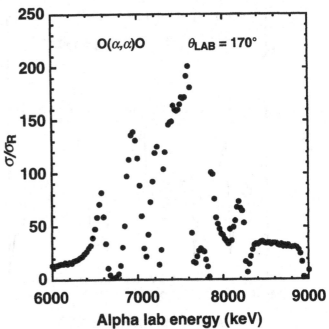

FIG. A7.74 Cheng *et al.* (1993).

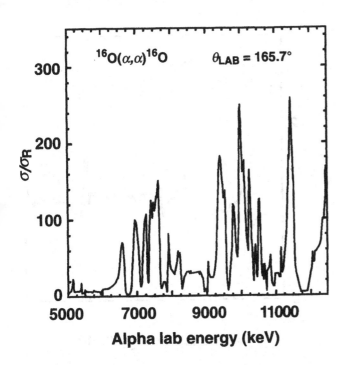

FIG. A7.75. John *et al.* (1969).

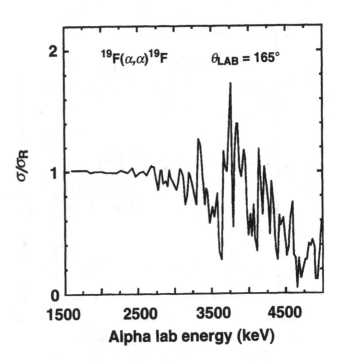

FIG. A7.77. Cheng *et al.* (1994a).

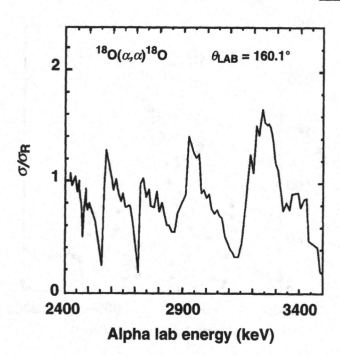

FIG. A7.76. Powers *et al.* (1964).

FIG. A7.78. Goldberg *et al.* (1954).

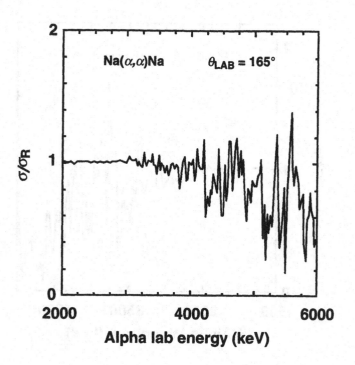

FIG. A7.79. Cheng *et al.* (1991).

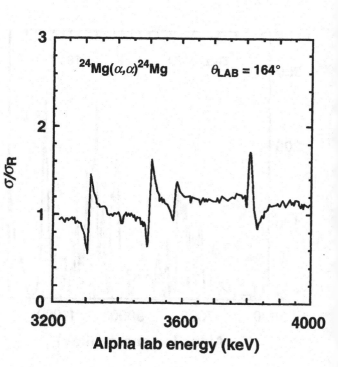

FIG. A7.81. Kaufmann *et al.* (1952).

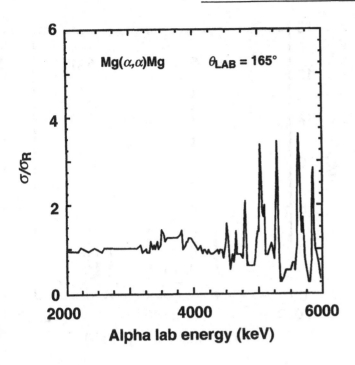

FIG. A7.80. Cheng *et al.* (1994a).

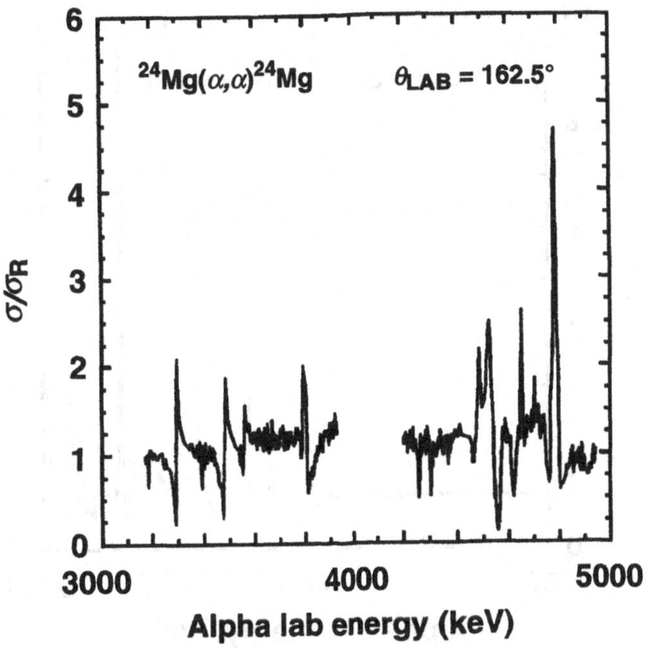

FIG. A7.82. Cseh *et al.* (1982).

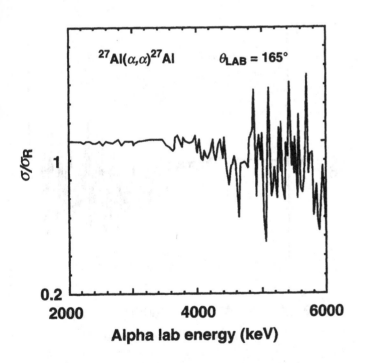

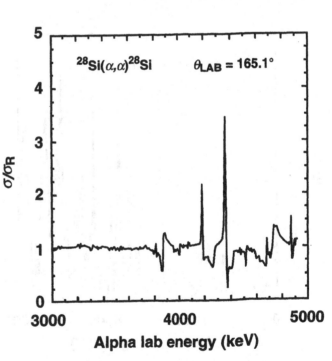

FIG. A7.83. Cheng *et al.* (1991).

FIG. A7.85. Leung *et al.* (1972).

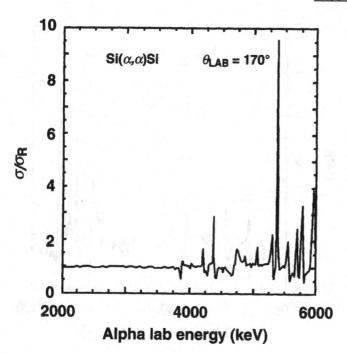

FIG. A7.84. Cheng *et al.* (1994a).

FIG. A7.86. Leung *et al.* (1972).

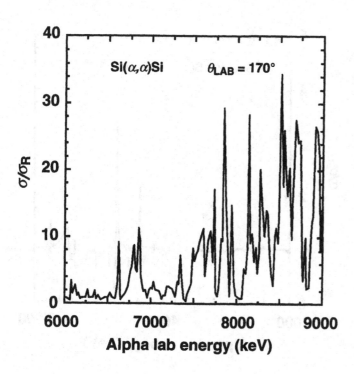

FIG. A7.87. Cheng *et al.* (1994a).

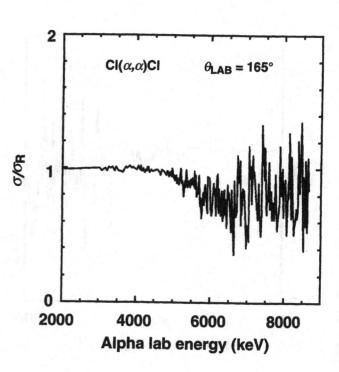

FIG. A7.89. Cheng *et al.* (1994a).

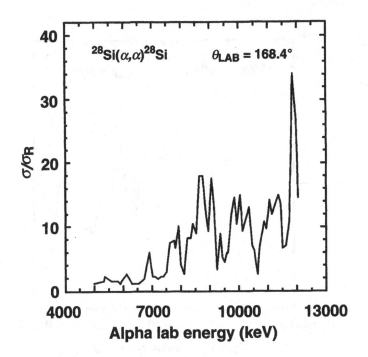

FIG. A7.88. Lawrie *et al.* (1986).

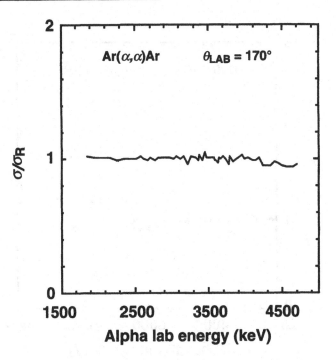

FIG. A7.90. Leavitt *et al.* (1986).

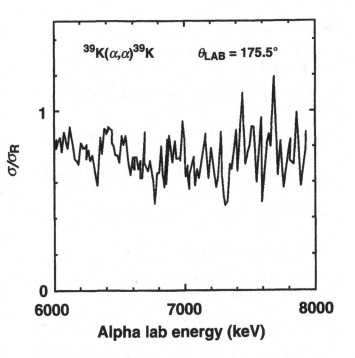

FIG. A7.91. Frekers *et al.* (1983).

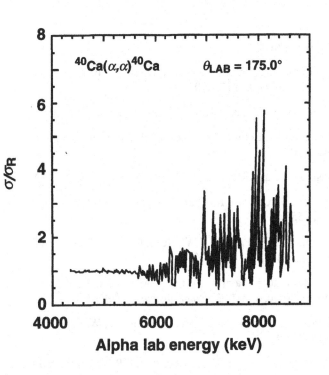

FIG. A7.93. Frekers *et al.* (1983).

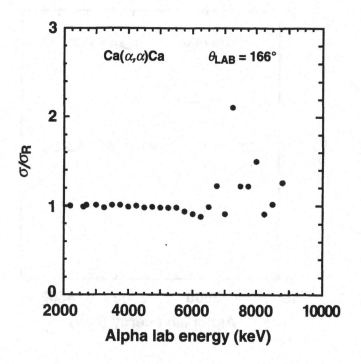

FIG. A7.92. Hubbard *et al.* (1990).

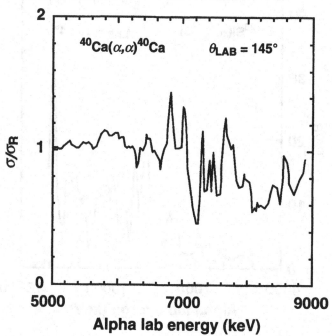

FIG. A7.94. Sellschop *et al.* (1987).

REFERENCES

Amirikas, R., Jamieson, D.N., and Dooley, S.P. (1993), *Nucl. Instrum. Methods* **B77**, 110.

Barnhard, A.C.L., and Kim, C.C. (1961), *Nucl. Phys.* **28**, 428.

Barnes, B.K., Belote, T.A., and Risser, J.R. (1965), *Phys. Rev.* **140**, B616.

Bashkin, S., and Richards, H.T. (1951), *Phys. Rev.* **84**, 1124.

Bashkin, S., Carlson, R.R., and Douglas, R.A. (1959), *Phys. Rev.* **114**, 1552.

Baumann, N.P., Prosser, F.W., Read, G.W., and Krone, R.W. (1956), *Phys. Rev.* **104**, 376.

Bogdoanović, I., Fazinić, S., Jakšić, M., Tadić, T., Valković, O., and Valković, V. (1993), *Nucl. Instrum. Methods* **B79**, 524.

Bohlen, H., Marquardt, N., and Von Oertzen, W. (1972), *Nucl. Phys.* **A179**, 504.

Cheng, H.-S., Lee, X.-Y., and Yang, F. (1991), *Nucl. Instrum. Methods* **B56–57**, 749.

Cheng, H.-S., Shen, H., Tang, J., and Yang, F. (1993), *Nucl. Instrum. Methods* **B83**, 449.

Cheng, H.-S., Shen, H., Yang, F., and Tang, J. (1994a), *Nucl.*

Instrum. Methods **B85**, 47.

Cheng, H.-S., Shen, H., Tang, J., and Yang, F. (1994b), *Acta Phys. Sin.* **43**, 1569.

Chow, H.C., Griffiths, G.M., and Hall, T.H. (1975), *Can. J. Phys.* **53**, 1672.

Claassen, R.S., Brown, J.S., Freier, G.O., and Stratton, W.R. (1951), *Phys. Rev.* **82**, 589.

Cseh, C, Koltay, E., Mate, Z., Somorjai, E., and Zolnai, L. (1982), *Nucl. Phys.* **A385**, 43.

Cohen-Ganouna, J., Lambert, M., and Schmouker, J. (1963), *Nucl. Phys.* **40**, 67.

Dearnaley, G. (1956), *Philos. Mag. Ser. 8* **1**, 821.

DeSimone, D.J., Haertling, C., Tesmer, J.R., and Wang, Y.Q. (2007), *Nucl. Instrum. Methods* **B261**, 405.

Feng, Y., Zhou, Z., Zhou, Y., and Zhou, G. (1994a), *Nucl. Instrum. Methods* **B86**, 225.

Feng, Y., Zhou, Z., Zhou, G., and Yang, F. (1994b), *Nucl. Instrum. Methods* **B94**, 11.

Frekers, D., Santo, R., and Langke, K. (1983), *Nucl. Phys.* **A394**, 189.

Freier, G., Lampi, E., Sleator, W., and Williams, J.H. (1949), *Phys. Rev.* **75**, 1345.

Freier, G.D., Famularo, K.F., Zipoy, D.M., and Leigh, J. (1958), *Phys. Rev.* **110**, 446.

Goldberg, E., Haberli, W., Galonsky, A.L., and Douglas, R.A. (1954), *Phys. Rev.* **93**, 799.

Goss, J.D., Blatt, S.L., Parsignault, D.R., Porterfield, CD., and Riffle, F.L. (1973), *Phys. Rev.* **C7**, 1837.

Hagedorn, F.B., Mozer, F.S., Webb, T.S., Fowler, W.A., and Lauritsen, C.C. (1957), *Phys. Rev.* **105**, 219.

Hubbard, K.M., Martin, J.A., Muenchhausen, R.E., Tesmer, J.R., and Nastasi, M. (1990), in *High Energy and Heavy Ion Beams in Materials Analysis* (Tesmer, J.R., Maggiore, C.J., Nastasi, M., Barbour, J.C., and Mayer, J.W., eds.), Materials Research Society, Pittsburgh, PA, p. 129.

IBANDL (Ion Beam Analysis Nuclear Data Library): http://www-nds.iaea.org/ibandl/.

Jackson, H.L., Galonsky, A.I., Eppling, F.J., Hill, R.W., Goldberg, E., and Cameron, J.R. (1953), *Phys. Rev.* **89**, 365.

John, J., Aldridge, J.P., and Davis, R.H. (1969), *Phys. Rev.* **181**, 1455.

Kaufmann, S., Goldberg, E., Koester, L.J., and Mooring, F.P. (1952), *Phys. Rev.* **88**, 673.

Kerr, G.W., Morris, J.M., and Risser, J.R. (1968), *Nucl. Phys.* **A110**, 637.

Knox, J.M., and Harmon, J.F. (1989), *Nucl. Instrum. Methods* **B44**, 40.

Lambert, M., and Durand, M. (1967), *Phys. Lett.* **24B**, 287.

Lambert, M., Midy, P., Drain, D., Amiel, M., Beaumevielle, H., Dauchy, A., and Meynadier, C. (1972), *J. Phys.* **33**, 155.

Laubenstein, R.A., Laubenstein, M.J.W., Koestler, L.J., and Mobley, R.C. (1951), *Phys. Rev.* **84**, 12.

Lawrie, J.J., Cowley, A.A., Whittal., D.M., Mills, S.J., and McMurray, W.R. (1986), *Z. Phys.* **A325**, 175.

Leavitt, J.A., Stoss, P., Cooper, D.B., Seerveld, J.L., McIntyre Jr., L.C., Davis, R.E., Gutierrez, S., and Reith, T.M. (1986), *Nucl. Instrum. Methods* **B15**, 296.

Leavitt, J.A., McIntyre Jr., L.C., Stoss, P., Oder, J.G., Ashbaugh, M.D., Dezfouly-Arjomandy, B., Yang, Z.M., and Lin, Z. (1989), *Nucl. Instrum. Methods* **B40–41**, 776.

Leavitt, J.A., McIntyre Jr., L.C., Ashbaugh, M.D., Oder, J.G., Lin, Z., and Dezfouly-Arjomandy, B. (1990), *Nucl. Instrum. Methods* **B44**, 260.

Leavitt, J.A., and McIntyre Jr., L.C. (1991), *Nucl Instrum. Methods* **B56–57**, 734.

Leavitt, J.A., McIntyre Jr., L.C, Champlin, R.S., Stoner Jr., J.O., Lin, Z., Ashbaugh, M.D., Cox, R.P., and Frank, J.D. (1994), *Nucl. Instrum. Methods* **B85**, 37.

Leung, M.K. (1972), Ph.D. dissertation, University of Kentucky, Lexington, KY.

Liu, Z., Li, B. Duan, D., and He, H. (1993), *Nucl. Instrum. Methods* **B74**, 439.

Malmberg, P.R. (1956), *Phys. Rev.* **101**, 114.

McIntyre Jr., L.C, Leavitt, J.A., Ashbaugh, M.D., Lin, Z., and Stoner Jr., J.O. (1992), *Nucl. Instrum. Methods* **B64**, 457.

Mooring, F.P., Koestler, L.J. Jr., Goldberg, E., Saxon, D., and Kaufmann, S.G. (1951), *Phys. Rev.* **84**, 703.

Mozer, F.S. (1956), *Phys. Rev.* **104**, 1386.

NNDC (National Nuclear Data Center), Brookhaven National Laboratory, Upton, NY. Several online databases are available at http://www.nndc.bnl.gov/.

Olness, J.W., Vorona, J., and Lewis, H.W. (1958a), *Phys. Rev.* **112**, 475.

Olness, J.W., Haeberli, W., and Lewis, H.W. (1958b), *Phys. Rev.* **112**, 1702.

Ott, W.R., and Weller, H.R. (1972), *Nucl. Phys.* **A198**, 505.

Overley, J.C., and Whaling, W. (1962), *Phys. Rev.* **128**, 315.

Powers, D., Blair, J.K., Ford Jr., J.L.C., and Willard, H.B. (1964), *Phys. Rev.* **134**, B1237.

Ramirez, J.J., Blue, R.A., and Weller, H.R. (1972), *Phys. Rev.* **C5**, 17.

Rauhala, E. (1985), *Nucl. Instrum. Methods* **B12**, 447.

Rauhala, E., and Luomajärvi, M. (1988), *Nucl. Instrum. Methods* **B33**, 628.

Rauhala, E. (1989), *Nucl. Instrum. Methods* **B40–41**, 790.

Rauhala, E. (1991), "Ion Backscattering Spectrometry", in *Elemental Analysis by Particle Accelerators* (Alfassi, Z., and Peisach, M., eds.), CRC Press, Boca Raton, FL.

Sellschop, J.P.F., Zucchiati, A., Mirman, L., Gering, M.Z.I., and DiSalvo, E. (1987), *J. Phys.* **G13**, 1129.

Smotrich, H., Jones, K.W., McDermott, L.C., and Benenson, R.E. (1961), *Phys. Rev.* **122**, 232.

Tang, J.Y., and Yang, F.J. (1995), *Nucl. Instrum. Methods* **B100**, 524.

Tautfest, G.W., and Rubin, S. (1956), *Phys. Rev.* **103**, 196.

Taylor, R.B., Fletcher, N.P., and Davis, R.H. (1965), *Nucl. Phys.* **65**, 318.

Tombrello, T.A., Jones, C.M., Phillips, G.C., and Weil, J.L. (1962), *Nucl. Phys.* **39**, 541.

Vorona, J., Olness, J.W., Haeberli, W., and Lewis, H.W. (1959), *Phys. Rev.* **116**, 1563.

Warters, W.D., Fowler, W.A., and Lauritsen, C.C. (1953), *Phys. Rev.* **91**, 917.

Webb, T.S., Hagedorn, F.B., Fowler, W.A., and Lauritsen, C.C. (1955), *Phys. Rev.* **99**, 138.

Wilson Jr., W.M., Moses, J.D., and Bilpuch, E.G. (1974), *Nucl. Phys.* **A227**, 277.

Zhou, Z.Y., Zhou, Y.Y., Zhang, Y., Xu, W.D., Zhao, G.Q., Tang, J.Y., and Yang, F.J. (1995), *Nucl. Instrum. Methods* **B100**, 524.

ACTUAL COULOMB BARRIERS

Compiled by

M. Bozoian

Los Alamos National Laboratory, Los Alamos, New Mexico, USA

The concept of the actual Coulomb barrier was introduced in an attempt to develop an approach for calculations of the deviation of backscattering cross sections from the Rutherford value. However, the validity of its physical meaning was perhaps not adequately defined, as pointed out by Gurbich (2004). Nevertheless, the concept has been in common use in ion beam analysis, and in many cases, it can provide rough guidance for ion beam analysis practitioners as to when they need to worry about the scattering process being non-Rutherford.

The following tables and figures present optical model calculations of E_{nr}^{cm}, the center-of-mass energy, for a 4% deviation from Rutherford backscattering (RBS) for various optical model parameter sets. Note that the ^1H results are based on both experiments and calculations, whereas the ^4He results are based solely on experiments. The ^4He table and figures also include 2% and 6% deviations from RBS.

The tables are self-explanatory. The accompanying figures are plots of the tabular data plus least-squares linear fits to the data. Each least-squares straight-line equation is included within the corresponding figure.

NOTE WELL: An experimentalist's prescription

To find E_{nr}^{lab}, the *laboratory* kinetic energy for departures from RBS, in terms of E_{nr}^{cm}, the projectile mass M_1, the target mass M_2, and the laboratory scattering angle θ_1, the following formula, valid within the stated inequalities, can be used:

For $0 < M_1/M_2 < 1$ and $150° < \theta_{lab} < 180°$,

$$E_{nr}^{lab} = E_{nr}^{cm}(M_1 + M_2)/M_2$$

Thus, one can read E_{nr}^{cm} from a figure or extrapolate the figure's straight-line fit at either end, providing that the stated inequalities are true, and then use the given formula to find E_{nr}^{lab}. Note that, if $M_1/M_2 = 1$, then $\theta_{lab} \leq 90°$ for $\theta_{cm} \leq 180°$, and if $M_1/M_2 > 1$, then $\theta_{lab} < 90°$. In both of these cases, no RBS occurs. See Appendix 4 and Chapter 3 for discussions on center-of-mass to laboratory relationships.

Finally, because cross-section ratios are usually preferred in ion beam analysis (i.e., the desired cross section is measured with respect to a known Rutherford cross section), there is no need to worry about transforming cross sections from the center-of-mass to the laboratory system, or vice versa, because the transforming factors will cancel in the cross-section ratios.

Table A8.1. [1]H-ion energy at which the scattering cross section deviates by 4% from its Rutherford value (Bozoian *et al.*, 1990).

Target atomic number (Z_2)	Target	E_{nr}^{cm} (data, calculation)
6	^{12}C	0.40, 0.54
7	^{14}N	0.64, 0.63
8	^{16}O	0.64, 0.81
10	^{20}Ne	1.10, 1.28
12	^{24}Mg	0.78, 1.63
14	^{28}Si	1.44, 1.63
15	^{31}P	1.21, 1.69
16	^{32}S	1.70, 1.81
18	^{40}Ar	1.78, 2.16
24	^{52}Cr	2.65, 2.83
26	^{56}Fe	2.46, 3.04
29	^{63}Cu	2.95, 3.43

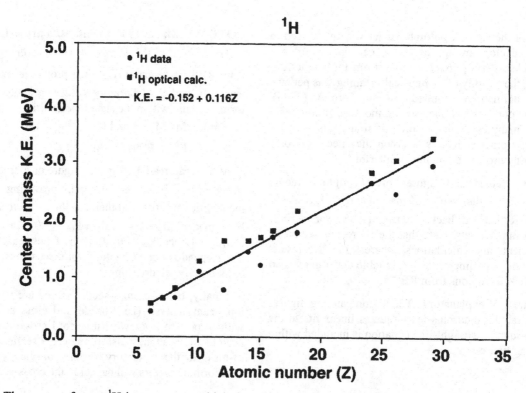

FIG. A8.1. The center-of-mass [1]H-ion energy at which the scattering cross section deviates by 4% from its Rutherford value versus atomic number for several elements. The solid line is a linear least-squares fit to the data.

Table A8.2. ^{2}H-ion energy at which the scattering cross section deviates by 4% from its Rutherford value (Bozoian, 1991a).

Target atomic number (Z_2)	Target	E_{nr}^{cm} (calculation)
6	^{12}C	1.01
7	^{14}N	0.96
8	^{16}O	1.03
12	^{24}Mg	1.40
14	^{28}Si	1.60
16	^{32}S	1.81
24	^{52}Cr	2.63
26	^{56}Fe	2.83
29	^{63}Cu	3.12

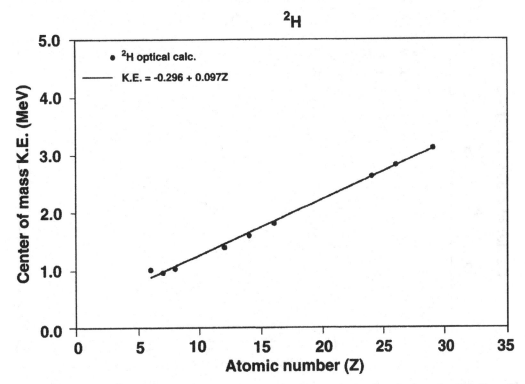

FIG. A8.2. The center-of-mass ^{2}H-ion energy at which the scattering cross section deviates by 4% from its Rutherford value versus atomic number for several elements. The solid line is a linear least-squares fit to the data.

Table A8.3. ^{3}H-ion energy at which the scattering cross section deviates by 4% from its Rutherford value (Bozoian, 1991a).

Target atomic number (Z_2)	Target	E_{nr}^{cm} (calculation)
6	^{12}C	0.77
8	^{16}O	0.92
20	^{40}Ca	2.15
32	^{74}Ge	3.33

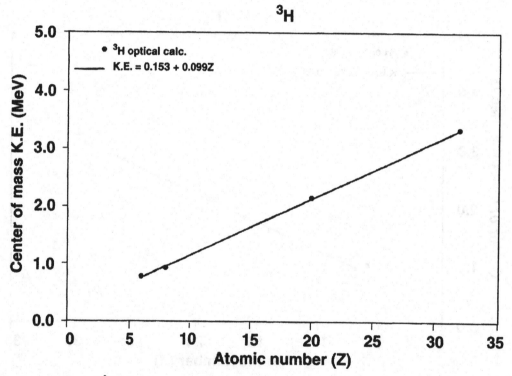

FIG. A8.3. The center-of-mass ^{3}H-ion energy at which the scattering cross section deviates by 4% from its Rutherford value versus atomic number for several elements. The solid line is a linear least-squares fit to the data.

Table A8.4. ^{3}He-ion energy at which the scattering cross section deviates by 4% from its Rutherford value (Bozoian, 1991a).

Target atomic number (Z_2)	Target	E_{nr}^{cm} (calculation)
6	^{12}C	1.41
7	^{14}N	1.64
8	^{16}O	1.87
9	^{19}F	2.08
12	^{24}Mg	2.78
13	^{27}Al	3.00
14	^{28}Si	3.24
20	^{40}Ca	4.58
22	^{48}Ti	4.96
29	^{63}Cu	6.44
32	^{72}Ge	7.03

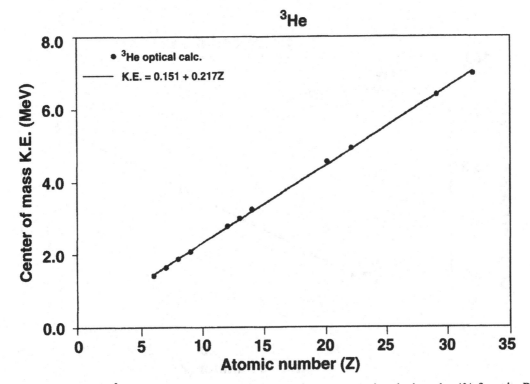

FIG. A8.4. The center-of-mass ^{3}He-ion energy at which the scattering cross section deviates by 4% from its Rutherford value versus atomic number for several elements. The solid line is a linear least-squares fit to the data.

Table A8.5. ^{4}He-ion energy at which the scattering cross section deviates by 2, 4 and 6% from its Rutherford value. Values are measurements and their associated error (Bozoian *et al.*, 1990).

Target atomic number (Z_2)	Target	E_{nr}^{cm}		
		2% (data, error)	4% (data, error)	6% (data, error)
6	^{12}C	1.61, 0.04	1.64, 0.02	1.67, 0.02
7	^{14}N	1.93, 0.06	1.96, 0.06	1.98, 0.06
8	^{16}O	1.89, 0.06	1.90, 0.06	1.92, 0.06
9	^{19}F	1.67, 0.06	1.67, 0.06	1.67, 0.06
12	^{24}Mg	2.74, 0.06	2.77, 0.06	2.79, 0.06
13	^{27}Al	3.10, 0.10	3.10, 0.10	3.10, 0.10
14	^{28}Si	3.32, 0.05	3.33, 0.05	3.35, 0.05
20	^{40}Ca	4.70, 0.30	5.10, 0.20	5.30, 0.20
22	^{48}Ti	5.10, 0.20	5.40, 0.20	5.70, 0.20
29	^{63}Cu	6.80, 0.20	7.20, 0.20	7.40, 0.20
32	^{72}Ge	7.50, 0.20	7.80, 0.10	8.00, 0.10

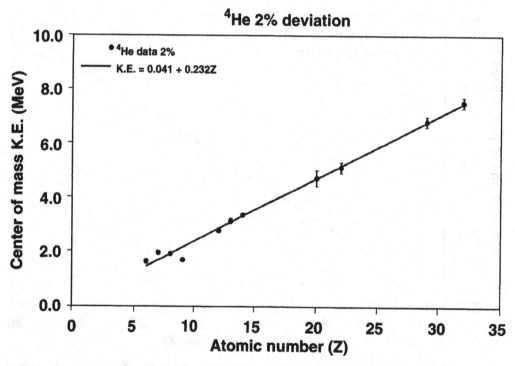

FIG. A8.5(a). The center-of-mass ^{4}He-ion energy at which the scattering cross section deviates by 2% from its Rutherford value versus atomic number for several elements. The solid line is a linear least-squares fit to the data.

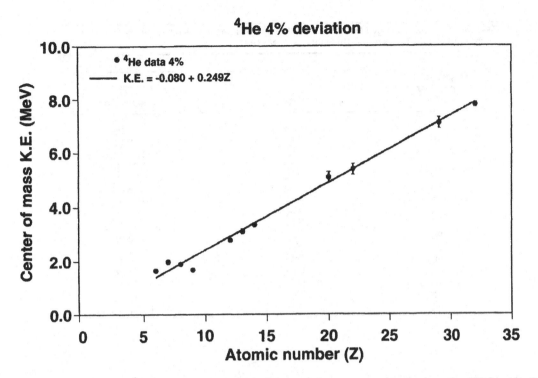

FIG. A8.5(b). The center-of-mass ^{4}He-ion energy at which the scattering cross section deviates by 4% from its Rutherford value versus atomic number for several elements. The solid line is a linear least-squares fit to the data.

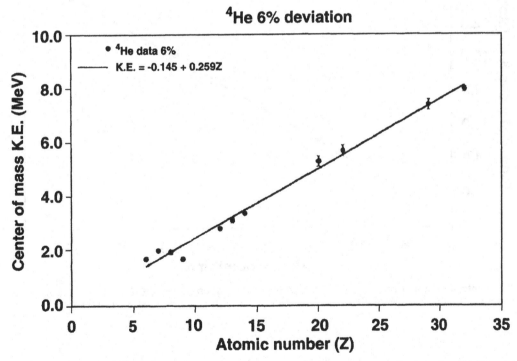

FIG. A8.5(c). The center-of-mass ^{4}He-ion energy at which the scattering cross section deviates by 6% from its Rutherford value versus atomic number for several elements. The solid line is a linear least-squares fit to the data.

Table A8.6. ^{6}Li-ion energy at which the scattering cross section deviates by 4% from its Rutherford value (Bozoian, 1991b; Hubbard *et al.*, 1991).

Target atomic number (Z_2)	Target	E_{nr}^{cm} (calculation)
4	^{9}Be	1.39
5	^{10}B	2.08
6	^{12}C	2.20, 2.20, 2.23
8	^{16}O	2.93, 2.94
12	^{24}Mg	4.32, 4.37
13	^{27}Al	4.69
14	^{28}Si	5.12, 5.16, 5.19
19	^{39}K	6.19
20	^{40}Ca	7.05, 7.05, 7.13, 7.22
20	^{48}Ca	6.93
27	^{59}Co	9.23
28	^{58}Ni	9.82
29	^{63}Cu	10.09
30	^{64}Zn	10.35
32	^{72}Ge	10.74
40	^{90}Zr	13.34, 13.46

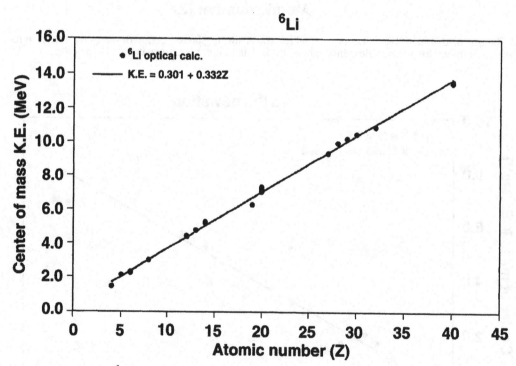

FIG. A8.6. The center-of-mass ^{6}Li-ion energy at which the scattering cross section deviates by 4% from its Rutherford value versus atomic number for several elements. The solid line is a linear least-squares fit to the data.

Table A8.7. ^{7}Li-ion energy at which the scattering cross section deviates by 4% from its Rutherford value (Bozoian, 1991b, Hubbard *et al.*, 1991).

Target atomic number (Z_2)	Target	E_{nr}^{cm} (calculated)
6	^{12}C	2.26
6	^{13}C	2.23
7	^{15}N	2.60
8	^{16}O	2.99
12	^{24}Mg	4.12, 4.35
12	^{25}Mg	4.36, 4.42
12	^{26}Mg	4.23
13	^{27}Al	4.77
20	^{40}Ca	6.85, 7.12, 7.15
20	^{48}Ca	6.89, 6.90
28	^{62}Ni	9.60
30	^{64}Zn	10.37
32	^{72}Ge	10.86
40	^{90}Zr	13.51

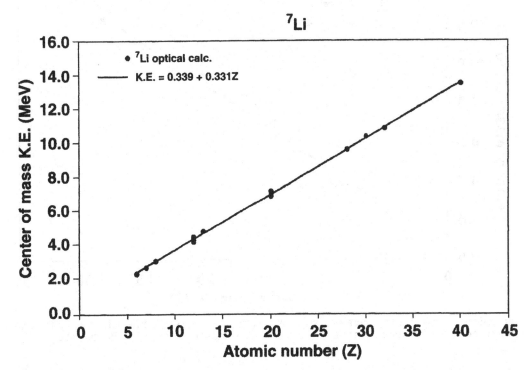

FIG. A8.7. The center-of-mass ^{7}Li-ion energy at which the scattering cross section deviates by 4% from its Rutherford value versus atomic number for several elements. The solid line is a linear least-squares fit to the data.

Table A8.8. ^{9}Be-ion energy at which the scattering cross section deviates by 4% from its Rutherford value (Bozoian, 1993).

Target atomic number (Z_2)	Target	E_{nr}^{cm} (calculated)
5	^{10}B	2.44
6	^{12}C	2.89, 2.93, 2.95, 3.07
6	^{13}C	2.88, 2.93
7	^{14}N	3.54
8	^{16}O	3.81, 3.98, 4.02, 4.14
12	^{26}Mg	5.64
14	^{28}Si	6.59, 6.64, 6.67, 6.79, 6.84, 7.11
20	^{40}Ca	9.20, 9.23, 9.30, 9.42, 9.64
28	^{58}Ni	12.48
28	^{60}Ni	12.27

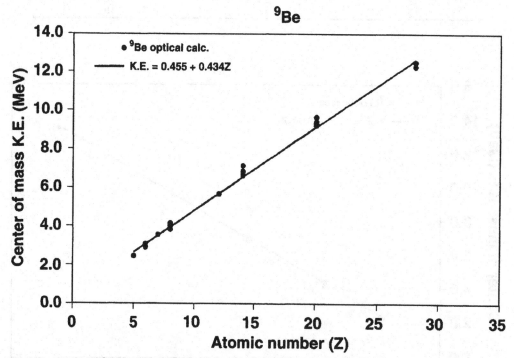

FIG. A8.8. The center-of-mass ^{9}Be-ion energy at which the scattering cross section deviates by 4% from its Rutherford value versus atomic number for several elements. The solid line is a linear least-squares fit to the data.

Table A8.9. ^{10}B-ion energy at which the scattering cross section deviates by 4% from its Rutherford value (Bozoian, 1993).

Target atomic number (Z_2)	Target	E_{nr}^{cm} (calculated)
6	^{12}C	3.87, 3.98, 3.99
6	^{13}C	3.86
7	^{14}N	4.56, 4.71
8	^{16}O	4.86, 5.09, 5.17, 5.25, 5.25, 5.39
10	^{20}Ne	6.28
12	^{24}Mg	7.16
12	^{25}Mg	7.12
13	^{27}Al	7.93
14	^{28}Si	8.76
19	^{39}K	10.68
20	^{40}Ca	12.30
28	^{60}Ni	15.38

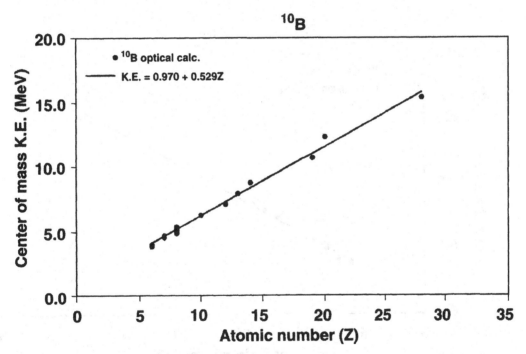

FIG. A8.9. The center-of-mass ^{10}B-ion energy at which the scattering cross section deviates by 4% from its Rutherford value versus atomic number for several elements. The solid line is a linear least-squares fit to the data.

Table A8.10. ^{11}B-ion energy at which the scattering cross section deviates by 4% from its Rutherford value (Bozoian, 1993).

Target atomic number (Z_2)	Target	E_{nr}^{cm} (calculated)
6	^{12}C	3.70, 3.88, 3.93, 3.97, 4.05
7	^{14}N	4.15
8	^{16}O	4.96, 4.97, 4.98, 5.32, 5.40
10	^{20}Ne	6.17
12	^{24}Mg	7.53
13	^{27}Al	7.92, 7.98
14	^{28}Si	8.52
20	^{40}Ca	12.34, 12.35
27	^{59}Co	15.10
28	^{60}Ni	15.33

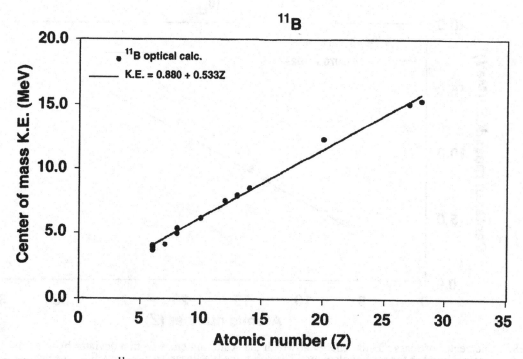

FIG. A8.10. The center-of-mass ^{11}B-ion energy at which the scattering cross section deviates by 4% from its Rutherford value versus atomic number for several elements. The solid line is a linear least-squares fit to the data.

Table A8.11. ^{12}C-ion energy at which the scattering cross section deviates by 4% from its Rutherford value (Bozoian, 1993).

Target atomic number (Z_2)	Target	E_{nr}^{cm} (calculated)
8	^{16}O	6.81
10	^{20}Ne	7.75
14	^{28}Si	10.82, 11.51
20	^{40}Ca	14.87

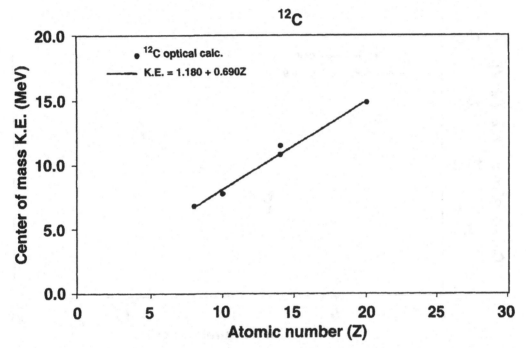

FIG. A8.11. The center-of-mass ^{12}C-ion energy at which the scattering cross section deviates by 4% from its Rutherford value versus atomic number for several elements. The solid line is a linear least-squares fit to the data.

125

Table A8.12. ^{14}N-ion energy at which the scattering cross section deviates by 4% from its Rutherford value (Bozoian, 1993).

Target atomic number (Z_2)	Target	E_{nr}^{cm} (calculated)
8	^{16}O	6.46, 6.78, 7.55, 7.79
12	^{24}Mg	10.36
13	^{27}Al	11.45
14	^{28}Si	12.28, 12.60, 13.52

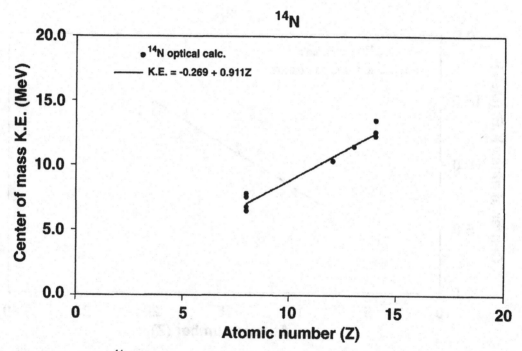

FIG. A8.12. The center-of-mass ^{14}N-ion energy at which the scattering cross section deviates by 4% from its Rutherford value versus atomic number for several elements. The solid line is a linear least-squares fit to the data.

Table A8.13. ^{16}O-ion energy at which the scattering cross section deviates by 4% from its Rutherford value (Bozoian, 1993).

Target atomic number (Z_2)	Target	E_{nr}^{cm} (calculated)
10	^{20}Ne	10.17
12	^{24}Mg	12.92, 13.27
12	^{26}Mg	13.14
14	^{28}Si	14.35, 14.77, 15.20
14	^{30}Si	14.88
20	^{40}Ca	20.40, 20.74, 20.79
20	^{48}Ca	20.49

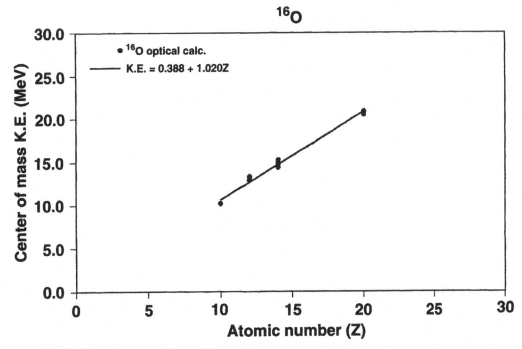

FIG. A8.13. The center-of-mass ^{16}O-ion energy at which the scattering cross section deviates by 4% from its Rutherford value versus atomic number for several elements. The solid line is a linear least-squares fit to the data.

REFERENCES

Bozoian, M., Hubbard, K.M., and Nastasi, M. (1990), *Nucl Instrum. Methods* **B51**, 311.

Bozoian, M. (1991a), *Nucl. Instrum. Methods* **B58**, 127.

Bozoian, M. (1991b), *Nucl. Instrum. Methods* **B56–57**, 740.

Bozoian, M. (1993), *Nucl Instrum. Methods* **B82**, 602.

Gurbich, A, (2004), *Nucl. Instrum. Methods* **B217**, 183.

Hubbard, K.M., Tesmer, J.R., Nastasi, M., and Bozoian, M. (1991), *Nucl. Instrum. Methods* **B58,** 121.

APPENDIX

9

EVALUATED ELASTIC SCATTERING CROSS SECTIONS

Compiled by

A. F. Gurbich

Institute of Physics and Power Engineering, Obninsk, Russia

The utilization of protons and ^{4}He beams with energies at which the elastic scattering cross sections are non-Rutherford has become very common. Consequently, the differential cross sections for elastic backscattering of protons and alpha particles from light nuclei have become among the most important data for ion beam analysis (IBA). The linear dependence of the recorded signal on the cross section results in obvious requirements for the accuracy of the data. It is evident that the concentration cannot be determined with an accuracy that exceeds the accuracy of the cross section. Thus, the importance of a precise knowledge of the differential cross sections cannot be overemphasized.

Cross sections are usually obtained by experimental means and are hence subject to statistical fluctuations, inconsistencies in sample preparation, and the pitfalls inherent in any experiment. Highly reliable data are available, and confidence is high where values from many different laboratories agree. However, even reliable experimental data usually remain discrete in both energy and angle.

Nuclear data evaluation is a routine procedure in preparing neutron data for different applications. It has also been successfully extended to charged-particle cross sections (see Chapter 3, Section 3.5). The evaluation of the cross sections for any particular reaction includes the elaboration of the most accurate possible cross sections through incorporation of all relevant experimental data within the framework of nuclear physics theory. Although nuclear physics theory cannot provide cross-section data with the required accuracy by calculations based on first principles, it does provide a powerful tool for data evaluation. Nuclear model parameters are adjusted using experimental information taken from different sources. When experiment and theory lock into a coherent unity, this provides confirmation that a reliable result has been obtained. Then, the required excitation functions for analytical purposes can be calculated for any scattering angle. The reliability of the evaluated cross sections has been verified by numerous comparisons with later measurements and benchmark experiments. Therefore, the use of evaluated data leads to substantial improvements in the accuracy of the analytical results obtained by IBA.

The evaluated cross sections in this appendix are presented in graphical form only, as plots in which theoretical data are compared with experimental results exist for only a very limited number of angles. A solid line in the plots shows results of the evaluation. The reference on the paper where the evaluation is described is made if available in the figure caption. The corresponding numerical data can be obtained by means of the online calculator SigmaCalc (http://www-nds.iaea.org/sigmacalc/). The SigmaCalc tool is also integrated into the Web site of the Ion Beam Analysis Nuclear Data Library (IBANDL, http://www-nds.iaea.org/ibandl/), where a comparison of the available experimental data with results of the evaluation can be easily made. Theoretical and methodological details of the evaluation procedure can be found elsewhere (Gurbich, 2007).

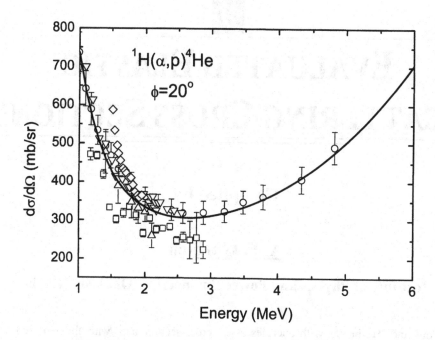

FIG. A9.1. □, Baglin *et al.* (1992); ○, Kim *et al.* (1999); △, Wang and Zhou (1988); ▽, Quillet *et al.* (1993); ◇, Nagata *et al.* (1985); solid line, Pusa *et al.* (2004).

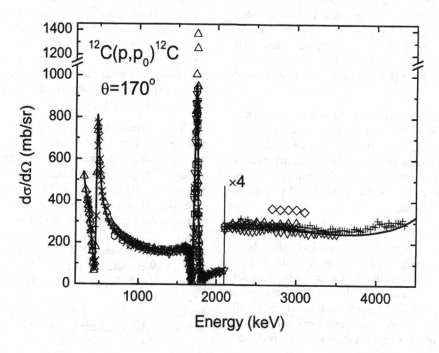

FIG. A9.2. ○, Rauhala (1985); △, Liu *et al.* (1993); □, Salomonovič (1993); ▽, Amirikas *et al.* (1993); ◇, Yang *et al.* (1991); +, Jackson *et al.* (1953); ×, Mazzoni *et al.* (1999); solid line, Gurbich (1998a).

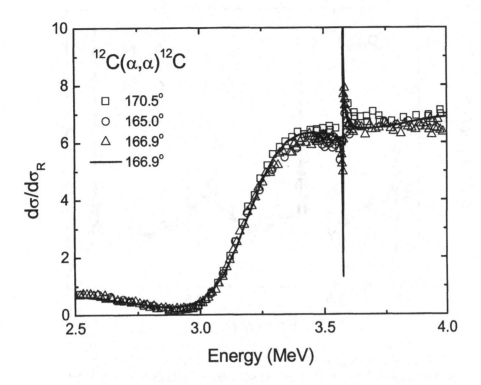

FIG. A9.3. □, Leavitt *et al.* (1989); ○, Feng *et al.* (1994); △, Hill (1953); solid line, Gurbich (2000).

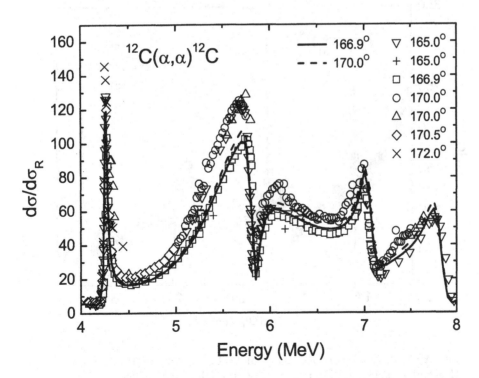

FIG. A9.4. ▽, Feng *et al.* (1994); +, Gosset (1989); □, Bittner and Moffat (1954); ○, Cheng *et al.* (1994a); △, Davies *et al.* (1994); ◇, Leavitt *et al.* (1989); ×, Somatri *et al.* (1996); solid and dashed lines, Gurbich (2000).

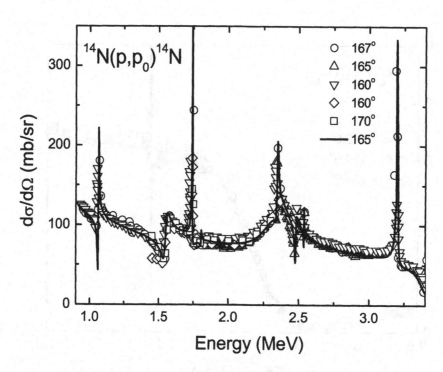

FIG. A9.5. ○, Olness *et al.* (1958b); △, Lambert and Durand (1967); ▽, Bashkin *et al.* (1959); ◇, Havranek *et al.* (1991); □, Rauhala (1985); solid line, Gurbich (2008).

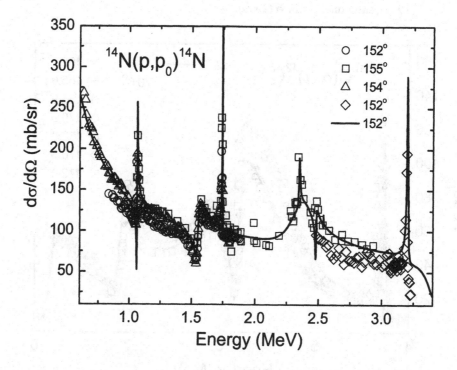

FIG. A9.6. ○, Tautfest and Rubin (1956); ○, Ferguson *et al.* (1959); △, Hagedorn *et al.* (1957); ◇, Jiang *et al.* (2005); solid line, Gurbich (2008).

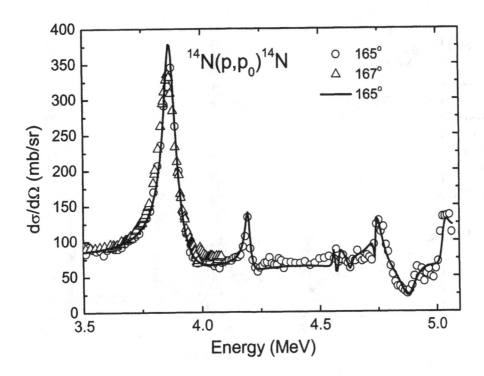

FIG. A9.7. ○, Bogdanović Radović *et al.* (2008); △, Olness *et al.* (1958a); solid line, Bogdanović Radović *et al.* (2008).

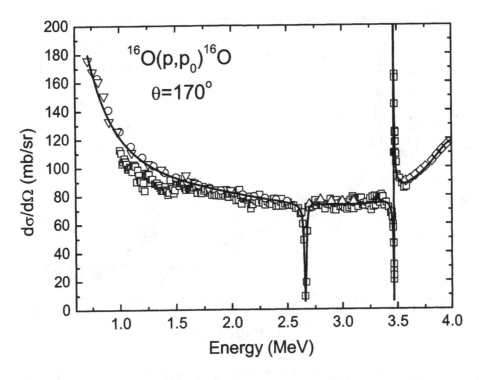

FIG. A9.8. □, Amirikas *et al.* (1993); ○, Luomajarvi *et al.* (1985); △, Yang *et al.* (1991); ▽, Chow *et al.* (1975); ◇, Eppling *et al.* (1953); solid line, Gurbich (1997).

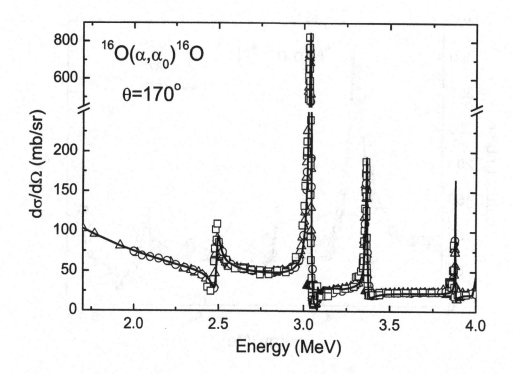

FIG. A9.9. △, Leavitt *et al.* (1990); ○, Cheng *et al.* (1993); □, Demarche and Terwagne (2006), solid line, evaluation (unpublished).

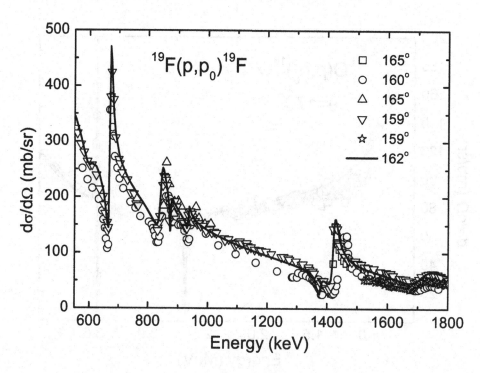

FIG. A9.10. □, Jesus *et al.* (2001); ○, Dearnaley (1956); △, Knox and Harmon (1989); ▽, Webb *et al.* (1955); ☆, Ouichaoui *et al.* (1986), solid line, evaluation (unpublished).

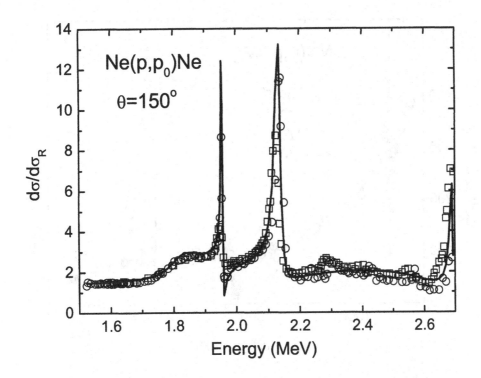

FIG. A9.11. ○, Lambert *et al.* (1972); □, Valter *et al.* (1960), solid line, evaluation (unpublished).

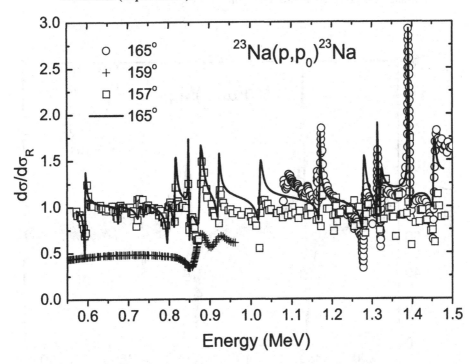

FIG. A9.12. ○, Vanhoy *et al.* (1987); +, Dearnaley (1956); □, Baumann *et al.* (1956), solid line, evaluation (unpublished).

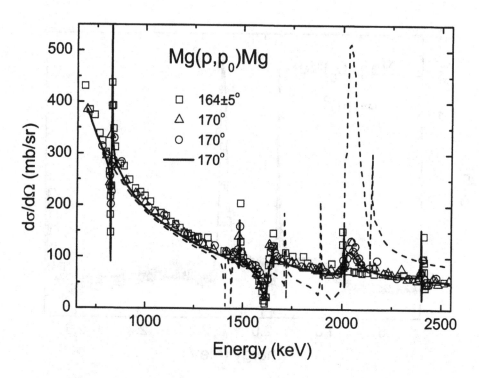

FIG. A9.13. □, Mooring *et al.* (1951); △, Rauhala and Luomajärvi (1988); ○, Zhang *et al.* (2003); solid line, Gurbich and Jeynes (2007); dashed line, theoretical cross-section for ^{26}Mg.

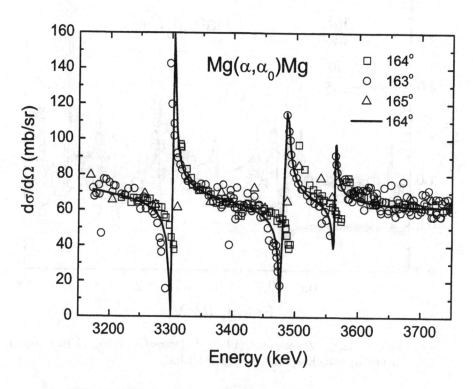

FIG. A9.14. □, Kaufman *et al.* (1952); ○, Cseh *et al.* (1982); △, Cheng *et al.* (1994b), solid line, evaluation (unpublished).

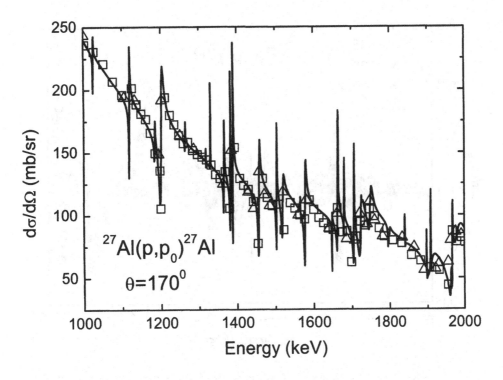

FIG. A9.15. △, Rauhala (1989); □, Chiari *et al.* (2001); solid line, Gurbich *et al.* (2002).

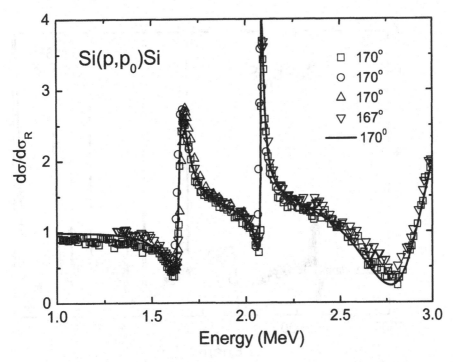

FIG. A9.16. □, Amirikas *et al.* (1993); ○, Rauhala (1985); △, Salomonovič (1993); ▽, Vorona *et al.* (1959); solid line, Gurbich (1998b).

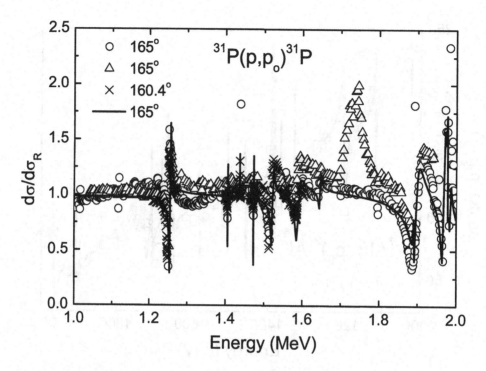

FIG. A9.17. O, Fang *et al.* (1988); △, Cohen-Ganouna *et al.* (1963); ×, Vernotte *et al.* (1973), solid line, evaluation (unpublished).

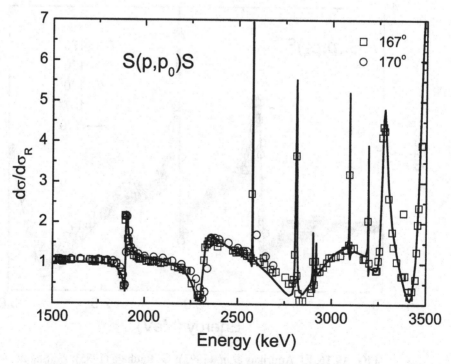

FIG. A9.18. □, Olness *et al.* (1958a); O, Rauhala and Luomajärvi (1988), solid line, evaluation (unpublished).

138

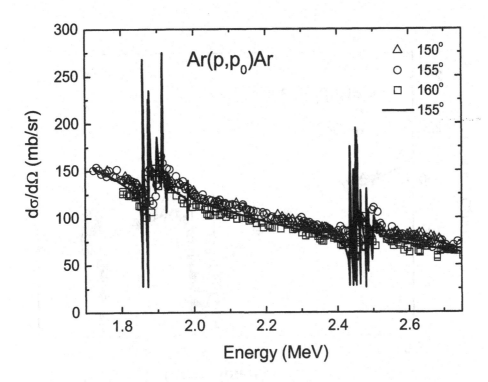

FIG. A9.19. △, Valter *et al.* (1959); ○, Frier *et al.* (1958); □, Barnhard and Kim (1961), solid line, evaluation (unpublished).

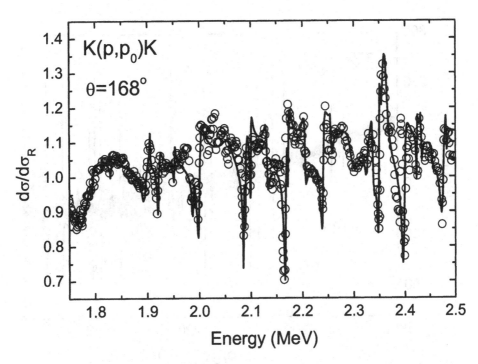

FIG. A9.20. ○, De Meijer *et al.* (1970) normalized against theory, solid line, evaluation (unpublished).

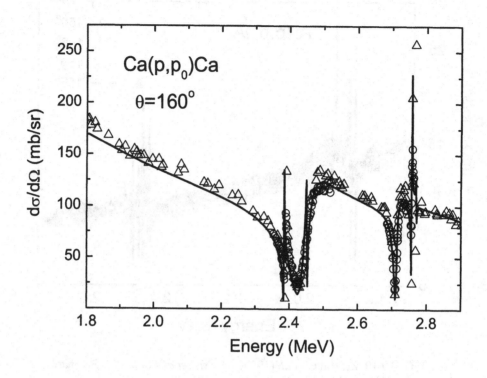

FIG. A9.21. ◯, Koltay *et al.* (1975); △, Wilson *et al.* (1974), solid line, evaluation (unpublished).

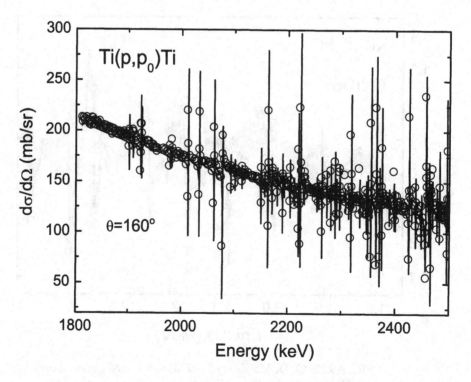

FIG. A9.22. ◯, Prochnow *et al.* (1972), solid line, evaluation (unpublished).

REFERENCES

Amirikas, R., Jamieson, D.N., and Dooley, S.P. (1993), *Nucl. Instrum. Methods* **B77**, 110.

Baglin, J.E.E., Kellock, A.J., Crockett, M.A., and Smith, A. (1992), *Nucl. Instrum. Methods* **B64**, 469.

Barnhard, A.C.L., and Kim, C.C. (1961), *Nucl. Phys.* **28**, 428.

Bashkin, S., Carlson, R.R., and Douglas, R.A. (1959), *Phys. Rev.* **114**, 1552.

Baumann, N.P., Prosser, F.W., Read, G.W., and Krone, R.W. (1956), *Phys. Rev.* **104**, 376.

Bittner, J.W., and Moffat, R.D. (1954), *Phys. Rev.* **96**, 374.

Bogdanović Radović, I., Siketić, Z., Jakšić, M., and Gurbich, A.F. (2008), *J. Appl. Phys.* **104**, 074905.

Cheng, H.-S., Shen, H., Tang, J., and Yang, F. (1993), *Nucl. Instrum. Methods* **B83**, 449.

Cheng, H.-S., Shen, H., Tang, J., and Yang, F. (1994a), *Acta Phys. Sin.* **43**, 1569.

Cheng, H.-S., Shen, H., Yang, F., and Tang, J. (1994b), *Nucl. Instrum. Methods* **B85**, 47.

Chiari, M., Giuntini, L., Mando, P.A., and Taccetti, N. (2001), *Nucl. Instrum. Methods* **B174**, 259.

Chow, H.C., Griffits, G.M., and Hall, T.H. (1975), *Can. J. Phys.* **53**, 1672.

Cohen-Ganouna, J., Lambert, M., and Schmouker, J. (1963), *Nucl. Phys.* **40**, 67.

Cseh, J., Koltay, E., Mate, Z., Somorjai, E., and Zolnai, L. (1982), *Nucl. Phys.* **A385**, 43.

Davies, J.A., Almeida, F.J.D., Haugen, H.K., Siegele, R., Forster, J.S., and Jackman, T.E. (1994), *Nucl. Instrum. Methods* **B85**, 28.

Dearnaley, G. (1956), *Philos. Mag. Ser. 8* **1**, 821.

Demarche J., and Terwagne, G. (2006), *J. Appl. Phys.* **100**, 124909.

De Meijer, R.J., Sieders, A.A., Landman, H.A.A., and De Roos, G. (1970), *Nucl. Phys.* **A155**, 109.

Eppling, F.J., Cameron, J.R., Davis, R.H., Divatia, A.S., Galonsky, A.L., Goldberg, E., and Hill, R.W. (1953), *Phys. Rev.* **91**, 438.

Fang, D.F., Bilpuch, E.G., Westerfeldt, C.R., and Mitchell, G.E. (1988), *Phys. Rev.* **C37** 28.

Feng, Y., Zhou, Z., Zhou, C., and Zhao, G. (1994), *Nucl. Instrum. Methods* **B86**, 225.

Ferguson, A.J., Clarke, R.L., and Gove, H.E. (1959), *Phys. Rev.* **115**, 1655.

Frier, G.D., Famularo, K.F., Zipoy, D.M., and Leigh, J. (1958), *Phys. Rev.* **110**, 446.

Gosset, C.R. (1989), *Nucl. Instrum. Methods* **B40–41**, 813.

Gurbich, A.F. (1997), *Nucl. Instrum. Methods* **B129**, 311.

Gurbich, A.F. (1998a), *Nucl. Instrum. Methods* **B136–138**, 60.

Gurbich, A.F. (1998b), *Nucl. Instrum. Methods* **B145**, 578.

Gurbich, A.F. (2000), *Nucl. Instrum. Methods* **B161–163**, 125.

Gurbich, A.F., Barradas, N.P., Jeynes, C., and Wendler, E. (2002), *Nucl. Instrum. Methods* **B190**, 237.

Gurbich, A.F. (2007), *Nucl. Instrum. Methods* **B261**, 401.

Gurbich, A.F., and Jeynes, C. (2007), *Nucl. Instrum. Methods* **B265**, 447.

Gurbich, A.F. (2008), *Nucl. Instrum. Methods* **B266**, 1193.

Hagedorn, F.B., Mozer, F.S., Webb, T.S., Fowler, W.A., and Lauritsen, C.C. (1957), *Phys. Rev.* **105**, 219.

Havranek, V., Hnatowic, V., and Kvitek, J. (1991), *Czech. J. Phys.* **41**, 921.

Hill, R.W. (1953), *Phys. Rev.* **90**, 845.

Jackson, H.L., Galonsky, A.I., Eppling, F.J., Hill, R.W., Goldberg, E., and Cameron, J.R. (1953), *Phys. Rev.* **89**, 365.

Jesus, A.P., Braizinha, B., Cruz, J., and Ribeiro, J.P. (2001), *Nucl. Instrum. Methods* **B174**, 229.

Jiang, W., Shutthanandan, V., Thevuthasan, S., Wang, C.M., and Weber, W.J. (2005), *Surf. Interface Anal.* **37**, 374.

Kaufman, S., Goldberg, E., Koester, L.J., and Mooring, F.P. (1952), *Phys. Rev.* **88**, 673.

Kim, C.-S., Kim, S.-K., and Choi, D. (1999), *Nucl. Instrum. Methods* **B155**, 229.

Knox, J.M., and Harmon, J.F. (1989), *Nucl. Instrum. Methods* **B44**, 41.

Koltay, E., Mesko, L., and Vegh, L. (1975), *Nucl. Phys.* **A249**, 173.

Lambert, M., and Durand, M. (1967), *Phys. Lett.* **24B**, 287.

Lambert, M., Midy, P., Drain, D., Amiel, M., Beaumevielle, H., Dauchy, A., and Meynadier, C. (1972), *J. Phys.* **33**, 155.

Leavitt, J.A., McIntyre Jr., L.C., Stoss, P., Oder, J.G., Ashbaugh, M.D., Dezfouly-Arjomandy, B., Yang, Z.-M., and Lin, Z. (1989), *Nucl. Instrum. Methods* **B40–41**, 776.

Leavitt, J.A., McIntyre Jr., L.C., Ashbaugh, M.D., Oder, J.G., Lin, Z., and Dezfouly-Arjomandy, B. (1990), *Nucl. Instrum. Methods* **B44**, 260.

Liu, Z., Li, B., Duan, Z., and He, H. (1993), *Nucl. Instrum. Methods* **B74**, 439.

Luomajarvi, M., Rauhala, E., and Hautala, M. (1985), *Nucl. Instrum. Methods* **B9**, 255.

Mazzoni, S., Chiari, M., Giuntini, L., Mando, P.A., and Taccetti, N. (1999), *Nucl. Instrum. Methods* **B159**, 191.

Mooring, F.P., Koster, L.J., Goldberg, E., Saxon, D., and Kaufmann, S.G. (1951), *Phys. Rev.* **84**, 703.

Nagata, S., Yamaguchi, S.. Fujino, Y., Hori, Y., Sugiyama, N., and Kamada, K. (1985), *Nucl. Instrum. Methods* **B6**, 533.

Olness, J.W., Haeberli, W., and Lewis, H.W. (1958a), *Phys. Rev.* **112**, 1702.

Olness, J.W., Vorona, J., and Lewis, H.W. (1958b), *Phys. Rev.* **112**, 475.

Ouichaoui, S., Beaumevieille, H., and Bendjaballah, N. (1986), *Nuovo Cim.* **A94**, 133.

Prochnow N.H., Newson, H.W., Bilpuch, E.G., and Mitchell, G.E. (1972), *Nucl. Phys.* **A194**, 353.

Pusa, P., Rauhala, E., Gurbich, A., and Nurmela, A. (2004), *Nucl. Instrum. Methods* **B222**, 686.

Quillet, V., Abel, F., and Schott, M. (1993), *Nucl. Instrum. Methods* **B83**, 47.

Rauhala, E. (1985), *Nucl. Instrum. Methods* **B12**, 447.

Rauhala, E., and Luomajärvi, M. (1988), *Nucl. Instrum. Methods* **B33**, 628.

Rauhala, E. (1989), *Nucl. Instrum. Methods* **B40–41**, 790.

Salomonovič, R. (1993), *Nucl. Instrum. Methods* **B82**, 1.

Somatri, R., Chailan, J.F., Chevarier, A., Chevarier, N., Ferro, G., Monteil, Y., Vincent, H., and Bouix, J. (1996), *Nucl. Instrum. Methods* **B113**, 284.

Tautfest, G.W., and Rubin, S. (1956), *Phys. Rev.* **103**, 196.

Valter, A.K., Malakhov, I.Ya., Sorokin, P.V., and Taranov, A.Ya. (1959), *Izv. Akad. Nauk SSSR, Ser. Fiz.* **23**, 846 (in Russian).

Valter, A.K., Deineko, A.S., Sorokin, P.V., and Taranov, A.Ya. (1960), *Izv. Akad. Nauk SSSR, Ser. Fiz.* **24**, 884 (in Russian).

Vanhoy, J.R., Bilpuch, E.G., Westerfeldt, C.R., and Mitchell, G.E. (1987), *Phys. Rev.* **C36**, 920.

Vernotte, J., Gales, S., Langevin, M., and Maison, J.M. (1973), *Nucl. Phys.* **A212**, 493.

Vorona, J., Olness, J.W., Haeberli, W., and Lewis, H.W. (1959), *Phys. Rev.* **116**, 1563.

Wang, H., and Zhou, G. (1988), *Nucl. Instrum. Methods* **B34**, 145.

Webb, T.S., Hagedorn, F.B., Fowler, W.A., and Lauritsen, C.C. (1955), *Phys. Rev.* **99**, 138.

Wilson Jr., W.M., Moses, J.D., and Bilpuch, E.G. (1974), *Nucl. Phys.* **A227**, 277.

Yang, G., Zhu, D., Xu, H., and Pan, H. (1991), *Nucl. Instrum. Methods* **B61**, 175.

Zhang, X., Li, G., Ding, B., and Liu, Z. (2003), *Nucl. Instrum. Methods* **B201**, 551.

DEUTERIUM-INDUCED NUCLEAR REACTION PARAMETERS

Compiled by

G. Vizkelethy

Sandia National Laboratories, Albuquerque, New Mexico, USA

This table gives the energies of protons and alphas from (d,p) and (d,α) reactions before (E_{out}) and after having passed through (E_{mylar}) an absorber foil of Mylar that stops the deuteron beam. E_d is the energy of the deuterons; X_{mylar} is the thickness of the absorber foil; and E_{min}^{α} and E_{min}^{p} are the minimum energies with which the α-particles and protons, respectively, can penetrate the absorber foils (G. Amsel, Universités Paris 7 et Paris 6, Paris, France, personal communication, 1995.). The table can also be produced using SIMNRA software (Mayer, M., *Nucl. Instrum. Methods* B194 (2002) 177).

(a) $E_d = 0.9$ MeV

     $E_{min}^{\alpha} = 2.9$ MeV $\Theta = 150°$ $X_{mylar} = 12$ μm

                                                           $E_{min}^{p} = 0.9$ MeV

Element	Reaction (d,α)	Reaction (d,p)	Q (MeV)	E_{out} (MeV)	E_{mylar} (MeV)
D		D(d,p)T	4.033	2.36	2.1
^{6}Li		^{6}Li(d,p$_0$)^{7}Li	5.027	4.36	4.2
		^{6}Li(d,p$_1$)^{7}Li	4.549	3.97	3.8
		^{6}Li(d,p$_2$)^{7}Li	0.397	0.67	0
	^{6}Li(d,α)^{4}He		22.36	9.61	8.6
^{7}Li		^{7}Li(d,p)^{8}Li	−0.192	0.29	0
	^{7}Li(d,α)^{5}He		14.163	6.82	5.48
^{9}Be		^{9}Be(d,p)^{10}B	4.585	4.38	4.2
	^{9}Be(d,α$_0$)^{7}Li		7.152	4.10	2.08
	^{9}Be(d,α$_1$)^{7}Li		6.674	3.83	1.68
	^{9}Be(d,α$_2$)^{7}Li		2.522	1.50	0

Element	Reaction (d,α)	Reaction (d,p)	Q (MeV)	E_{out} (MeV)	E_{mylar} (MeV)
^{10}B		$^{10}B(d,p_0)^{11}B$	9.237	8.58	8.48
		$^{10}B(d,p_1)^{11}B$	7.107	6.69	6.56
		$^{10}B(d,p_2)^{11}B$	4.777	4.64	4.50
		$^{10}B(d,p_3)^{11}B$	4.207	4.14	4.00
		$^{10}B(d,p_4)^{11}B$	2.477	2.63	2.40
		$^{10}B(d,p_5)^{11}B$	2.427	2.59	2.35
		$^{10}B(d,p_6)^{11}B$	1.937	2.17	1.90
		$^{10}B(d,p_7)^{11}B$	1.247	1.54	1.23
		$^{10}B(d,p_8)^{11}B$	0.667	1.09	0.58
	$^{10}B(d,\alpha_0)^8Be$		17.819	11.04	10.30
	$^{10}B(d,\alpha_1)^{8B}e$		14.919	9.22	8.20
^{11}B		$^{11}B(d,p_0)^{12}B$	1.138	1.52	1.16
		$^{11}B(d,p_1)^{12}B$	0.188	0.72	0
	$^{11}B(d,\alpha_0)^9Be$		8.022	5.22	3.60
	$^{11}B(d,\alpha_1)^9Be$		6.272	4.10	2.10
	$^{11}B(d,\alpha_2)^9Be$		5.592	3.67	1.44
	$^{11}B(d,\alpha_3)^9Be$		4.982	3.29	0.76
	$^{11}B(d,\alpha_4)^9Be$		3.282	2.22	0
	$^{11}B(d,\alpha_5)^9Be$		1.262	1.00	0
^{13}B	$^{13}B(d,\alpha)^{11}Be$		5.167	3.73	1.50
^{12}C		$^{12}C(d,p)^{13}C$	2.719	2.95	2.72
^{13}C		$^{13}C(d,p)^{14}C$	5.947	5.89	5.77
	$^{13}C(d,\alpha_0)^{11}B$		5.167	3.73	1.30
	$^{13}C(d,\alpha_1)^{11}B$		3.037	2.30	0
	$^{13}C(d,\alpha_2)^{11}B$		0.707	0.78	0
	$^{13}C(d,\alpha_3)^{11}B$		0.137	0.44	0
^{14}N		$^{14}N(d,p_0)^{15}N$	8.615	8.39	8.25
		$^{14}N(d,p_1)^{15}N$	3.335	3.58	3.40
		$^{14}N(d,p_2)^{15}N$	3.305	3.56	3.37
		$^{14}N(d,p_3)^{15}N$	2.285	2.64	2.40
		$^{14}N(d,p_4)^{15}N$	1.455	1.99	1.60
		$^{14}N(d,p_5)^{15}N$	1.305	1.76	1.47
		$^{14}N(d,p_6)^{15}N$	1.045	1.53	1.17
	$^{14}N(d,\alpha_0)^{12}C$		13.579	9.84	9.05
	$^{14}N(d,\alpha_1)^{12}C$		9.146	6.67	5.35
	$^{14}N(d,\alpha_2)^{12}C$		5.923	4.40	2.44
	$^{14}N(d,\alpha_3)^{12}C$		3.949	3.02	0.20
	$^{14}N(d,\alpha_4)^{12}C$		3.479	2.70	0
	$^{14}N(d,\alpha_5)^{12}C$		2.739	2.19	0
	$^{14}N(d,\alpha_6)^{12}C$		2.479	2.10	0
^{15}N		$^{15}N(d,p)^{16}N$	0.267	0.83	0.08
	$^{15}N(d,\alpha)^{13}C$		7.683	5.80	4.35

Element	Reaction (d,α)	Reaction (d,p)	Q (MeV)	E_{out} (MeV)	E_{mylar} (MeV)
^{16}O		$^{16}O(d,p_0)^{17}O$	1.919	2.36	2.12
		$^{16}O(d,p_1)^{17}O$	1.048	1.58	1.23
	$^{16}O(d,\alpha_0)^{14}N$		3.116	2.61	0
	$^{16}O(d,\alpha_1)^{14}N$		0.804	0.97	0
^{17}O		$^{17}O(d,p_0)^{18}O$	5.842	5.99	5.90
		$^{17}O(d,p_1)^{18}O$	3.860	4.17	4.00
	$^{17}O(d,\alpha_0)^{15}N$		9.812	7.68	6.50
	$^{17}O(d,\alpha_1)^{15}N$		4.532	3.72	1.30
^{18}O		$^{18}O(d,p_0)^{19}O$	1.731	2.24	1.96
		$^{18}O(d,p_1)^{19}O$	1.635	2.15	1.87
		$^{18}O(d,p_2)^{19}O$	0.262	0.90	0.27
	$^{18}O(d,\alpha_0)^{16}N$		4.237	3.58	1.10
	$^{18}O(d,\alpha_1)^{16}N$		4.117	3.49	0.95
	$^{18}O(d,\alpha_2)^{16}N$		3.942	3.36	0.77
	$^{18}O(d,\alpha_3)^{16}N$		3.845	3.29	0.52
	$^{18}O(d,\alpha_4)^{16}N$		0.707	0.97	0
	$^{18}O(d,\alpha_5)^{16}N$		0.257	0.65	0
^{19}F		$^{19}F(d,p_0)^{20}F$	4.379	4.92	4.70
		$^{19}F(d,p_1)^{20}F$	3.729	4.31	4.03
		$^{19}F(d,p_2)^{20}F$	3.549	4.15	3.90
		$^{19}F(d,p_3)^{20}F$	3.389	4.00	3.74
		$^{19}F(d,p_4)^{20}F$	3.319	3.94	3.68
		$^{19}F(d,p_5)^{20}F$	3.069	3.71	3.44
		$^{19}F(d,p_6)^{20}F$	2.409	3.10	2.80
		$^{19}F(d,p_7)^{20}F$	2.329	3.02	2.70
		$^{19}F(d,p_8)^{20}F$	2.179	2.89	2.56
		$^{19}F(d,p_9)^{20}F$	1.509	2.27	1.86
		$^{19}F(d,p_{10})^{20}F$	1.409	2.18	1.79
		$^{19}F(d,p_{11})^{20}F$	0.889	1.71	1.16
		$^{19}F(d,p_{12})^{20}F$	0.849	1.67	1.11
		$^{19}F(d,p_{13})^{20}F$	0.789	1.62	1.05
		$^{19}F(d,p_{14})^{20}F$	0.699	1.54	0.93
		$^{19}F(d,p_{15})^{20}F$	0.419	1.28	0.56
		$^{19}F(d,p_{16})^{20}F$	0.299	1.17	0.35
	$^{19}F(d,\alpha_0)^{17}O$		10.038	8.25	6.50
	$^{19}F(d,\alpha_1)^{17}O$		9.167	7.57	5.60
	$^{19}F(d,\alpha_2)^{17}O$		6.980	5.89	3.36

(b) $E_d = 1.2 \, \text{MeV}$ $X_{mylar} = 19 \, \mu m$

  $E_{min}^{\alpha} = 4.0 \, \text{MeV}$ $\Theta = 150°$ $E_{min}^{p} = 1.06 \, \text{MeV}$

Element	Reaction (d,α)	Reaction (d,p)	Q (MeV)	E_{out} (MeV)	E_{mylar} (MeV)
D		D(d,p)T	4.033	2.306	1.87
^{6}Li		^{6}Li(d,p$_0$)^{7}Li	5.027	4.44	4.15
		^{6}Li(d,p$_1$)^{7}Li	4.549	4.05	3.75
		^{6}Li(d,p$_2$)^{7}Li	0.397	0.80	0
	^{6}Li(d,α)^{4}He		22.36	9.87	7.75
^{7}Li		^{7}Li(d,p)^{8}Li	−0.192	0.43	0
	^{7}Li(d,α)^{5}He		14.163	6.72	4.48
^{9}Be		^{9}Be(d,p)^{10}B	4.585	4.522	4.27
	^{9}Be(d,α_0)^{7}Li		7.152	4.108	0.04
	^{9}Be(d,α_1)^{7}Li		6.674	3.837	0
	^{9}Be(d,α_2)^{7}Li		2.522	1.54	0
^{10}B		^{10}B(d,p$_0$)^{11}B	9.237	8.71	8.57
		^{10}B(d,p$_1$)^{11}B	7.107	6.83	6.67
		^{10}B(d,p$_2$)^{11}B	4.777	4.79	4.58
		^{10}B(d,p$_3$)^{11}B	4.207	4.29	4.04
		^{10}B(d,p$_4$)^{11}B	2.477	2.80	2.46
		^{10}B(d,p$_5$)^{11}B	2.427	2.75	2.40
		^{10}B(d,p$_6$)^{11}B	1.937	2.33	1.97
		^{10}B(d,p$_7$)^{11}B	1.247	1.75	1.215
		^{10}B(d,p$_8$)^{11}B	0.667	1.26	0.51
	^{10}B(d,α_0)^{8}Be		17.819	11.00	9.6
	^{10}B(d,α_1)^{8}Be		14.919	9.19	7.56
^{11}B		^{11}B(d,p$_0$)^{12}B	1.138	1.70	1.15
		^{11}B(d,p$_1$)^{12}B	0.188	0.90	0
	^{11}B(d,α_0)^{9}Be		8.022	5.25	2.48
	^{11}B(d,α_1)^{9}Be		6.272	4.15	0.4
	^{11}B(d,α_2)^{9}Be		5.592	3.72	0
	^{11}B(d,α_3)^{9}Be		4.982	3.34	0
	^{11}B(d,α_4)^{9}Be		3.282	2.29	0
	^{11}B(d,α_5)^{9}Be		1.262	1.09	0
^{13}B	^{13}B(d,α)^{11}Be		5.167	3.81	0
^{12}C		^{12}C(d,p)^{13}C	2.719	3.13	2.80
^{13}C		^{13}C(d,p)^{14}C	5.947	6.07	5.90
	^{13}C(d,α_0)^{11}B		5.167	3.81	0
	^{13}C(d,α_1)^{11}B		3.037	2.39	0
	^{13}C(d,α_2)^{11}B		0.707	0.90	0
	^{13}C(d,α_3)^{11}B		0.137	0.55	0

Element	Reaction		Q (MeV)	E_{out} (MeV)	E_{mylar} (MeV)
	(d,α)	(d,p)			
^{14}N		$^{14}N(d,p_0)^{15}N$	8.615	8.56	8.41
		$^{14}N(d,p_1)^{15}N$	3.335	3.78	3.50
		$^{14}N(d,p_2)^{15}N$	3.305	3.75	3.47
		$^{14}N(d,p_3)^{15}N$	2.285	2.83	2.48
		$^{14}N(d,p_4)^{15}N$	1.455	2.10	1.64
		$^{14}N(d,p_5)^{15}N$	1.305	1.96	1.48
		$^{14}N(d,p_6)^{15}N$	1.045	1.73	1.20
	$^{14}N(d,\alpha_0)^{12}C$		13.579	9.88	8.34
	$^{14}N(d,\alpha_1)^{12}C$		9.146	6.74	4.52
	$^{14}N(d,\alpha_2)^{12}C$		5.923	4.48	1.15
	$^{14}N(d,\alpha_3)^{12}C$		3.949	3.12	0
	$^{14}N(d,\alpha_4)^{12}C$		3.479	2.80	0
	$^{14}N(d,\alpha_5)^{12}C$		2.739	2.30	0
	$^{14}N(d,\alpha_6)^{12}C$		2.479	2.12	0
^{15}N		$^{15}N(d,p)^{16}N$	0.267	1.03	0
	$^{15}N(d,\alpha)^{13}C$		7.683	5.88	3.36
^{16}O		$^{16}O(d,p_0)^{17}O$	1.919	2.57	2.20
		$^{16}O(d,p_1)^{17}O$	1.048	1.79	1.27
	$^{16}O(d,\alpha_0)^{14}N$		3.116	2.73	0
	$^{16}O(d,\alpha_1)^{14}N$		0.804	1.11	0
^{17}O		$^{17}O(d,p_0)^{18}O$	5.842	6.19	6.00
		$^{17}O(d,p_1)^{18}O$	3.860	4.37	4.17
	$^{17}O(d,\alpha_0)^{15}N$		9.812	7.78	5.90
	$^{17}O(d,\alpha_1)^{15}N$		4.532	3.85	0
^{18}O		$^{18}O(d,p_0)^{19}O$	1.731	2.45	2.1
		$^{18}O(d,p_1)^{19}O$	1.635	2.37	1.95
		$^{18}O(d,p_2)^{19}O$	0.262	1.13	0.17
	$^{18}O(d,\alpha_0)^{16}N$		4.237	3.72	0
	$^{18}O(d,\alpha_1)^{16}N$		4.117	3.68	0
	$^{18}O(d,\alpha_2)^{16}N$		3.942	3.49	0
	$^{18}O(d,\alpha_3)^{16}N$		3.845	3.42	0
	$^{18}O(d,\alpha_4)^{16}N$		0.707	1.12	0
	$^{18}O(d,\alpha_5)^{16}N$		0.257	0.81	0

Element	Reaction (d,α)	Reaction (d,p)	Q (MeV)	E_{out} (MeV)	E_{mylar} (MeV)
^{19}F		$^{19}F(d,p_0)^{20}F$	4.379	5.13	4.82
		$^{19}F(d,p_1)^{20}F$	3.729	4.53	4.20
		$^{19}F(d,p_2)^{20}F$	3.549	4.37	4.00
		$^{19}F(d,p_3)^{20}F$	3.389	4.22	3.85
		$^{19}F(d,p_4)^{20}F$	3.319	4.16	3.80
		$^{19}F(d,p_5)^{20}F$	3.069	3.93	3.55
		$^{19}F(d,p_6)^{20}F$	2.409	3.32	2.87
		$^{19}F(d,p_7)^{20}F$	2.329	3.25	2.82
		$^{19}F(d,p_8)^{20}F$	2.179	3.11	2.70
		$^{19}F(d,p_9)^{20}F$	1.509	2.50	1.93
		$^{19}F(d,p_{10})^{20}F$	1.409	2.41	1.83
		$^{19}F(d,p_{11})^{20}F$	0.889	1.93	1.30
		$^{19}F(d,p_{12})^{20}F$	0.849	1.90	1.20
		$^{19}F(d,p_{13})^{20}F$	0.789	1.84	1.10
		$^{19}F(d,p_{14})^{20}F$	0.699	1.76	0.94
		$^{19}F(d,p_{15})^{20}F$	0.419	1.51	0.57
		$^{19}F(d,p_{16})^{20}F$	0.299	1.40	0.34
	$^{19}F(d,\alpha_0)^{17}O$		10.038	8.37	5.85
	$^{19}F(d,\alpha_1)^{17}O$		9.167	7.70	4.97
	$^{19}F(d,\alpha_2)^{17}O$		6.980	6.02	2.36

APPENDIX

11

PARTICLE-PARTICLE NUCLEAR REACTION CROSS SECTIONS

Compiled by

L. Foster

Los Alamos National Laboratory, Los Alamos, New Mexico, USA

G. Vizkelethy

Sandia National Laboratories, Albuquerque, New Mexico, USA

M. Lee, J. R. Tesmer, M. Nastasi, and Y.Q. Wang

Los Alamos National Laboratory, Los Alamos, New Mexico, USA

The data presented in this appendix are a partial listing of the most relevant reactions discussed in Section 6.5 of Chapter 6. The data were digitized from the original references given in the figure captions. All values of cross section, scattering angle, and ion energy are presented in laboratory units, unless otherwise noted. The original data were assumed to be in center-of-mass units unless explicitly stated to be in laboratory units. Center-of-mass values of cross section, scattering angle, and ion energy were converted into laboratory units using the conversion formulas given in Appendix 4. The reader should note that the absolute utility of the cross sections presented in this appendix is uncertain at best. This is compounded by the fact that, in many instances, a user's scattering geometry is not identical to those used to obtain the present data. It is recommended that the cross-section data as a function of incident-ion energy be used only as a guide in determining the appropriate energy range to perform nuclear reaction analysis (NRA). The most reliable analysis requires mapping of the cross section as a function of ion energy

for the scattering geometry employed and the careful use of well-calibrated standards. The use of standards in particle–particle nuclear reaction analysis is discussed in detail in Chapter 6.

The comprehensive database of nuclear reaction cross sections of specific interest for NRA is maintained by the International Atomic Energy Agency (IAEA) as the Ion Beam Analysis Nuclear Data Library (IBANDL, http://www-nds.iaea.org/ibandl). IBANDL originally incorporated the SigmaBase and NRABase (archival copies of which are available at http://www.mfa.kfki.hu/sigmabase/ and http://www.mfa.kfki.hu/sigmabase/programs/nrabase2.html, respectively) and now contains many more cross sections, including many of those initially compiled by Jarjis (1979). It is continually updated, and the cross sections can be downloaded in formats suitable for use with simulation programs and for plotting.

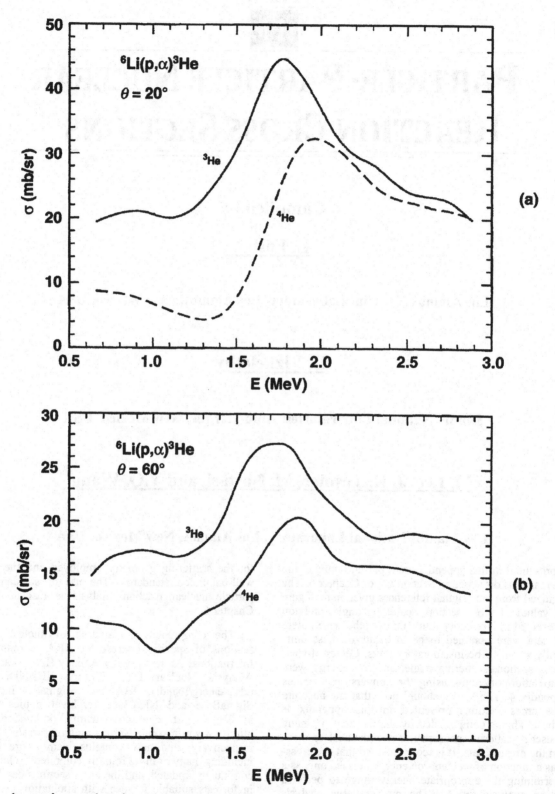

FIG. A11.1. The ^{6}Li(p,α)^{3}He and ^{6}Li(p,^{3}He)^{4}He reactions between 0.5 MeV and 3 MeV at (a) θ = 20° and (b) θ = 60°. From Marion, J.B., Weber, G., and Mozer, F.S. (1956), *Phys. Rev.* **104**, 1402.

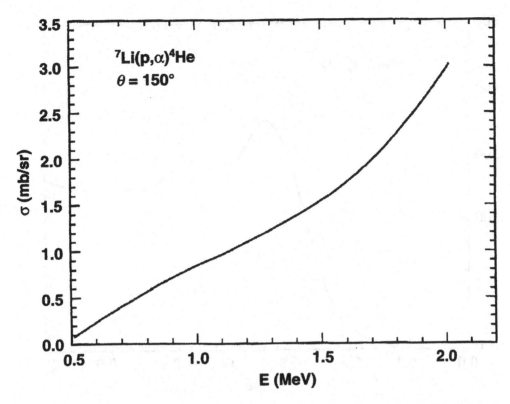

FIG. A11.2. The ^{7}Li(p,α)^{4}He reaction at $\theta = 150°$ in the energy range of 0.5–2 MeV. From Maurel, B., Amsel, G., and Dieumegard, D (1977), Chapter 4, Page 133, in *Ion Beam Handbook for Material Analysis*, Mayer, J.W., and Rimini, E. (eds.), Academic Press, New York.

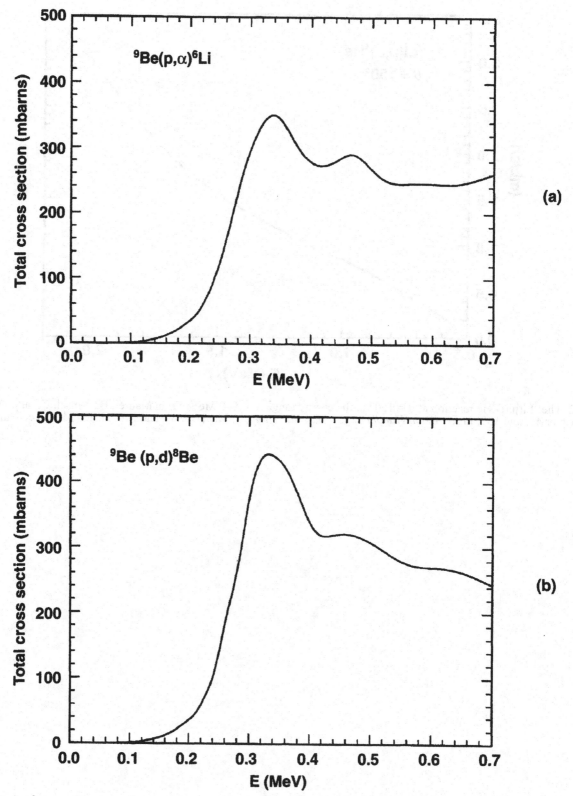

FIG. A11.3. The (a) ^{9}Be(p,α)^{6}Li and (b) ^{9}Be(p,d)^{8}Be reactions at $\theta = 90°$ between 0.1 MeV and 0.7 MeV. From Sierk, A.J., and Tombrello, T.A. (1973), *Nucl. Phys.* **A210**, 341.

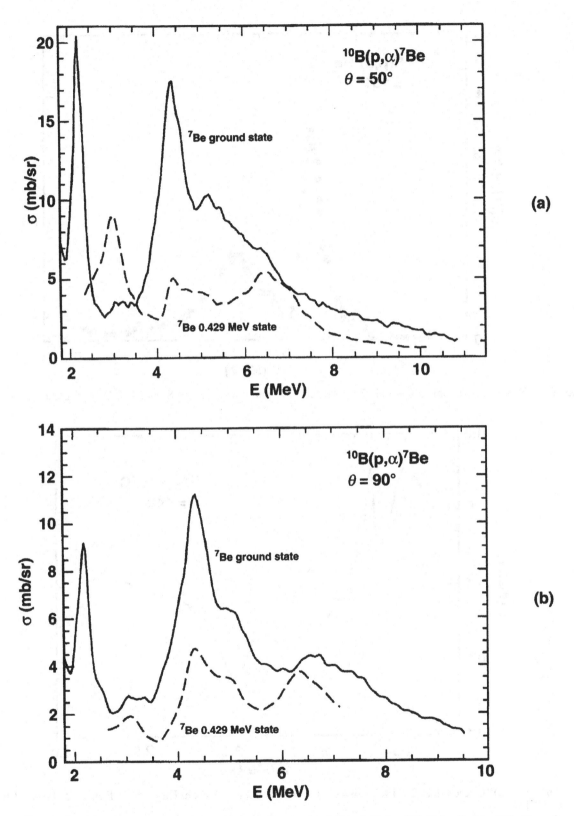

FIG. A11.4. The ^{10}B(p,α)^{7}Be reaction between 2 MeV and 11 MeV at (a) $\theta = 50°$ and (b) $\theta = 90°$. From Jenkin, J.G., Earwaker, L.G., and Titterton, E.W. (1964), *Nucl. Phys.* **50**, 517.

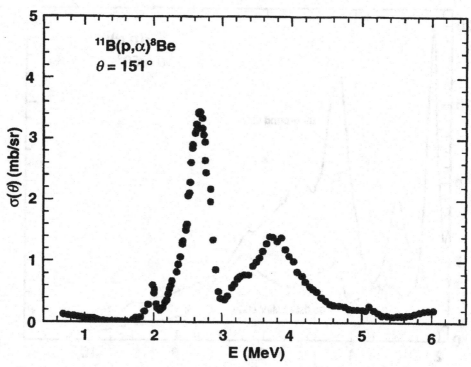

FIG. A11.5. The ^{11}B(p,α)^{8}Be reaction at $\theta = 151°$ between 0.8 MeV and 6 MeV. From Symons, G.D., and Treacy, P.B. (1963), *Nucl. Phys.* **46**, 93.

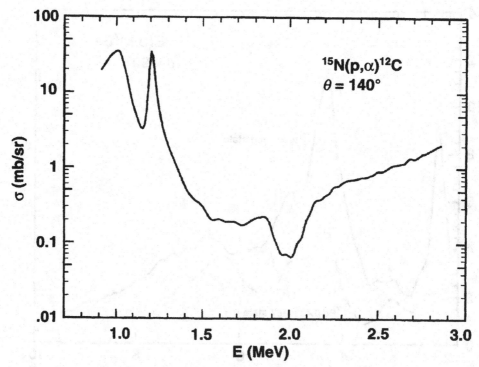

FIG. A11.6. The ^{15}N(p,α)^{12}C reaction at $\theta = 140°$ between 0.9 MeV and 3 MeV. From Hagendorn, F.B., and Marion, J.B. (1957), *Phys. Rev.* **108**, 1015.

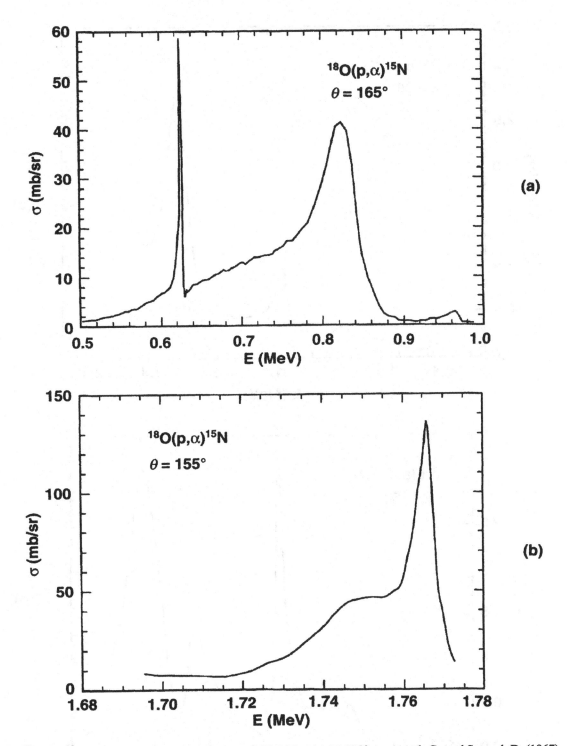

FIG. A11.7. The $^{18}O(p,\alpha)^{15}N$ reaction at (a) $\theta = 165°$ between 0.5 MeV and 1 MeV [from Amsel, G., and Samuel, D. (1967), *Anal. Chem.* **39**, 1689] and (b) at $\theta = 155°$ between 1.7 MeV and 1.775 MeV [from Alkemade, P.F.A., Stap, C.A.M., Habraken, F.H.P.M., and van der Weg, W.F. (1988), *Nucl. Instrum. Methods* **B35**, 135].

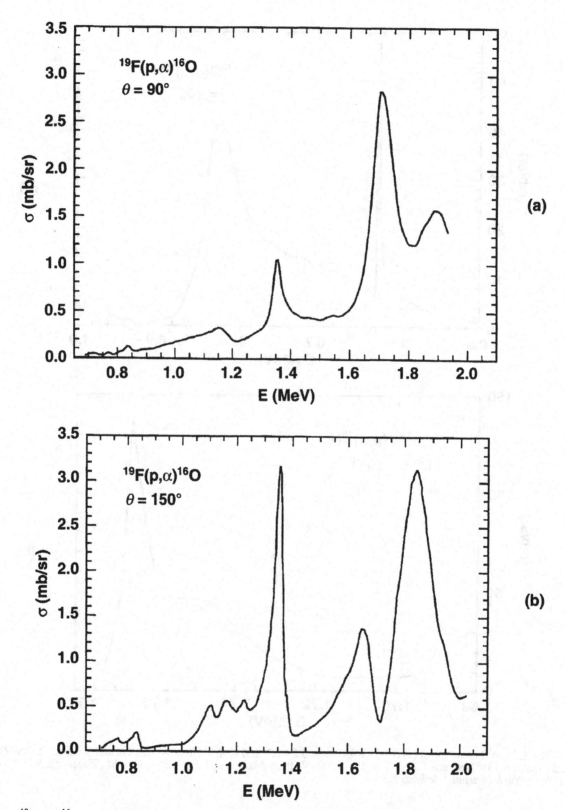

FIG. A11.8. The $^{19}F(p,\alpha)^{16}O$ reaction between 0.5 MeV and 2 MeV at (a) $\theta = 90°$ and (b) $\theta = 150°$. From Dieumegard, D., Dubreuil, D., and Amsel, G. (1979), *Nucl. Instrum. Methods* **166**, 431.

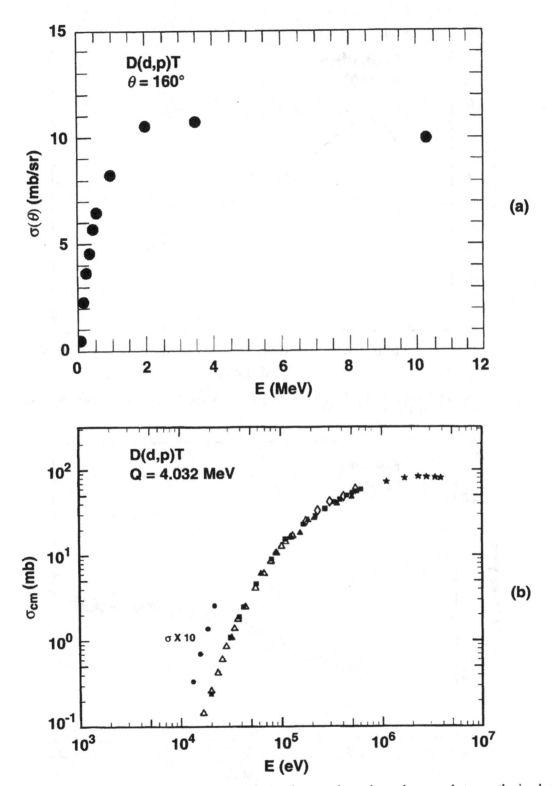

FIG. A11.9. The D(d,p)T reaction, which is always present when a deuteron beam is used, occurs between the implanted deuterium and the beam. Shown are (a) the differential cross section at θ = 160°, and (b) the total cross section. From Jarmie, N., and Seagrove, J. (1957), Los Alamos Scientific Laboratory Report LA-2014, Los Alamos National Laboratory, Los Alamos, NM.

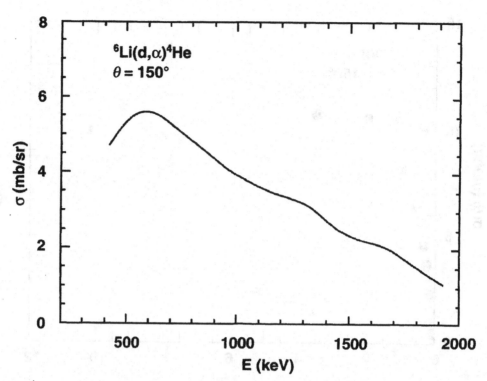

FIG. A11.10. The ^{6}Li(d,α)^{4}He reaction measured at $\theta = 150°$ between 0.5 MeV and 2 MeV. From Maurel, B., Amsel, G., and Dieumegard, D. (1981), *Nucl. Instrum. Methods* **191**, 349.

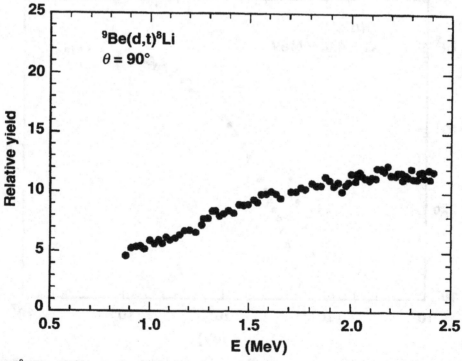

FIG. A11.11. The ^{9}Be(d,t)^{8}Be reaction measured at $\theta = 90°$ from 1.5 MeV to 2.5 MeV. From Biggerstaff, J.A., Hood, R.F., Scott, H., and McEllistrem, M.T. (1962), *Nucl. Phys.* **36**, 631.

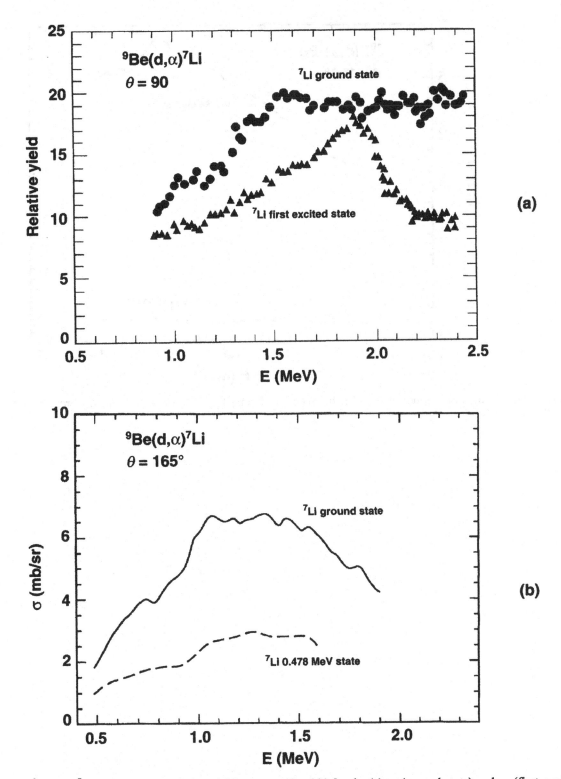

FIG. A11.12. The ^{9}Be(d,α)^{7}Li reaction. The relative yield curves at $\theta = 90°$ for the (a) α_0 (ground-state) and α_1 (first-excited-state) reactions. (b) The cross section for $\theta = 165°$ showing ^{7}Li in the ground state and the 0.478 MeV state. From Biggerstaff, J.A., Hood, R.F., Scott, H., and McEllistrem, M.T. (1962), *Nucl. Phys.* **36**, 631.

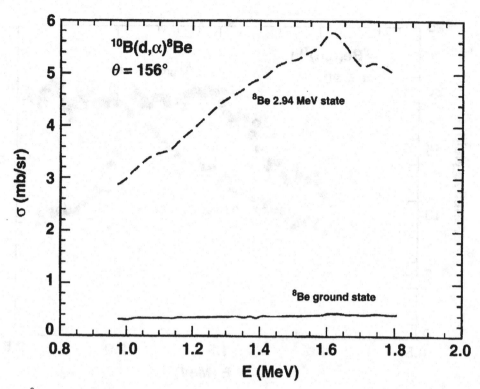

FIG. A11.13. The ^{10}B(d,α)^{8}Be reaction at $\theta = 156°$ between 1 MeV and 1.8 MeV for both α_0 and α_1. From Purser, K.H., and Wildenthal, B.H. (1963), *Nucl. Phys.* **44**, 22.

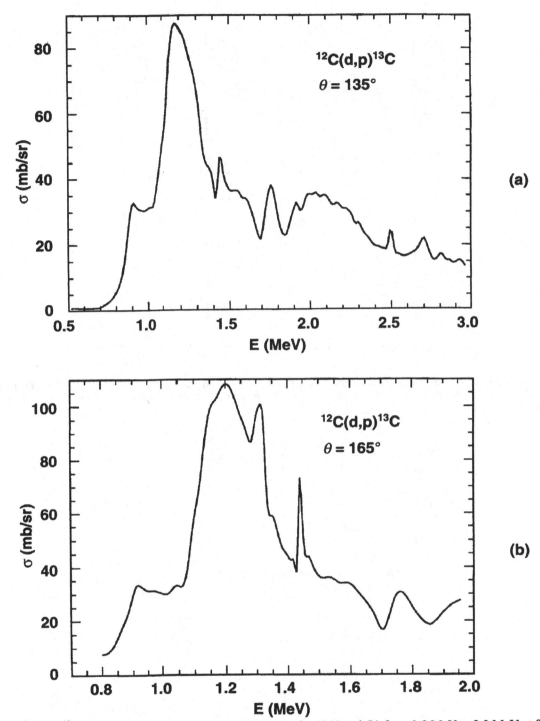

FIG. A11.14. The $^{12}C(d,p)^{13}C$ reaction (a) from 0.5 MeV to 3.0 MeV at θ =135° and (b) from 0.8 MeV to 2.0 MeV at θ =165°. From McEllistrem, M.T., Jones, K.W., Chiba, R., Douglas, R.A., Herring,D.F., and Silverstein, E.A. (1956), Phys. Rev. 104, 1008.

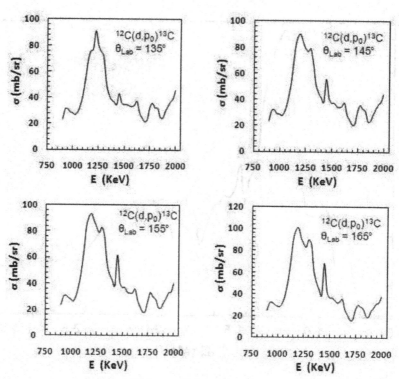

FIG. A11.15. Differential cross-section curves for the ^{12}C(d, p$_0$)^{13}C reaction at laboratory angles of 135–165°. From Kokkoris, M., Misaelides, P., Kossionides, S., Zarkadas, Ch., Lagoyannis, A., Vlastou, R., Papadopoulos, C.T., and Kontos, A. (2006), Nucl. Instrum. Methods B249, 77.

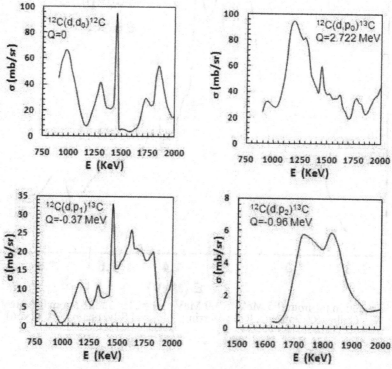

FIG. A11.16. Differential cross-section curves for the ^{12}C(d, d$_0$)^{12}C, ^{12}C(d, p$_0$)^{13}C, ^{12}C(d, p$_1$)^{13}C, and ^{12}C(d, p$_2$)^{13}C reactions at a laboratory angle of 150°. From Kokkoris, M., Misaelides, P., Kossionides, S., Zarkadas, Ch., Lagoyannis, A., Vlastou, R., Papadopoulos, C.T., and Kontos, A. (2006), Nucl. Instrum. Methods B249, 77.

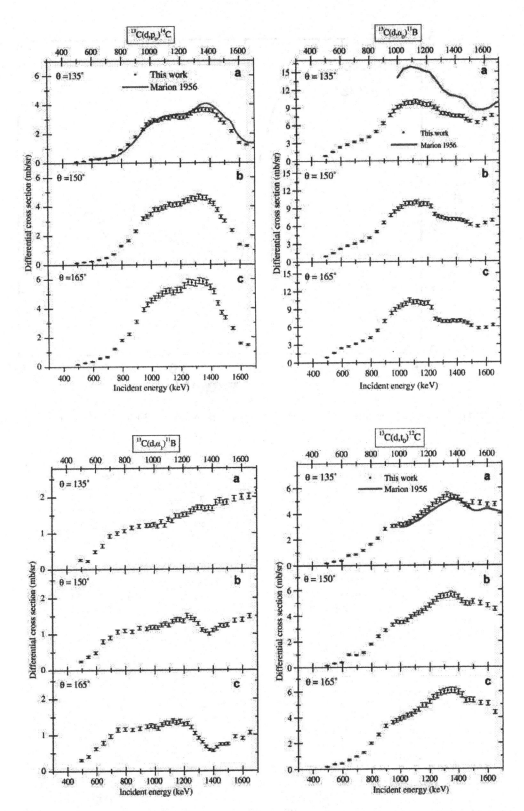

FIG. A11.17. Differential cross-section curves for the $^{13}C(d,p_0)^{14}C$, $^{13}C(d,\alpha_0)^{11}B$, $^{13}C(d,\alpha_1)^{11}B$, and $^{13}C(d,t_0)^{14}C$ reactions at laboratory angles of 135°, 150°, and 165°. From Colaux, J.L., Thomé, T., and Terwagne, T. (2007), *Nucl. Instrum. Methods* **B254**, 25.

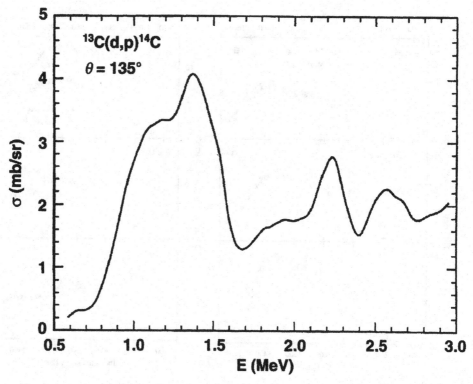

FIG. A11.18. The $^{13}C(d,p)^{14}C$ reaction measured at $\theta = 135°$ in the energy range of 0.6–3 MeV.From Marion, J.B., and Weber, *Phys. Rev.* **103**, 167.

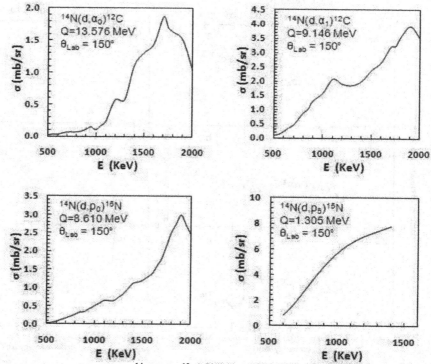

FIG. A11.19. Differential cross-section curves for the $^{14}N(d,\alpha_0)^{12}C$, $^{14}C(d,\alpha_1)^{12}C$, $^{14}N(d,p_0)^{15}N$, and $^{14}N(d,p_5)^{15}N$ reactions at a laboratory angle of 150°. From Pellegrino, S., Beck, L., and Trouslard,Ph.. (2004), *Nucl. Instrum. Methods* **B219–220**, 140.

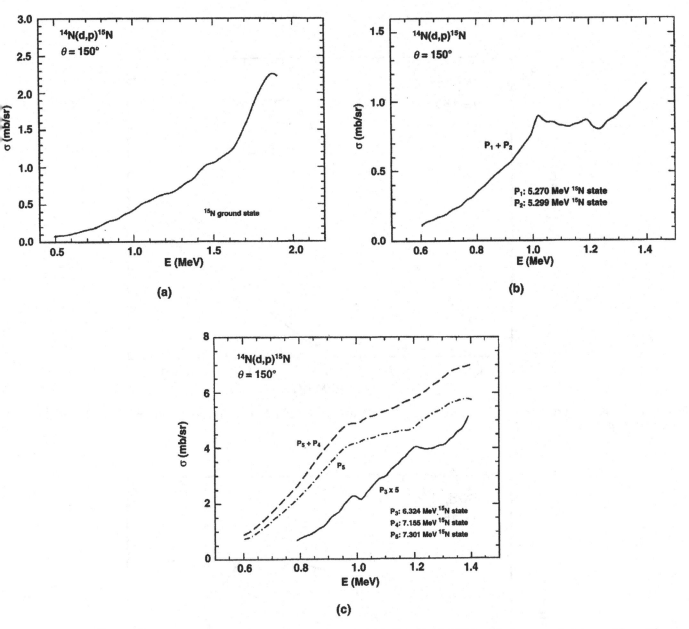

FIG. A11.20. The ^{14}N(d,p)^{15}N reaction measured at $\theta = 150°$ from 0.6 MeV to 1.4 MeV for (a) the ground state, p$_0$ [from Simpson, .C.B., and Earwaker, L.G. (1984), *Vacuum* **34**, 899]; (b) the p$_1$ and p$_2$ states; and (c) the p$_3$, p$_4$, and p$_5$ states. [(b) and (c) from Amsel, i., and David, D. (1969), *Rev. Phys. Appl.* **4**, 383.]

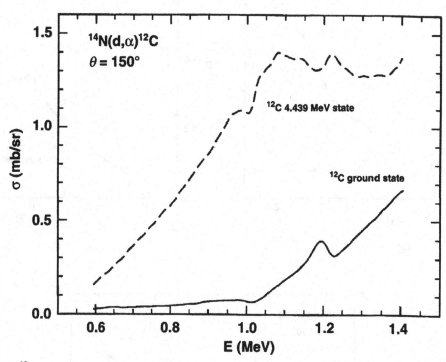

FIG. A11.21. The ^{14}N(d,α)^{12}C reaction measured at $\theta = 150°$ from 0.6 MeV to 1.4 MeV (α_0 and α_1). From Amsel, G., and David, D. (1969), *Rev. Phys. Appl.* **4**, 383.

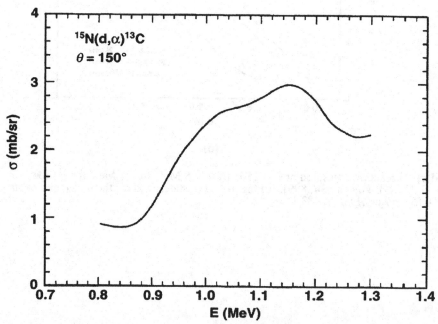

FIG. A11.22. The ^{15}N(d,α)^{13}C reaction measured at $\theta = 150°$ in the 0.8–1.3 MeV range. From Sawicki, J.A., Davies, J.A., and Jackman, T.E. (1986), *Nucl. Instrum. Methods* **B15**, 530.

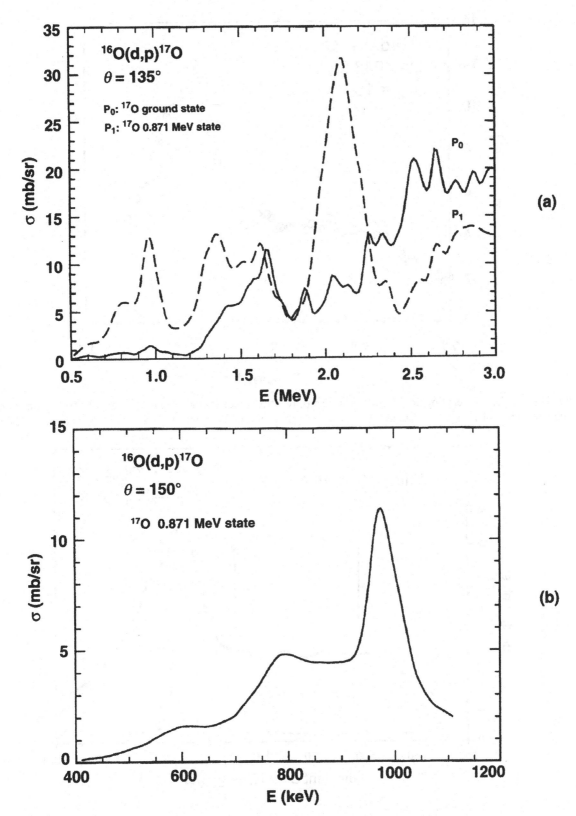

FIG. A11.23. The $^{16}O(d,p)^{17}O$ reaction measured at (a) $\theta = 135°$ from 0.5 MeV to 3 MeV [from Jarjis, R.A. (1979), Internal Report, University of Manchester, Manchester, UK] and at (b) $\theta = 150°$ between 0.4 MeV and 1.1 MeV [from Amsel, G., and Samuel, D. 1967), *Anal. Chem.* **39**, 1689].

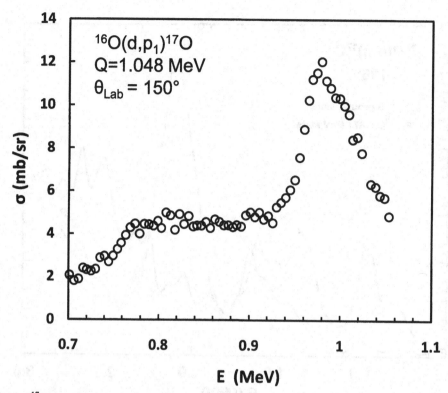

FIG. A11.24. The $^{16}O(d,p)^{17}O$ reaction cross sections for the p_1 reaction product at a laboratory angle of 150°. From Jiang, W., Shutthanandan, V., Thevuthasan, S., McCready, D.E., and Weber, W.J. (2003), Nucl. Instrum. Methods Phys. Res. B207, 453.

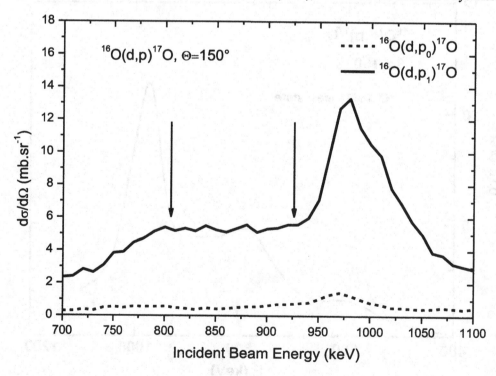

FIG. A11.25. The $^{16}O(d,p)^{17}O$ reaction cross sections for both p_0 and p_1 reaction products. This reaction has a plateau below 900 keV down to 800 keV. From Gurbich, A.S., and Molotdsov, S. (2004), *Nucl. Instrum. Methods* **B226**, 637.

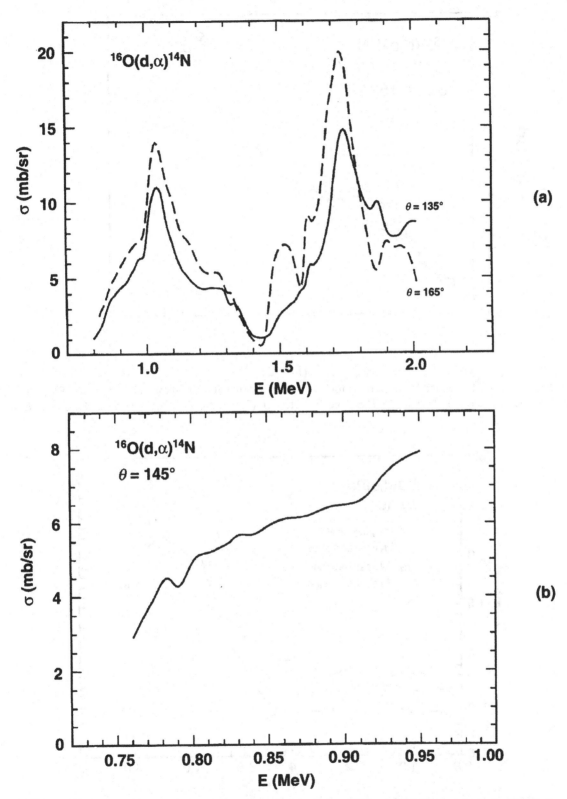

FIG. A11.26. The ^{16}O(d,α)^{14}N reaction in the 0.8–2 MeV energy range (a) at θ = 135° and 165° [from Amsel, G. (1964), *Ann. Phys.* **9**, 297] and (b) at θ = 145° between 0.75 MeV and 0.95 MeV [from Turos, A., Wielunski, L., and Batcz, A. (1973), *Nucl. Instrum. Methods* **111**, 05].

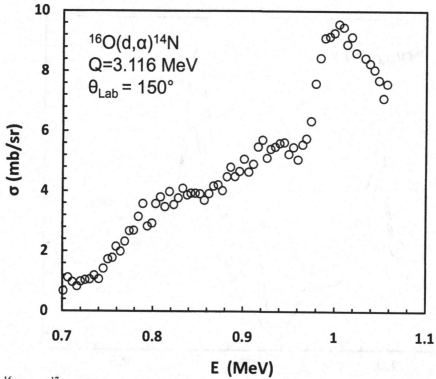

FIG. A11.27. The $^{16}O(d, \alpha)^{17}N$ reaction cross sections for α_0 reaction product at a laboratory angle of 150°. From Jiang, W., Shutthanandan, V., Thevuthasan, S., McCready, D.E., and Weber, W.J. (2003), *Nucl. Instrum. Methods Phys. Res.* **B207**, 453.

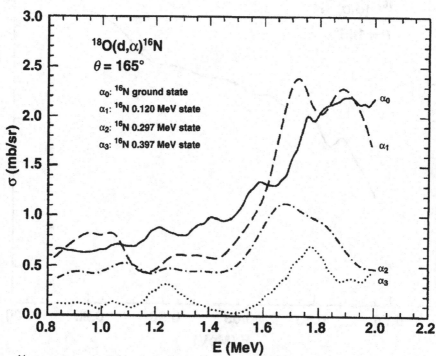

FIG. A11.28. The $^{16}O(d,\alpha)^{14}N$ reaction at different energy states measured at $\theta = 165°$ from 0.7 MeV to 2.2 MeV. From Seller, F., Jones, C.H., Anzick, W.J., Herring, D.F., and Jones, K.W. (1963), *Nucl. Phys.* **45**, 647.

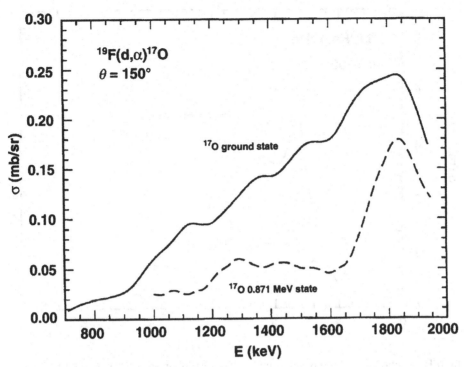

FIG. A11.29. The $^{19}F(d,\alpha)^{17}$ reaction measured at $\theta = 150°$ between 0.8 MeV and 2 MeV for both α_0 and α_1. From Maurel, B., Amsel, C., and Dieumegard, D. (1981), *Nucl. Instrum. Methods* **191**, 349.

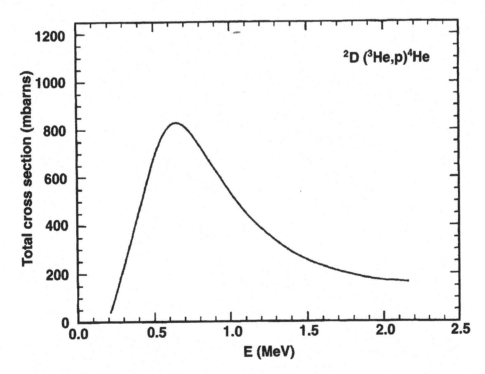

FIG. A11.30. The total cross section for the D(^{3}He,p)^{4}He reaction. From Moller, W., and Besenbacher, F. (1980), *Nucl. Instrum. Methods* **168**, 111.

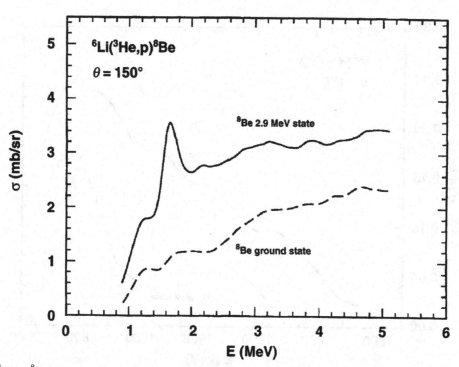

FIG. A11.31. The $^6\text{Li}(^3\text{He,p})^8\text{Be}$ reaction for p_0 and p_1 measured at $\theta = 150°$ between 1 MeV and 5 MeV. From Schiffer, J.P., Bonner, T.W., Davis, R.H., and Prosser Jr., F.W. (1956), *Phys. Rev.* **104**, 1064.

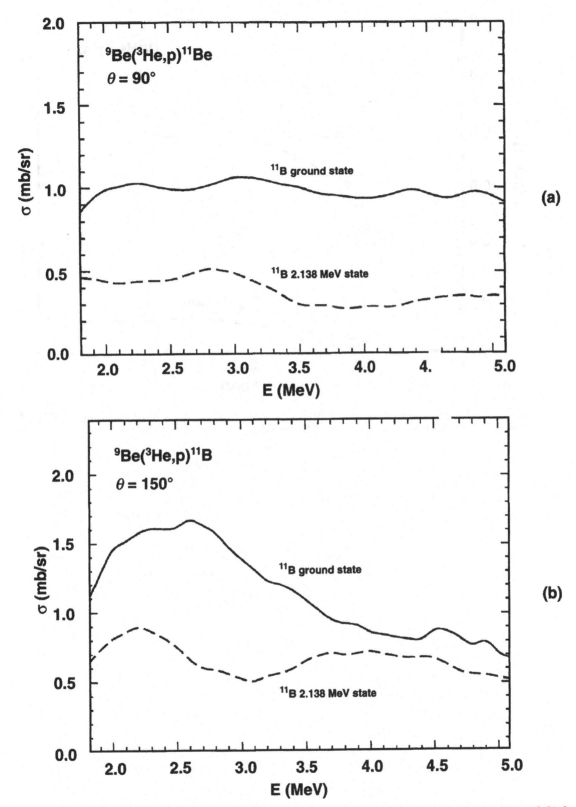

FIG. A11.32. The $^9Be(^3He,p)^{11}B$ reaction for p_0 and p_1 measured between 1.8 MeV and 5.1 MeV at (a) $\theta = 90°$ and (b) $\theta = 150°$. From Wolicki, E.A., Holmgren, H.D., Johnston, R.L., and Geer Illsley, E. (1959), *Phys. Rev.* **116**, 1585.

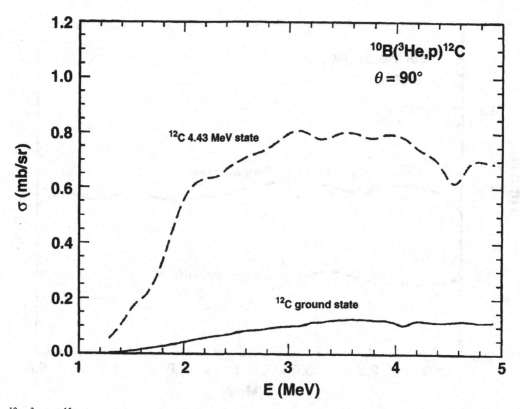

FIG. A11.33. The ^{10}B(^{3}He,p)^{15}N reaction measured at θ = 90° for p$_0$ and p$_1$ from 0.5 MeV to 5 MeV. From Schiffer, J.P., Bonner, T.W., Davis, R.H., and Prosser Jr., F.W. (1956), *Phys. Rev.* **104**, 1064.

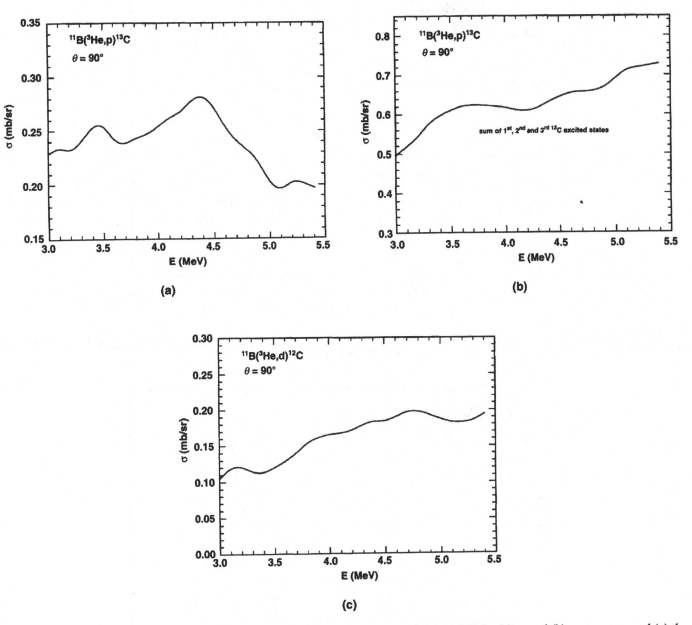

FIG. A11.34. The ^{11}B(^{3}He,p)^{13}C reaction measured at $\theta = 90°$ between 3 MeV and 5.5 MeV for (a) p_0 and (b) $p_1 + p_2 + p_3$ and (c) the ^{11}B(^{3}He,d)^{12}C reaction for d_0 measured at $\theta = 90°$ between 3 MeV and 5.5 MeV. From Holmgren, H.D., Wolicki, E.A., and Johnston, R.L. (1959), *Phys. Rev.* **114**, 1281.

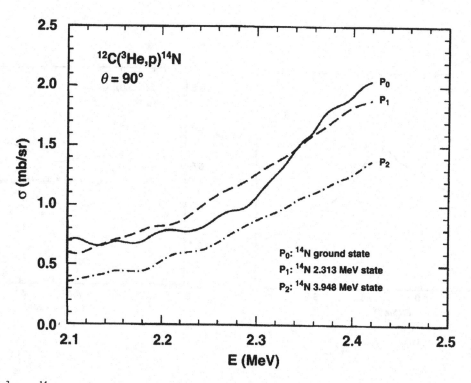

FIG. A11.35. The $^{12}C(^{3}He,p)^{14}N$ reaction for p_0–p_2 measured at $\theta = 90°$ from 2.1 MeV to 2.4 MeV. From Tong, S.Y., Lennard, W.N. Alkemada, P.F.A., and Mitchell, I.V. (1990), *Nucl. Instrum. Methods* **B45**, 91.

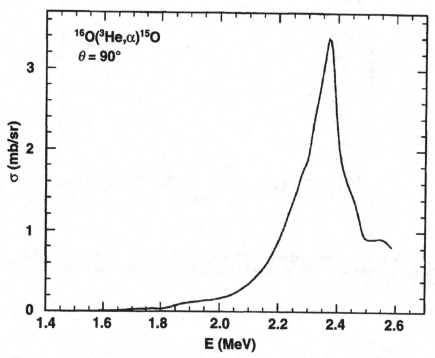

FIG. A11.36. The $^{16}O(^{3}He,\alpha)^{15}O$ reaction measured at $\theta = 90°$ in the energy range between 1.6 MeV and 2.6 MeV. From Abel, F., Amsel, G., d'Artemare, E., Ortega, C., Siejka, J., and Vizkelethy, G. (1990), *Nucl. Instrum. Methods* **B45**, 10.

PARTICLE INDUCED GAMMA EMISSION DATA

Compiled by

J. Räisänen

University of Helsinki, Helsinki, Finland

Contributors: J.-P. Hirvonen and R. Lappalainen

2.1 Depth profiling with resonance reactions

Table A12.1 is based on a similar table for Z = 3–9 by ·licheff *et al.* (1972) and on Table A12.2 of the previous ·tion of this handbook. The (p,γ) data of Table A12.2 are ·sed on the work of Butler (1959) and on Table A12.1 of the ·ond edition of the Ion Beam Handbook (Mayer and Rimini, ·77). All data have been updated using the information ·ovided in the references of Table 12.1 and the Ion Beam ·alysis Nuclear Data Library (IBANDL) database ·tp://www-nds.iaea.org/ibandl/). The range of bombarding ·ergies has been limited to 3 MeV.

The tables list reactions, resonance energies, energies of ·itted gamma rays, cross sections, and resonance widths. The ·mber of gamma-ray energies is limited to those that are of the ·eatest aid in identifying the resonance. Therefore, they are ·nerally the three highest energies, but where an exceptionally ·tense low-energy gamma ray occurs, it has been used to ·bstitute one of the high-energy lines.

Where the identification of the gamma-ray transition in a ·mpound nucleus is definite, the gamma-ray energy listed was ·otained by taking the difference between the excitation energies of the two states involved in the transition, instead of using the directly measured value.

The cross section given is the total cross section in millibarns (mb) or the resonance strength (eV) at the resonance peak. Where more than one primary gamma ray is emitted, the tabulated value of the cross section is the sum of all such individual primary gamma-ray cross sections. For those resonances that are too narrow for such cross-section measurements, the integrated cross section, $\int \sigma \, dE$, has been tabulated where this measurement has been made. In these instances the abbreviation "eV-b" for "electronvolt barn" has been inserted in the cross-section column.

The width column gives the measured full width at half-maximum of the resonance in the laboratory system of coordinates. In most cases, this is the observed overall experimental width. However, for certain narrow well-known resonances, the tabulated width is the actual intrinsic resonance width; that is, it is a processed number that results after the factors such as beam width and Doppler broadening have been removed from the actual experimental value.

Table A12.1. (p,γ) and (p,αγ) resonances by element.

Element	Reaction	E_p (keV)	σ (mb)/ Strength (eV)	Resonance width (keV)	E_γ (MeV)	Relative intensity
Li	$^7Li(p,\gamma)^8Be$	441	5.9 mb	12.2	17.64	63
					14.75	37
		1030		168	18.15	40
					15.25	60
		2060		310	16.15	100
Be	$^9Be(p,\gamma)^{10}B$	319	0.14 eV	120	6.15	21
					5.15	55
					4.75	11
					1.02	58
					0.72	84
		992	10.4 eV	82	7.5	~100
		1083	0.78 eV	2.65	6.85	85
					4.45	4.5
					3.01	9.5
					2.4	15
					0.72	96
B	$^{11}B(p,\gamma)^{12}C$	163	0.157 mb	5.2	16.11	3.5
					11.68	96.5
					4.43	96.5
		675	0.050 mb	322	12.15	~100
	$^{10}B(p,\gamma)^{11}C$	1146	0.0055 mb	450	9.7	
		1180	0.0075 mb	570	9.4	
	$^{11}B(p,\gamma)^{12}C$	1390	0.053 mb	1270	17.23	66
					12.80	34
					4.43	34
C	$^{12}C(p,\gamma)^{13}N$	457	0.127 mb	35	2.36	100
	$^{13}C(p,\gamma)^{14}N$	551	1.44 mb	25	8.06	80
		1152	0.56 mb	4.1	8.62	23
					4.67	24
					2.42	40
					2.31	62
		1320	0.062 mb	440	8.71	~90
		1462	0.074 mb	17	5.83	18
					5.10	53
					3.08	84
		1540	0.037 mb	9	8.98	~100
	$^{12}C(p,\gamma)^{13}N$	1689	0.035 mb	67	3.51	95
					2.36	5
					1.14	5
	$^{13}C(p,\gamma)^{14}N$	1748	340 mb	0.075	9.17	86
					6.45	6
					2.73	9
					2.74	

Table A12.1. (p,γ) and (p,αγ) resonances by element (continued).

Element	Reaction	E_p (keV)	σ (mb)/ Strength (eV)	Resonance width (keV)	E_γ (MeV)	Relative intensity		
N	^{14}N(p,γ)^{15}O	278	0.014 eV	1.06	6.79	23		
					6.18	58		
					1.38	58		
	^{15}N(p,γ)^{16}O	335	0.007 mb	110	12.44	~100		
	^{15}N(p,αγ)^{12}C	335	0.03 mb	110	4.43	100		
		429	1560 mb	0.12	4.43	100		
	^{15}N(p,γ)^{16}O	429	0.001 mb	0.103	6.40	60		
					6.13	60		
		710		40	6.72			
	^{15}N(p,αγ)^{12}C	897	800 mb	1.7	4.43	100		
		1028	15 mb	140	4.43	100		
	^{15}N(p,αγ)^{12}C	1028	1 mb	140	13.09	~100		
	^{14}N(p,γ)^{15}O	1058	0.37 mb	3.9	8.28	53		
					5.24	42		
					3.04	42		
	^{15}N(p,αγ)^{12}C	1210	425 mb	22.5	4.43	100		
	^{14}N(p,γ)^{15}O	1550	0.09 eV	34	6.18	36		
					5.18	64		
	^{15}N(p,αγ)^{12}C	1640	340 mb	68	4.43	100		
	^{14}N(p,γ)^{15}O	1742	0.06 eV	8	8.92	50		
					5.18	39		
		1806	0.52 mb	4.2	8.98	94		
	^{15}N(p,αγ)^{12}C	1979	35 mb	23	4.43	100		
O	^{18}O(p,γ)^{19}F	630	0.10 eV	2.0	8.39	42		
		841	1.4 eV	48	8.68	30		
		1167	0.29 eV	0.05	6.32	47		
					2.58	47		
					0.197	>50		
					0.110	>35		
		1398	0.08 eV	3.6	9.32	30		
		1684	0.025 eV	8	8.24	32		
		1768	1.2 eV	3.8	9.67	22		
		1928	2.8 eV	0.3	9.62	41		
F	^{19}F(p,αγ)^{16}O				γ_1	γ_1	γ_2	γ_3
					6.13		6.72	7.12
		224	0.2 mb	0.94		100		
		340.5	102 mb	2.22		96.5	0.5	3
		484	32 mb	0.86		79	1	20
		594	7 mb	24		~100		
		668	57 mb	6.4		81	0.3	19
		832	19 mb	6.5				
		872	661 mb	4.3		68	24	8
		935	180 mb	7.7		76	3	21
		1088	13 mb	0.14				
		1136	15 mb	2.5				
		1280	29 mb	19		74	8	18
		1347	89 mb	4.7		55	14	31
		1371	300 mb	11.8		87	8	5
		1692		35				
	^{19}F(p,γ)^{22}Ne	1350			1.63			

Table A12.1. (p,γ) and (p,αγ) resonances by element (continued).

Element	Reaction	E_p (keV)	σ (mb)/ Strength (eV)	Resonance width (keV)	E_γ (MeV)	Relative intensity
Ne	^{22}Ne(p,γ)^{23}Na	638	5.6 eV	0.065	9.406	73
					0.440	12
		851	14 eV	0.006	9.606	27
					9.166	42
					5.693	21
					0.440	>47
	^{20}Ne(p,γ)^{21}Na	1169	1.6 eV	0.016	3.544	92
					1.828	6
					0.332	8
Na	^{23}Na(p,γ)^{24}Mg	308.75	0.84 eV	<0.002	10.618	28
					7.748	46
					4.239	35
					1.369	>60
		512	0.73 eV	<0.05	10.813	71
					7.943	9
					1.369	>80
		592	1.9 eV	0.03	10.9	
		677	5.1 eV	<0.07	10.970	11
					8.100	45
					7.104	21
					4.239	35
					1.369	>52
		1011	0.37 eV	<0.5		
	^{23}Na(p,αγ)^{20}Ne	1011	55 eV	<0.1	1.634	100
		1164	160 eV	1.2	1.634	100
	^{23}Na(p,γ)^{24}Mg	1164	1.8 eV	1.4		
Mg	^{24}Mg(p,γ)^{25}Al	419	0.042 eV	<0.044	2.674	25
					2.222	31
					0.452	43
	^{26}Mg(p,γ)^{27}Al	454	0.0715 eV	<0.08	7.864	48
					7.695	16
		823	0.64 eV	1.3	0.843	>62
		840	1.9 eV	0.240	2.610	77
					0.452	83
		1549	5.8 eV	0.018	9.761	20
					8.748	30
					7.552	27
					1.013	>40
Al	^{27}Al(p,γ)^{28}Si	406	0.00863 eV	<0.04	7.360	72
					2.835	>72
					1.779	>80
		632	2.6 eV	0.0067	10.416	74
					7.581	15
					1.779	>94
		992	24 eV	0.070	10.763	76
					4.744	10
					1.779	>93
Si	^{29}Si(p,γ)^{30}P	417	1.04 eV	~0.1	5.9	
	^{30}Si(p,γ)^{31}P	620	4.4 eV	0.068	7.897	93

Table A12.1. (p,γ) and (p,αγ) resonances by element (continued).

Element	Reaction	E_p (keV)	σ (mb)/ Strength (eV)	Resonance width (keV)	E_γ (MeV)	Relative intensity
P	^{31}P(p,γ)^{32}S	541	0.480 eV	<<1	7.159	80
					2.230	~100
		811	1.0 eV	<0.42	7.420	57
					4.956	38
					2.230	>84
		1251	4.3 eV	1.50	7.846	32
					3.852	42
					2.230	>90
S	^{34}S(p,γ)^{35}Cl	929	1 eV	0.014	7.274	72
					6.055	20
					1.219	24
		1211	9.7 eV	0.0105	4.385	97
					3.163	87
	^{32}S(p,p'γ)^{32}S	3094	61 eV	0.34	2.23	
Cl	^{37}Cl(p,γ)^{38}Ar	766	4.3 eV	<1	8.820	44
					2.168	>60
Ar	^{40}Ar(p,γ)^{41}K	1087	2.1 eV	0.008	8.86	
Ca	^{40}Ca(p,γ)^{41}Sc	1842	0.280 mb		2.883	
Ti	^{48}Ti(p,γ)^{49}V	1007			7.582	
		1013	<1.2 eV		7.650	
Cr	^{52}Cr(p,γ)^{53}Mn	1005	0.89 eV	<0.1	2.5–4.7	
Ni	^{58}Ni(p,γ)^{59}Cu	1424	1.7 eV-b	<0.045	4.82	27
					4.33	50

New data are based on the following nuclear data compilations and references therein:

For A = 5–7,   Tilley, D.R., Cheves, C.M., Godwin, J.L., Hale, G.M., Hofmann, H.M., Kelley, J.H., Sheu,. C.G., and Weller, H.R. (2002), *Nucl. Phys.* **A708**, 3.

For A = 5–10,   Ajzenberg-Selove, F. (1988), *Nucl. Phys.* **A490**, l.

For A = 11, 12, Ajzenberg-Selove, F. (1990), *Nucl. Phys.* **A506**, l.

For A = 13–15   Ajzenberg-Selove, F. (1991), *Nucl. Phys.* **A523**, l.

For A = 16, 17, Tilley, D.R., Weller, H.R., and Cheves, C.M. (1993), *Nucl. Phys.* **A564**, l.

For A = 18, 19, Tilley, D.R., Weller, H.R., Cheves, C.M., and Chasteler, R.M.: (1995), *Nucl. Phys.* **A595**, l.

For A = 20,   Tilley, D.R., Cheves, C.M., Kelley, J.H., Raman, S., and Weller, H.R. (1998), *Nucl. Phys.* **A636**, 249.

For A = 21–44, Endt, P.M. (1990), *Nucl. Phys.* **A521**, l; Supplement (1998), *Nucl. Phys.* **A633**, 1.

For ^{15}N(p,αγ)^{12}C, Horn, K.M., and Lanford, W.A. (1988), Nucl. Instrum. Methods **B34**, 1.

Table A12.2. Selected strong (p,γ) resonances versus bombarding energy.

Proton energy (keV)	Reaction	Gamma-ray energy (MeV)	σ (mb)/ Strength (eV)	Width (keV)
163	^{11}B(p,γ)^{12}C	16.11, 11.68, 4.43	0.157 mb	5.2
224	^{19}F(p,αγ)^{16}O	7.12, 6.72, 6.13	0.2 mb	0.94
278	^{14}N(p,γ)^{15}O	6.79, 6.18, 1.38	0.014 eV (ω$_γ$)	1.06
309	^{23}Na(p,γ)^{24}Mg	10.6, 7.7, 4.2	0.84 eV (S)	<0.002
319	^{9}Be(p,γ)^{10}B	6.15, 5.15, 4.75, 1.02, 0.72	0.14 eV (ω$_γ$)	120
340	^{19}F(p,αγ)^{16}O	7.12, 6.72, 6.13	102 mb	2.22
335	^{15}N(p,γ)^{16}O	12.44, 6.40	0.007 mb	110
335	^{15}N(p,αγ)^{12}C	4.43	0.03 mb	110
374	^{23}Na(p,γ)^{24}Mg	6.26	0.011 eV (S)	<0.02
390	^{25}Mg(p,γ)^{26}Al	6.26	0.88 eV (S)	0.0012
406	^{27}Al(p,γ)^{28}Si	7.36, 2.84, 1.78	0.00863 eV (ω$_γ$)	<0.04
417	^{29}Si(p,γ)^{30}P	5.9	1.04 eV (S)	~0.1
419	^{24}Mg(p,γ)^{25}Al	2.67, 2.22, 0.45	0.042 eV (ω$_γ$)	<0.044
429	^{15}N(p,αγ)^{12}C	4.43	1560 mb	0.12
429	^{15}N(p,γ)^{16}O	6.40, 6.13	0.001 mb	0.103
441	^{7}Li(p,γ)^{8}Be	17.64, 14.75	5.9 mb	12.2
454	^{26}Mg(p,γ)^{27}Al	7.86, 7.70, 0.84	2.5 mb	<0.08
457	^{12}C(p,γ)^{13}N	2.36	0.0715 eV (ω$_γ$)	35
484	^{19}F(p,αγ)^{16}O	7.12, 6.72, 6.13	32 mb	0.86
512	^{23}Na(p,γ)^{24}Mg	10.81, 7.94, 1.37	0.73 eV (S)	<0.05
541	^{31}P(p,γ)^{32}S	7.16, 2.23	0.480 eV (S)	<<1
551	^{13}C(p,γ)^{14}N	8.06	1.44 mb	25
592	^{23}Na(p,γ)^{24}Mg	10.9, 8.0	1.9 eV (S)	0.03
594	^{19}F(p,αγ)^{16}O	7.12, 6.72, 6.13	7 mb	24
620	^{30}Si(p,γ)^{31}P	7.90	4.4 eV (S)	0.068
630	^{18}O(p,γ)^{19}F	8.39	0.10 eV (ω$_γ$)	2.0
632	^{27}Al(p,γ)^{28}Si	10.42, 7.58, 1.78	2.6 eV (S)	0.0067
638	^{22}Ne(p,γ)^{23}Na	9.41, 0.44	5.6 eV (S)	0.065
668	^{19}F(p,γ)^{20}Ne	11.88, 1.63	0.5 mb	6.4
668	^{19}F(p,αγ)^{16}O	7.12, 6.72, 6.13	57 mb	6.0
675	^{11}B(p,γ)^{12}C	12.15, 4.43	0.05 mb	322

Table A12.2. Selected strong (p,γ) resonances versus bombarding energy (continued).

Proton energy (keV)	Reaction	Gamma-ray energy (MeV)	σ (mb)/ Strength (eV)	Width (keV)
677	^{23}Na(p,γ)^{24}Mg	11.0, 8.1, 7.1	5.1 eV (S)	<0.07
710	^{15}N(p,γ)^{16}O	6.72		40
766	^{37}Cl(p,γ)^{38}Ar	8.82, 2.17	4.3 eV (S)	<1
811	^{31}P(p,γ)^{32}S	7.39	1.0 eV (S)	<0.42
823	^{24}Mg(p,γ)^{25}Al	2.61, 0.45	0.64 eV (ω$_γ$)	1.3
832	^{19}F(p,αγ)^{16}O	7.12, 6.72, 6.13	19 mb	
840	^{26}Mg(p,γ)^{27}Al		1.9 eV (S)	0.240
841	^{18}O(p,γ)^{19}F	8.7	1.4 eV (ω$_γ$)	48
851	^{22}Ne(p,γ)^{23}Na	9.61, 9.17, 5.70	14 eV (S)	0.006
872	^{19}F(p,αγ)^{16}O	7.12, 6.72, 6.13	661 mb	4.3
897	^{15}N(p,αγ)^{12}C	4.43	800 mb	1.7
929	^{34}S(p,γ)^{35}Cl	7.27, 6.05, 1.22	1 eV (S)	0.014
935	^{19}F(p,αγ)^{16}O	7.12, 6.72, 6.13	180 mb	7.7
992	^{9}Be(p,γ)^{10}B	7.5, 6.8, 5.8	10.4 eV (ω$_γ$)	82
992	^{27}Al(p,γ)^{28}Si	10.76, 4.74, 1.78	24 eV (S)	0.070
1005	^{52}Cr (p,γ)^{53}Mn	2.5-4.7	0.89 eV (S)	<0.10
1007	^{48}Ti(p,γ)^{49}V	7.58	<1.2 eV (S)	
1011	^{23}Na(p,αγ)^{20}Ne	1.634	55 eV (S)	<0.1
1011	^{23}Na(p,γ)^{24}Mg	11.0, 8.1, 7.1	0.37 eV (S)	<0.5
1013	^{48}Ti(p,γ)^{49}V	7.58	<1.2 eV (S)	
1028	^{15}N(p,γ)^{16}O	13.09	1 mb	140
1028	^{15}N(p,αγ)^{12}C	4.43	15 mb	140
1030	^{7}Li(p,γ)^{8}Be	18.15, 15.25, 0.478		168
1058	^{14}N(p,γ)^{15}O	8.28, 5.24, 3.04	0.37 mb	3.9
1083	^{9}Be(p,γ)^{10}B	6.85, 4.45, 3.01, 2.4, 0.72	0.78 eV (ω$_γ$)	2.65
1087	^{40}Ar(p,γ)^{41}K	8.86	2.1 eV (S)	0.008
1088	^{19}F(p,γ)^{20}Ne	12.28, 8.84, 1.63	>0.05 mb	0.14
1090	^{19}F(p,αγ)^{16}O	7.12, 6.72, 6.13	>13 mb	0.7
1136	^{19}F(p,αγ)^{16}O	7.12, 6.72, 6.13	15 mb	
1146	^{10}B(p,γ)^{11}C	9.7	0.0055 mb	450
1152	^{13}C(p,γ)^{14}N	8.62, 4.67, 2.42, 2.31	0.56 mb	4.1
1164	^{23}Na(p,αγ)^{20}Ne	1.634	160 eV (S)	1.2
1164	^{23}Na(p,γ)^{24}Mg	11.0, 8.1, 7.1	1.8 eV (S)	1.4
1167	^{18}O(p,γ)^{19}F	6.32, 2.58, 0.197, 0.110	0.29 eV (ω$_γ$)	0.05
1169	^{20}Ne(p,γ)^{21}Na	3.54, 1.83, 0.33	1.6 eV (S)	0.016

Table A12.2. Selected strong (p,γ) resonances versus bombarding energy (continued).

Proton energy (keV)	Reaction	Gamma-ray energy (MeV)	σ (mb)/ Strength (eV)	Width (keV)
1180	^{10}B(p,γ)^{11}C	9.4	0.0075 mb	570
1210	^{15}N(p,αγ)^{12}C	4.43	425 mb	22.5
1211	^{34}S(p,γ)^{35}Cl	4.38, 3.16	9.7 eV (S)	0.0105
1251	^{31}P(p,γ)^{32}S	7.85, 3.85, 2.23	4.3 eV (S)	1.50
1280	^{19}F(p,αγ)^{16}O	7.12, 6.72, 6.13	29 mb	
1320	^{13}C(p,γ)^{14}N	8.71	0.062 mb	440
1347	^{19}F(p,αγ)^{16}O	7.12, 6.72, 6.13	89 mb	4.7
1350	^{19}F(p,γ)^{20}Ne	1.63		
1371	^{18}O(p,γ)^{19}F	7.12, 6.72, 6.13	300 mb	11.8
1390	^{11}B(p,γ)^{12}C	17.23, 12.80, 4.43	0.053 mb	1270
1398	^{18}O(p,γ)^{19}F	9.32	0.08 eV (ω$_γ$)	3.6
1424	^{58}Ni(p,γ)^{59}Cu	4.82, 4.33	1.7 eV-b	<0.045
1431	^{19}F(p,γ)^{20}Ne	12.60, 1.63	0.19 mb	15.7
1462	^{13}C(p,γ)^{14}N	5.83, 5.10, 3.08	0.074 mb	17
1540	^{13}C(p,γ)^{14}N	8.98	0.037 mb	9
1549	^{26}Mg(p,γ)^{27}Al	9.76, 8.75, 7.55, 1.01	5.8 eV (S)	0.018
1550	^{14}N(p,γ)^{15}O	6.18, 5.18	0.09 eV (ω$_γ$)	34
1599	^{60}Ni(p,γ)^{61}Cu	6.38, 5.00	2.3 eV-b	<1
1605	^{60}Ni(p,γ)^{61}Cu	6.39, 5.01	2.0 eV-b	<1
1620	^{60}Ni(p,γ)^{61}Cu	6.40, 5.02	1.8 eV-b	<1
1640	^{15}N(p,αγ)^{12}C	4.43	340 mb	68
1684	^{18}O(p,γ)^{19}F	8.24	0.025 eV (ω$_γ$)	8
1689	^{12}C(p,γ)^{13}N	3.51, 2.36, 1.14	0.035 mb	67
1692	^{19}F(p,αγ)^{16}O	7.12, 6.72, 6.13		35
1742	^{14}N(p,γ)^{15}O	8.92, 5.18	0.06 eV (ω$_γ$)	8
1748	^{13}C(p,γ)^{14}N	9.17, 6.45, 2.73	340 mb	0.075
1768	^{18}O(p,γ)^{19}F	9.67	1.2 eV (ω$_γ$)	3.8
1806	^{14}N(p,γ)^{15}O	8.98	0.52 mb	4.2
1842	^{40}Ca(p,γ)^{41}Sc	2.88	0.280 mb	
1844	^{58}Ni(p,γ)^{59}Cu	5.23	2.1 eV-b	<0.1
1928	^{18}O(p,γ)^{19}F	9.62	2.8 eV (ω$_γ$)	0.3
1979	^{15}N(p,αγ)^{12}C	4.43	35 mb	23
2060	^{7}Li(p,γ)^{8}Be	16.15		310
2120	^{13}C(p,γ)^{14}N	5.10, 4.39	0.20 mb	45
3000	^{15}N(p,αγ)^{12}C	4.43	750 mb	45
3094	^{32}S(p,p'γ)^{32}S	2.23	61 eV (ω$_γ$)	0.34

Table A12.3. Useful (α,γ) and $(\alpha,p\gamma)$ resonances for depth profiling.

Element	Reaction	E_α (keV)	σ (mb)/ Strength (eV)	Resonance width (keV)	E_γ (MeV)	Relative intensity
Li	$^7Li(\alpha,\gamma)^{11}B$	814	0.3 eV	0.0018	6.74	9
					4.75	87
					4.44	91
		953	6.9 eV	4	9.27	18
					4.83	70
					4.44	73
					2.53	12
B	$^{10}B(\alpha,p\gamma)^{13}C$	1507	6 eV	25	3.85	54
					3.68	42
		1645	8 eV	22	3.85	59
					3.68	39
N	$^{14}N(\alpha,\gamma)^{18}F$	1529	2.6 eV	<1.2	4.52	54
					2.47	32
					1.08	62
		1618	1.4 mb	<0.8	4.59	52
					2.54	29
					1.08	59
Ne	$^{20}Ne(\alpha,\gamma)^{24}Mg$	1928	2.2 eV	<1	9.56	85

See references for nuclear data compilations in Table A12.1

Table A12.4. Standard samples for resonance reactions. For most of the metals, high purity foils or sheets are available, and evaporated films can be easily made. These metals include Be, Mg, Al, Ti, V, Cr, Mn, Fe, Co, Ni, Cu, Zn, Y, Zr, Nb, Mo, Ru, Pd, Ag, Cd, In, Sn, Sb, Hf, Ta, W, Pt, Au, and Pb. This table list includes typical standards for non-metallic light elements and metals which can not be used in elemental form.

Element	Stable isotopes	Standards
H	^{1}H 99.985 ^{2}H 0.015	hydrogenated Ta[1] polypropyle $(C_3H_6)_n$, polyester, (Mylar) $(C_{10}H_8O_4)_n$, Kapton® $(C_{22}H_{10}O_5N_2)_n$
He	^{4}He ~100	implanted sample (Ta)
Li	^{6}Li 7.5, ^{7}Li 92.5	LiF[2]
B	^{10}B 19.9, ^{11}B 80.1	B[2]
C	^{12}C 98.9, ^{13}C 1.10	high purity graphite or evaporated film
N	^{14}N 99.63, ^{15}N 0.37	TiN, TaN, NbN[3], N_2O, NO^4, Si_3N_4, AlN[5]
O	^{16}O 99.76, ^{17}O 0.0374 ^{18}O 0.2039	Al_2O_3, MgO, SiO_2[5], Ta_2O_5[6]
F	^{19}F 100	CaF_2, LiF[2]
Ne	^{20}Ne 90.48, ^{21}Ne 0.27 ^{22}Ne 9.25	implanted isotopic sample (Ta)
Na	^{23}Na 100	fresh cleaved NaCl (hygroscopic !)
Si	^{28}Si 92.2, ^{29}Si 4.67 ^{30}Si 3.10	Si (channeling possible at certain angles !)
P	^{31}P 100	InP[2]
S	^{32}S 95.02, ^{33}S 0.75 ^{34}S 4.21, ^{36}S 0.02	ZnS, PbS[2]
Cl	^{35}Cl 75.77, ^{37}Cl 24.23	fresh cleaved NaCl, $PbCl_2$ (powder pellet)
Ar	^{36}Ar 0.34, ^{38}Ar 0.06 ^{40}Ar 99.60	implanted isotopic sample (Ta)
K	^{39}K 93.26, ^{40}K 0.012 ^{41}K 6.73	KI, KCl[2]
Ca	^{40}Ca 96.94, ^{42}Ca 0.65 ^{43}Ca 0.135, ^{44}Ca 2.09	CaO, $CaCO_3$[2]

1. B. Hjorvarsson, J. Ryden, T. Ericsson and E. Karlsson. (1989), *Nucl. Instr. Meth.* B42, 257.
2. Powder pellet or evaporated film.
3. Powder pellet or deposited film (CVD, sputtering, thermal nitriding, etc.).
4. Frozen gas (J.A. Davies *et ah* (1983), *Nucl. Instr. Meth.* **218,** 141.
5. Sintered ceramic or powder pellet.
6. Films by anodic or thermal oxidation (^{16}O, ^{17}O and ^{18}O) (G. Amsel *et al.* (1978), *Nucl. Instr. Meth.* **149,** 713.

Table A12.5. Summary of sensitivities (atomic fraction) for depth profiling of the light elements with resonances leading to gamma-ray emission.

Over 1%	Between 1 and 0.1%	Better than 0.1%
O	Be, B, C, N, ^{18}O, Ne, Na, Mg, P, S, Cl, Ar, Ti, Cr, Ni	^{13}C, ^{15}N, F, ^{22}Ne, Al

A12.2 Elemental analysis using particle-induced gamma-ray emission (PIGE)

The available particle-induced thick-target gamma-ray yields and relevant references are provided in Table A12.6. The data search was extended until June 2009. As most applications deal with use of protons as bombarding particles, they are treated in more detail for elemental analysis at the beginning. For more special applications, the relevant data for other light ions are provided, including hydrogen isotopes and α-particles.

Especially the use of deuterons as bombarding particles has gained more interest recently. Because of the use of heavy ions in specialized applications of ion beam characterization techniques, the data available for heavy ions are now also summarized. For each ion estimates of typical sensitivities obtainable are provided for guidance. In practical work, a useful atlas of spectra can be found for many ions from the original publications cited in Table A12.6.

Table A12.6. Particle-induced gamma-ray yields available in the literature. Only those references in which two or more elements were studied are included.

Bombarding particle	Energy (MeV)	Measurement angle	Element(s)	Atlas of spectra	Reference(s)
Protons	2, 2.5	135°	Li–Cl	No	Bird *et al.* (1978)
	1.2–2.9	90°	Rh, Pd, Ag, Pt, Au	No	Deconninck and Demortier (1975)
	0.6–3.2	90°	Ti–Zn	Yes	Demortier (1978)
	2, 2.5	135°	F–Si, Ti–Zn, Mo, Ag, Au	No	Kenny *et al.* (1980)
	1, 1.7, 2.4	55°	Li–Sc	Yes	Anttila *et al.* (1981)
	1.7, 2.4	55°	Ga–Pb	Yes	Räisänen and Hänninen (1983)
	2.4–4.2	55°	Li–Sc	Yes	Kiss *et al.* (1985)
	7, 9	55°	Li–Pb	Yes	Räisänen *et al.* (1987)
	0.4–1.5	90°	Li–F	No	Golicheff *et al.* (1972)
	4.5	45°	F–Co	No	Gihwala and Peisach (1982)
	1.0–4.1	90°	Li, B, F–P	No	Savidou *et al.* (1999)
	3.0–5.7	135°	Li, F	No	Caciolli *et al.* (2006)

Table A12.6. Particle-induced gamma-ray yields available in the literature. Only the references in which two or more elements were studied are included (continued).

Bombarding particle	Energy (MeV)	Measurement angle	Element(s)	Atlas of spectra	Reference(s)
Deuterons	0.7–3.4	135°	Li–Ca	Yes	Kiss *et al.* (1994), Elekes *et al.* (2000)
Tritons	2, 3, 3.5	90°	Li, O–Mg, Si–Au	No	Borderie and Barrandon (1978)
^{4}He ions	2.4	55°	Li–Si	Yes	Lappalainen *et al.* (1983)
	3.5	90°	Li–Na, Ti–Fe	No	Borderie and Barrandon (1978)
	5	90°	Li–Bi	Yes	Giles and Peisach (1979)
	5.6–10	30.6°, 109.9°	Be, B, N, F, Na, Mg, Al binned yields only	No	Heaton *et al.* (1995)
	5, 10	55°	Ti–Zn	Yes	Kocsonya *et al.* (2006)
^{7}Li ions	12, 18	55°	Li–Pb	No	Räisänen (1990)
^{12}C ions	22, 28	55°	Li–Ca, Ti–Zn	Yes	Seppälä *et al.* (1998)
^{14}N ions	28	55°	Li–F	Yes	Seppälä *et al.* (1998)
^{16}O ions	28, 33	55°	Li–F	Yes	Seppälä *et al.* (1998)
^{55}Cl ions	55	90°	Li, B, F–Si, Ti–Th	No	Borderie *et al.* (1979)

A12.2.1 Proton-induced gamma-ray emission

Table A12.7. Useful proton-induced reactions for the elemental analysis of the light elements ($Z \leq 30$).

Element	E_γ (keV)	Reaction	Remarks
Li	429	^{7}Li(p,nγ)^{7}Be	E_p should be >3 MeV.
	478	^{7}Li(p,p'γ)^{7}Li	Provides best sensitivity.
Be	3562	^{9}Be(p,$\alpha\gamma$)^{6}Li	
B	429	^{10}B(p,$\alpha\gamma$)^{7}Be	
	718	^{10}B(p,p'γ)^{10}B	
	2125	^{11}B(p,p'γ)^{11}B	For ^{11}B detection at E_p > 3 MeV.
C	3089	^{13}C(p,p'γ)^{13}C	For ^{13}C detection at E_p > 3.5 MeV.
	4439	^{12}C(p,p'γ)^{12}C	For ^{12}C detection at E_p > 5.4 MeV.
N	2313	^{14}N(p,p'γ)^{14}N	E_p should be >4 MeV.
	4439	^{15}N(p,$\alpha\gamma$)^{12}C	For ^{15}N detection.
O	495	^{16}O(p,γ)^{17}F	For ^{16}O detection at low energies.
	871	^{17}O(p,p'γ)^{17}O	For ^{17}O detection.
	1982	^{18}O(p,p'γ)^{18}O	For ^{18}O detection.
	6129	^{16}O(p,p'γ)^{16}O	E_p should be >7.5 MeV.
F	110	^{19}F(p,p'γ)^{19}F	Absorption problems possible.
	197	^{19}F(p,p'γ)^{19}F	Absorption problems possible.
	6129	^{19}F(p,$\alpha\gamma$)^{16}O	

Table A12.7. Useful proton-induced reactions for the elemental analysis of the light elements (Z ≤ 30) (continued).

Element	E_γ (keV)	Reaction	Remarks
Na	440	$^{23}Na(p,p'\gamma)^{23}Na$	
	1634	$^{23}Na(p,\alpha\gamma)^{20}Ne$	Overlapping with 1636 keV gamma line.
	1636	$^{23}Na(p,p'\gamma)^{23}Na$	
Mg	390	$^{25}Mg(p,p'\gamma)^{25}Mg$	
	585	$^{25}Mg(p,p'\gamma)^{25}Mg$	
	975	$^{25}Mg(p,p'\gamma)^{25}Mg$	
	1369	$^{24}Mg(p,p'\gamma)^{24}Mg$	For ^{24}Mg detection.
	1612	$^{25}Mg(p,p'\gamma)^{25}Mg$	E_p should be >3 MeV.
	1809	$^{26}Mg(p,p'\gamma)^{26}Mg$	For ^{26}Mg detection at E_p > 3MeV.
Al	844	$^{27}Al(p,p'\gamma)^{27}Al$	
	1014	$^{27}Al(p,p'\gamma)^{27}Al$	
	1369	$^{27}Al(p,\alpha\gamma)^{24}Mg$	
	1779	$^{27}Al(p,\gamma)^{28}Si$	Only at E_p < 3 MeV.
Si	755	$^{29}Si(p,p'\gamma)^{29}Si$	E_p should be >3 MeV.
	1273	$^{29}Si(p,p'\gamma)^{29}Si$	
	1779	$^{28}Si(p,p'\gamma)^{28}Si$	For ^{28}Si detection.
	2233	$^{30}Si(p,\gamma)^{31}P$	For ^{30}Si detection.
	2235	$^{30}Si(p,p'\gamma)^{30}Si$	Overlapping with 2233 keV gamma line.
P	1266	$^{31}P(p,p'\gamma)^{31}P$	
	1779	$^{31}P(p,\alpha\gamma)^{28}Si$	
	2230	$^{31}P(p,\gamma)^{32}S$	Overlapping with 2233 keV gamma line.
	2233	$^{31}P(p,p'\gamma)^{31}P$	
S	841	$^{33}S(p,p'\gamma)^{33}S$	For ^{33}S detection.
	1219	$^{34}S(p,\gamma)^{35}Cl$	For ^{34}S detection.
	2230	$^{32}S(p,p'\gamma)^{32}S$	E_p should be >4.9 MeV.
Cl	1219	$^{35}Cl(p,p'\gamma)^{35}Cl$	
	1410	$^{37}Cl(p,n\gamma)^{37}Ar$	E_p should be >3.5 MeV.
	1611	$^{37}Cl(p,n\gamma)^{37}Ar$	E_p should be >3.5 MeV.
	1763	$^{35}Cl(p,p'\gamma)^{35}Cl$	
	2126	$^{37}Cl(p,\alpha\gamma)^{38}Ar$	
K	1294	$^{41}K(p,p'\gamma)^{41}K$	E_p should be >3.5 MeV.
	1943	$^{41}K(p,n\gamma)^{41}Ca$	E_p should be >3.5 MeV.
	2010	$^{41}K(p,n\gamma)^{41}Ca$	E_p should be >3.5 MeV.
	2168	$^{41}K(p,\alpha\gamma)^{38}Ar$	
	2522	$^{39}K(p,n\gamma)^{41}Ca$	For ^{39}K detection at E_p > 4 MeV.

Table A12.7. Useful proton-induced reactions for the elemental analysis of the light elements ($Z \leq 30$) (continued).

Element	E_γ (keV)	Reaction	Remarks
Ca	371	$^{48}Ca(p,n\gamma)^{48}Sc$	For ^{48}Ca detection.
	373	$^{43}Ca(p,p'\gamma)^{43}Ca$	For ^{43}Ca detection.
	1157	$^{44}Ca(p,p'\gamma)^{44}Ca$	For ^{44}Ca detection.
	1525	$^{42}Ca(p,p'\gamma)^{42}Ca$	For ^{42}Ca detection.
Sc	364	$^{45}Sc(p,p'\gamma)^{45}Sc$	
	718	$^{45}Sc(p,p'\gamma)^{45}Sc$	
Ti	308	$^{47}Ti(p,\gamma)^{48}V$ $^{48}Ti(p,n\gamma)^{48}V$	For ^{47}Ti detection.
	889	$^{46}Ti(p,p'\gamma)^{46}Ti$	For ^{46}Ti detection.
	983	$^{48}Ti(p,p'\gamma)^{48}Ti$	For ^{48}Ti detection.
V	319	$^{51}V(p,p'\gamma)^{51}V$	
	749	$^{51}V(p,n\gamma)^{51}Cr$	
	808	$^{51}V(p,n\gamma)^{51}Cr$	Useful at $E_p > 3.5$ MeV.
Cr	379	$^{52}Cr(p,\gamma)^{53}Mn$	Strongest line from Cr.
	565	$^{53}Cr(p,p'\gamma)^{53}Cr$	For ^{53}Cr detection.
	783	$^{50}Cr(p,p'\gamma)^{50}Cr$	For ^{50}Cr detection.
Mn	411	$^{55}Mn(p,n\gamma)^{55}Fe$	
	931	$^{55}Mn(p,n\gamma)^{55}Fe$	Strongest line from Mn.
Fe	847	$^{56}Fe(p,p'\gamma)^{56}Fe$	Strongest line from Fe.
	1408	$^{54}Fe(p,p'\gamma)^{54}Fe$	For ^{54}Fe detection at $E_p > 7$ MeV.
Co	339	$^{59}Co(p,n\gamma)^{59}Ni$	
	1332	$^{59}Co(p,\gamma)^{60}Ni$	
Ni	1172	$^{62}Ni(p,p'\gamma)^{62}Ni$	For ^{62}Ni detection at $E_p > 3.5$ MeV.
	1332	$^{60}Ni(p,p'\gamma)^{60}Ni$	Strongest line from Ni.
	1454	$^{58}Ni(p,p'\gamma)^{58}Ni$	For ^{58}Ni detection at $E_p > 3.5$ MeV.
Cu	962	$^{63}Cu(p,p'\gamma)^{63}Cu$	For $E_p > 3$ MeV.
	992	$^{63}Cu(p,\gamma)^{64}Zn$	Best at $E_p < 3$ MeV.
	865	$^{65}Cu(p,n\gamma)^{65}Zn$	For ^{65}Cu detection at $E_p > 3.5$ MeV.
Zn	360	$^{67}Zn(p,n\gamma)^{67}Ga$	For ^{67}Zn detection.
	508	$^{70}Zn(p,n\gamma)^{70}Ga$	For ^{70}Zn detection at $E_p > 3.5$ MeV.
	874	$^{68}Zn(p,p'\gamma)^{68}Zn$	For ^{68}Zn detection.
	992	$^{64}Zn(p,p'\gamma)^{64}Zn$	For ^{64}Zn detection.
	1039	$^{66}Zn(p,p'\gamma)^{66}Zn$	For ^{66}Zn detection.

Table A12.8a. Absolute thick-target gamma-ray yields for light elements (Z < 21) with 1.0–4.2 MeV protons. The detection angle is 55° with respect to the beam direction. Data from Anttila *et al.* (1981), Kiss *et al.* (1985).

Element	E_γ (keV)	Yield ($N_\gamma/\mu C$ sr)						Reaction
		1.0 MeV	1.7 MeV	2.4 MeV	3.1 MeV	3.8 MeV	4.2 MeV	
Li	429	-	-	-	$9.2 \cdot 10^6$	$2.6 \cdot 10^7$	$3.3 \cdot 10^7$	^{7}Li(p, n$_1\gamma$)^{7}Be
	478	$6.5 \cdot 10^5$	$8.6 \cdot 10^6$	$2.6 \cdot 10^7$	$5.6 \cdot 10^7$	$8.9 \cdot 10^7$	$1.1 \cdot 10^8$	^{7}Li(p, p$_1\gamma$)^{7}Li
Be	415	$0.4 \cdot 10^3$	$0.6 \cdot 10^3$	$0.6 \cdot 10^3$	-	-	-	^{9}Be(p, γ_{3-2})^{10}B
	718	$1.2 \cdot 10^3$	$3.5 \cdot 10^3$	$5.3 \cdot 10^3$	-	-	-	^{9}Be (p, γ_1)^{10}B
	1022	$0.4 \cdot 10^3$	$0.8 \cdot 10^3$	$1.3 \cdot 10^3$	-	-	-	^{9}Be(p, γ_{2-1})^{10}B
	3562	-	$0.1 \cdot 10^3$	$2.5 \cdot 10^4$	$2.5 \cdot 10^6$	$5.1 \cdot 10^6$	$6.2 \cdot 10^6$	^{9}Be(p, $\alpha_1\gamma$)^{6}Li
	7477	$3.3 \cdot 10^3$	$3.0 \cdot 10^3$	-	-	-	-	^{9}Be (p,γ)^{10}B
B	429	$1.4 \cdot 10^4$	$9.1 \cdot 10^5$	$3.5 \cdot 10^6$	$7.2 \cdot 10^6$	$8.3 \cdot 10^6$	$8.3 \cdot 10^6$	^{10}B(p, $\alpha_1\gamma$)^{7}Be
	718	-	$0.4 \cdot 10^4$	$1.2 \cdot 10^5$	$1.3 \cdot 10^6$	$2.6 \cdot 10^6$	$4.7 \cdot 10^6$	^{10}B(p, p$_1\gamma$)^{10}B
	2125	-	-	-	$4.8 \cdot 10^6$	$1.3 \cdot 10^7$	$1.5 \cdot 10^7$	^{11}B(p, p$_1\gamma$)^{11}B
C	1635	6.0	9.0	$2.6 \cdot 10$	$3.5 \cdot 10$	-	-	^{13}C(p, γ_{2-1})^{14}N
	2313	$1.7 \cdot 10$	$2.6 \cdot 10$	$3.0 \cdot 10$	$1.4 \cdot 10^2$	-	-	^{13}C(p, γ_1)^{14}N
	2366	$2.6 \cdot 10^2$	$2.2 \cdot 10^2$	$2.3 \cdot 10^2$	$2.7 \cdot 10^2$	-	-	^{12}C(p, γ_1)^{13}N
	3089	-	-	-	-	$6.2 \cdot 10^3$	$4.1 \cdot 10^4$	^{13}C(p, p$_1\gamma$)^{13}C
	3511	-	$2.3 \cdot 10^2$	$2.1 \cdot 10^2$	-	-	-	^{12}C(p, γ_2)^{13}N
	8062	$2.5 \cdot 10$	$2.5 \cdot 10$	-	-	-	-	^{13}C(p,γ)^{14}N
N	2313	-	-	-	-	$5.5 \cdot 10^4$		^{14}N(p, p$_1\gamma$)^{14}N
	4439	$1.2 \cdot 10^2$	$7.7 \cdot 10^2$	$5.0 \cdot 10^3$	$5.0 \cdot 10^4$	$6.0 \cdot 10^4$		^{15}N(p, $\alpha_1\gamma$)^{12}C
	6793	-	-	$4.0 \cdot 10^2$	-	-		^{14}N(p,γ)^{15}O

Table A12.8a. Absolute thick-target gamma-ray yields for light elements (Z < 21) with 1.0–4.2 MeV protons. The detection angle is 55° with respect to the beam direction. Data from Anttila *et al.* (1981), Kiss *et al.* (1985) (continued).

Element	E_γ (keV)	Yield ($N_\gamma/\mu C$ sr)						Reaction
		1.0 MeV	1.7 MeV	2.4 MeV	3.1 MeV	3.8 MeV	4.2 MeV	
	2981	-	-	-	-	-	$3.6 \cdot 10^4$	^{27}Al(p, p$_5\gamma$)^{27}Al
	3004	-	-	-	-	-	$7.7 \cdot 10^4$	^{27}Al(p, p$_6\gamma$)^{27}Al
Si	677	4.0	$2.1 \cdot 10$	$3.1 \cdot 10$	-	-		^{29}Si(p, γ_1)^{30}P
	709	3.0	$1.9 \cdot 10$	$4.4 \cdot 10$	-	-		^{29}Si(p, γ_2)^{30}P
	755	-	-	-	$1.0 \cdot 10^5$	$6.3 \cdot 10^5$		^{29}Si(p, pγ_{2-1})^{29}Si
	1266	$1.2 \cdot 10$	$9.6 \cdot 10$	$1.9 \cdot 10^2$				^{30}Si(p, γ_1)^{31}P
	1273	-	-	$1.8 \cdot 10^3$	$1.2 \cdot 10^5$	$7.6 \cdot 10^5$		^{29}Si(p, p$_1\gamma$)^{29}Si
	1384	-	$5.2 \cdot 10$	$9.1 \cdot 10$	-	-		^{28}Si(p, γ_1)^{29}P
	1779	-	-	$2.3 \cdot 10^2$	$1.2 \cdot 10^6$	$7.2 \cdot 10^6$		^{28}Si(p, p$_1\gamma$)^{28}Si
	2028	-	-	-	$2.8 \cdot 10^3$	$5.3 \cdot 10^4$		^{29}Si(p, p$_2\gamma$)^{29}Si
	2233 and 2235	-	-	$1.3 \cdot 10^{2\,b}$	$4.7 \cdot 10^3$	$3.4 \cdot 10^4$		^{30}Si(p, γ_2)^{31}P ^{30}Si(p, p$_1\gamma$)^{30}Si
	4343	-	$1.6 \cdot 10^2$	$1.8 \cdot 10^2$	-	-		^{28}Si(p,γ)^{29}P
P	1266	-	$7.2 \cdot 10^2$	$3.8 \cdot 10^4$	$1.6 \cdot 10^6$	$5.2 \cdot 10^6$	$2.3 \cdot 10^7$	^{31}P(p, p$_1\gamma$)^{31}P
	1779	-	-	$2.0 \cdot 10^3$	$2.1 \cdot 10^5$	$6.5 \cdot 10^5$	$1.6 \cdot 10^6$	^{31}P(p, $\alpha_1\gamma$)^{28}Si
	2230 and 2233	$1.3 \cdot 10^2$	$1.7 \cdot 10^3$	$3.5 \cdot 10^3$	$1.2 \cdot 10^4$	$4.0 \cdot 10^5$	$9.5 \cdot 10^5$	^{31}P(p, γ_1)^{32}S ^{31}P(p, p$_2\gamma$)^{31}P
S	811	$2.1 \cdot 10$	$3.4 \cdot 10$	$4.5 \cdot 10$	$1.2 \cdot 10^2$	-	-	^{32}S(p, γ_1)^{33}Cl
	841	-	9.1	$1.4 \cdot 10^2$	$8.7 \cdot 10^3$	$1.7 \cdot 10^4$	$2.9 \cdot 10^4$	^{33}S(p, p$_1\gamma$)^{33}S
	1219	$1.0 \cdot 10$	$3.7 \cdot 10$	$6.5 \cdot 10$	$4.3 \cdot 10^3$	$4.6 \cdot 10^4$	$2.8 \cdot 10^5$	^{34}S(p, γ_1)^{35}Cl
	2035	$2.0 \cdot 10$	$2.3 \cdot 10$	-	-	-	-	^{32}S(p, γ_{2-1})^{33}Cl
	2127	-	-	-	-	$2.1 \cdot 10^4$	$4.8 \cdot 10^4$	^{34}S(p, p$_1\gamma$)^{34}S
	2230	-	-	-	$5.3 \cdot 10^3$	$1.5 \cdot 10^5$	$8.9 \cdot 10^5$	^{32}S(p, p$_1\gamma$)^{32}S
Cl	670	$0.1 \cdot 10^2$	$2.1 \cdot 10^2$	$5.2 \cdot 10^2$	$3.8 \cdot 10^2$	-		^{37}Cl(p, γ_{3-2})^{38}Ar
	1219	-	-	$3.5 \cdot 10^3$	$2.2 \cdot 10^5$	$1.5 \cdot 10^6$		^{35}Cl(p, p$_1\gamma$)^{35}Cl
	1410	-	-	-	-	$2.1 \cdot 10^5$		^{37}Cl(p, n$_1$)^{37}Ar
	1611	-	-	-	-	$1.9 \cdot 10^5$		^{37}Cl(p, n$_2\gamma$)^{37}Ar
	1642	$0.3 \cdot 10^2$	$4.9 \cdot 10^2$	$1.0 \cdot 10^3$				^{37}Cl(p, γ_{2-1})^{38}Ar
	1727	-	-	-	$3.8 \cdot 10^3$	$6.9 \cdot 10^4$		^{37}Cl(p, p$_1\gamma$)^{37}Cl
	1763	-	-	-	$5.2 \cdot 10^4$	$6.8 \cdot 10^5$		^{35}Cl(p, p$_2\gamma$)^{35}Cl
	1970	$2.2 \cdot 10^2$	$4.5 \cdot 10^2$	$1.1 \cdot 10^3$	$3.3 \cdot 10^3$	-		^{35}Cl(p, γ_1)^{36}Ar
	2127	-	-	$1.2 \cdot 10^3$	$3.1 \cdot 10^4$	$1.8 \cdot 10^5$		^{37}Cl(p, $\alpha_1\gamma$)^{38}Ar
	2168	$1.6 \cdot 10^2$	$1.6 \cdot 10^3$	$2.9 \cdot 10^3$	$3.9 \cdot 10^3$	$5.3 \cdot 10^3$		^{37}Cl(p, γ_1)^{38}Ar
	2208	$1.2 \cdot 10^2$	$2.1 \cdot 10^2$	$4.5 \cdot 10^2$	-	-		^{35}Cl(p, γ_{2-1})^{36}Ar
K	313	4.0	$5.6 \cdot 10$	-	-	-	-	^{41}K(p, γ_{2-1})^{42}Ca
	755	-	$6.1 \cdot 10$	-	-	-	-	^{39}K(p, γ_{2-1})^{40}Ca
	899	7.0	$1.1 \cdot 10$	$2.5 \cdot 10^2$	-	-	-	^{41}K(p, γ_{3-1})^{42}Ca
	980	-	-	$5.2 \cdot 10^2$	$1.1 \cdot 10^4$	$5.5 \cdot 10^4$	$9.3 \cdot 10^4$	^{41}K(p, p$_1\gamma$)^{41}K
	1294	-	-	-	-	$3.0 \cdot 10^4$	$1.2 \cdot 10^5$	^{41}K(p, p$_2\gamma$)^{41}K
	1525	$4.3 \cdot 10$	$5.4 \cdot 10^2$	$1.3 \cdot 10^3$	-	-	-	^{41}K(p, γ_1)^{42}Ca
	1943	-	-	-	-	$8.0 \cdot 10^4$	$1.4 \cdot 10^5$	^{41}K(p, n$_1\gamma$)^{41}Ca

Table A12.8a. Absolute thick-target gamma-ray yields for light elements (Z < 21) with 1.0–4.2 MeV protons. The detection angle is 55° with respect to the beam direction. Data from Anttila *et al.* (1981), Kiss *et al.* (1985) (continued).

Element	E_γ (keV)	Yield $(N_\gamma/\mu C\ sr)$						Reaction
		1.0 MeV	1.7 MeV	2.4 MeV	3.1 MeV	3.8 MeV	4.2 MeV	
	2010	-	-	-	-	$3.5 \cdot 10^4$	$1.2 \cdot 10^5$	^{41}K(p, $n_2\gamma$)^{41}Ca
	2168	-	$2.0 \cdot 10^2$	$2.6 \cdot 10^3$	$2.5 \cdot 10^4$	$9.5 \cdot 10^4$	$1.6 \cdot 10^5$	^{41}K(p, $\alpha_1\gamma$)^{38}Ar
	2522	-	-	-	-	-	$8.2 \cdot 10^4$	^{39}K(p, $n_1\gamma$)^{41}Ca
	2576	-	-	-	-	-	$1.4 \cdot 10^4$	^{41}K(p, $n_1\gamma$)^{41}Ca
	2604	-	-	-	-	-	$9.7 \cdot 10^3$	^{41}K(p, $n_1\gamma$)^{41}Ca
Ca	364	-	$4.3 \cdot 10$	$2.4 \cdot 10^2$	-	-		^{44}Ca(p, γ_{2-1})^{45}Sc
	371	-	-	-	$3.7 \cdot 10^3$	$1.6 \cdot 10^4$		^{48}Ca(p, $n\gamma_{2-1}$)^{48}Sc
	373	-	$3.2 \cdot 10$	$8.7 \cdot 10^2$	-	-		^{43}Ca(p, $p_1\gamma$)^{43}Ca
	520	-	-	-	$7.3 \cdot 10^2$	$3.6 \cdot 10^3$		^{48}Ca(p, $n\gamma_{3-2}$)^{48}Sc
	531	-	$1.5 \cdot 10$	$6.9 \cdot 10$	-	-		^{44}Ca(p, $p\gamma_{3-1}$)^{44}Ca
	543	-	$1.2 \cdot 10$	$4.6 \cdot 10$	-	-		^{44}Ca(p, γ_3)^{45}Sc
	720	-	$2.9 \cdot 10$	$1.0 \cdot 10^2$	-	-		^{44}Ca(p, γ_4)^{45}Sc
	779	-	-	-	$1.2 \cdot 10^3$	$6.5 \cdot 10^2$		^{48}Ca(p, $n\gamma_{4-2}$)^{48}Sc
	1157	-	-	-	$6.9 \cdot 10^3$	$8.6 \cdot 10^3$		^{44}Ca(p, $p_1\gamma$)^{44}Ca
	1525	-	-	-	$5.9 \cdot 10^2$	$7.7 \cdot 10^3$		^{42}Ca(p, $p_1\gamma$)^{42}Ca
Sc	364	-	$1.4 \cdot 10^3$	$3.6 \cdot 10^4$	$1.3 \cdot 10^5$	$3.6 \cdot 10^5$	$5.1 \cdot 10^5$	^{45}Sc(p, $p\gamma_{2-1}$)^{45}Sc
	431	-	-		$1.8 \cdot 10^4$	$5.8 \cdot 10^4$	-	^{45}Sc(p, $p\gamma_{6-3}$)^{45}Sc
	531	-	-	$2.5 \cdot 10^4$	$1.1 \cdot 10^5$	$3.4 \cdot 10^5$	$3.6 \cdot 10^5$	^{45}Sc(p, $p\gamma_{3-1}$)^{45}Sc
	543	-	-	-	$9.8 \cdot 10^4$	$2.5 \cdot 10^5$	$2.3 \cdot 10^5$	^{45}Sc(p, $p_3\gamma$)^{45}Sc
	720	-	-	$2.5 \cdot 10^4$	$2.5 \cdot 10^5$	$7.2 \cdot 10^5$	$6.0 \cdot 10^5$	^{45}Sc(p, $p_4\gamma$)^{45}Sc
	889	$0.6 \cdot 10^3$	$9.8 \cdot 10^3$	$6.4 \cdot 10^4$	$1.9 \cdot 10^5$	$3.4 \cdot 10^5$	$2.8 \cdot 10^5$	^{45}Sc(p, γ_1)^{46}Ti
	962	-	-	-	$1.9 \cdot 10^5$	$3.7 \cdot 10^5$	$2.6 \cdot 10^5$	^{45}Sc(p, $p\gamma_{6-1}$)^{45}Sc
	974	-	-	-	$9.3 \cdot 10^4$	$3.3 \cdot 10^5$	$2.9 \cdot 10^5$	^{45}Sc(p, $p_6\gamma$)^{45}Sc
	1049	$0.1 \cdot 10^3$	$1.7 \cdot 10^3$	-	$2.9 \cdot 10^4$	$3.6 \cdot 10^4$	-	^{45}Sc(p, γ_{5-2})^{46}Ti
	1121	$0.4 \cdot 10^3$	$6.1 \cdot 10^3$	$4.2 \cdot 10^4$	$1.2 \cdot 10^5$	$1.5 \cdot 10^5$	-	^{45}Sc(p, γ_{2-1})^{46}Ti
	1236	-	-	-	$1.0 \cdot 10^5$	$4.7 \cdot 10^5$	$4.0 \cdot 10^5$	^{45}Sc(p, $p_8\gamma$)^{45}Sc
	1409	-	-	-	-	$2.1 \cdot 10^5$	$2.3 \cdot 10^5$	^{45}Sc(p, $p_{10}\gamma$)^{45}Sc
	1662	-	-	-	-	$4.9 \cdot 10^4$	$7.5 \cdot 10^4$	^{45}Sc(p, $p_{13}\gamma$)^{45}Sc

Table A12.8b. Absolute thick-target gamma-ray yields for heavy elements (Z > 30) with 1.7 MeV and 2.4 MeV protons. The detection angle is 55° with respect to the beam direction. Data from Räisänen and Hänninen (1983).

Element	E_γ (keV)[1]	Yield E_p ($\mu C\ sr)^{-1}$ 1.7 MeV	2.4 MeV	Reaction
Ga	175	320	1.2×10^4	^{69}Ga(p,γ)^{70}Ge, ^{71}Ga(p, nγ)^{71}Ge
	318	160	1.5×10^3	^{69}Ga(p, p'γ)^{69}Ga
	327		3.0×10^3	^{71}Ga(p, nγ)^{71}Ge
	391		1.7×10^3	^{71}Ga(p, nγ)^{71}Ge, ^{71}Ga(p, p'γ)^{71}Ga
	500		4.3×10^3	^{71}Ga(p, nγ)^{71}Ge
	668	80	3.0×10^3	^{69}Ga(p,γ)^{70}Ge
	1040	450	1.2×10^4	^{69}Ga(p,γ)^{70}Ge
	1708	670	2.0×10^3	^{69}Ga(p,γ)^{70}Ge
Ge	147	70	1.6×10^3	^{70}Ge(p,γ)^{71}As
	199	50	2.1×10^3	^{74}Ge(p,γ)^{75}As
	254	45	1.1×10^3	^{72}Ge(p,γ)^{73}As, ^{73}Ge(p, nγ)^{73}As
	265	70	2.3×10^3	^{74}Ge(p,γ)^{75}As
	280	40	2.0×10^3	^{74}Ge(p,γ)^{75}As
	361	40	1.9×10^3	^{72}Ge(p,γ)^{73}As, ^{73}Ge(p, nγ)^{73}As
	563	60	1.4×10^3	^{76}Ge(p, p'γ)^{76}Ge
	596	1.9×10^3	5.6×10^3	^{74}Ge(p, p'γ)^{74}Ge, ^{74}Ge(p,γ)^{75}As
As	112	30	4.8×10^3	^{75}As(p, nγ)^{75}Se
	199	1.3×10^3	4.3×10^3	^{75}As(p, p'γ)^{75}As
	280	1.9×10^3	8.8×10^3	^{75}As(p, p'γ)^{75}As
	287		7.4×10^3	^{75}As(p, nγ)^{75}Se
	559	220	1.2×10^3	^{75}As(p,γ)^{76}Se
	573	95	2.3×10^3	^{75}As(p, p'γ)^{75}As
Se	239	1.6×10^3	4.1×10^3	^{77}Se(p, p'γ)^{77}Se
	276	80	2.6×10^3	^{80}Se(p,γ)^{81}Br
	559	130	2.1×10^3	^{76}Se(p, p'γ)^{76}Se
	614	120	2.4×10^3	^{78}Se(p, p'γ)^{78}Se
	666	80	2.7×10^3	^{80}Se(p, p'γ)^{80}Se
Br	190		1.3×10^3	^{81}Br(p, nγ)^{81}Kr
	217	1.4×10^3	6.2×10^3	^{79}Br(p, p'γ)^{79}Br
	276	1.2×10^3	6.2×10^3	^{81}Br(p, p'γ)^{81}Br
	306	310	2.1×10^3	^{79}Br(p, p'γ)^{79}Br
	523	75	1.6×10^3	^{79}Br(p, p'γ)^{79}Br
	617	55	3.9×10^3	^{79}Br(p,γ)^{80}Kr
Rb	130	130	780	^{85}Rb(p, p'γ)^{85}Rb
	151	35	1.6×10^3	^{85}Rb(p, p'γ)^{85}Rb
	232		5.0×10^3	^{85}Rb(p, nγ)^{85}Sr
	485	15	350	^{87}Rb(p, nγ)^{87}Sr
	1077	40	790	^{85}Rb(p,γ)^{86}Sr
	1229		470	^{87}Rb(p, nγ)^{87}Sr
	1854		150	^{85}Rb(p,γ)^{86}Sr

1. Values from Lederer, C.M., and Shirley, V.S. (eds.) (1978), *Table of Isotopes*, Wiley, New York.

Table A12.8b. Absolute thick-target gamma-ray yields for heavy elements (Z > 30) with 1.7 MeV and 2.4 MeV protons. The detection angle is 55° with respect to the beam direction. Data from Räisänen and Hänninen (1983) (continued).

Element	E_γ (keV)[1]	Yield E_p $(\mu C\ sr)^{-1}$ 1.7 MeV	2.4 MeV	Reaction
Sr	232		50	$^{87}Sr(p,\gamma)^{88}Y$
	793	1	20	$^{86}Sr(p,\gamma)^{87}Y$
	909	4	110	$^{88}Sr(p,\gamma)^{89}Y$
	1313	2	50	$^{88}Sr(p,\gamma)^{89}Y$
	1507	5	220	$^{88}Sr(p,\gamma)^{89}Y$
	1745		35	$^{88}Sr(p,\gamma)^{89}Y$
Y	420	7	60	$^{89}Y(p,\gamma)^{90}Zr$
	562	9	100	$^{89}Y(p,\gamma)^{90}Zr$
	891		35	$^{89}Y(p,\gamma)^{90}Zr$
	2186	40	650	$^{89}Y(p,\gamma)^{90}Zr$
	2319	9	80	$^{89}Y(p,\gamma)^{90}Zr$
Zr	656		60	$^{92}Zr(p,\gamma)^{93}Nb$
	744		20	$^{92}Zr(p,\gamma)^{93}Nb$
	919		10	$^{94}Zr(p,\ p'\gamma)^{94}Zr$
	1082		50	$^{90}Zr(p,\gamma)^{91}Nb$
	1208		75	$^{90}Zr(p,\gamma)^{91}Nb$
Nb	449		15	$^{93}Nb(Zr(p,\gamma)^{94}Mo$
	703	2	230	$^{93}Nb(p,\gamma)^{94}Mo$
	742		50	$^{93}Nb(p,\gamma)^{94}Mo,\ ^{93}Nb(p,\ p'\gamma)^{93}Nb$
	850		60	$^{93}Nb(p,\gamma)^{94}Mo$
	871	3	270	$^{93}Nb(p,\gamma)^{94}Mo$
Mo	204	210	940	$^{95}Mo(p,\ p'\gamma)^{95}Mo$
	481	2	60	$^{97}Mo(p,\ p'\gamma)^{97}Mo$
	535	20	930	$^{100}Mo(p,\ p'\gamma)^{100}Mo$
	778		110	$^{96}Mo(p,\ p'\gamma)^{96}Mo$
	787		150	$^{98}Mo(p,\ p'\gamma)^{98}Mo$
Ru	90	380	1.1×10^3	$^{99}Ru(p,\ p'\gamma)^{99}Ru$
	127	310	1.3×10^3	$^{101}Ru(p,\ p'\gamma)^{101}Ru$
	358	610	7.4×10^3	$^{104}Ru(p,\ p'\gamma)^{104}Ru$
	475	130	3.8×10^3	$^{102}Ru(p,\ p'\gamma)^{102}Ru$
	540	15	700	$^{100}Ru(p,\ p'\gamma)^{100}Ru$
Rh	295	1.6×10^3	1.6×10^4	$^{103}Rh(p,\ p'\gamma)^{103}Rh$
	357	880	1.2×10^4	$^{103}Rh(p,\ p'\gamma)^{103}Rh$
Pd	280	20	170	$^{105}Pd(p,\ p'\gamma)^{105}Pd$
	374	250	4.0×10^3	$^{110}Pd(p,\ p'\gamma)^{110}Pd$
	434	210	4.8×10^3	$^{108}Pd(p,\ p'\gamma)^{108}Pd$
	442		980	$^{105}Pd(p,\ p'\gamma)^{105}Pd$
	556	7	480	$^{102}Pd(p,\ p'\gamma)^{102}Pd,\ ^{104}Pd(p,\ p'\gamma)^{104}Pd$

1. Values from Lederer, C.M., and Shirley, V.S. (eds.) (1978), Wiley, New York.

Table A12.8b. Absolute thick-target gamma-ray yields for heavy elements (Z > 30) with 1.7 MeV and 2.4 MeV protons. The detection angle is 55° with respect to the beam direction. Data from Räisänen and Hänninen (1983) (continued).

Element	E_γ (keV)[1]	Yield E_p (μC sr)$^{-1}$		Reaction
		1.7 MeV	2.4 MeV	
Ag	311	810	7.3×10^3	^{109}Ag(p, p'γ)^{109}Ag
	325	610	6.0×10^3	^{107}Ag(p, p'γ)^{107}Ag
	415	220	4.2×10^3	^{109}Ag(p, p'γ)^{109}Ag
	423	180	3.6×10^3	^{107}Ag(p, p'γ)^{107}Ag
Cd	298	85	770	^{113}Cd(p, p'γ)^{113}Cd
	342	45	480	^{111}Cd(p, p'γ)^{111}Cd
	558	15	980	^{114}Cd(p, p'γ)^{114}Cd
	617	5	450	^{112}Cd(p, p'γ)^{112}Cd
	658		140	^{110}Cd(p, p'γ)^{110}Cd
In	115		60	^{115}In(p, nγ)^{115}Sn
	497		30	^{115}In(p, nγ)^{115}Sn
Sn				
Sb	160	40	320	^{123}Sb(p, p'γ)^{123}Sb
	382	1	50	^{123}Sb(p, p'γ)^{123}Sb
	508		30	^{121}Sb(p, p'γ)^{121}Sb
	542	0.3	25	^{123}Sb(p, p'γ)^{123}Sb
	573	0.6	55	^{121}Sb(p, p'γ)^{121}Sb
Te	666		570	^{126}Te(p, p'γ)^{126}Te
I	145	30	240	^{127}I(p, p'γ)^{127}I
	172	10	220	^{127}I(p, p'γ)^{127}I
	203	340	2.2×10^3	^{127}I(p, p'γ)^{127}I
	375	10	210	^{127}I(p, p'γ)^{127}I
	418	4	120	^{127}I(p, p'γ)^{127}I
	629		130	^{127}I(p, p'γ)^{127}I
Cs	161	160	620	^{133}Cs(p, p'γ)^{133}Cs
	303		410	^{133}Cs(p, p'γ)^{133}Cs
	384		170	^{133}Cs(p, p'γ)^{133}Cs
Ba	221		30	^{135}Ba(o, p'γ)^{135}Ba
	279	9	75	^{137}Ba(p, p'γ)^{137}Ba
	481	4	55	^{135}Ba(p, p'γ)^{135}Ba
	605		10	^{134}Ba(p, p'γ)^{134}Ba
La				
Ce				
Pr	145	7	55	^{141}Pr(p, p'γ)^{141}Pr
Nd	130	400	4.4×10^3	^{150}Nd(p, p'γ)^{148}Nd
	302	60	1.0×10^3	^{148}Nd(p, p'γ)^{148}Nd
	454	7	400	^{146}Nd(p, p'γ)^{146}Nd
	696		25	^{144}Nd(p, p'γ)^{144}Nd

1. Values from Lederer, C.M., and Shirley, V.S. (eds.) (1978), Wiley, New York.

Table A12.8b. Absolute thick-target gamma-ray yields for heavy elements (Z > 30) with 1.7 MeV and 2.4 MeV protons. The detection angle is 55° with respect to the beam direction. Data from Räisänen and Hänninen (1983) (continued).

Element	E_γ (keV)[1]	Yield E_p ($\mu C \, sr)^{-1}$ 1.7 MeV	2.4 MeV	Reaction
Sm	82	2.4×10^3	1.5×10^4	^{154}Sm(p, p'γ)^{154}Sm
	122	3.2×10^3	3.4×10^4	^{152}Sm(p, p'γ)^{152}Sm, ^{147}Sm(p,p'γ)^{147}Sm
	197	20	220	^{147}Sm(p, p'γ)^{147}Sm
	334	40	990	^{150}Sm(p, p'γ)^{150}Sm
	550	6	70	^{148}Sm(p, p'γ)^{148}Sm
Eu	83	1.1×10^4	3.0×10^4	^{153}Eu(p, p'γ)^{153}Eu
	111	490	3.0×10^3	^{151}Eu(p, p'γ)^{151}Eu
	193	1.1×10^3	6.2×10^3	^{153}Eu(p, p'γ)^{153}Eu
	197	230	1.6×10^3	^{151}Eu(p, p'γ)^{151}Eu
	286	25	400	^{151}Eu(p, p'γ)^{151}Eu
	307	170	2.7×10^3	^{151}Eu(p, p'γ)^{151}Eu
Gd	80	3.9×10^3	9.0×10^3	^{158}Gd(p, p'γ)^{158}Gd
	89	6.8×10^3	1.7×10^4	^{156}Gd(p, p'γ)^{156}Gd
	123	620	2.2×10^3	^{154}Gd(p, p'γ)^{154}Gd
	131	110	440	^{157}Gd(p, p'γ)^{157}Gd
	146	130	560	^{155}Gd(p, p'γ)^{155}Gd
	344		20	^{152}Gd(p, p'γ)^{152}Gd
Tb				
Dy	81	3.3×10^3	8.5×10^3	^{162}Dy(p, p'γ)^{162}Dy
	167	580	3.3×10^3	^{163}Dy(p, p'γ)^{163}Dy, ^{164}Dy(p, p'γ)^{164}Dy
Ho	95	4.0×10^4	1.2×10^3	^{165}Ho(p, p'γ)^{165}Ho
	115	2.3×10^3	1.6×10^4	^{165}Ho(p, p'γ)^{165}Ho
	210	110	760	^{165}Ho(p, p'γ)^{165}Ho
Er				
Tm	110	1.4×10^4	1.8×10^5	^{169}Tm(p, p'γ)^{169}Tm
	118	1.4×10^3	1.4×10^4	^{169}Tm(p, p'γ)^{169}Tm
Yb	181	20	150	^{172}Yb(p, p'γ)^{172}Yb, ^{173}Yb(p, p'γ)^{173}Yb
Lu	114	3.8×10^3	1.6×10^4	^{175}Lu(p, p'γ)^{175}Lu
	138	180	2.4×10^3	^{175}Lu(p, p'γ)^{175}Lu
	185	90	680	^{176}Lu(p, p'γ)^{176}Lu
	251	120	1.6×10^3	^{175}Lu(p, p'γ)^{175}Lu
Hf	93	1.2×10^4	4.2×10^4	^{178}Hf(p, p'γ)^{178}Hf, ^{180}Hf(p, p'γ)^{180}Hf
	113	2.5×10^3	9.0×10^3	^{177}Hf(p, p'γ)^{177}Hf
	123	1.2×10^3	4.9×10^3	^{179}Hf(p, p'γ)^{179}Hf
	146	40	280	^{179}Hf(p, p'γ)^{179}Hf
	250	110	930	^{177}Hf(p, p'γ)^{177}Hf
Ta	136	3.1×10^4	1.6×10^5	^{181}Ta(p, p'γ)^{181}Ta
	165	80	1.5×10^3	^{181}Ta(p, p'γ)^{181}Ta
	302	55	1.1×10^3	^{181}Ta(p, p'γ)^{181}Ta

1. Values from Lederer, C.M., and Shirley, V.S. (eds.) (1978), Wiley, New York.

Table A12.8b. Absolute thick-target gamma-ray yields for heavy elements (Z > 30) with 1.7 MeV and 2.4 MeV protons. The detection angle is 55° with respect to the beam direction. Data from Räisänen and Hänninen (1983) (continued).

Element	E_γ (keV)[1]	Yield E_p ($\mu C\ sr$)$^{-1}$		Reaction
		1.7 MeV	2.4 MeV	
W	100	5.0×10^3	1.5×10^4	$^{182}W(p, p'\gamma)^{182}W$
	111	6.1×10^3	2.1×10^4	$^{184}W(p, p'\gamma)^{184}W$
	123	5.4×10^3	2.2×10^4	$^{186}W(p, p'\gamma)^{186}W$
	292	4	60	$^{183}W(p, p'\gamma)^{183}W$
Pt	211	110	1.2×10^3	$^{195}Pt(p, p'\gamma)^{195}Pt$
	239	40	430	$^{195}Pt(p, p'\gamma)^{195}Pt$
	329	60	1.9×10^3	$^{194}Pt(p, p'\gamma)^{194}Pt$
	356	20	930	$^{196}Pt(p, p'\gamma)^{196}Pt$
	407	2	110	$^{198}Pt(p, p'\gamma)^{198}Pt$
Au	191	20	240	$^{197}Au(p, p'\gamma)^{197}Au$
	269	1	23	$^{197}Au(p, p'\gamma)^{197}Au$
	279	100	1.3×10^3	$^{197}Au(p, p'\gamma)^{197}Au$
	548		60	$^{197}Au(p, p'\gamma)^{197}Au$
Hg	158	70	460	$^{199}Hg(p, p'\gamma)^{199}Hg$
	208	15	160	$^{199}Hg(p, p'\gamma)^{199}Hg$
	368	7	310	$^{200}Hg(p, p'\gamma)^{200}Hg$
	412		110	$^{198}Hg(p, p'\gamma)^{198}Hg$
Tl	204	65	670	$^{205}Tl(p, p'\gamma)^{205}Tl$
	279	9	230	$^{203}Tl(p, p'\gamma)^{203}Tl$
	416		17	$^{205}Tl(p, p'\gamma)^{205}Tl$
Pb				

1. Values from Lederer, C.M., and Shirley, V.S. (eds.) (1978), Wiley, New York.

Table A12.8c. Absolute thick-target gamma-ray yields with 7 MeV and 9 MeV protons. Data from Räisänen et al. (1987).

Isotope	E_γ (keV)[a]	Reaction[b]	Absolute γ-ray yield[c] ($\mu C\ sr)^{-1}$		Isotope	E_γ (keV)[a]	Reaction[b]	Absolute γ-ray yield[c] ($\mu C\ sr)^{-1}$	
			7 MeV	9 MeV				7 MeV	9 MeV
^{7}Li[d]	429	2	4.33 (7)	2.23 (8)	^{31}P[j]	1266	1	9.74 (7)	1.74 (8)
	478	1	3.59 (8)	2.09 (9)		1779	3	5.22 (7)	8.10 (7)
^{9}Be	3562	3	1.95 (8)	1.86 (8)		2230; 2235	4; 1	7.01 (7)	1.33 (8)
^{10}B	718	1	1.16 (8)	1.54 (8)	^{32}S[k]	2230	1	6.17 (7)	1.48 (8)
^{11}B	2000	2	2.25 (7)	9.49 (7)		4282	1	5.99 (6)	3.66 (7)
	2125	1	2.79 (8)	4.27 (8)	^{35}Cl[l]	1219	1	5.67 (7)	8.55 (8)
^{12}C	4439	1	7.53 (8)	2.89 (9)		1763	1	9.51 (7)	1.62 (8)
^{14}N[e]	1635	1	1.26 (7)	3.12 (7)		2230	3	5.13 (7)	1.06 (8)
	2313	1	1.82 (7)	4.24 (7)		3163	1	2.27 (7)	5.71 (7)
	5106	1	2.21 (6)	2.47 (7)	^{39}K[m]	1970	3	1.37 (7)	3.12 (7)
^{16}O[f]	6129	1	1.15 (7)	1.80 (9)	^{39}K; ^{41}K	2523	1; 2	1.57 (7)	2.46 (7)
	6919	1		8.29 (7)	^{39}K	2814	1	3.98 (7)	1.03 (8)
^{19}F[d]	1236	1; 2	3.44 (7)	3.00 (8)		3019	1	1.80 (7)	2.92 (7)
	1346; 1357	1; 1	1.35 (8)	7.70 (8)		3598	1	6.99 (6)	1.51 (7)
	6129	3	1.94 (8)	9.43 (8)	^{40}Ca[n]	755	1	5.68 (5)	2.55 (7)
^{23}Na[g]	440	1	7.31 (8)	6.67 (8)		3736	1	7.40 (7)	2.65 (8)
	1633	1; 3	4.75 (8)	4.60 (8)		3904	1	8.59 (7)	2.84 (8)
	3915	1	1.51 (7)	3.24 (7)	^{47}Ti; ^{48}Ti	308	4; 2	3.88 (7)	1.46 (8)
^{24}Mg[h]	1369	1	7.28 (8)	9.28 (8)	^{46}Ti	889	1	9.57 (6)	2.57 (7)
^{26}Mg	1809	1	7.81 (7)	8.58 (7)	^{48}Ti	983	1	6.15 (7)	1.27 (8)
^{24}Mg	2754	1	1.12 (7)	7.38 (7)		1312	1	4.34 (6)	1.72 (7)
	4239	1	5.20 (7)	9.25 (7)		1437	1	6.97 (6)	1.71 (7)
^{27}Al	844	1	2.06 (8)	3.78 (8)	^{51}V	319	1	1.46 (8)	2.46 (8)
	1014	1	4.97 (8)	8.89 (8)		749	2	3.58 (8)	5.38 (8)
	1369	3	3.92 (8)	7.62 (8)		808	2	1.47 (8)	2.20 (8)
	1720	1	1.33 (8)	2.38 (8)		1148	2	8.21 (7)	1.23 (8)
	2210	1	4.17 (8)	8.00 (8)		1164	2	3.29 (8)	4.85 (8)
	2981	1	7.32 (7)	1.01 (8)		1480	2	1.01 (8)	1.72 (8)
	3004	1	2.04 (8)	5.44 (8)	^{52}Cr	936	1	5.05 (6)	3.95 (7)
^{28}Si[i]	1779	1	2.08 (8)	4.68 (8)		1334	1	2.68 (6)	4.62 (7)
^{29}Si; ^{30}Si	2028	1; 4	4.87 (6)	1.08 (7)		1434	1	1.10 (8)	3.26 (8)
^{30}Si	2235	1; 4	4.30 (6)	1.15 (7)		1531	1	5.06 (6)	3.75 (7)
^{28}Si	2839	1	4.87 (6)	2.16 (7)		1728	1	2.66 (6)	2.39 (7)
	3200	1	2.78 (6)	2.91 (7)	^{55}Mn	411	2	1.04 (8)	1.26 (8)

a. From Lederer, C.M. and Shirley, V.S. (eds.) (1978), Table of Isotopes, Wiley, New York. The gamma-ray energies that can lead to an ambgiuous analysis are as follows: 288 (Cd), 292 (W), 297 (Gd), 307 (Cd), 308 (Ti), 319 (V), 321 (I), 326 (Gd), 328 (Pt, Hf), 336 (Mo), 339 (Co), 365 (Ag, Ta), 375 (I), 382 (Sb), 411 (Mn, I), 429 (Li), 343 (Pd), 583 (Au), 588 (Y), 700 (Sn), 702 (Ag), 804 (Mn), 808 (V, Zn), 833 (Ru), 834 (Ge), 844 (Al), 847 (Fe, Co), 931 (Mn), 934 (In), 936 (Cr), 1039 (Zn), 1040 (Ge), 1230 (Sn), 1236 (F), 1237 (In), 1238 (Fe), 1312 (Ti), 1314 (Mn), 1315 (Zn), 1327 (Cu), 1332 (Cu, Co, Ni), 1334 (Cr), 1338 (Co), 1369 (Mg, Al), 1408 (Mn, Fe), 1434 (Cr), 1437 (Ti), 1477 (Nb), 1480 (V), 1507 (Y), 1510 (Mo), 1627 (Y), 1632 (Na), 1635 (N), 1779 (P, Si), 1809 (Mg), 1811 (Fe), 2230 (S, Cl), 2235 (P, S), 6126 (O, F).

b. 1, (p,p'); 2, (p,n); 3 (p,α); 4 (p,γ).

c. (n) means x10[n].

d. LiF target.

e. TaN target.

f. MgO, SiO_2, amd Sm_2O_3 targets.

g. NaCl target.

h. MgO target.

i. Si and SiO_2 targets.

j. InP target.

k. PbS target.

l. NcCl and $PbCl_2$ targets.

m. KI target.

n. CaO target.

Table A12.8c. Absolute thick-target gamma-ray yields with 7 MeV and 9 MeV protons. Data from Räisänen *et al.* (1987) (continued).

Isotope	E_γ (keV)[a]	Reaction[b]	Absolute γ-ray yield[c] ($\mu C\ sr)^{-1}$ 7 MeV	9 MeV
	804	2	6.83 (7)	1.01 (8)
	931	2	3.45 (8)	4.73 (8)
	1314	2	2.11 (8)	3.09 (8)
	1408	2	7.39 (7)	9.90 (7)
^{56}Fe	847	1	6.01 (8)	7.61 (8)
	1238	1	3.67 (7)	1.01 (8)
^{54}Fe	1408	1	2.94 (7)	4.64 (7)
^{56}Fe	1811	1	4.49 (7)	7.13 (7)
	2113	1	2.71 (7)	4.29 (7)
^{59}Co	339	2	1.01 (8)	1.88 (8)
	847	3	1.02 (7)	2.52 (7)
	878	2	1.13 (7)	1.93 (7)
	998	2	2.40 (7)	4.77 (7)
	1332; 1338	4; 2	1.83 (7)	3.36 (7)
	1428	2	1.90 (7)	3.80 (7)
^{60}Ni	826	1	3.19 (7)	8.22 (7)
^{58}Ni	1005	1	2.26 (7)	1.61 (8)
	1321	1	2.59 (7)	1.18 (8)
^{60}Ni	1332	1	1.21 (8)	3.29 (8)
^{58}Ni	1454; 1448	1	3.12 (8)	8.77 (8)
	1584	1	1.02 (7)	5.02 (7)
^{63}Cu	962	1	1.37 (7)	3.39 (7)
	1065	2	5.09 (6)	1.54 (7)
	1327	1	5.78 (6)	1.54 (7)
	1332	3	4.51 (6)	1.39 (7)
^{64}Zn	808	1	5.15 (7)	1.66 (8)
	992	1	2.52 (8)	7.60 (8)
^{66}Zn	1039	1	9.73 (7)	2.13 (8)
^{64}Zn	1315	1	1.17 (7)	7.50 (7)
	1799	1	2.21 (7)	7.27 (7)
^{72}Ge	834	1	3.63 (7)	7.84 (7)
^{70}Ge	1040	1	7.22 (7)	2.13 (8)
	1708	1	9.44 (6)	2.87 (7)
^{89}Y	588	2	1.35 (8)	3.25 (8)
	770	2	1.54 (7)	4.41 (7)
	863	2	2.06 (7)	8.64 (7)
	1155	2	1.39 (7)	4.25 (7)
	1507	1	1.81 (7)	4.07 (7)
	1627	2	1.71 (7)	5.95 (7)
	1833	2	1.42 (7)	5.08 (7)
^{90}Zr	1761	1	1.27 (6)	1.53 (7)
	2186	1		4.18 (7)
^{93}Nb	685; 687	2; 1	3.29 (7)	1.01 (8)
	1363	2	3.18 (7)	9.30 (7)
	1477	2	1.11 (8)	2.84 (8)
	1520	2	1.97 (7)	4.48 (7)

Isotope	E_γ (keV)[a]	Reaction[b]	Absolute γ-ray yield[c] ($\mu C\ sr)^{-1}$ 7 MeV	9 MeV
^{95}Mo	336	2	1.13 (7)	5.28 (7)
	627	2	1.21 (7)	3.34 (7)
^{92}Mo	1510	1	6.58 (6)	5.41 (7)
^{104}Ru	358	1	1.70 (7)	4.01 (7)
^{96}Ru	833	1	4.75 (6)	1.94 (7)
^{108}Pd	434	1	2.41 (6)	8.99 (6)
^{107}Ag	365	2	2.13 (7)	9.00 (7)
^{109}Ag	614	2	3.28 (7)	5.07 (7)
	624	2	2.75 (7)	4.69 (7)
^{107}Ag; ^{109}Ag	702	2; 1	9.20 (6)	4.44 (7)
^{114}Cd	210	2	1.04 (8)	3.07 (8)
	288	2	6.94 (7)	1.84 (8)
	307	2	3.26 (7)	1.71 (8)
^{114}Cd; ^{111}Cd	536, 537	2	2.91 (7)	1.43 (8)
^{111}Cd	1102	2	5.81 (6)	1.55 (7)
^{113}In; ^{115}In[o]	497	2	1.05 (8)	4.28 (8)
^{115}In	934	1	2.43 (7)	9.50 (7)
	1078	1	3.27 (7)	1.49 (8)
	1237	2	1.64 (6)	7.18 (6)
^{119}Sn	700	2	1.63 (7)	5.36 (7)
^{118}Sn	1230	1	3.29 (6)	3.02 (7)
^{116}Sn	1294	1	1.95 (7)	1.11 (8)
^{121}Sb	212	2	1.84 (8)	5.84 (8)
	230	2	6.56 (7)	1.76 (8)
	244	2	4.55 (7)	1.36 (8)
^{123}Sb	281	2	3.52 (7)	1.13 (8)
	382	1	3.32 (7)	1.10 (8)
^{127}I	321	2	2.77 (7)	9.60 (7)
	375	1	3.42 (7)	1.15 (8)
	411	2	857 (6)	1.15(8)
^{147}Sm[p]	777	2	7.37 (5)	2.05 (6)
^{156}Gd; ^{157}Gd	297	1; 2	4.87 (6)	3.36 (7)
	326	2	5.22 (5)	7.78 (6)
	598	2	3.55 (5)	2.70 (6)
^{179}Hf; ^{178}Hf	325; 328;	2; 1; 1	1.80 (6)	9.29 (6)
^{180}Hf	328			
^{181}Ta[q]	365	2	7.02 (6)	3.10 (7)
	661	2	9.46 (5)	3.04 (6)
^{183}W	292	1	5.14 (5)	6.00 (6)
^{195}Pt	262	2	4.33 (5)	4.69 (7)
^{194}Pt	328	1	1.88 (7)	3.44 (7)
^{197}Au	548	1	1.64 (6)	4.03 (6)
	583	1	5.48 (5)	2.56 (6)
^{207}Pb; ^{208}Pb	570; 571	1; 2	1.37 (5)	6.62 (6)
^{208}Pb	970	2	1.21 (5)	2.56 (6)

a. From Lederer, C.M. and Shirley, V.S. (eds.) (1978), *Table of Isotopes*, Wiley, New York. The gamma-ray energies that can lead to an ambgiuous analysis are as follows: 288 (Cd), 292 (W), 297 (Gd), 307 (Cd), 308 (Ti), 319 (V), 321 (I), 326 (Gd), 328 (Pt, Hf), 336 (Mo), 339 (Co), 365 (Ag, Ta), 375 (I), 382 (Sb), 411 (Mn, I), 429 (Li), 343 (Pd), 583 (Au), 588 (Y), 700 (Sn), 702 (Ag), 804 (Mn), 808 (V, Zn), 833 (Ru), 834 (Ge), 844 (Al), 847 (Fe, Co), 931 (Mn), 934 (In), 936 (Cr), 1039 (Zn), 1040 (Ge), 1230 (Sn), 1236 (F), 1237 (In), 1238 (Fe), 1312 (Ti), 1314 (Mn), 1315 (Zn), 1327 (Cu), 1332 (Cu, Co, Ni), 1334 (Cr), 1338 (Co), 1369 (Mg, Al), 1408 (Mn, Fe), 1434 (Cr), 1437 (Ti), 1477 (Nb), 1480 (V), 1507 (Y), 1510 (Mo), 1627 (Y), 1632 (Na), 1635 (N), 1779 (P, Si), 1809 (Mg), 1811 (Fe), 2230 (S, Cl), 2235 (P, S), 6126 (O, F).

b. 1, (p,p'); 2, (p,n); 3 (p,α); 4 (p,γ).

c. (n) means $\times 10^n$.

o. In and InP targets.

p. Sm_2O_3 target.

q. Ta and TaN targets.

Table A12.8d. Relative neutron yields induced with 7 MeV and 9 MeV protons. Data from Räisänen *et al.* (1987).

Target	Relative neutron yields[a]	
	7 MeV	9 MeV
LiF	8.8 (4)	1.6 (5)
Be	2.7 (5)	4.9 (5)
B	4.7 (4)	8.9 (4)
C	8.2 (2)	2.7 (3)
TaN	3.5 (2)	5.0 (3)
MgO	1.5 (3)	6.6 (3)
SiO_2	4.3 (2)	1.9 (3)
Sm_2O_3	9.7 (2)	1.0 (4)
NaCl	1.5 (4)	4.0 (4)
Al	2.8 (3)	1.8 (4)
Si	5.8 (2)	3.4 (3)
InP	5.3 (3)	2.9 (4)
PbS	1.4 (2)	1.7 (3)
$PbCl_2$	6.7 (3)	1.6 (4)
KI	4.0 (3)	2.0 (4)
CA	7.5 (2)	2.1 (3)
Ti	3.0 (4)	9.6 (4)
V	7.7 (4)	1.2 (5)
Fe	7.6 (3)	4.4 (4)
Co	4.9 (4)	1.1 (5)
Ni	2.0 (3)	1.1 (4)
Cu	3.2 (4)	8.4 (4)
Zn	1.1 (4)	4.4 (4)
Ge	2.6 (4)	8.8 (4)
Y	1.2 (4)	5.5 (4)
Zr	9.4 (3)	4.0 (4)
Nb	1.5 (4)	5.7 (4)
Mo	1.3 (4)	5.2 (4)
Ru	1.1 (4)	5.1 (4)
Pd	9.0 (3)	4.1 (4)
Ag	8.6 (3)	3.9 (4)
Cd	7.1 (3)	3.7 (4)
In	7.0 (3)[b]	3.5 (4)
Sn	5.6 (3)[b]	3.1 (4)
Sb	5.4 (3)[b]	3.1 (4)
Gd	1.0 (3)[b]	1.2 (4)
Yb	5.2 (2)	6.1 (3)
Hf	5.4 (2)	7.3 (3)
Ta	3.5 (2)	4.5 (3)[c]
W	8.6 (2)	5.7 (3)[c]
Pt	2.0 (2)	5.3 (3)[c]
Au	1.7 (2)	2.6 (3)[c]
Pb	1.0 (2)	4.0 (3)[c]

a. Yield obtained in the geometry used in the original article per μC.
b. The values given in Elwyn *et al.* (1966) are 7.0 (3) for In, 6.0 (3) for Sn, 5.6 (3) for Sb, 1.1 (3) for Gd, where (n) means $x \times 10^n$.
c. The values given in Elwyn *et al.* (1966) are: 4.5 (3) for Ta, 3.7 (3) for W, 2.6 (3) for Pt, 2.3 (3) for Au, 1.8 (3) for Pb where (n) means $\times 10^n$.

Table A12.9. Sensitivities obtainable with proton-induced PIGE.

a. Sensitivity in atomic fraction when proton energy is limited to <9 MeV. The crude grouping is based on Table A12.5c of the previous edition of the Ion Beam Handbook.

Over 1%	Between 1 and 0.1%	Better than 0.1%
Pd, Sm, Gd, Hf, W, Au, Pb	S, K, Sc, Ti, Cr, Co, Cu, Ge, Y, Zr, Mo, Ru, Ag, Sn, I, Ta, Pt	Li, Be, B, C, N, O, F, Na, Mg, Al, Si, P, Cl, Ca, V, Mn, Fe, Ni, Zn, Nb, Cd, In, Sb

b. Detection limits for some elements detectable in typical biomedical and organic samples by external-beam PIGE. The values were compiled from Räisänen (1987), Räisänen and Lapatto (1988), and Räisänen (1989).

Element	E_p (MeV)[a]	Reaction	E_γ (keV)	Detection limit (ppm by weight)
Li	1.8	$^7Li(p,p'\gamma)^7Li$	478	0.15
B	1.8	$^{10}B(p,\alpha\gamma)^7Be$	429	0.18
C	7.6	$^{12}C(p,p'\gamma)^{12}C$	4439	100
N	4.1	$^{14}N(p,p'\gamma)^{14}N$	2313	10
O	7.6	$^{16}O(p,p'\gamma)^{16}O$	6129	300
F	2.4	$^{19}F(p,\alpha\gamma)^{16}O$	6129	0.5
Na	2.4	$^{23}Na(p,p'\gamma)^{23}Na$	440	0.3
Mg	3.0	$^{25}Mg(p,p'\gamma)^{25}Mg$	585	15
Al[b]	3.1	$^{27}Al(p,p'\gamma)^{27}Al$	1014	40
P	3.0	$^{31}P(p,p'\gamma)^{31}P$	1266	15
Ca	2.4	$^{40}Ca(p,p'\gamma)^{40}Ca$	3736	4000
S	5.0	$^{32}S(p,p'\gamma)^{32}S$	2230	100
Cl	3.0	$^{35}Cl(p,p'\gamma)^{35}Cl$	1219	100

a. Energy on the exit foil.
b. Sample is ash diluted in graphite, and values are for an in-vacuum measurement (Boni *et al.*, 1990).

A12.2.2 Deuteron- and triton-induced gamma-ray emission

Table A12.10. Absolute thick-target gamma-ray yields induced with 1.8 MeV deuterons. Data from Elekes *et al.* (2000) and Kiss *et al.* (1994).

Element	E_γ (keV)	Yield$_{meas}$ ($N_\gamma/sr/\mu C$)	Reaction	Remarks
Li	429	1.10(6)	$^6Li(d,n\gamma)^7Be$	
	478	2.31(6)	$^6Li(d,p\gamma)^7Li$	
B	953	1.09(7)	$^{11}B(d,p\gamma)^{12}B$	
	1674	1.36(7)	$^{11}B(d,p\gamma)^{12}B$	
C	3089	1.55(7)	$^{12}C(d,p\gamma)^{13}C$	The line is Doppler broadened.
	3684	1.14(6)	$^{12}C(d,p\gamma)^{13}C$	
	3853	6.92(4)	$^{12}C(d,p\gamma)^{13}C$	
N	1885	2.56(6)	$^{14}N(d,p\gamma)^{15}N$	
	2297	2.15(6)	$^{14}N(d,p\gamma)^{15}N$	
	5240	5.46(5)	$^{14}N(d,n\gamma)^{15}O$	
	8310	4.30(6)	$^{14}N(d,p\gamma)^{15}N$	The gamma-ray energy is above the majority of lines induced from other elements.
O	871	1.22(7)	$^{16}O(d,p\gamma)^{17}O$	The presence of fluorine must be taken into account, and in such cases, the yield ratios should be applied.
F	656	2.02(6)	$^{19}F(d,p\gamma)^{20}F$	
	823	2.30(5)	$^{19}F(d,p\gamma)^{20}F$	
	871	1.16(6)	$^{19}F(d,\alpha\gamma)^{17}O$	Yield from F is lower than that from O.
	984	3.57(5)	$^{19}F(d,p\gamma)^{20}F$	
	1057	2.29(6)	$^{19}F(d,p\gamma)^{20}F$	
	1309	4.25(5)	$^{19}F(d,p\gamma)^{20}F$	
Na	351	6.61(5)	$^{23}Na(d,\alpha\gamma)^{21}Ne$	
	472	4.68(6)	$^{23}Na(d,p\gamma)^{24}Na$	
	1369	2.39(6)	$^{23}Na(d,n\gamma)^{24}Mg$	

Table A12.10. Absolute thick-target gamma-ray yields induced with 1.8 MeV deuterons. Data from Elekes *et al.* (2000) and Kiss *et al.* (1994) (continued).

Element	E_γ (keV)	Yield$_{meas}$ $(N_\gamma/sr/\mu C)$	Reaction	Remarks
Mg	390	3.98(5)	$^{24}Mg(d,p\gamma)^{25}Mg$	
	452	8.53(5)	$^{24}Mg(d,n\gamma)^{25}Al$	
	585	1.66(6)	$^{24}Mg(d,p\gamma)^{25}Mg$	Overlapping from aluminum.
	975	4.61(5)	$^{24}Mg(d,p\gamma)^{25}Mg$	
	1809	2.09(5)	$^{25}Mg(d,p\gamma)^{26}Mg$	
Al	585	1.22(5)	$^{27}Al(d,\alpha\gamma)^{25}Mg$	Al yield is lower than Mg yield.
	975	1.96(5)	$^{27}Al(d,\alpha\gamma)^{25}Mg$	Sum of 975 keV and 983 keV gamma
	1014	9.94(4)	$^{27}Al(d,p\gamma)^{28}Al$	
	2272	8.27(4)	$^{27}Al(d,p\gamma)^{28}Al$	
	2839	1.09(5)	$^{27}Al(d,p\gamma)^{28}Al$	
Si	755	5.76(4)	$^{28}Si(d,p\gamma)^{29}Si$	
	1273	4.96(5)	$^{28}Si(d,p\gamma)^{29}Si$	
	1384	1.07(5)	$^{28}Si(d,n\gamma)^{29}P$	
	2028	2.26(5)	$^{28}Si(d,p\gamma)^{29}Si$	
	2426	1.25(5)	$^{28}Si(d,p\gamma)^{29}Si$	
	4934	6.42(5)	$^{28}Si(d,p\gamma)^{29}Si$	
S	841	2.18(5)	$^{32}S(d,p\gamma)^{33}S$	
Cl	671	2.11(4)	$^{37}Cl(d,p\gamma)^{38}Cl$	
	788	5.66(4)	$^{35}Cl(d,p\gamma)^{36}Cl$	
	1165	7.13(4)	$^{35}Cl(d,p\gamma)^{36}Cl$	
	1951	1.45(5)	$^{35}Cl(d,p\gamma)^{36}Cl$	Sum of 1951, 1959, and 1970 keV yields.
	2167	7.79(4)	$^{37}Cl(d,n\gamma)^{38}Ar$	
K	770	4.29(4)	$^{39}K(d,p\gamma)^{40}K$	
	2070	9.21(3)	$^{39}K(d,p\gamma)^{40}K$	Sum of 2070 and 2074 keV peaks.
	3737	2.29(4)	$^{39}K(d,n\gamma)^{40}Ca$	

Table A12.11. Detection limits for the transition elements in steel samples by 5 MeV deuterons. Data from Peisach *et al.* (1994).

Element	E_γ (keV)	Reaction	Detection limit (mg/g)
Ti	153	$^{48}Ti(d,n\gamma)^{49}V$	0.51
V	1434	$^{51}V(d,n\gamma)^{52}Cr$	0.04
Cr	379	$^{52}Cr(d,n\gamma)^{53}Mn$	0.21
Mn	847	$^{55}Mn(d,n\gamma)^{56}Fe$	0.11
Fe	352	$^{56}Fe(d,p\gamma)^{57}Fe$	0.35
Co	1332	$^{59}Co(d,n\gamma)^{60}Ni$	0.12
Ni	339	$^{58}Ni(d,p\gamma)^{59}Ni$	0.26
Cu	992	$^{63}Cu(d,n\gamma)^{64}Zn$	0.20
Zn	360	$^{66}Zn(d,n\gamma)^{67}Ga$	1.30

Table A12.12. Gamma-ray yields measured at 90° and sensitivity values obtained by triton analysis of a niobium sample. Coulomb excitations are marked by CE. The data were obtained from Borderie and Barrandon (1978) and Borderie (1980).

Element	Reaction	Energy (MeV)	E_γ (keV)	Gamma-ray yield $[(\mu C\,sr)^{-1}]$	Detection limit (ppm)
Li	$^{7}Li(t,t'\gamma)^{7}Li$	2	478	3.6×10^6	2 (at 3 MeV)
C	$^{12}C(t,n\gamma)^{14}N$	2	1632	2.5×10^6	5 (at 3 MeV)
O	$^{16}O(t,p\gamma)^{18}O$	2	1982	1.1×10^6	1 (at 3 MeV)
F	$^{19}F(t,p\gamma)^{21}F$	2	280	3.8×10^5	5 (at 3 MeV)
Na	$^{23}Na(t,\alpha\gamma)^{22}Ne$	2	1275	4.2×10^5	30
Mg	$^{24}Mg(t,n\gamma)^{26}Al$	2	417	2.6×10^5	5 (at 3 MeV)
Si	$^{28}Si(t,n\gamma)^{30}P$	3	709	4.1×10^5	20
S	$^{31}S(t,n\gamma)^{34}Cl$	3	461	4.4×10^5	15
Ti	$^{48}Ti(t,n\gamma)^{50}V$	3.5	226	1.7×10^6	2
V	$^{51}V(t,d\gamma)^{52}V$	3	124	6.3×10^4	50
Cr	$^{52}Cr(t,n\gamma)^{54}Mn$	3.5	157	3.7×10^5	15
Mn	$^{55}Mn(t,\alpha\gamma)^{54}Cr$	3.5	835	8.0×10^5	15
Fe	$^{56}Fe(t,n\gamma)^{58}Co$	3.5	367	1.1×10^5	70
Co	$^{59}Co(t,d\gamma)^{60}Co$	3.5	945	6.0×10^4	220
Ni	$^{58}Ni(t,d\gamma)^{59}Ni$	3.5	460	3.1×10^4	290
Cu	$^{63}Cu(t,n\gamma)^{65}Zn$	3.5	115	2.1×10^4	260
Zn	$^{66}Zn(t,n\gamma)^{68}Ga$	3.5	175	2.0×10^4	350
Ga	$^{69}Ga(t,n\gamma)^{71}Ge$	3.5	831	3.4×10^4	420
Ge	$^{72}Ge(t,p\gamma)^{74}Ge$	3.5	596	2.8×10^4	460
As	$^{75}As\ (CE)$	3.5	279	3.8×10^4	275
Se	$^{77}Se\ (CE)$	3.5	239	1.3×10^4	680
Mo	$^{95}Mo\ (CE)$	3.5	204	3.9×10^3	2300
Pd	$^{110}Pd\ (CE)$	3.5	374	1.3×10^4	840
Ag	$^{109}Ag\ (CE)$	3.5	311	2.7×10^4	420
Cd	$^{114}Cd\ (CE)$	3.5	556	3.3×10^3	4400
Ta	$^{181}Ta\ (CE)$	3.5	136	1.2×10^5	90
W	$^{184}W\ (CE)$	3.5	111	4.7×10^4	220
Re	$^{187}Re\ (CE)$	3.5	134	4.3×10^4	250
Pt	$^{194}Pt\ (CE)$	3.5	328	7.7×10^3	2300
Au	$^{197}Au\ (CE)$	3.5	279	7.4×10^3	2100

A12.2.3 Alpha-particle-induced gamma-ray emission

Table A12.13a. Absolute thick-target gamma-ray yields for the light elements with a 2.4 MeV $^4\text{He}^+$ beam. The detection angle is 55° with respect to the beam direction. Data from Lappalainen *et al.* (1983).

Element	E_γ (keV)	Absolute γ-yield $\gamma/(\mu C \cdot Sr)$	Detection limit[b]
Li	478	9.6×10^5	0.45 ppm
Be	4439	1.6×10^6	0.42 ppm
B	170	2.5×10^4	
	3088	3.1×10^3	
	3684	4.9×10^4	
	3854	3.1×10^4	10 ppm
C	a		
N	937	5.9	
	1041	5.6	
	1080	13	
	2125	4.5	
	2471	4.4	
	2542	7.8	
	4525	5.9	1.1%
O	351	15	
	1634	16	2.0%
F	110	9.2×10^2	
	197	1.6×10^4	22 ppm
	1275	6.2×10^3	
Na	440	1.7×10^3	260 ppm
	1130	15	
	1809	610	
Mg	1273	8.8	
	1779	15	1.3%
	2839	0.6	
Al	2235	15	3.2%
Si	2230	0.3	65%

a. No separate gamma-ray was observable.
b. The detection limits are based on the volcanic stone measurement.

Table A12.13b. Prompt gamma rays generated by 5 MeV alpha particles. The notation x(m,n) means reaction with production of light ion x and gamma-ray transition between states m and n in the residual nucleus. $-m_e$ and $-2m_e$ mean single and double escape peaks, respectively. The states are counted starting from 0 for the ground state. c/mC and bg/mC mean the net gamma-ray and background counts/accumulated charge, respectively. Sensitivity values correspond to the net count in the peak equal to three times the standard deviation of the peak background. Data from Giles and Peisach (1979).

Element	Assignment	E_γ (keV)	Relative intensity (%)	c/mC	bg/mC	‰
Lithium	^{7}Li α(1,0)	478	100.0	72000	6300	0.03
Boron	^{10}B p(3,2)	170	100.0	4700	3600	1.2
	^{10}B p(2,1)	598	——————— not possible to integrate ———————			
	^{10}B α(1,0)	718	66.0	3100	1800	1.3
	^{10}B p(3,1)	768	0.8	38	710	66
	^{11}B $-2m_e$	1291	3.6	170	830	16
	^{11}B $-m_e$	1802	0.4	18	280	88
	^{11}B n(1,0)	2313	25.5	1200	570	1.9
	^{10}B $-2m_e$	2662	18.7	880	1100	
	^{10}B $-2m_e$	2832	8.3	390	570	5.9
	^{10}B p(1,0)	3086	4.9	230	890	12
	^{10}B $-m_e$	3173	12.3	580	1200	5.6
	^{10}B $-m_e$	3343	4.5	210	470	9.9
	^{10}B p(2,0)	3684	27.7	1300	210	1.1
	^{10}B p(3,0)	3854	9.8	460	18	0.87
Nitrogen	^{14}N pl(1,0)	871	100.0	1500	40	0.04
Oxygen	^{18}O n(1,0)	351	100.0	120	12	12
	^{18}O n(2,1)	1395	5.7	6.8	3.3	11
	^{17}O n(1,0)	1634	1.8	2.2	1.6	24
	^{18}O n(3,1)	2438	1.0	1.2	0.94	34
	^{17}O n(2,1)	2614	0.1	0.15	0.4	170
Fluorine	^{19}F n(2,1)	74	2.5	430	3500	7.9
	^{19}F α(1,0)	110	56.5	9600	3700	0.37
	^{19}F α(2,0)	197	100.0	17000	2500	0.17
	^{19}F n(1,0)	583	21.2	3600	660	0.42
	^{19}F n(4,3)	637	0.2	32	500	40
	^{19}F n(3,0)	891	5.2	890	550	1.5
	^{19}F n(3,1)	1236	1.4	240	180	3.3
	^{19}F p(1,0)	1275	18.8	3200	280	0.30
	^{19}F n(1,0)	1280	——————— not resolved ———————			
	^{19}F α(4,1)	1349	1.3	220	170	3.4
	^{19}F α(5,2)	1357	0.3	50	82	11
	^{19}F n(6,1)	1369	0.4	63	100	9.3
	^{19}F n(7,1)	1400	0.1	21	74	23
	^{19}F α(4,0)	1459	0.3	47	150	15
	^{19}F n(4,0)	1528	1.1	180	85	3.0
	^{19}F $-m_e$	1570	0.1	9.6	56	45
	^{19}F p(2,1)	2081	1.4	240	41	1.5
	^{19}F $-2m_e$	2160	0.1	16	35	21
	^{19}F $-m_e$	2671	0.04	6.7	18	37
	^{19}F p(3,1)	3182	0.2	34	7	4.5

Table A12.13b. Prompt gamma rays generated by 5 MeV alpha particles. The notation x(m,n) means reaction with production of light ion x and gamma-ray transition between states m and n in the residual nucleus. $-m_e$ and $-2m_e$ mean single and double escape peaks, respectively. The states are counted starting from 0 for the ground state. c/mC and bg/mC mean the net gamma-ray and background counts/accumulated charge, respectively. Sensitivity values correspond to the net count in the peak equal to three times the standard deviation of the peak background. Data from Giles and Peisach (1979) (continued).

Element	Assignment	E_γ (keV)	Relative intensity (%)	c/mC	bg/mC	‰
Sodium	^{23}Na n(2,0)	417	66.2	860	320	0.54
	^{23}Na α(1,0)	440	100.0	1300	280	0.32
	^{23}Na $-2m_e$	787	7.7	100	440	5.4
	^{23}Na n(3,1)	830	2.0	26	150	12
	^{23}Na p(4,2)	1003	8.5	110	170	3.0
	^{23}Na p(2,1)	1130	51.5	670	240	0.59
	^{23}Na $-m_e$	1298	2.8	36	220	11
	^{23}Na p(7,2)	1412	2.6	34	190	11
	^{23}Na p(3,1)	1780	2.8	37	74	6.1
	^{23}Na p(1,0)	1809	100.0	1300	94	0.19
	^{23}Na p(8,2)	1897	0.1	1.5	16	71
	^{23}Na $-2m_e$	1916	0.1	1.8	17	60
	^{23}Na p(4,1)	2132	2.3	30	31	4.9
	^{23}Na $-m_e$	2427	0.4	4.9	12	19
	^{23}Na p(5,1)	2511	2.3	30	15	3.3
	^{23}Na p(6,1)	2524	1.5	19	9.0	4.0
	^{23}Na p(7,1)	2541	1.2	16	9.7	5.1
	^{23}Na p(2,0)	2938	2.1	27	5.5	2.3
	^{23}Na p(9,1)	3092	0.2	2.3	1.5	14
Magnesium	^{25}Mg α(1,0)	585	58.6	170	660	5.6
	^{26}Mg n(2,1)	755	9.0	26	520	32
	^{24}Mg p(1,0)	844	93.1	270	950	4.3
	^{24}Mg p(2,0)	1015	55.2	160	650	5.9
	^{26}Mg n(1,0)	1273	100.0	290	280	2.1
	^{25}Mg n(1,0)	1779	58.6	170	86	2.0
	^{26}Mg n(4,1)	1794	5.5	16	56	18
	^{26}Mg n(2,0)	2028	10.7	31	27	6.2
	^{25}Mg $-m_e$	2328	1.4	4.2	14	32
	^{26}Mg n(3,0)	2426	12.4	36	20	4.6
	^{25}Mg n(2,1)	2839	8.3	24	16	6.2
Aluminum	^{27}Al n(1,0)	677	4.0	56	220	25
	^{27}Al n(2,0)	709	27.1	380	340	4.6
	^{27}Al α(1,0)	844	6.5	91	270	17
	^{27}Al α(2,0)	1015	5.5	77	280	21
	^{27}Al $-2m_e$	1214	15.7	220	300	7.4
	^{27}Al p(2,1)	1263	36.4	510	330	3.4
	^{27}Al p(5,2)	1311	4.5	63	220	22
	^{27}Al p(6,2)	1332	0.9	13	200	100
	^{27}Al n(3,0)	1454	3.1	43	210	32
	^{27}Al n(3,1)	1534	4.9	69	240	21
	^{27}Al n(4,1)	1552	2.1	30	240	50
	^{27}Al $-m_e$	1725	7.9	110	410	10
	^{27}Al p(1,0)	2236	100.0	1400	170	0.87

Table A12.13b. Prompt gamma rays generated by 5 MeV alpha particles. The notation x(m,n) means reaction with production of light ion x and gamma-ray transition between states m and n in the residual nucleus. $-m_e$ and $-2m_e$ mean single and double escape peaks, respectively. The states are counted starting from 0 for the ground state. c/mC and bg/mC mean the net gamma-ray and background counts/accumulated charge, respectively. Sensitivity values correspond to the net count in the peak equal to three times the standard deviation of the peak background. Data from Giles and Peisach (1979) (continued).

Element	Assignment	E_γ (keV)	Relative Intensity (%)	c/mC	bg/mC	‰
	^{27}Al $-2m_e$	2476	6.9	96	110	10
	^{27}Al p(5,1)	2574	0.4	6.2	63	120
	^{27}Al p(6,1)	2595	7.1	100	94	9.2
	^{27}Al $-2m_e$	2748	1.0	14	98	68
	^{27}Al $-m_e$	2987	4.2	59	94	16
	^{27}Al $-m_e$	3259		———— obscured by Compton edge ————		
	^{27}Al p(2,0)	3498	11.4	160	39	3.8
	^{27}Al p(3,0)	3770	2.9	40	26	12
Silicon	^{29}Si p(1,0)	78	100.0	97	140	12
	(^{28}Si+^{30}Si) $-2m_e$	1213	5.6	5.4	8.1	50
	^{28}Si p(1,0)	1266	7.6	7.4	4.3	26
	^{29}Si α(1,0)	1273	3.2	3.1	3.9	60
	(^{28}Si+^{30}Si) $-m_e$	1724	2.8	2.7	4.5	75
	^{28}Si α(1,0)	1779	4.3	4.2	6.1	56
	^{30}Si α(1,0)	2235	29.9	29	1.9	4.6
	^{28}Si p(2,0)			——— not resolved ———		
Phosphorus	^{31}P $-2m_e$	1105	12.1	5.7	9.6	2.3
	^{31}P p(2,1)	1176	15.7	7.4	7.9	1.6
	^{31}P α(1,0)	1266	3.4	1.6	5.1	6.1
	^{31}P $-m_e$	1616	4.3	2	7	5.6
	^{31}P p(1,0)	2127	100.0	47	2.6	0.15
	^{31}P $-2m_e$	2282	1.1	0.52	0.77	7.1
	^{31}P $-m_e$	2793	1.0	0.49	0.54	6.4
	^{31}P p(2,0)	3304	2.1	1.0	1.4	4.8
Sulphur	-					
Chlorine	^{35}Cl $-2m_e$	1146	11.6	3.7	5.0	30
	^{35}Cl p(2,1)	1210	3.4	1.1	3.1	77
	^{37}Cl p(1,0)	1461	1.8	0.56	3.6	170
	^{35}Cl p(3,1)	1643	11.3	3.6	6.5	35
	^{35}Cl $-m_e$	1657	6.9	2.2	4.5	49
	^{35}Cl p(1,0)	2168	100.0	32	1.5	1.9
Potassium	^{39}K p(2,1)	313	20.0	2.8	8.6	17
	^{39}K p(3,1)	899	11.4	1.6	3.9	19
	^{39}K p(1,0)	1524	100.0	14	2.4	1.7
Calcium	-					
Scandium	^{55}Sc α(2,1)	364	100.0	43	650	31
Titanium	^{47}Ti α(1,0)	159	100.0	390	91	2.3
	^{48}Ti n(1,0)	749	4.1	16	9.9	19
	^{46}Ti α(1,0)	889	2.8	11	3.6	4.4
	^{48}Ti α(1,0)	983	10.3	40	3.4	4.4
Vanadium	^{51}V n(1,0)	54	0.8	5.1	110	190
	^{51}V n(2,0)	157	5.2	32	130	34
	^{51}V n(3,2)	207	1.6	10	43	61
	^{51}V α(1,0)	320	100.0	620	36	0.91

Table A12.13b. Prompt gamma rays generated by 5 MeV alpha particles. The notation x(m,n) means reaction with production of light ion x and gamma-ray transition between states m and n in the residual nucleus. $-m_e$ and $-2m_e$ mean single and double escape peaks respectively. The states are counted starting from 0 for the ground state. c/mC and bg/mC mean the net gamma-ray and background counts/accumulated charge, respectively. Sensitivity values correspond to the net count in the peak equal to three times the standard deviation of the peak background. Data from Giles and Peisach (1979) (continued).

Element	Assignment	E_γ (keV)	Relative intensity (%)	c/mC	bg/mC	‰
	^{51}V α(2,1)	609	0.1	0.75	1.9	170
	^{51}V α(2,0)	929	0.4	2.5	1.3	43
Chromium	^{53}Cr α(1,0)	564	97.6	8.2	13	42
	^{50}Cr α(1,0)	783	100.0	8.4	5.1	26
	^{54}Cr α(1,0)	835	41.7	3.5	4.0	54
Manganese	^{55}Mn α(1,0)	126	100.0	3700	610	0.63
	^{55}Mn α(2,1)	858	0.1	3.6	1.6	34
Iron	^{57}Fe α(2,1)	122	100.0	70	44	9.0
	^{57}Fe α(2,0)	137	12.3	8.6	34	64
	^{57}Fe α(3,2)	230	2.3	1.6	18	250
	^{57}Fe α(3,1)	353		——— obscured by ^{18}O ———		
	^{57}Fe α(3,0)	367	3.6	2.5	10	120
	^{58}Fe α(1,0)	811	0.8	0.54	1.1	180
	^{56}Fe α(1,0)	847	84.3	59	1.7	2.1
Cobalt	-					
Nickel	-					
Copper	^{63}Cu α(1,0)	670	100.0	9.6	2.4	15
	^{65}Cu α(1,0)	771	18.8	1.8	0.70	46
	^{63}Cu α(2,0)	962	24.0	2.3	1.1	43
	^{65}Cu α(2,0)	1116	5.8	0.56	1.4	200
Zinc	^{67}Zn α(1,0)	93	17.2	6.2	40	97
	^{67}Zn α(2,0)	185	100.0	36	7.1	11
	^{64}Zn α(1,0)	992	17.2	6.2	0.85	14
	^{66}Zn α(1,0)	1039	5.6	2.0	0.71	41
	^{68}Zn α(1,0)	1077	2.5	0.90	0.63	84
Gallium	-					
Bromine	^{79}Br α(2,0)	217	100.0	130	43	1.6
	^{79}Br α(3,0)	261	9.2	12	22	13
	^{81}Br α(1,0)	276	84.6	110	26	1.4
	^{79}Br α(4,0)	306	25.4	33	17	4.0
	^{79}Br α(5,0)	397	1.9	2.5	7.0	33
	^{79}Br α(6,0)	523	12.3	16	5.2	4.6
	^{81}Br α(3,0)	538	3.3	4.3	6.9	19
	^{79}Br α(7,0)	606	0.7	0.93	5.0	76
Rubidium	^{85}Rb α(1,0)	151	100.0	82	2400	46
Strontium	-					
Yttrium	-					
Zirconium	^{96}Zr α(1,0)	1594	100.0	0.26	0.54	270
Niobium	-					
Molybdenum	^{95}Mo α(1,0)	204	100.0	57	30	9.0
	^{97}Mo α(1,0)	481	3.3	1.9	2.7	83
	^{100}Mo α(1,0)	536	40.4	23	2.1	6.0
	^{96}Mo α(1,0)	778	2.6	1.5	0.50	44
	^{95}Mo α(2,0)	786	2.6	1.5	0.60	50

Table A12.13b. Prompt gamma rays generated by 5 MeV alpha particles. The notation x(m,n) means reaction with production of light ion x and gamma-ray transition between states m and n in the residual nucleus. $-m_e$ and $-2m_e$ mean single and double escape peaks, respectively. The states are counted starting from 0 for the ground state. c/mC and bg/mC mean the net gamma-ray and background counts/accumulated charge, respectively. Sensitivity values correspond to the net count in the peak equal to three times the standard deviation of the peak background. Data from Giles and Peisach (1979) (continued).

Element	Assignment	E_γ (keV)	Relative Intensity (%)	c/mC	bg/mC	‰
	^{98}Mo α(2,0)	787		— not resolved —		
	^{94}Mo α(1,0)	871		— obscured by ^{14}N —		
Ruthenium	^{99}Ru α(1,0)	89	40.4	38	330	45
	^{101}Ru α(1,0)	127	51.1	48	250	31
	^{104}Ru α(1,0)	358	100.0	94	57	7.6
	^{102}Ru α(1,0)	475	26.6	25	73	33
	^{100}Ru α(1,0)	540	3.9	3.7	39	160
Rhodium	^{103}Rh α(3,0)	295	100.0	1000	130	1.1
	^{103}Rh α(4,0)	358	69.0	690	31	0.77
Palladium	^{105}Pd α(1,0)	281	5.9	8.8	31	60
	^{110}Pd α(1,0)	374	100.0	150	33	3.6
	^{108}Pd α(1,0)	434	86.7	130	40	4.5
	^{110}Pd α(2,1)	440	8.7	13	14	28
	^{104}Pd α(1,0)	556	7.3	11	3.9	17
	^{102}Pd α(1,0)	557				
Silver	^{107}Ag α(2,1)	98	5.3	20	150	58
	^{109}Ag α(3,2)	103	4.7	18	180	70
	^{109}Ag α(2,0)	312	100.0	380	69	2.1
	^{107}Ag α(1,0)	325	81.6	310	41	2.0
	^{109}Ag α(3,0)	415	36.8	140	16	2.7
	^{109}Ag α(2,0)	423	31.6	120	11	2.6
Cadmium	^{113}Cd α(2,0)	299	100.0	25	14	14
	^{111}Cd α(2,0)	342	68.0	17	9.5	17
	^{114}Cd α(1,0)	558	80.0	20	2.9	8.1
	^{113}Cd α(6,0)	584	13.2	3.3	1.7	38
	^{112}Cd α(1,0)	617	32.8	8.2	2.2	17
	^{110}Cd α(1,0)	658	8.8	2.2	1.0	43
Indium		-				
Tin		-				
Tellurium		-				
Barium		-				
Lanthanum		-				
Cerium		-				
Praseodymium		-				
Neodymium		-				
Erbium	^{167}Er α(2,0)	178	100.0	16	140	61
	^{170}Er α(2,1)	182	22.5	3.6	100	230
	^{166}Er α(2,1)	184		not resolved		
	^{168}Er α(2,1)	184		not resolved		
Hafnium	^{178}Hf α(1,0)	93	100.0	120	190	11
	^{180}Hf α(1,0)	93		not resolved		
	^{177}Hf α(1,0)	113	34.2			
	^{179}Hf α(1,0)	123	23.3	41	120	25

Table A12.13b. Prompt gamma rays generated by 5 MeV alpha particles. The notation x(m,n) means reaction with production of light ion x and gamma-ray transition between states m and n in the residual nucleus. $-m_e$ and $-2m_e$ mean single and double escape peaks, respectively. The states are counted starting from 0 for the ground state. c/mC and bg/mC mean the net gamma-ray and background counts/accumulated charge, respectively. Sensitivity values correspond to the net count in the peak equal to three times the standard deviation of the peak background. Data from Giles and Peisach (1979) (continued).

Element	Assignment	E_γ (keV)	Relative intensity (%)	c/mC	bg/mC	‰
Tantalum	^{181}Ta α(1,0)	136	100.0	1200	200	1.1
	^{181}Ta α(2,1)	165	5.2	62	78	14
	^{181}Ta α(2,0)	301	3.3	39	16	9.7
Tungsten	^{183}W α(2,0)	99		not resolved		
	^{182}W α(1,0)	100	27.5	110	210	12
	^{184}W α(1,0)	111	65.0	260	150	4.5
	^{186}W α(1,0)	122	100.0	400	100	2.4
Rhenium	^{185}Re α(1,0)	125	59.1	650	420	3.0
	^{187}Re α(1,0)	134	100.0	1100	210	1.2
	^{185}Re α(2,1)	159	4.1	45	63	17
	^{187}Re α(3,1)	167	5.3	58	53	12
	^{185}Re α(2,0)	285	0.5	5.7	15	65
	^{187}Re α(3,0)	301	0.6	6.4	14	55
Iridium	^{193}Ir α(1,0)	73	100.0	510	260	3.0
	^{193}Ir α(3,1)	107	1.9	9.9	97	94
	^{191}Ir α(2,0)	129	56.9	290	150	3.9
	^{193}Ir α(2,0)	139	88.2	450	89	2.0
	^{191}Ir α(4,0)	179	1.3	6.4	32	84
	^{193}Ir α(3,0)	180		not resolved		
	^{191}Ir α(5,2)	214	2.5	13	24	36
	^{193}Ir α(4,2)	219	3.7	19	19	21
	^{191}Ir α(5,0)	343	2.7	14	5.7	16
	^{193}Ir α(4,0)	358	0.6	2.9	4.3	67
	^{193}Ir α(5,0)	362	3.1	16	4.6	13
Platinum	^{195}Pt α(1,0)	99	5.9	4.1	72	200
	^{195}Pt α(2,0)	130	4.6	3.2	59	230
	^{195}Pt α(5,1)	140	9.6	6.7	54	100
	^{195}Pt α(4,0)	211	92.9	65	35	8.6
	^{195}Pt α(5,0)	239	32.9	23	18	17
	^{192}Pt α(1,0)	317	4.0	2.8	9.6	100
	^{194}Pt α(1,0)	329	100.0	70	10	4.3
	^{196}Pt α(1,0)	356	42.9	30	9.0	9.5
	^{198}Pt α(1,0)	407	4.3	3.0	2.6	52
Gold	^{197}Au α(1,0)	77	100.0	190	140	5.7
	^{197}Au α(2,1)	192	7.9	15	36	39
	^{197}Au α(2,0)	269	1.1	2.0	6.8	120
	^{197}Au α(3,0)	279	36.8	70	15	5.0
	^{197}Au α(6,0)	548	0.6	1.2	1.5	97
Mercury	^{199}Hg α(1,0)	158	100.0	16	65	42
	^{199}Hg α(2,0)	208	16.9	2.7	28	160
	^{200}Hg α(1,0)	368	21.9	3.5	16	94
	^{198}Hg α(1,0)	412	9.4	1.5	15	220
	^{202}Hg α(1,0)	439	11.9	1.9	7.7	120
Lead	-					
Bismuth	-					

Table A12.13c Absolute gamma-ray yields [$(\mu C\ sr)^{-1}$] for elements Ti–Zn induced with 5 MeV and 10 MeV α-particles. The measurement angle is 55°.(Kocsonya *et al.*, 2006). Coulomb excitations are marked by CE.

Element	Energy (keV)	Reaction	5 MeV	10 MeV
Ti	320.6	$^{48}Ti(\alpha,p)^{51}V$	1.1×10^2	2.9×10^5
	749.8	$^{48}Ti(\alpha,n)^{51}Cr$	1.5×10^3	2.6×10^6
	808.8	$^{48}Ti(\alpha,n)^{51}Cr$	53	6.4×10^5
	984.1	^{48}Ti (CE)	4.7×10^3	3.3×10^5
	1166.2	$^{48}Ti(\alpha,n)^{51}Cr$	47	9.3×10^5
Cr	378.3	$^{50}Cr(\alpha,d)^{52}Mn$	12	7.3×10^4
	412.1	$^{52}Cr(\alpha,n)^{55}Fe$	8.5	3.4×10^5
	784.0	^{50}Cr (CE)	7.8×10^2	2.9×10^4
	847.4	$^{52}Cr(\alpha,\gamma)^{56}Fe$	61	4.5×10^5
	1239.0	$^{52}Cr(\alpha,\gamma)^{56}Fe$	25	1.1×10^5
	1434.7	^{52}Cr (CE)	92	1.4×10^5
Mn	366.5	$^{55}Mn(\alpha,pn)^{57}Fe$	27	1.8×10^5
	811.5	$^{55}Mn(\alpha,p)^{58}Fe$	12	1.3×10^5
	859.0	^{55}Mn(CE)	4.5×10^2	2.7×10^4
	1238.2	$^{55}Mn(\alpha,t)^{56}Fe$	22	4.1×10^4
Fe	847.2	^{56}Fe (CE)	6.9×10^3	5.8×10^5
	1332.7	$^{56}Fe(\alpha,\gamma)^{60}Ni$	30	6.6×10^4
Co	1173.2	$^{59}Co(\alpha,p)^{62}Ni$	16	3.5×10^5
Ni	1332.8	^{60}Ni (CE)	10	6.6×10^4
	1454.7	^{58}Ni (CE)	10	9.3×10^4
Cu	302.3	$^{65}Cu(\alpha,pn)^{67}Zn$	14	1.0×10^4
	670.4	^{63}Cu (CE)	930	3.5×10^4
	771.7	^{65}Cu (CE)	180	9.4×10^3
	962.5	^{63}Cu (CE)	360	4.4×10^4
Zn	301.8	^{67}Zn (CE)	350	1.2×10^4
	992.0	^{64}Zn (CE)	440	1.4×10^5
	1039.7	^{66}Zn (CE)	150	1.1×10^5

Table A12.14 Summary of sensitivities (atomic fraction) obtainable with 5 MeV alpha particles. Data from Giles and Peisach (1979).

Over 1%	Between 1 and 0.1%	Better than 0.1%
Sc, Cr, Cu, Zn, Rb, Zr, Er, Hf, Hg	O, Mg, Si, Cl, K, Ti, Fe, Br, Mo, Ru, Pd, Ag, Cd, Ta, W, Re, Ir, Pt, Au	Li, B, N, F, Na, Al, P, V, Mn, Rh

A12.2.4 Heavy-ion-induced gamma-ray emission

Table A12.15 Reactions leading to the highest gamma-ray yields following 18 MeV ^{7}Li-ion bombardment. The two strongest gamma-ray lines are listed for each element when available. The data were compiled from Räisänen (1990).

Element	E_γ (keV)	Reaction	Gamma-ray yield at 55° [$(\mu C\ sr)^{-1}$]
Li	No significant lines		
Be	No significant lines		
B	No significant lines		
C	871	$^{12}C(^7Li,pn\gamma)^{17}O$, $^{13}C(^7Li,p2n\gamma)^{17}O$	7.6×10^5
	937	$^{12}C(^7Li,n\gamma)^{18}F$, $^{13}C(^7Li,2n\gamma)^{18}F$	5.6×10^5
N	1233	$^{14}N(^7Li,2n\gamma)^{19}Ne$	4.7×10^5
	1633	$^{14}N(^7Li,n\gamma)^{20}Ne$, $^{15}N(^7Li,2n\gamma)^{20}Ne$	1.9×10^5
O	350	$^{16}O(^7Li,pn\gamma)^{21}Ne$	5.0×10^6
	937	$^{16}O(^7Li,\alpha n\gamma)^{18}F$	2.4×10^6
F	350	$^{19}F(^7Li,\alpha n\gamma)^{21}Ne$	4.3×10^6
	1636	$^{19}F(^7Li,p2n\gamma)^{23}Na$	3.3×10^6
Na	585	$^{23}Na(^7Li,\alpha n\gamma)^{25}Mg$	1.1×10^6
	1369	$^{23}Na(^7Li,pn\gamma)^{28}Al$	9.3×10^5
Mg	1779	$^{24}Mg(^7Li,p2n\gamma)^{28}Si$	1.8×10^6
	1809	$^{24}Mg(^7Li,\alpha p\gamma)^{26}Mg$, $^{25}Mg(^7Li,\alpha pn\gamma)^{26}Mg$	2.6×10^6
Al	1323	$^{27}Al(^7Li,pn\gamma)^{32}P$	2.8×10^6
	2028	$^{27}Al(^7Li,\alpha n\gamma)^{29}Si$	1.9×10^6
Si	1966	$^{28}Si(^7Li,pn\gamma)^{33}S$, $^{29}Si(^7Li,p2n\gamma)^{33}S$	1.5×10^6
	2233	$^{28}Si(^7Li,\alpha\gamma)^{31}P$, $^{29}Si(^7Li,\alpha n\gamma)^{31}P$	2.8×10^6
	+ 2235	$^{28}Si(^7Li,\alpha p\gamma)^{30}Si$, $^{29}Si(^7Li,\alpha pn\gamma)^{30}Si$	

Table A12.15 Reactions leading to the highest gamma-ray yields following 18 MeV ^{7}Li-ion bombardment. The two strongest gamma-ray lines are listed for each element when available. The data were compiled from Räisänen (1990) (continued).

Element	E_γ (keV)	Reaction	Gamma-ray yield at 55° [$(\mu C\ sr)^{-1}$]
P	788	^{31}P(^{7}Li,pnγ)^{36}Cl	1.3×10^6
	1966, 1970	^{31}P(^{7}Li,αnγ)^{36}Ar, ^{31}P(^{7}Li,2nγ)^{36}Ar	9.4×10^5
S	1611, 1601	Several reactions on $^{32-34}$S	3.0×10^6
Cl	892	^{35}Cl(^{7}Li,pnγ)^{40}K	1.1×10^6
	1614	^{35}Cl(^{7}Li,pnγ)^{40}K	8.1×10^5
K	350	^{39}K(^{7}Li,pnγ)^{44}Sc, ^{40}K(^{7}Li,p2nγ)^{44}Sc	8.4×10^5
	2163	^{40}K(^{7}Li,γ)^{47}Ti, ^{41}K(^{7}Li,nγ)^{47}Ti	2.8×10^5
Ca	1227	^{40}Ca(^{7}Li,pnγ)^{45}Ti	2.9×10^5
	1525	^{40}Ca(^{7}Li,αpγ)^{42}Ca, ^{42}Ca(^{7}Li,^{7}Li′)^{42}Ca	2.9×10^5
Ti	936	Several reactions on $^{46-48}$Ti	8.1×10^5
	1434	Several reactions on $^{46-48}$Ti	1.5×10^6
V	847	^{50}V(^{7}Li,nγ)^{56}Fe, ^{51}V(^{7}Li,2nγ)^{56}Fe	1.3×10^6
	1316	^{50}V(^{7}Li,2nγ)^{55}Fe	9.0×10^5
Cr	851	Several reactions on ^{50,52,54}Cr	1.7×10^6
	1238	^{50}Cr(^{7}Li,pγ)^{56}Fe, ^{52}Cr(^{7}Li,p2nγ)^{56}Fe	8.1×10^5
Mn	851	^{55}Mn(^{7}Li,αpγ)^{57}Mn	7.1×10^5
	1332	^{55}Mn(^{7}Li,2nγ)^{60}Ni	8.2×10^5
Fe	847	^{54}Fe(^{7}Li,αpγ)^{56}Fe; ^{56}Fe(^{7}Li,^{7}Li′)^{56}Fe	4.1×10^5
	1332	Several reactions on $^{54,56-58}$Fe	6.1×10^5
Co	992	^{59}Co(^{7}Li,2nγ)^{64}Zn	8.9×10^5
	1315	^{59}Co(^{7}Li,2nγ)^{64}Zn	4.7×10^5
Ni	1173	Several reactions on $^{58,60-62}$Ni	2.7×10^5
	1332	Several reactions on ^{58,60,61}Ni	5.5×10^5
Cu	1017	^{63}Cu(^{7}Li,2nγ)^{68}Ge	4.8×10^5
	1040	Several reactions on ^{63,65}Cu	5.5×10^5
Zn	398	Several reactions on $^{64,66-68}$Zn	2.6×10^5
	1040	Several reactions on ^{64,66}Zn	3.3×10^5
Ge	596	Several reactions on $^{72-74}$Ge	3.5×10^5
	834	^{70}Ge(^{7}Li,αpγ)^{72}Ge, ^{72}Ge(^{7}Li,^{7}Li′)^{72}Ge	3.2×10^5
Br	793	^{79}Br(^{7}Li,2nγ)^{84}Sr	5.2×10^4
Y	703	^{89}Y(^{7}Li,2nγ)^{94}Mo	2.0×10^5
	871	^{89}Y(^{7}Li,2nγ)^{94}Mo	8.5×10^4
Zr	No significant lines		
Nb	744	^{93}Nb(^{7}Li,^{7}Li′)^{93}Nb	6.8×10^4
	950	^{93}Nb(^{7}Li,^{7}Li′)^{93}Nb	7.4×10^4
Mo	535	^{100}Mo(^{7}Li,^{7}Li′)^{100}Mo	9.8×10^4
	787	^{98}Mo(^{7}Li,^{7}Li′)^{98}Mo	7.6×10^4

Table A12.15 Reactions leading to the highest gamma-ray yields following 18 MeV ^{7}Li-ion bombardment. The two strongest gamma ray lines are listed for each element when available. The data were compiled from Räisänen (1990) (continued).

Element	E_γ (keV)	Reaction	Gamma-ray yield at 55° $[(\mu C \ sr)^{-1}]$
Ru	358	^{104}Ru(^{7}Li,^{7}Li')^{104}Ru	3.0×10^5
	475	^{102}Ru(^{7}Li,^{7}Li')^{102}Ru	3.3×10^5
Pd	434	^{108}Pd(^{7}Li,^{7}Li')^{108}Pd	3.1×10^5
	557	^{102}Pd(^{7}Li,^{7}Li')^{102}Pd	2.7×10^5
Ag	415	^{109}Ag(^{7}Li,^{7}Li')^{109}Ag	2.3×10^5
	423	^{107}Ag(^{7}Li,^{7}Li')^{107}Ag	2.2×10^5
Cd	558	^{114}Cd(^{7}Li,^{7}Li')^{114}Cd	1.6×10^5
	617	^{112}Cd(^{7}Li,^{7}Li')^{112}Cd	1.3×10^5
In	1133	^{115}In(^{7}Li,^{7}Li')^{115}In	2.1×10^4
	1291	^{115}In(^{7}Li,^{7}Li')^{115}In	1.3×10^4
Sn	1172	^{120}Sn(^{7}Li,^{7}Li')^{120}Sn	1.4×10^4
Sb	573	^{121}Sb(^{7}Li,^{7}Li')^{121}Sb	1.1×10^4
	1030	^{123}Sb(^{7}Li,^{7}Li')^{123}Sb	1.4×10^4
I	No significant lines		
Sm	No significant lines		
Gd	No significant lines		
Yb	No significant lines		
Hf	No significant lines		
Ta	301	^{181}Ta(^{7}Li,^{7}Li')^{181}Ta	5.2×10^4
W	No significant lines		
Re	No significant lines		
Pt	328	^{194}Pt(^{7}Li,^{7}Li')^{194}Pt	1.5×10^5
	333	^{196}Pt(^{7}Li,^{7}Li')^{196}Pt	1.1×10^5
Au	279	^{197}Au(^{7}Li,^{7}Li')^{197}Au	5.0×10^4
	548	^{197}Au(^{7}Li,^{7}Li')^{197}Au	6.5×10^4
Pb	No significant lines		

Table A12.16 Gamma-ray yields with ^{12}C, ^{14}N, and ^{16}O ion bombardment at 55°. Data from Seppälä *et al.* (1998).

Element	E_γ (keV)	E_{ion} (MeV)	Reaction	Gamma-ray yield [$(\mu C\ sr)^{-1}$]
Li	871	28	^{7}Li(^{12}C,pγ)^{17}O	7.0×10^4
	1236	28	^{6}Li(^{14}N,pγ)^{19}F	7.7×10^4
	2256	33	^{6}Li(^{16}O,2pγ)^{20}F	7.7×10^5
Be	6163	28	^{9}Be(^{12}C,p2nγ)^{18}F	1.3×10^6
	937	28	^{9}Be(^{14}N,αnγ)^{18}F	1.1×10^6
	1636	33	^{9}Be(^{16}O,pnγ)^{23}Na	3.0×10^6
B	1419, 1429	28	^{10}B(^{12}C,2pnγ)^{19}F, ^{10}B(^{12}C,2pnγ)^{19}F	8.6×10^5
	440	28	^{10}B(^{14}N,pγ)^{23}Na	6.7×10^5
	1369	33	^{10}B(^{16}O,2p2nγ)^{22}Na, ^{10}B(^{16}O,2nγ)^{24}Mg	6.8×10^4
C	440	28	^{12}C(^{12}C,pγ)^{23}Na	1.3×10^6
	1636	28	^{12}C(^{12}C,pγ)^{23}Na	1.3×10^6
	3333	28	^{12}C(^{12}C,$\alpha\gamma$)^{20}Ne	2.9×10^6
	350	28	^{12}C(^{14}N,αpγ)^{21}Ne	8.6×10^5
	583, 585	28	^{12}C(^{14}N,$\alpha\gamma$)^{22}Na, ^{12}C(^{14}N,pγ)^{25}Mg	3.1×10^5
	1369	28	^{12}C(^{14}N,$\alpha\gamma$)^{22}Na, ^{12}C(^{14}N,pnγ)^{24}Mg	1.4×10^6
	2788, 2794, 2798	28	^{12}C(^{14}N,αpγ)^{21}Ne, ^{12}C(^{14}N,αnγ)^{21}Na	9.9×10^6
	417	33	^{12}C(^{16}O,pnγ)^{26}Al	1.0×10^5
	440	33	^{12}C(^{16}O,αpγ)^{23}Na	1.7×10^5
	1369, 1371	33	^{12}C(^{16}O,αpnγ)^{22}Na, ^{12}C(^{16}O,pnγ)^{26}Al	8.1×10^5
N	350	28	^{14}N(^{12}C,αpγ)^{21}Ne	1.4×10^5
	1369	28	^{14}N(^{12}C,$\alpha\gamma$)^{22}Na, ^{16}O(^{12}C,pnγ)^{24}Mg	2.6×10^5
	2794	28	^{14}N(^{12}C,αpγ)^{21}Ne	1.1×10^5
	417	28	^{14}N(^{14}N,pnγ)^{26}Al	2.2×10^4
	440	28	^{14}N(^{14}N,αpγ)^{23}Na	6.4×10^4
	1636	28	^{14}N(^{14}N,αpγ)^{23}Na	2.2×10^4
	1809	28	^{14}N(^{14}N,2pγ)^{26}Mg	2.2×10^4
	585	33	^{14}N(^{16}O,αpγ)^{25}Mg	2.4×10^4
	975	33	^{14}N(^{16}O,αpγ)^{25}Mg	9.0×10^3
	1779, 1780	33	^{14}N(^{16}O,pnγ)^{28}Si, ^{16}O(^{12}C,$\alpha\gamma$)^{26}Al	4.8×10^4

Table A12.16 Gamma-ray yields with ^{12}C, ^{14}N, and ^{16}O ion bombardment at 55°. Data from Seppälä *et al.* (1998) (continued).

Element	E_γ (keV)	E_{ion} (MeV)	Reaction	Gamma-ray yield $[(\mu C\ sr)^{-1}]$
O	440	28	$^{16}O(^{12}C,\alpha p\gamma)^{23}Na$	1.9×10^5
	1369, 1371	28	$^{16}O(^{12}C,\alpha pn\gamma)^{22}Na$, $^{16}O(^{12}C,\alpha\gamma)^{24}Mg$ $^{16}O(^{12}C,p\gamma)^{27}Al$	6.2×10^5
	1821	28	$^{16}O(^{12}C,pn\gamma)^{26}Al$	1.8×10^5
	583, 585	28	$^{16}O(^{14}N,2\alpha\gamma)^{22}Na$, $^{16}O(^{14}N,\alpha p\gamma)^{25}Mg$	1.3×10^5
	1779, 1780	28	$^{16}O(^{14}N,pn\gamma)^{28}Si$, $^{16}O(^{14}N,\alpha\gamma)^{26}Al$	7.9×10^5
	2887, 2904	28	$^{16}O(^{14}N,2p\gamma)^{28}Al$, $^{16}O(^{14}N,\alpha 2p\gamma)^{24}Na$	2.9×10^6
	709	33	$^{16}O(^{16}O,pn\gamma)^{30}P$	6.5×10^4
	1369	33	$^{16}O(^{16}O,\alpha p\gamma)^{22}Na$	1.5×10^5
	2233, 2235	33	$^{16}O(^{16}O,p\gamma)^{31}P$, $^{16}O(^{16}O,2p\gamma)^{30}Si$	2.1×10^5
F	1809	28	$^{19}F(^{12}C,\alpha p\gamma)^{26}Mg$	6.5×10^4
	2258	28	$^{19}F(^{14}N,p2n\gamma)^{30}P$	1.9×10^5
	2258, 2260	33	$^{19}F(^{16}O,\alpha n\gamma)^{30}P$	1.9×10^5
Na	2249	28	$^{23}Na(^{12}C,^{12}C')^{23}Na$	3.4×10^5
Mg	2243, 2246	28	$^{24}Mg(^{12}C,2pn\gamma)^{33}S$, $^{25}Mg(^{12}C,p\gamma)^{36}Cl$	1.9×10^5
Al	1611	28	$^{27}Al(^{12}C,pn\gamma)^{37}Ar$	5.0×10^5
Si	2167	28	$^{28}Si(^{12}C,2p\gamma)^{38}Ar$	8.1×10^5
P	2167	28	$^{31}P(^{12}C,\alpha p\gamma)^{38}Ar$	4.1×10^4
S	1525	28	$^{32}S(^{12}C,2p\gamma)^{42}Ca$	4.7×10^4
Cl	1525	28	$^{35}Cl(^{12}C,\alpha p\gamma)^{42}Ca$	1.4×10^4
K	891	28	$^{40}K(^{12}C,^{12}C')^{40}K$	2.4×10^4
Ca	783	28	$^{40}Ca(^{12}C,2p\gamma)^{50}Cr$	1.0×10^5
Ti	984	28	$^{48}Ti(^{12}C,^{12}C')^{48}Ti$	1.1×10^5
V	320	28	$^{50}V(^{12}C,^{12}C')^{50}V$	8.1×10^4
Cr	1434	28	$^{52}Cr(^{12}C,^{12}C')^{52}Cr$	4.3×10^4
Mn	858	28	$^{55}Mn(^{12}C,^{12}C')^{55}Mn$	4.1×10^4
Fe	847	28	$^{56}Fe(^{12}C,^{12}C')^{56}Fe$	1.8×10^5
Co	1190	28	$^{59}Co(^{12}C,^{12}C')^{59}Co$	3.0×10^4
Ni	1333	28	$^{60}Ni(^{12}C,^{12}C')^{60}Ni$	1.1×10^4
Cu	962	28	$^{63}Cu(^{12}C,^{12}C')^{63}Cu$	2.5×10^4
Zn	992	28	$^{64}Zn(^{12}C,^{12}C')^{64}Zn$	5.9×10^4

Table A12.17 Measured gamma-ray yields at 90° and calculated detection limits for 55 MeV ^{35}Cl ions. Coulomb excitations are marked by CE. The data were collected from Borderie *et al.* (1979).

Element	Reaction	E_γ (keV)	Gamma-ray yield $[(\mu C\ sr)^{-1}]$	Detection limit (ppm)
Li	^{7}Li(^{35}Cl,nγ)^{41}Ca	2606	7.1×10^4	8
B	^{10}B(^{35}Cl,npγ)^{43}Sc	350	3.4×10^3	175
F	^{19}F (CE)	197	1.3×10^5	7
Na	^{23}Na (CE)	440	1.8×10^5	17
Mg	^{24}Mg (CE)	1369	1.5×10^3	350
Al	^{27}Al (CE)	844	2.3×10^3	310
Ti	^{47}Ti (CE)	159	1.4×10^4	105
V	^{51}V (CE)	319	3.6×10^4	35
Mn	^{55}Mn (CE)	126	1.7×10^5	10
Fe	^{56}Fe (CE)	847	1.2×10^4	60
Cu	^{63}Cu (CE)	669	2.8×10^3	
Ga	^{69}Ga (CE)	319	6.6×10^3	230
Ge	^{74}Ge (CE)	596	3.0×10^4	85
As	^{75}As (CE)	279	9.6×10^4	20
Se	^{77}Se (CE)	239	2.2×10^4	75
Br	^{79}Br (CE)	217	3.8×10^4	50
Nb	^{93}Nb (CE)	744	8.6×10^2	
Mo	^{100}Mo (CE)	535	7.5×10^4	330
Ru	^{101}Ru (CE)	127	9.3×10^3	230
Rh	^{103}Rh (CE)	297	1.3×10^5	15
Pd	^{108}Pd (CE)	434	3.2×10^4	70
Ag	^{109}Ag (CE)	311	5.3×10^4	35
Cd	^{111}Cd (CE)	299	4.7×10^3	395
Sb	^{123}Sb (CE)	160	2.1×10^3	1100
Te	^{120}Te (CE)	560	7.2×10^3	420
I	^{127}I (CE)	203	2.5×10^4	90
Nd	^{150}Nd (CE)	131	6.5×10^4	40
Sm	^{147}Sm (CE)	122	1.4×10^5	20
Eu	^{153}Eu (CE)	83	1.7×10^5	20
Gd	^{156}Gd (CE)	89	9.7×10^4	35
Tb	^{159}Tb (CE)	79	1.1×10^5	40
Dy	^{163}Dy (CE)	73	1.3×10^5	35
Ho	^{165}Ho (CE)	95	2.9×10^5	10
Er	^{166}Er (CE)	80	3.5×10^5	10
Yb	^{171}Yb (CE)	78	1.9×10^5	35
Hf	^{180}Hf (CE)	93	2.2×10^5	15
Ta	^{181}Ta (CE)	136	2.8×10^5	10
W	^{184}W (CE)	111	1.1×10^5	35
Re	^{187}Re (CE)	134	1.1×10^5	35
Pt	^{194}Pt (CE)	328	2.1×10^4	140
Au	^{197}Au (CE)	279	1.9×10^4	170
Th	^{232}Th (CE)	113	2.6×10^3	2700

Table A12.18. Typical gamma-ray background peaks. 4n and 4n+2 refer to the natural radioactive decay chains. Data from Giles and Peisach (1979).

E_γ (keV)	Assignment	Origin or natural decay chain	E_γ (keV)	Assignment	Origin or natural decay chain
57	Ta Kα X-rays	Chamber lining	844	^{27}Mg	^{27}Al(n, p)^{27}Mg by fast neutrons
61				^{27}Al α,α'(1,0)	Scattered-alpha excitation of chamber
66	Ta Kβ X-rays	Chamber lining	861	^{208}Tl	4n
74	^{208}Bi; ^{212}Pb	^{209}Bi(n, 2n) ^{208}Bi; 4n unresolved Kα X-ray	871	^{14}N α,p(1,0)	Nitrogen on target surface
86	^{208}Bi; ^{212}Pb	^{209}Bi(n, 2n) ^{208}Bi; 4n unresolved Kβ X-ray	885	^{73}Ge n,α(1,0)	Neutron bombardment of detector
91			894	^{72}Ge n,n'(4,2)	Neutron bombardment of detector
110	^{19}F n,n'(1,0)	Neutron bombardment of detector	911	^{228}Ac	4n
129	^{228}Ac	4n	935	^{214}Bi	(4n + 2)
136	^{181}Ta, α,α'(1,0)	Collimators and chamber lining	969	^{228}Ac	4n
186	^{226}Ra	(4n + 2)	1015	^{27}Al α,α'(2,0)	Scattered-alpha excitation of chamber
197	^{19}F n,n'(2,0)	Neutron bombardment of detector	1120	^{214}Bi	(4n + 2)
239	^{212}Pb; ^{214}Pb	4n; (4n + 2)	1155	^{214}Bi	(4n + 2)
277	^{208}Tl; ^{228}Ac	4n; 4n	1214	(2236-2m$_e$)	^{27}Al
285	^{214}Bi	(4n + 2)	1223		
296	^{210}Tl; ^{214}Pb	(4n + 2); (4n + 2)	1238	^{214}Bi	(4n + 2)
301	^{181}Ta α,α' (2,0)	Collimators and chamber lining	1275		
322			1369	^{24}Na	^{27}Al(n, α)^{24}Na by fast neutrons
328	^{228}Ac	4n	1378	^{214}Bi	(4n + 2)
339	^{228}Ac	4n	1406		
351	^{18}O α,n(1,0)	Oxygen on target surface	1408	^{214}Bi	(4n + 2)
352	^{214}Pb	(4n + 2)	1461	^{40}K	0.012 atom% of natural K
417	^{23}Na α,n(2,0)	Sodium contamination of target	1464	^{72}Ge n,n'(3,0)	Neutron bombardment of detector
440	^{23}Na α,α'(1,0)	Sodium contamination of target	1509	^{214}Bi	(4n + 2)
463	^{228}Ac	4n	1588	^{228}Ac	4n
478	^{7}Be	Cosmic-ray produced	1592	(2614-2m$_e$)	^{208}Bi; ^{208}Tl
511	β^+; ^{208}Tl	Various; 4n	1693	(2204-m$_e$)	^{214}Bi
563	^{76}Ge n,n'(1,0)	Neutron bombardment of detector	1725	(2236-m$_e$)	^{27}Al
583	^{208}Tl	4n	1732	(2754-2m$_e$), ^{214}Bi	^{24}Na; (4n + 2)
596	^{74}Ge n,n'(1,0)	Neutron bombardment of detector	1756		
604			1764	^{214}Bi	(4n + 2)
608	^{74}Ge n,n'(1,0)	Neutron bombardment of detector	1809	^{23}Na α,p(1,0)	Sodium contamination of targets
609	^{214}Bi	(4n + 2)	1850	^{214}Bi	(4n + 2)
666	^{214}Bi	(4n + 2)	2103	(2614-m$_e$)	^{208}Bi; ^{208}Tl
691	^{72}Ge n,n'(1,0)	Neutron bombardment of detector	2204	^{214}Bi	(4n + 2)
718			2236	^{27}Al, α,p(1,0)	Scattered-alpha excitation of chamber
727	^{212}Bi	4n	2243	(2754-m$_e$)	^{24}Na
769	^{214}Bi	(4n + 2)	2614	^{208}Bi; ^{208}Tl	^{209}Bi(n, 2n)^{208}Bi by fast neutrons; 4n
795	^{210}Tl; ^{228}Ac	(4n + 2); 4n	2754	^{24}Na	^{27}Al(n,α)^{24}Na by fast neutrons
806					
834	^{72}Ge n,n'(2,0)	Neutron bombardment of detector			
835	^{54}Mn	^{54}Fe(n, p)^{54}Mn by fast neutrons			

REFERENCES

Anttila, A., Hänninen, R., and Räisänen, J. (1981), *J. Radioanal. Chem.* **62**, 293.

Bird, J.R., Scott, M.D., Russel, L.H., and Kenny, M.J. (1978), *Aust. J. Phys.* **31**, 209.

Borderie, B., and Barrandon, J.N. (1978), *Nucl. Instrum. Methods* **156**, 483.

Borderie, B., Barrandon, J.N., Delaunay, B., and Basutcu, M. (1979), *Nucl. Instrum. Methods* **163**, 441.

Borderie, B. (1980), *Nucl. Instrum. Methods* **175**, 465.

Boni, C., Caridi, A., Cereda, E., Braga Marcazzan, G.M., and Redaelli, P. (1990), *Nucl. Instrum. Methods* B **49**, 106.

Butler, J. W. (1959), NRL Report 5282, U.S. Naval Research Laboratory, Washington, DC.

Caciolli, A., Chiari, M., Climent-Font, A., Fernández-Jiménez, M.T., García-López, G., Lucarelli, F., Nava, S., and Zucchiatti, A. (2006), *Nucl. Instrum. Methods* **B249**, 98.

Deconninck, G., and Demortier, G. (1975), *J. Radioanal. Chem.* **24**, 437.

Demortier, G. (1978), *J. Radioanal. Chem.* **45**, 459.

Elwyn, A.J., Marinov, A., and Schiffer, J.P. (1966), *Phys. Rev.* **146**, 957.

Elekes, Z., Kiss, Á.Z., Biron, I., Calligaro, T., and Salomon, J. (2000), *Nucl. Instrum. Methods* **B168**, 305.

Giles, I.S., and Peisach, M. (1979), *J. Radioanal. Chem.* **50**, 307.

Gihwala, D., and Peisach, M. (1982), *J. Radioanal. Chem.* **70**, 287.

Golicheff, L., Loeuillet, M., and Engelmann, C. (1972), *J. Radioanal. Chem.* **12**, 233.

Heaton, R:K., Lee, H.W., Robertson, B.C., Norman, E.B., Lesko, K.T., and Sur, B. (1995), *Nucl. Instrum. Methods* **A364**, 317.

Horn, K.M., and Lanford, W.A. (1988), *Nucl. Instrum. Methods* **B34**, 1.

Kenny, M.J., Bird, J.R., and Clayton, E. (1980), *Nucl. Instrum. Methods* **168**, 115.

Kiss, Á.Z., Koltay, E., Nyakó, B., Somorjai, E., Anttila, A., and Räisänen, J. (1985), *J. Radioanal. Nucl. Chem.* **89**, 123.

Kiss, Á.Z., Biron, I., Calligaro, T., and Salomon, J. (1994), *Nucl. Instrum. Methods* **B85**, 118.

Kocsonya, A., Szökefalvi-Nagy, Z., Torri, A., Rauhala, E., and Räisänen, J. (2006), *Nucl. Instrum. Methods* **B251**, 367.

Lappalainen, R., Anttila, A., and Räisänen, J. (1983), *Nucl. Instrum. Methods* **212**, 441.

Mayer, J.W., and Rimini, E. (eds.) (1977), *Ion Beam Handbook for Materials Analysis*, Academic Press, New York.

Peisach, M., Pineda, C.A., and Pillay, A.E. (1994), *Nucl. Instrum. Methods* **B85**, 142.

Räisänen, J., and Hänninen, R. (1983), *Nucl. Instrum. Methods* **205**, 259.

Räisänen, J., Witting, T., and Keinonen, J. (1987), *Nucl. Instrum. Methods* **B28**, 199.

Räisänen, J. (1987), *Biol. Trace Elem. Res.* **12**, 55.

Räisänen, J., and Lapatto, R. (1988), *Nucl. Instrum. Methods* **B30**, 90.

Räisänen, J. (1989), *Nucl. Instrum. Methods* **B40–41**, 638.

Räisänen, J. (1990), *Nucl. Instrum. Methods* **A299**, 387.

Savidou, A., Aslanoglou, X., Paradellis, T., and Pilakouta, M. (1999), *Nucl. Instrum. Methods* **B152**, 12.

Seppälä, A., Räisänen, J., Rauhala, E., and Szökefalvi-Nagy, Z. (1998), *Nucl. Instrum. Methods* **B143**, 233.

APPENDIX

13

HYDROGEN NUCLEAR REACTION DATA

Compiled by

W. A. Lanford

State University of New York, Albany, New York, USA

CONTENTS

Table A13.1 Nuclear reaction cross section characteristics[1,2].

Reaction	Units	^{7}Li + H	^{15}N + H	^{15}N + H	^{19}F + H	^{19}F + H
Resonance energy	MeV	3.070	6.385	13.35	6.418	16.44
Cross section (σ) at resonance	mbarn	4.800	1650.000	1050.00	88.000	440.00
Resonance width (Γ)	keV	81.000	1.800	25.40	44.000	86.00
σ·Γ	mbarn x keV	389.000	2970.000	26700.00	3870.000	37800.00
Relative yield		0.130	1.000	9.00	1.300	12.70
dE/dx in Si	MeV/micron	0.442	1.450	1.35	1.940	1.91
Γ/(dE/dx)	nanometers	183.000	1.200	18.80	22.700	45.00
Energy of next resonance	MeV	7.110	13.350	18.00	9.100	17.60
Approximate analysis depth	microns	9.140	4.800	3.44	1.380	0.61
Gamma-ray energy	MeV	17.7, 14.7	4.430	4.43	6.13, 6.98, 7.12	6.13, 6.98, 7.12

1. The resonance data for ^{7}Li are from Trocellier, P., and Engelmann, Ch. (1986), *J. Radioanal. Nucl. Chem.* **100**, 117.

2. The resonance data for ^{15}N and ^{19}F are from Xiong, F., Rauch, F., Shi, C., Zhou, Z., Livi, R.P., and Tombrello, T.A. (1987), *Nucl. Instrum. Methods Phys. Res.* **B27**, 432.

Table A13.2. dE/dx for ^{7}Li, ^{15}N and ^{19}F at their resonance energies.

Element	Atomic number	Density	^{7}Li at 3.07 MeV (MeV/mg/cm^2)	^{15}N at 6.385 MeV (MeV/mg/cm^2)	^{15}N at 13.35 MeV (MeV/mg/cm^2)	^{19}F at 6.418 MeV (MeV/mg/cm^2)	^{19}F at 16.44 MeV (MeV/mg/cm^2)
H	1		8.715E+00	2.838E+01	2.166E+01	4.145E+01	3.222E+01
He	2		3.508E+00	1.148E+01	9.233E+00	1.606E+01	1.370E+01
Li	3	0.54	2.606E+00	8.514E+00	7.399E+00	1.166E+01	1.087E+01
Be	4	1.82	2.330E+00	7.693E+00	7.055E+00	1.022E+01	1.032E+01
B	5	2.47	2.528E+00	8.307E+00	7.227E+00	1.143E+01	1.062E+01
C	6	3.52	2.574E+00	8.453E+00	7.397E+00	1.160E+01	1.086E+01
N	7		2.607E+00	8.537E+00	7.112E+00	1.192E+01	1.048E+01
O	8		2.425E+00	7.965E+00	6.923E+00	1.087E+01	1.016E+01
F	9		2.096E+00	6.915E+00	6.266E+00	9.046E+00	9.178E+00
Ne	10		2.043E+00	6.747E+00	6.198E+00	8.799E+00	9.077E+00
Na	11	1.01	2.117E+00	6.987E+00	6.442E+00	9.204E+00	9.422E+00
Mg	12	1.74	1.990E+00	6.535E+00	6.022E+00	8.682E+00	8.795E+00
Al	13	2.70	1.775E+00	5.858E+00	5.556E+00	7.633E+00	8.099E+00
Si	14	2.33	1.896E+00	6.242E+00	5.677E+00	8.421E+00	8.297E+00
P	15	1.82	1.843E+00	6.055E+00	5.458E+00	8.288E+00	7.976E+00
S	16	2.07	1.779E+00	5.820E+00	4.924E+00	8.231E+00	7.218E+00
Cl	17		1.883E+00	6.162E+00	5.222E+00	8.699E+00	7.655E+00
Ar	18		1.649E+00	5.416E+00	4.834E+00	7.464E+00	7.059E+00
K	19	0.91	1.875E+00	6.133E+00	5.456E+00	8.222E+00	8.010E+00
Ca	20	1.53	1.823E+00	5.978E+00	5.354E+00	7.942E+00	7.839E+00
Sc	21	2.99	1.713E+00	5.613E+00	4.872E+00	7.655E+00	7.148E+00
Ti	22	4.51	1.571E+00	5.178E+00	4.737E+00	6.909E+00	6.921E+00
V	23	6.09	1.599E+00	5.263E+00	4.566E+00	7.195E+00	6.698E+00
Cr	24	7.19	1.456E+00	4.825E+00	4.477E+00	6.330E+00	6.538E+00
Mn	25	7.47	1.374E+00	4.550E+00	4.264E+00	5.936E+00	6.221E+00
Fe	26	7.87	1.373E+00	4.531E+00	4.261E+00	5.926E+00	6.211E+00
Co	27	8.9	1.283E+00	4.243E+00	4.037E+00	5.510E+00	5.877E+00
Ni	28	8.91	1.250E+00	4.160E+00	4.116E+00	5.277E+00	5.971E+00
Cu	29	8.93	1.137E+00	3.776E+00	3.814E+00	4.778E+00	5.516E+00
Zn	30	7.13	1.145E+00	3.797E+00	3.783E+00	4.801E+00	5.481E+00
Ga	31	5.91	1.092E+00	3.606E+00	3.537E+00	4.651E+00	5.119E+00
Ge	32	5.32	1.071E+00	3.547E+00	3.593E+00	4.536E+00	5.184E+00
As	33	5.77	1.058E+00	3.501E+00	3.417E+00	4.588E+00	4.945E+00
Se	34	4.81	1.003E+00	3.321E+00	3.319E+00	4.311E+00	4.794E+00
Br	35		1.041E+00	3.422E+00	3.464E+00	4.500E+00	4.988E+00
Kr	36		1.048E+00	3.434E+00	3.340E+00	4.572E+00	4.828E+00
Rb	37	1.63	1.153E+00	3.783E+00	3.570E+00	4.952E+00	5.203E+00
Sr	38	2.58	1.146E+00	3.749E+00	3.431E+00	5.089E+00	4.986E+00
Y	39	4.48	1.118E+00	3.669E+00	3.404E+00	4.934E+00	4.946E+00
Zr	40	6.51	1.126E+00	3.704E+00	3.405E+00	5.031E+00	4.952E+00
Nb	41	8.58	1.158E+00	3.828E+00	3.601E+00	5.009E+00	5.248E+00
No	42	10.22	1.037E+00	3.440E+00	3.249E+00	4.502E+00	4.730E+00
Tc	43	11.5	1.044E+00	3.455E+00	3.302E+00	4.527E+00	4.799E+00
Ru	44	12.36	1.036E+00	3.431E+00	3.271E+00	4.458E+00	4.759E+00
Rh	45	12.42	1.006E+00	3.332E+00	3.211E+00	4.307E+00	4.667E+00
Pd	46	12.00	9.752E-01	3.244E+00	3.169E+00	4.155E+00	4.600E+00
Ag	47	10.5	9.602E-01	3.187E+00	3.032E+00	4.139E+00	4.406E+00
Cd	48	8.65	9.329E-01	3.082E+00	2.998E+00	3.993E+00	4.350E+00
In	49	7.29	9.298E-01	3.066E+00	2.984E+00	3.968E+00	4.329E+00
Sn	50	5.76	9.053E-01	2.985E+00	2.855E+00	3.907E+00	4.147E+00
Sb	51	6.69	8.942E-01	2.955E+00	2.866E+00	3.839E+00	4.160E+00
Te	52	6.25	8.879E-01	2.932E+00	2.782E+00	3.850E+00	4.040E+00
I	53		8.977E-01	2.947E+00	2.859E+00	3.920E+00	4.137E+00
Xe	54		8.820E-01	2.894E+00	2.825E+00	3.832E+00	4.086E+00
Cs	55	1.99	9.601E-01	3.151E+00	2.994E+00	4.111E+00	4.359E+00
Ba	56	3.59	9.256E-01	3.037E+00	2.891E+00	4.028E+00	4.199E+00
La	57	6.17	9.380E-01	3.083E+00	2.866E+00	4.071E+00	4.180E+00
Ce	58	6.77	8.289E-01	2.729E+00	2.732E+00	3.611E+00	3.937E+00
Pr	59	6.78	8.666E-01	2.859E+00	2.828E+00	3.680E+00	4.097E+00
Nd	60	7.00	8.349E-01	2.753E+00	2.731E+00	3.552E+00	3.953E+00

Table A13.2. dE/dx for ^{7}Li, ^{15}N and ^{19}F at their resonance energies (continued).

Element	Atomic number	Density	^{7}Li at 3.07 MeV (MeV/mg/cm^2)	^{15}N at 6.385 MeV (MeV/mg/cm^2)	^{15}N at 13.35 MeV (MeV/mg/cm^2)	^{19}F at 6.418 MeV (MeV/mg/cm^2)	^{19}F at 16.44 MeV (MeV/mg/cm^2)
Pm	61		8.362E-01	2.755E+00	2.709E+00	3.587E+00	3.920E+00
Sm	62	7.54	8.612E-01	2.844E+00	2.809E+00	3.623E+00	4.074E+00
Eu	63	5.25	7.640E-01	2.523E+00	2.613E+00	3.192E+00	3.760E+00
Gd	64	7.89	7.704E-01	2.541E+00	2.541E+00	3.281E+00	3.670E+00
Tb	65	8.27	7.389E-01	2.441E+00	2.503E+00	3.106E+00	3.608E+00
Dy	66	8.53	7.273E-01	2.405E+00	2.477E+00	3.041E+00	3.571E+00
Ho	67	8.80	6.587E-01	2.181E+00	2.308E+00	2.731E+00	3.315E+00
Er	68	9.04	7.032E-01	2.327E+00	2.413E+00	2.916E+00	3.478E+00
Tm	69	9.32	6.543E-01	2.165E+00	2.276E+00	2.733E+00	3.269E+00
Yb	70	6.97	6.562E-01	2.164E+00	2.207E+00	2.774E+00	3.174E+00
Lu	71	9.84	6.336E-01	2.094E+00	2.184E+00	2.670E+00	3.137E+00
Hf	72	13.2	6.221E-01	2.060E+00	2.157E+00	2.617E+00	3.098E+00
Ta	73	16.66	6.286E-01	2.092E+00	2.191E+00	2.635E+00	3.151E+00
W	74	19.25	6.258E-01	2.083E+00	2.083E+00	2.652E+00	3.013E+00
Re	75	21.03	6.200E-01	2.062E+00	2.137E+00	2.583E+00	3.078E+00
Os	76	22.58	6.085E-01	2.025E+00	2.101E+00	2.532E+00	3.027E+00
Ir	77	22.55	5.951E-01	1.981E+00	2.082E+00	2.481E+00	2.994E+00
Pt	78	21.47	5.751E-01	1.917E+00	2.015E+00	2.433E+00	2.890E+00
Au	79	19.28	6.070E-01	2.022E+00	2.043E+00	2.568E+00	2.944E+00
Hg	80	14.26	5.900E-01	1.949E+00	2.000E+00	2.504E+00	2.872E+00
Tl	81	11.87	5.901E-01	1.948E+00	2.017E+00	2.496E+00	2.898E+00
Pb	82	11.34	6.185E-01	2.042E+00	2.022E+00	2.649E+00	2.915E+00
Bi	83	9.8	6.470E-01	2.134E+00	2.077E+00	2.794E+00	3.001E+00
Po	84	9.24	6.363E-01	2.097E+00	2.090E+00	2.763E+00	3.009E+00
At	85		6.462E-01	2.122E+00	2.113E+00	2.787E+00	3.046E+00
Rn	86		6.182E-01	2.030E+00	2.025E+00	2.669E+00	2.917E+00
Fr	87		6.369E-01	2.088E+00	2.021E+00	2.766E+00	2.919E+00
Ra	88	5	6.498E-01	2.129E+00	2.040E+00	2.849E+00	2.946E+00
Ac	89		6.751E-01	2.217E+00	2.110E+00	2.962E+00	3.055E+00
Th	90	11.2	6.432E-01	2.117E+00	2.035E+00	2.804E+00	2.942E+00
Pa	91	15.4	6.568E-01	2.207E+00	2.090E+00	2.933E+00	3.028E+00
U	92	18.7	5.880E-01	1.948E+00	1.962E+00	2.521E+00	2.828E+00

Table A13.3. Hydrogen bearing materials.

Common name	Chemical name	Composition
Celluloid	Cellulose acetate	$(C_9H_{13}O_7)_n$
Kapton	Polyamide film	$(C_{22}H_{10}N_2O_4)_n$
Kel-F	Chlorotrifluoro-ethylene polymer	$(CF_2CHCl)_n$
Lucite (Plexiglas)	methyl methacrylate	$(C_5H_8O_2)_n$
Mica	Muscovite	$(KF)_2(Al_2O_3)_3(SiO_2)_6(H_2O)$
Mylar	Polyester film	$(C_{10}H_8O_4)_n$
Nylon	polyamides	$(C_{12}H_{22}N_2O_2)_n$
Polythylene		$(CH_2)_n$
Polystyrene		$(CH)_n$

PROTON–PROTON SCATTERING CROSS SECTIONS

Compiled by

P. Reichart and G. Dollinger

Universität der Bundeswehr München, Neubiberg, Germany

CONTENTS

A14.1 PROTON–PROTON SCATTERING CROSS-SECTION DATA

Table A14.1 shows an arbitrary excerpt of the available proton–proton (pp) scattering data in the energy range E_{in} = 1.4–50 MeV with given absolute errors in percent. Some of the data set are plotted in Fig. 9.5 (Chapter 9) for the laboratory system. Equations from Section 9.5.3 (Chapter 9) were used to transform the measured data into the laboratory or center-of-mass (CM) system. These data and further data, including inelastic scattering data, are available from the original sources (Allred et al., 1952; Berdoz et al., 1986; Blair et al., 1948; Burkig et al., 1959; Dayton and Schrank, 1956; Imai et al., 1975; Jarmie et al., 1970; Jarmie et al., 1971; Jarmie and Jett, 1974; Jarmie et al., 1974; Jeong et al., 1960; Johnston and Young, 1959; Kikuchi et al., 1960; Knecht et al. 1959; Panofsky and Fillmore, 1950; Slobodrian et al., 1968; Slobodrian et al., 1976; Wassmer and Muhry, 1973; Worthington et al., 1953; Yntema and White, 1954), collections (Bystricky et al., 1980), or databases such as EXFOR/CSISRS (IAEA Nuclear Data Services, 2009, NNDC Databases, 2009).

Table A14.1. Cross-section data for proton–proton scattering in the laboratory and center-of-mass (CM) system (arbitrary excerpt of available data at proton energies that are useful for hydrogen analysis). Data taken from original sources are marked bold; all other values were calculated with relativistic equations and the assumption of a symmetric CM cross section about θ_{cm} = 90°. Additionally, the second-to-last column lists the percentage deviation of the measured data from nonlinear regression with Eq. (A14.2) using the values in Table A14.3.

Energy E_{in} (MeV)	θ (deg)	$d\sigma/d\Omega$ (mb/sr)	θ_{cm} (deg)	$d\sigma/d\Omega_{cm}$ (mb/sr)	Error (%)	Fit dev. (%)	Data set
1.397	6.0	80161.4	12	20136	0.34		Knecht et al. (1959)
1.397	7.0	41927.9	14	10553	0.32		
1.397	8.0	23626.2	16	5960.3	0.27		
1.397	10.0	8966.5	20	2274.6	0.26	2.0	
1.397	12.0	4048.8	24	1034.1	0.23	5.4	
1.397	15.0	1626.5	30	420.68	0.2	5.2	
1.397	17.5	976.0	35	255.67	0.17	3.4	
1.397	20.0	719.8	40	191.39	0.16	16.2	
1.397	25.0	570.8	50	157.36	0.14	34.2	
1.397	30.0	541.2	60	156.15	0.14	38.3	
1.397	35.0	526.1	70	160.5	0.14	36.7	
1.397	55.0	368.2	110	160.5	0.14		
1.397	60.0	312.3	120	156.15	0.14		
1.397	65.0	266.0	130	157.36	0.14		
1.397	70.0	261.8	140	191.39	0.16		
1.397	72.5	307.4	145	255.67	0.17		
1.397	75.0	435.4	150	420.68	0.2		
1.397	78.0	859.7	156	1034.1	0.23		
1.397	80.0	1579.4	160	2274.6	0.26		
1.397	82.0	3316.9	164	5960.3	0.27		
1.397	83.0	5142.5	166	10553	0.32		
1.397	84.0	8416.1	168	20136	0.34		

Table A14.1. Cross-section data for proton–proton scattering in the laboratory and center-of-mass (CM) system (arbitrary excerpt of available data at proton energies that are useful for hydrogen analysis). Data taken from original sources are marked bold; all other values were calculated with relativistic equations and the assumption of a symmetric CM cross section about $\theta_{cm} = 90°$. Additionally, the second-to-last column lists the percentage deviation of the measured data from nonlinear regression with Eq. (A14.2) using the values in Table A14.3 (continued).

Energy E_{in} (MeV)	θ (deg)	$d\sigma/d\Omega$ (mb/sr)	θ_{cm} (deg)	$d\sigma/d\Omega_{cm}$ (mb/sr)	Error (%)	Fit dev. (%)	Data set
1.855	7.0	22863.4	14	5753.21	0.27		Worthington et al. (1953)
1.855	8.0	12895.6	16	3252.45	0.29		
1.855	10.0	4958.7	20	1257.6	0.21	1.1	
1.855	12.0	2321.4	24	592.77	0.25	5.7	
1.855	15.0	1076.0	30	278.24	0.21	8.2	
1.855	17.5	754.1	35	197.5	0.26	2.7	
1.855	20.0	630.6	40	167.63	0.2	5.5	
1.855	25.0	565.7	50	155.94	0.23	15.1	
1.855	30.0	554.3	60	159.92	0.21	16.3	
1.855	35.0	538.3	70	164.2	0.25	15.4	
1.855	40.0	513.5	80	167.53	0.24	14.2	
1.855	45.0	475.1	90	167.93	0.28	14.8	
1.855	50.0	430.8	100	167.53	0.24		
1.855	55.0	376.7	110	164.2	0.25		
1.855	60.0	319.8	120	159.92	0.21		
1.855	65.0	263.6	130	155.94	0.23		
1.855	70.0	229.3	140	167.63	0.2		
1.855	72.5	237.5	145	197.5	0.26		
1.855	75.0	287.9	150	278.24	0.21		
1.855	78.0	492.8	156	592.77	0.25		
1.855	80.0	873.1	160	1257.6	0.21		
1.855	82.0	1809.8	164	3252.45	0.29		
1.855	83.0	2803.2	166	5753.21	0.27		
2.425	6.0	25553.0	12	6415.29	0.26		Worthington et al. (1953)
2.425	8.0	7532.2	16	1899.17	0.34		
2.425	10.0	2970.3	20	753.11	0.22	1.5	
2.425	12.0	1480.2	24	377.87	0.29	6.6	
2.425	15.0	791.2	30	204.53	0.15	9.0	
2.425	17.5	621.5	35	162.73	0.22	4.9	
2.425	20.0	561.5	40	149.22	0.28	0.1	
2.425	25.0	532.4	50	146.73	0.26	5.4	
2.425	30.0	522.3	60	150.64	0.23	6.2	
2.425	35.0	507.1	70	154.67	0.21	5.2	
2.425	40.0	481.6	80	157.08	0.2	4.4	
2.425	45.0	443.8	90	156.86	0.24	5.1	
2.425	50.0	401.3	100	156.04	0.28		
2.425	55.0	354.5	110	154.5	0.37		
2.425	60.0	300.2	120	150.12	0.49		
2.425	65.0	248.0	130	146.73	0.26		

Table A14.1. Cross-section data for proton–proton scattering in the laboratory and center-of-mass (CM) system (arbitrary excerpt of available data at proton energies that are useful for hydrogen analysis). Data taken from original sources are marked bold; all other values were calculated with relativistic equations and the assumption of a symmetric CM cross section about $\theta_{cm} = 90°$. Additionally, the second-to-last column lists the percentage deviation of the measured data from nonlinear regression with Eq. (A14.2) using the values in Table A14.3 (continued).

Energy E_{ln} (MeV)	θ (deg)	$d\sigma/d\Omega$ (mb/sr)	θ_{cm} (deg)	$d\sigma/d\Omega_{cm}$ (mb/sr)	Error (%)	Fit dev. (%)	Data set
2.425	70.0	204.1	140	149.22	0.28		Worthington *et al.* (1953)
2.425	72.5	195.6	145	162.73	0.22		(continued)
2.425	75.0	211.6	150	204.53	0.15		
2.425	78.0	314.1	156	377.87	0.29		
2.425	80.0	522.8	160	753.11	0.22		
2.425	82.0	1056.6	164	1899.17	0.34		
2.425	84.0	2680.6	168	6415.29	0.26		
3.037	6.0	16124.9	**12**	**4046.98**	**0.3**		Worthington et al. (1953)
3.037	7.0	8370.5	**14**	**2105.01**	**0.33**		
3.037	8.0	4804.4	**16**	**1211.01**	**0.4**		
3.037	10.0	1953.7	**20**	**495.19**	**0.3**	0.0	
3.037	12.0	1045.5	**24**	**266.8**	**0.35**	5.5	
3.037	15.0	630.2	**30**	**162.86**	**0.49**	7.7	
3.037	17.5	529.8	**35**	**138.68**	**0.29**	4.4	
3.037	20.0	495.3	**40**	**131.6**	**0.2**	0.8	
3.037	25.0	479.2	**50**	**132.03**	**0.25**	2.4	
3.037	30.0	472.1	**60**	**136.15**	**0.26**	2.1	
3.037	35.0	453.6	**70**	**138.32**	**0.28**	1.9	
3.037	40.0	431.3	**80**	**140.67**	**0.23**	0.8	
3.037	45.0	398.5	**90**	**140.84**	**0.28**	1.1	
3.037	50.0	361.6	**100**	**140.6**	**0.33**		
3.037	55.0	320.1	**110**	**139.53**	**0.39**		
3.037	60.0	271.5	**120**	**135.79**	**0.54**		
3.037	65.0	223.1	130	132.03	0.25		
3.037	70.0	179.9	140	131.6	0.2		
3.037	72.5	166.7	145	138.68	0.29		
3.037	75.0	168.5	150	162.86	0.49		
3.037	78.0	221.7	156	266.8	0.35		
3.037	80.0	343.7	160	495.19	0.3		
3.037	82.0	1171.0	164	2105.01	0.33		
3.037	83.0	1971.3	166	4046.98	0.3		
3.527	6.0	11929.1	**12**	**2993.18**	**0.28**		Worthington et al. (1953)
3.527	7.0	6273.8	**14**	**1577.34**	**0.33**		
3.527	8.0	3599.2	**16**	**907**	**0.46**		
3.527	10.0	1501.8	**20**	**380.55**	**0.28**	0.9	
3.527	12.5	759.3	**25**	**194.1**	**0.45**	5.3	
3.527	15.0	545.3	**30**	**140.9**	**0.21**	6.3	

Table A14.1. Cross-section data for proton–proton scattering in the laboratory and center-of-mass (CM) system (arbitrary excerpt of available data at proton energies that are useful for hydrogen analysis). Data taken from original sources are marked bold; all other values were calculated with relativistic equations and the assumption of a symmetric CM cross section about $\theta_{cm} = 90°$. Additionally, the second-to-last column lists the percentage deviation of the measured data from nonlinear regression with Eq. (A14.2) using the values in Table A14.3 (continued).

Energy E_{in} (MeV)	θ (deg)	$d\sigma/d\Omega$ (mb/sr)	θ_{cm} (deg)	$d\sigma/d\Omega_{cm}$ (mb/sr)	Error (%)	Fit dev. (%)	Data set
3.527	17.5	475.4	35	124.42	0.25	3.8	Worthington *et al.* (1953)
3.527	20.0	453.9	40	120.56	0.2	1.5	(continued)
3.527	25.0	440.7	50	121.39	0.21	1.1	
3.527	30.0	434.4	60	125.26	0.25	0.3	
3.527	35.0	418.8	70	127.69	0.26	0.5	
3.527	40.0	396.3	80	129.25	0.24	1.2	
3.527	45.0	365.8	90	129.26	0.28	0.9	
3.527	50.0	331.4	100	128.85	0.28		
3.527	55.0	292.6	110	127.54	0.45		
3.527	60.0	250.5	120	125.26	0.25		
3.527	65.0	205.1	130	121.39	0.21		
3.527	70.0	164.8	140	120.56	0.2		
3.527	72.5	149.6	145	124.42	0.25		
3.527	75.0	145.8	150	140.9	0.21		
3.527	77.5	167.9	155	194.1	0.45		
3.527	80.0	264.1	160	380.55	0.28		
3.527	82.0	504.5	164	907	0.46		
3.527	83.0	768.2	166	1577.34	0.33		
3.527	84.0	1250.4	168	2993.18	0.28		
3.899	6.0	9750.7	12	**2446.1**	**0.47**		Worthington *et al.* (1953)
3.899	7.0	5118.4	14	**1286.6**	**0.6**		
3.899	8.0	2962.9	16	**746.49**	**0.34**		
3.899	10.0	1269.4	20	**321.62**	**0.35**	1.1	
3.899	12.5	668.9	25	**170.95**	**0.56**	4.8	
3.899	15.0	494.3	30	**127.69**	**0.27**	5.1	
3.899	17.5	438.7	35	**114.79**	**0.35**	2.8	
3.899	20.0	423.7	40	**112.52**	**0.19**	1.3	
3.899	25.0	413.1	50	**113.79**	**0.31**	0.6	
3.899	30.0	406.0	60	**117.05**	**0.33**	0.1	
3.899	35.0	390.3	70	**119**	**0.36**	0.8	
3.899	40.0	368.8	80	**120.27**	**0.29**	1.4	
3.899	45.0	341.3	90	**120.62**	**0.26**	1.4	
3.899	50.0	310.5	100	**120.74**	**0.25**		
3.899	55.0	273.6	110	**119.24**	**0.36**		
3.899	60.0	233.4	120	**116.74**	**0.53**		
3.899	65.0	192.3	130	**113.79**	**0.31**		
3.899	70.0	153.8	140	**112.52**	**0.19**		
3.899	72.5	138.0	145	**114.79**	**0.35**		
3.899	75.0	132.1	150	**127.69**	**0.27**		
3.899	77.5	147.9	155	**170.95**	**0.56**		

Table A14.1. Cross-section data for proton–proton scattering in the laboratory and center-of-mass (CM) system (arbitrary excerpt of available data at proton energies that are useful for hydrogen analysis). Data taken from original sources are marked bold; all other values were calculated with relativistic equations and the assumption of a symmetric CM cross section about $\theta_{cm} = 90°$. Additionally, the second-to-last column lists the percentage deviation of the measured data from nonlinear regression with Eq. (A14.2) using the values in Table A14.3 (continued).

Energy E_{ln} (MeV)	θ (deg)	$d\sigma/d\Omega$ (mb/sr)	θ_{cm} (deg)	$d\sigma/d\Omega_{cm}$ (mb/sr)	Error (%)	Fit dev. (%)	Data set
3.899	80.0	223.2	160	321.62	0.35		Worthington et al. (1953)
3.899	82.0	415.2	164	746.49	0.34		(continued)
3.899	83.0	626.6	166	1286.6	0.6		
3.899	84.0	1021.7	168	2446.1	0.47		
4.203	6.0	8411.6	**12**	**2109.83**	**0.48**		Worthington et al. (1953)
4.203	7.0	4407.6	**14**	**1107.75**	**0.82**		
4.203	8.0	2553.9	**16**	**643.36**	**1.11**		
4.203	12.5	603.2	**25**	**154.15**	**0.45**	3.4	
4.203	15.0	460.0	**30**	**118.82**	**0.32**	4.4	
4.203	17.5	413.8	**35**	**108.25**	**0.35**	2.4	
4.203	20.0	401.1	**40**	**106.52**	**0.3**	1.1	
4.203	25.0	392.7	**50**	**108.15**	**0.32**	0.2	
4.203	30.0	385.2	**60**	**111.04**	**0.33**	0.4	
4.203	35.0	371.2	**70**	**113.16**	**0.29**	1.4	
4.203	40.0	349.7	**80**	**114.03**	**0.29**	1.8	
4.203	45.0	323.2	**90**	**114.2**	**0.31**	1.7	
4.203	50.0	293.2	**100**	**114**	**0.34**		
4.203	55.0	259.6	**110**	**113.16**	**0.29**		
4.203	60.0	222.0	120	111.04	0.33		
4.203	65.0	182.7	130	108.15	0.32		
4.203	70.0	145.6	140	106.52	0.3		
4.203	72.5	130.1	145	108.25	0.35		
4.203	75.0	122.9	150	118.82	0.32		
4.203	77.5	133.3	155	154.15	0.45		
4.203	82.0	357.8	164	643.36	1.11		
4.203	83.0	539.4	166	1107.75	0.82		
4.203	84.0	881.2	168	2109.83	0.48		
4.21	**18.65**	26.0	37.3	**107.8**	28	2.7	Slobodrian et al. (1976)
4.21	**19.99**	116.5	40.0	**107.1**	28	1.8	
4.21	**21.25**	176.3	42.5	**106.6**	20	0.7	
4.21	**22.53**	222.4	45.1	**106.9**	35	0.2	
4.21	**23.73**	255.3	47.5	**107.5**	48	0.0	
4.21	**24.95**	280.4	49.9	**108.1**	19	0.1	
4.21	**26.09**	297.9	52.2	**108.6**	39	0.3	
4.21	**27.2**	312.5	54.5	**109.5**	21	0.1	
4.21	**28.41**	320.3	56.9	**109.2**	43	0.6	
4.21	**29.72**	330.8	59.5	**110.4**	19	0.1	
4.21	**31**	336.9	62.1	**111.1**	19	0.4	

Table A14.1. Cross-section data for proton–proton scattering in the laboratory and center-of-mass (CM) system (arbitrary excerpt of available data at proton energies that are useful for hydrogen analysis). Data taken from original sources are marked bold; all other values were calculated with relativistic equations and the assumption of a symmetric CM cross section about $\theta_{cm} = 90°$. Additionally, the second-to-last column lists the percentage deviation of the measured data from nonlinear regression with Eq. (A14.2) using the values in Table A14.3 (continued).

Energy E_{in} (MeV)	θ (deg)	$d\sigma/d\Omega$ (mb/sr)	θ_{cm} (deg)	$d\sigma/d\Omega_{cm}$ (mb/sr)	Error (%)	Fit dev. (%)	Data set
4.21	**32.24**	339.6	64.5	**111.4**	19	0.4	Slobodrian *et al.* (1976)
4.21	**33.46**	343.1	67.0	**112.4**	19	1.1	(continued)
4.21	**34.66**	343.8	69.4	**112.9**	23	1.4	
4.21	**35.84**	342.5	71.7	**113.1**	36	1.4	
4.21	**37.01**	340.4	74.1	**113.3**	54	1.5	
4.21	**38.16**	337.0	76.4	**113.3**	54	1.4	
4.21	**39.3**	333.8	78.7	**113.5**	39	1.5	
4.21	**40.43**	329.7	80.9	**113.6**	80	1.5	
4.21	**41.55**	325.3	83.2	**113.7**	47	1.6	
4.21	**42.68**	319.2	85.4	**113.4**	48	1.2	
4.21	**43.79**	313.5	87.6	**113.3**	20	1.1	
4.21	**44.9**	306.9	89.9	**113**	41	0.8	
4.21	**46.01**	301.7	92.1	**113.3**	25		
4.21	**47.13**	295.2	94.3	**113.3**	25		
4.21	47.3	294.7	94.6	113.4	48		
4.21	48.4	288.7	96.8	113.7	47		
4.21	49.5	281.3	99.1	113.6	80		
4.21	50.6	273.6	101.3	113.5	39		
4.21	51.8	265.3	103.6	113.3	54		
4.21	52.9	257.0	105.9	113.3	54		
4.21	54.1	247.8	108.3	113.1	36		
4.21	55.3	238.2	110.6	112.9	23		
4.21	56.5	227.3	113.0	112.4	19		
4.21	57.7	214.7	115.5	111.4	19		
4.21	58.9	203.0	117.9	111.1	19		
4.21	60.2	189.4	120.5	110.4	19		
4.21	61.5	173.8	123.1	109.2	43		
4.21	62.7	161.2	125.5	109.5	21		
4.21	63.9	146.5	127.8	108.6	39		
4.21	65.0	131.1	130.1	108.1	19		
4.21	66.2	112.9	132.5	107.5	48		
4.21	67.4	93.0	134.9	106.9	35		
4.21	68.7	69.4	137.5	106.6	20		
4.21	70.0	43.4	140.0	107.1	28		
4.21	71.3	9.9	142.7	107.8	28		
4.978	**8**	1829.6	16.0	**461.9**	**0.63**		Imai *et al.* (1975)
4.978	**9**	1192.7	18.0	**301.9**	**0.45**		
4.978	**10**	838.7	20.0	**212.9**	**0.4**	2.6	
4.978	**11**	635.3	22.0	**161.8**	**0.4**	1.0	
4.978	**12**	520.0	24.0	**132.9**	**0.4**	0.6	
4.978	**13**	451.3	26.0	**115.8**	**0.4**	1.6	

235

Table A14.1. Cross-section data for proton–proton scattering in the laboratory and center-of-mass (CM) system (arbitrary excerpt of available data at proton energies that are useful for hydrogen analysis). Data taken from original sources are marked bold; all other values were calculated with relativistic equations and the assumption of a symmetric CM cross section about $\theta_{cm} = 90°$. Additionally, the second-to-last column lists the percentage deviation of the measured data from nonlinear regression with Eq. (A14.2) using the values in Table A14.3 (continued).

Energy E_{in} (MeV)	θ (deg)	$d\sigma/d\Omega$ (mb/sr)	θ_{cm} (deg)	$d\sigma/d\Omega_{cm}$ (mb/sr)	Error (%)	Fit dev. (%)	Data set
4.978	14	410.2	28.0	105.7	0.4	2.0	Imai *et al.* (1975)
4.978	15	385.8	30.0	99.86	0.4	2.1	(continued)
4.978	16	370.0	32.0	96.24	0.4	1.7	
4.978	18	355.0	36.0	93.33	0.4	0.8	
4.978	20	350.0	40.0	93.11	0.4	0.3	
4.978	22.5	346.7	45.1	93.81	0.4	0.3	
4.978	25	344.3	50.1	94.98	0.4	0.3	
4.978	30	337.6	60.1	97.46	0.4	0.7	
4.978	35	323.3	70.1	98.68	0.4	1.2	
4.978	40	303.0	80.1	98.9	0.4	1.1	
4.978	45	282.8	90.1	100	0.4	2.0	
4.978	49.9	254.7	99.9	98.9	0.4		
4.978	54.9	226.8	109.9	98.68	0.4		
4.978	59.9	195.3	119.9	97.46	0.4		
4.978	64.9	160.9	129.9	94.98	0.4		
4.978	67.4	143.9	134.9	93.81	0.4		
4.978	70.0	127.7	140.0	93.11	0.4		
4.978	72.0	115.6	144.0	93.33	0.4		
4.978	74.0	106.4	148.0	96.24	0.4		
4.978	75.0	103.6	150.0	99.86	0.4		
4.978	76.0	102.5	152.0	105.7	0.4		
4.978	77.0	104.5	154.0	115.8	0.4		
4.978	78.0	110.8	156.0	132.9	0.4		
4.978	79.0	123.8	158.0	161.8	0.4		
4.978	80.0	148.3	160.0	212.9	0.4		
4.978	81.0	189.4	162.0	301.9	0.4		
4.978	82.0	257.8	164.0	461.9	0.4		
6.141	6	3896.2	12.0	976.3	3.1		Slobodrian *et al.* (1968)
6.141	6	3948.9	12.0	989.5	3.1		
6.141	7	2056.3	14.0	516.3	3.2		
6.141	7	2094.9	14.0	526	3.2		
6.141	8	1230.6	16.0	309.7	3.2		
6.141	8	1243.7	16.0	313	2.9		
6.141	9	844.1	18.0	213	2.8		
6.141	9	851.3	18.0	214.8	2.8		
6.141	10	603.8	20.0	152.8	0.7	2.8	
6.141	10	608.5	20.0	154	0.7	2.0	
6.141	11	479.3	22.0	121.7	0.6	0.0	
6.141	11	482.5	22.0	122.5	0.6	0.7	
6.141	12	405.8	24.0	103.4	0.5	1.6	
6.141	12	408.1	24.0	104	0.5	2.2	

Table A14.1. Cross-section data for proton–proton scattering in the laboratory and center-of-mass (CM) system (arbitrary excerpt of available data at proton energies that are useful for hydrogen analysis). Data taken from original sources are marked bold; all other values were calculated with relativistic equations and the assumption of a symmetric CM cross section about $\theta_{cm} = 90°$. Additionally, the second-to-last column lists the percentage deviation of the measured data from nonlinear regression with Eq. (A14.2) using the values in Table A14.3 (continued).

Energy E_{ln} (MeV)	θ (deg)	$d\sigma/d\Omega$ (mb/sr)	θ_{cm} (deg)	$d\sigma/d\Omega_{cm}$ (mb/sr)	Error (%)	Fit dev. (%)	Data set
6.141	14	333.7	28.0	**85.74**	0.5	2.0	Slobodrian et al. (1968)
6.141	14	335.6	28.0	**86.21**	0.5	2.5	(continued)
6.141	16	310.5	32.0	**80.52**	0.5	2.2	
6.141	16	312.2	32.0	**80.97**	0.5	2.7	
6.141	18	300.9	36.1	**78.89**	0.5	1.4	
6.141	18	302.4	36.1	**79.29**	0.5	1.9	
6.141	20	297.6	40.1	**78.99**	0.5	1.0	
6.141	20	299.2	40.1	**79.41**	0.5	1.5	
6.141	25	293.2	50.1	**80.7**	0.5	0.8	
6.141	25	294.5	50.1	**81.06**	0.5	1.2	
6.141	30	284.3	60.1	**81.94**	0.4	1.0	
6.141	30	285.2	60.1	**82.2**	0.4	1.3	
6.141	35	271.4	70.1	**82.74**	0.5	1.4	
6.141	35	272.7	70.1	**83.12**	0.5	1.8	
6.141	40	254.3	80.1	**82.95**	0.6	1.4	
6.141	40	256.0	80.1	**83.51**	0.6	2.1	
6.141	49.9	213.6	99.9	82.95	0.6		
6.141	49.9	215.0	99.9	83.51	0.6		
6.141	54.9	190.0	109.9	82.74	0.5		
6.141	54.9	190.9	109.9	83.12	0.5		
6.141	59.9	164.0	119.9	81.94	0.4		
6.141	59.9	164.5	119.9	82.2	0.4		
6.141	64.9	136.5	129.9	80.7	0.5		
6.141	64.9	137.1	129.9	81.06	0.5		
6.141	69.9	108.1	139.9	78.99	0.5		
6.141	69.9	108.7	139.9	79.41	0.5		
6.141	71.9	97.5	143.9	78.89	0.5		
6.141	71.9	98.0	143.9	79.29	0.5		
6.141	74.0	88.8	148.0	80.52	0.5		
6.141	74.0	89.3	148.0	80.97	0.5		
6.141	76.0	83.0	152.0	85.74	0.5		
6.141	76.0	83.4	152.0	86.21	0.5		
6.141	78.0	86.0	156.0	103.4	0.5		
6.141	78.0	86.5	156.0	104	0.5		
6.141	79.0	92.9	158.0	121.7	0.6		
6.141	79.0	93.5	158.0	122.5	0.6		
6.141	80.0	106.1	160.0	152.8	0.7		
6.141	80.0	107.0	160.0	154	0.7		
6.141	81.0	133.3	162.0	213	2.8		
6.141	81.0	134.4	162.0	214.8	2.8		
6.141	82.0	172.4	164.0	309.7	3.2		
6.141	82.0	174.3	164.0	313	2.9		

Table A14.1. Cross-section data for proton–proton scattering in the laboratory and center-of-mass (CM) system (arbitrary excerpt of available data at proton energies that are useful for hydrogen analysis). Data taken from original sources are marked bold; all other values were calculated with relativistic equations and the assumption of a symmetric CM cross section about $\theta_{cm} = 90°$. Additionally, the second-to-last column lists the percentage deviation of the measured data from nonlinear regression with Eq. (A14.2) using the values in Table A14.3 (continued).

Energy E_{ln} (MeV)	θ (deg)	$d\sigma/d\Omega$ (mb/sr)	θ_{cm} (deg)	$d\sigma/d\Omega_{cm}$ (mb/sr)	Error (%)	Fit dev. (%)	Data set
6.141	83.0	251.7	166.0	516.3	3.2		Slobodrian *et al.* (1968)
6.141	83.0	256.4	166.0	526	3.2		(continued)
6.141	84.0	408.2	168.0	976.3	3.1		
6.141	84.0	413.7	168.0	989.5	3.1		
6.968	**8**	986.2	16.0	**248.1**	**0.4**		Imai et al. (1975)
6.968	**9**	656.6	18.0	**165.6**	**0.4**		
6.968	**10**	487.0	20.0	**123.2**	**0.4**	4.6	
6.968	**11**	392.1	22.0	**99.52**	**0.4**	2.5	
6.968	**12**	336.8	24.0	**85.8**	**0.4**	1.3	
6.968	**13**	306.6	26.0	**78.41**	**0.4**	0.1	
6.968	**14**	288.2	28.0	**74.02**	**0.4**	0.3	
6.968	**15**	276.4	30.1	**71.31**	**0.4**	0.0	
6.968	**16**	269.0	32.1	**69.74**	**0.4**	0.5	
6.968	**18**	264.7	36.1	**69.38**	**0.4**	0.3	
6.968	**20**	261.5	40.1	**69.38**	**0.4**	1.0	
6.968	**22.5**	260.0	45.1	**70.16**	**0.4**	1.1	
6.968	**25**	258.3	50.1	**71.08**	**0.4**	0.8	
6.968	**30**	252.3	60.1	**72.71**	**0.4**	0.3	
6.968	**35**	242.4	70.1	**73.9**	**0.4**	1.4	
6.968	**40**	225.1	80.1	**73.43**	**0.4**	0.6	
6.968	**45**	209.6	90.1	**74.09**	**0.4**	1.4	
6.968	49.9	189.1	99.9	73.43	0.4		
6.968	54.9	169.8	109.9	73.9	0.4		
6.968	59.9	145.6	119.9	72.71	0.4		
6.968	64.9	120.2	129.9	71.08	0.4		
6.968	67.4	107.5	134.9	70.16	0.4		
6.968	69.9	95.0	139.9	69.38	0.4		
6.968	71.9	85.8	143.9	69.38	0.4		
6.968	73.9	76.9	147.9	69.74	0.4		
6.968	74.9	73.8	149.9	71.31	0.4		
6.968	76.0	71.6	152.0	74.02	0.4		
6.968	77.0	70.6	154.0	78.41	0.4		
6.968	78.0	71.4	156.0	85.8	0.4		
6.968	79.0	76.0	158.0	99.52	0.4		
6.968	80.0	85.6	160.0	123.2	0.4		
6.968	81.0	103.6	162.0	165.6	0.4		
6.968	82.0	138.1	164.0	248.1	0.4		

Table A14.1. Cross-section data for proton–proton scattering in the laboratory and center-of-mass (CM) system (arbitrary excerpt of available data at proton energies that are useful for hydrogen analysis). Data taken from original sources are marked bold; all other values were calculated with relativistic equations and the assumption of a symmetric CM cross section about $\theta_{cm} = 90°$. Additionally, the second-to-last column lists the percentage deviation of the measured data from nonlinear regression with Eq. (A14.2) using the values in Table A14.3 (continued).

Energy E_{lin} (MeV)	θ (deg)	$d\sigma/d\Omega$ (mb/sr)	θ_{cm} (deg)	$d\sigma/d\Omega_{cm}$ (mb/sr)	Error (%)	Fit dev. (%)	Data set
8.03	8	754.9	16.0	**189.8**	**0.4**		Imai et al. (1975)
8.03	9	516.5	18.0	**130.2**	**0.4**		
8.03	10	391.4	20.0	**98.95**	**0.4**	4.3	
8.03	11	320.4	22.0	**81.27**	**0.4**	2.8	
8.03	12	282.6	24.0	**71.95**	**0.4**	1.1	
8.03	13	259.7	26.1	**66.39**	**0.4**	0.5	
8.03	14	246.4	28.1	**63.25**	**0.4**	0.4	
8.03	15	239.7	30.1	**61.8**	**0.4**	0.1	
8.03	16	234.0	32.1	**60.65**	**0.4**	0.8	
8.03	18	230.4	36.1	**60.35**	**0.4**	1.1	
8.03	20	230.4	40.1	**61.09**	**0.4**	0.7	
8.03	22.5	228.5	45.1	**61.64**	**0.4**	1.1	
8.03	25	226.3	50.1	**62.26**	**0.4**	1.0	
8.03	30	221.1	60.1	**63.68**	**0.4**	0.1	
8.03	35	210.7	70.1	**64.22**	**0.4**	0.6	
8.03	40	197.0	80.1	**64.24**	**0.4**	0.4	
8.03	45	183.4	90.1	**64.85**	**0.4**	1.3	
8.03	49.9	165.5	99.9	64.24	0.4		
8.03	54.9	147.5	109.9	64.22	0.4		
8.03	59.9	127.5	119.9	63.68	0.4		
8.03	64.9	105.3	129.9	62.26	0.4		
8.03	67.4	94.4	134.9	61.64	0.4		
8.03	69.9	83.6	139.9	61.09	0.4		
8.03	71.9	74.6	143.9	60.35	0.4		
8.03	73.9	66.9	147.9	60.65	0.4		
8.03	74.9	64.0	149.9	61.8	0.4		
8.03	75.9	61.2	151.9	63.25	0.4		
8.03	76.9	59.8	153.9	66.39	0.4		
8.03	78.0	59.8	156.0	71.95	0.4		
8.03	79.0	62.0	158.0	81.27	0.4		
8.03	80.0	68.7	160.0	98.95	0.4		
8.03	81.0	81.5	162.0	130.2	0.4		
8.03	82.0	105.7	164.0	189.8	0.4		
8.97	6	2244.0	12.0	**564.1**	3.0		Slobodrian et al. (1968)
8.97	6	2271.5	12.0	**571**	3.0		
8.97	7	1265.7	14.0	**318.8**	2.0		
8.97	7	1278.8	14.0	**322.1**	2.0		
8.97	8	757.4	16.0	**191.2**	0.8		
8.97	8	764.9	16.0	**193.1**	0.8		
8.97	9	518.7	18.0	**131.3**	0.5		

Table A14.1. Cross-section data for proton–proton scattering in the laboratory and center-of-mass (CM) system (arbitrary excerpt of available data at proton energies that are useful for hydrogen analysis). Data taken from original sources are marked bold; all other values were calculated with relativistic equations and the assumption of a symmetric CM cross section about $\theta_{cm} = 90°$. Additionally, the second-to-last column lists the percentage deviation of the measured data from nonlinear regression with Eq. (A14.2) using the values in Table A14.3 (continued).

Energy E_{ln} (MeV)	θ (deg)	$d\sigma/d\Omega$ (mb/sr)	θ_{cm} (deg)	$d\sigma/d\Omega_{cm}$ (mb/sr)	Error (%)	Fit dev. (%)	Data set
8.97	9	521.9	18.0	**132.1**	0.6		Slobodrian *et al.* (1968)
8.97	10	388.4	20.0	**98.61**	0.7	3.4	(continued)
8.97	10	391.7	20.0	**99.43**	0.7	2.5	
8.97	11	322.1	22.1	**82.02**	0.6	0.7	
8.97	11	324.5	22.1	**82.64**	0.6	0.0	
8.97	12	284.9	24.1	**72.82**	0.7	1.2	
8.97	12	287.3	24.1	**73.44**	0.7	2.0	
8.97	14	246.1	28.1	**63.41**	0.6	0.7	
8.97	14	248.0	28.1	**63.91**	0.6	1.5	
8.97	16	237.1	32.1	**61.67**	0.7	1.7	
8.97	16	239.1	32.1	**62.19**	0.7	2.5	
8.97	18	231.4	36.1	**60.83**	0.6	0.5	
8.97	18	233.2	36.1	**61.31**	0.6	1.2	
8.97	20	229.6	40.1	**61.08**	0.6	0.0	
8.97	20	231.4	40.1	**61.55**	0.6	0.8	
8.97	25	226.3	50.1	**62.43**	0.5	0.0	
8.97	25	228.1	50.1	**62.92**	0.5	0.8	
8.97	30	219.1	60.1	**63.25**	0.6	0.2	
8.97	30	220.9	60.1	**63.77**	0.7	1.1	
8.97	35	208.1	70.1	**63.52**	0.7	0.3	
8.97	35	209.8	70.1	**64.03**	0.7	1.0	
8.97	40	195.9	80.1	**63.92**	0.8	0.7	
8.97	40	198.0	80.1	**64.61**	0.9	1.8	
8.97	45	183.6	90.1	**64.92**	0.8	2.2	
8.97	49.9	164.8	99.9	**63.92**	0.8		
8.97	49.9	166.6	99.9	64.61	0.9		
8.97	54.9	146.2	109.9	63.52	0.7		
8.97	54.9	147.4	109.9	64.03	0.7		
8.97	59.9	127.0	119.9	63.25	0.6		
8.97	59.9	128.0	119.9	63.77	0.7		
8.97	64.9	106.0	129.9	62.43	0.5		
8.97	64.9	106.8	129.9	62.92	0.5		
8.97	69.9	83.9	139.9	61.08	0.6		
8.97	69.9	84.6	139.9	61.55	0.6		
8.97	71.9	75.5	143.9	60.83	0.6		
8.97	71.9	76.1	143.9	61.31	0.6		
8.97	73.9	68.3	147.9	61.67	0.7		
8.97	73.9	68.9	147.9	62.19	0.7		
8.97	75.9	61.6	151.9	63.41	0.6		
8.97	75.9	62.1	151.9	63.91	0.6		
8.97	77.9	60.8	155.9	72.82	0.7		

Table A14.1. Cross-section data for proton–proton scattering in the laboratory and center-of-mass (CM) system (arbitrary excerpt of available data at proton energies that are useful for hydrogen analysis). Data taken from original sources are marked bold; all other values were calculated with relativistic equations and the assumption of a symmetric CM cross section about $\theta_{cm} = 90°$. Additionally, the second-to-last column lists the percentage deviation of the measured data from nonlinear regression with Eq. (A14.2) using the values in Table A14.3 (continued).

Energy E_{in} (MeV)	θ (deg)	$d\sigma/d\Omega$ (mb/sr)	θ_{cm} (deg)	$d\sigma/d\Omega_{cm}$ (mb/sr)	Error (%)	Fit dev. (%)	Data set
8.97	77.9	61.4	155.9	73.44	0.7		Slobodrian *et al.* (1968)
8.97	78.9	62.9	157.9	82.02	0.6		(continued)
8.97	78.9	63.4	157.9	82.64	0.6		
8.97	80.0	68.8	160.0	98.61	0.7		
8.97	80.0	69.4	160.0	99.43	0.7		
8.97	81.0	82.5	162.0	131.3	0.5		
8.97	81.0	83.0	162.0	132.1	0.6		
8.97	82.0	106.9	164.0	191.2	0.8		
8.97	82.0	108.0	164.0	193.1	0.8		
8.97	83.0	156.1	166.0	318.8	2.0		
8.97	83.0	157.8	166.0	322.1	2.0		
8.97	84.0	237.0	168.0	564.1	3.0		
8.97	84.0	239.9	168.0	571	3.0		
9.918	**10**	296.2	20.1	**74.83**	**0.8**	0.9	Jarmie et al. (1970)
9.918	**12.5**	212.6	25.1	**54.18**	**0.41**	1.1	
9.918	**15**	194.8	30.1	**50.2**	**0.39**	1.1	
9.918	**17.5**	191.7	35.1	**50.04**	**0.37**	0.9	
9.918	**20**	189.0	40.1	**50.09**	**0.69**	0.3	
9.918	**25**	185.9	50.1	**51.11**	**0.411**	0.5	
9.918	**30**	180.3	60.1	**51.91**	**0.339**	0.1	
9.918	**35**	172.4	70.1	**52.53**	**0.34**	1.0	
9.918	**40**	161.6	80.1	**52.69**	**0.34**	1.2	
9.918	**45**	148.4	90.2	**52.46**	**0.36**	0.7	
9.918	**50**	135.6	100.1	**52.78**	**0.36**		
9.918	54.9	120.7	109.9	52.53	0.34		
9.918	59.9	104.0	119.9	51.91	0.339		
9.918	64.9	86.5	129.9	51.11	0.411		
9.918	69.9	68.6	139.9	50.09	0.69		
9.918	72.4	60.2	144.9	50.04	0.37		
9.918	74.9	52.0	149.9	50.2	0.39		
9.918	77.4	46.9	154.9	54.18	0.41		
9.918	79.9	52.0	159.9	74.83	0.80		
9.918	**6**	1544.3	12.0	**386.2**	**3.86**		Slobodrian et al. (1968)
9.918	**6**	1559.9	12.0	**390.1**	**3.9**		
9.918	**7**	852.0	14.0	**213.5**	**2.29**		
9.918	**7**	860.0	14.0	**215.5**	**2.31**		
9.918	**8**	526.7	16.0	**132.3**	**0.91**		
9.918	**8**	531.9	16.0	**133.6**	**0.92**		

Table A14.1. Cross-section data for proton–proton scattering in the laboratory and center-of-mass (CM) system (arbitrary excerpt of available data at proton energies that are useful for hydrogen analysis). Data taken from original sources are marked bold; all other values were calculated with relativistic equations and the assumption of a symmetric CM cross section about $\theta_{cm} = 90°$. Additionally, the second-to-last column lists the percentage deviation of the measured data from nonlinear regression with Eq. (A14.2) using the values in Table A14.3 (continued).

Energy E_{in} (MeV)	θ (deg)	$d\sigma/d\Omega$ (mb/sr)	θ_{cm} (deg)	$d\sigma/d\Omega_{cm}$ (mb/sr)	Error (%)	Fit dev. (%)	Data set
9.918	9	373.8	18.0	**94.14**	**0.75**		Slobodrian et al. (1968)
9.918	9	377.6	18.0	**95.11**	**0.76**		(continued)
9.918	10	290.5	20.1	**73.39**	**0.61**	1.5	
9.918	10	293.7	20.1	**74.2**	**0.62**	0.4	
9.918	11	244.6	22.1	**61.99**	**0.54**	0.3	
9.918	11	247.4	22.1	**62.71**	**0.54**	0.9	
9.918	12	219.5	24.1	**55.83**	**0.4**	0.6	
9.918	12	221.5	24.1	**56.34**	**0.4**	1.5	
9.918	14	198.3	28.1	**50.86**	**0.28**	1.3	
9.918	14	199.5	28.1	**51.17**	**0.28**	1.9	
9.918	16	189.1	32.1	**48.97**	**0.27**	0.3	
9.918	16	190.4	32.1	**49.3**	**0.27**	0.3	
9.918	18	189.0	36.1	**49.48**	**0.22**	0.2	
9.918	18	190.0	36.1	**49.72**	**0.22**	0.7	
9.918	20	187.7	40.1	**49.74**	**0.33**	0.3	
9.918	20	189.2	40.1	**50.13**	**0.33**	0.4	
9.918	25	182.5	50.1	**50.16**	**0.39**	1.8	
9.918	25	184.2	50.1	**50.64**	**0.39**	0.8	
9.918	30	176.4	60.1	**50.8**	**0.43**	1.6	
9.918	30	178.4	60.1	**51.35**	**0.43**	0.5	
9.918	35	168.8	70.1	**51.43**	**0.47**	0.7	
9.918	35	170.8	70.1	**52.02**	**0.48**	0.4	
9.918	40	159.6	80.1	**52.05**	**0.5**	0.4	
9.918	40	161.6	80.1	**52.7**	**0.5**	1.6	
9.918	45	148.5	90.2	**52.49**	**0.56**	1.1	
9.918	45	150.5	90.2	**53.22**	**0.56**	2.5	
9.918	49.9	134.1	99.9	52.05	0.5		
9.918	49.9	135.8	99.9	52.7	0.5		
9.918	54.9	118.2	109.9	51.43	0.47		
9.918	54.9	119.6	109.9	52.02	0.48		
9.918	59.9	101.7	119.9	50.8	0.43		
9.918	59.9	102.8	119.9	51.35	0.43		
9.918	64.9	84.9	129.9	50.16	0.39		
9.918	64.9	85.7	129.9	50.64	0.39		
9.918	69.9	68.1	139.9	49.74	0.33		
9.918	69.9	68.6	139.9	50.13	0.33		
9.918	71.9	61.2	143.9	49.48	0.22		
9.918	71.9	61.5	143.9	49.72	0.22		
9.918	73.9	54.0	147.9	48.97	0.27		
9.918	73.9	54.4	147.9	49.3	0.27		
9.918	75.9	49.2	151.9	50.86	0.28		

Table A14.1. Cross-section data for proton–proton scattering in the laboratory and center-of-mass (CM) system (arbitrary excerpt of available data at proton energies that are useful for hydrogen analysis). Data taken from original sources are marked bold; all other values were calculated with relativistic equations and the assumption of a symmetric CM cross section about $\theta_{cm} = 90°$. Additionally, the second-to-last column lists the percentage deviation of the measured data from nonlinear regression with Eq. (A14.2) using the values in Table A14.3 (continued).

Energy E_{in} (MeV)	θ (deg)	$d\sigma/d\Omega$ (mb/sr)	θ_{cm} (deg)	$d\sigma/d\Omega_{cm}$ (mb/sr)	Error (%)	Fit dev. (%)	Data set
9.918	75.9	49.5	151.9	51.17	0.28		Slobodrian *et al.* (1968)
9.918	77.9	46.4	155.9	55.83	0.4		(continued)
9.918	77.9	46.9	155.9	56.34	0.4		
9.918	78.9	47.3	157.9	61.99	0.54		
9.918	78.9	47.9	157.9	62.71	0.54		
9.918	79.9	51.0	159.9	73.39	0.61		
9.918	79.9	51.5	159.9	74.2	0.62		
9.918	81.0	58.9	162.0	94.14	0.75		
9.918	81.0	59.5	162.0	95.11	0.76		
9.918	82.0	73.7	164.0	132.3	0.91		
9.918	82.0	74.4	164.0	133.6	0.92		
9.918	83.0	104.1	166.0	213.5	2.29		
9.918	83.0	105.1	166.0	215.5	2.31		
9.918	84.0	161.5	168.0	386.2	3.86		
9.918	84.0	163.1	168.0	390.1	3.9		
13.6	**10**	180.5	20.1	**45.5**	**1**	0.5	Jarmie *et al.* (1970)
13.6	**12.5**	141.9	25.1	**36.09**	**0.66**	0.5	
13.6	**15**	137.2	30.1	**35.28**	**0.61**	0.0	
13.6	**17.5**	135.4	35.1	**35.28**	**0.67**	1.4	
13.6	**20**	136.0	40.1	**35.99**	**0.45**	1.2	
13.6	**25**	134.0	50.2	**36.79**	**0.48**	1.2	
13.6	**30**	130.1	60.2	**37.43**	**0.61**	0.2	
13.6	**35**	123.1	70.2	**37.49**	**0.48**	0.3	
13.6	**40**	115.3	80.2	**37.58**	**0.48**	0.2	
13.6	**50**	96.2	100.2	**37.48**	**0.5**		
13.6	**55**	85.1	110.2	**37.2**	**0.51**		
13.6	59.8	75.0	119.8	37.43	0.61		
13.6	64.8	62.3	129.8	36.79	0.48		
13.6	69.9	49.3	139.9	35.99	0.45		
13.6	72.4	42.5	144.9	35.28	0.67		
13.6	74.9	36.5	149.9	35.28	0.61		
13.6	77.4	31.3	154.9	36.09	0.66		
13.6	79.9	31.6	159.9	45.5	1		
14.16	**9**	191.8	18.1	**48.2**	**9.793**		Kikuchi *et al.* (1960)
14.16	**10**	162.3	20.1	**40.9**	**5.4**	4.0	
14.16	**11**	147.5	22.1	**37.3**	**6.4**	0.7	
14.16	**12**	139.1	24.1	**35.3**	**4.6**	0.4	
14.16	**13**	133.8	26.1	**34.1**	**5.1**	0.0	

Table A14.1. Cross-section data for proton–proton scattering in the laboratory and center-of-mass (CM) system (arbitrary excerpt of available data at proton energies that are useful for hydrogen analysis). Data taken from original sources are marked bold; all other values were calculated with relativistic equations and the assumption of a symmetric CM cross section about $\theta_{cm} = 90°$. Additionally, the second-to-last column lists the percentage deviation of the measured data from nonlinear regression with Eq. (A14.2) using the values in Table A14.3 (continued).

Energy E_{in} (MeV)	θ (deg)	$d\sigma/d\Omega$ (mb/sr)	θ_{cm} (deg)	$d\sigma/d\Omega_{cm}$ (mb/sr)	Error (%)	Fit dev. (%)	Data set
14.16	**15**	129.9	30.1	**33.4**	**2**	1.0	Kikuchi *et al.* (1960)
14.16	**18**	129.4	36.1	**33.8**	**1.3**	1.8	(continued)
14.16	**20**	130.4	40.1	**34.5**	**0.9**	1.2	
14.16	**22.5**	129.7	45.2	**34.9**	**1.4**	1.4	
14.16	**25**	127.5	50.2	**35**	**1**	1.9	
14.16	**30**	122.7	60.2	**35.3**	**0.9**	1.9	
14.16	**35**	117.6	70.2	**35.8**	**1**	0.7	
14.16	**40**	110.1	80.2	**35.9**	**1.2**	0.4	
14.16	**45**	102.4	90.2	**36.2**	**1.2**	0.3	
14.16	**54**	83.7	108.2	**35.7**	**1.3**		
14.16	**55**	81.5	110.2	**35.6**	**1**		
14.16	**57**	76.9	114.2	**35.4**	**1.2**		
14.16	59.8	70.7	119.8	35.3	0.9		
14.16	64.8	59.2	129.8	35	1		
14.16	67.3	53.5	134.8	34.9	1.4		
14.16	69.9	47.2	139.9	34.5	0.9		
14.16	71.9	41.8	143.9	33.8	1.3		
14.16	74.9	34.6	149.9	33.4	2		
14.16	76.9	30.7	153.9	34.1	5.1		
14.16	77.9	29.4	155.9	35.3	4.6		
14.16	78.9	28.5	157.9	37.3	6.4		
14.16	79.9	28.4	159.9	40.9	5.4		
14.16	80.9	30.2	161.9	48.2	9.793		
18.2	14.9	97.4	**30**	**25**	**1**	2.1	Yntema and White, 1954
18.2	17.9	99.7	**36**	**25.98**	**1**	1.2	
18.2	19.9	100.4	**40**	**26.5**	**0.8**	0.7	
18.2	24.9	99.6	**50**	**27.27**	**0.7**	0.1	
18.2	29.9	95.6	**60**	**27.42**	**0.6**	0.0	
18.2	34.9	90.5	**70**	**27.47**	**0.5**	0.0	
18.2	39.9	83.9	**80**	**27.29**	**0.5**	0.6	
18.2	44.9	77.5	**90**	**27.32**	**0.5**	0.6	
18.2	39.9	83.9	80	27.29	0.5		
18.2	34.9	90.5	70	27.47	0.5		
18.2	29.9	95.6	60	27.42	0.6		
18.2	24.9	99.6	50	27.27	0.7		
18.2	19.9	100.4	40	26.5	0.8		
18.2	17.9	99.7	36	25.98	1		
18.2	14.9	97.4	30	25	1		

Table A14.1. Cross-section data for proton–proton scattering in the laboratory and center-of-mass (CM) system (arbitrary excerpt of available data at proton energies that are useful for hydrogen analysis). Data taken from original sources are marked bold; all other values were calculated with relativistic equations and the assumption of a symmetric CM cross section about $\theta_{cm} = 90°$. Additionally, the second-to-last column lists the percentage deviation of the measured data from nonlinear regression with Eq. (A14.2) using the values in Table A14.3 (continued).

Energy E_{in} (MeV)	θ (deg)	$d\sigma/d\Omega$ (mb/sr)	θ_{cm} (deg)	$d\sigma/d\Omega_{cm}$ (mb/sr)	Error (%)	Fit dev. (%)	Data set
19.8	7.0	239.5	14	59.7	3.35		Burkig et al. (1959)
19.8	8.0	152.5	16	38.1	1.575		
19.8	9.0	118.9	18	29.8	1.678		
19.8	9.9	103.8	20	26.1	1.533	1.7	
19.8	10.9	96.4	22	24.3	1.646	2.2	
19.8	11.9	92.5	24	23.4	1.282	1.6	
19.8	12.9	88.9	26	22.6	1.327	1.3	
19.8	14.9	92.0	30	23.6	1.059	1.6	
19.8	17.9	91.0	36	23.7	1.055	1.2	
19.8	19.9	89.9	40	23.7	1.266	2.8	
19.8	24.9	90.6	50	24.8	2.419	0.1	
19.8	29.9	83.7	60	24	1.25	4.1	
19.8	34.9	81.4	70	24.7	1.619	1.3	
19.8	39.9	75.1	80	24.4	2.459	2.5	
19.8	44.8	69.8	90	24.6	1.22	1.7	
19.8	49.9	62.8	100	24.4	2.459		
19.8	54.9	56.7	110	24.7	1.619		
19.8	59.9	47.9	120	24	1.25		
19.8	64.9	41.8	130	24.8	2.419		
19.8	69.9	32.3	140	23.7	1.266		
19.8	71.9	29.2	144	23.7	1.055		
19.8	74.9	24.3	150	23.6	1.059		
19.8	76.9	20.2	154	22.6	1.327		
19.8	77.9	19.4	156	23.4	1.282		
19.8	78.9	18.5	158	24.3	1.646		
19.8	79.9	18.0	160	26.1	1.533		
19.8	81.0	18.6	162	29.8	1.678		
19.8	82.0	21.1	164	38.1	1.575		
19.8	83.0	29.0	166	59.7	3.35		
25.63	5	442.60	10.1	109.6	1.8		Jeong et al. (1960)
25.63	6	227.00	12.1	56.31	1.2		
25.63	7	133.56	14.1	33.2	0.9		
25.63	8	95.35	16.1	23.76	0.8		
25.63	9	79.64	18.1	19.9	0.8		
25.63	9.5	74.73	19.1	18.7	0.8		
25.63	10	71.74	20.1	17.98	0.8	2.6	
25.63	11	68.91	22.1	17.33	0.8	2.6	
25.63	12	67.70	24.2	17.09	0.8	1.7	
25.63	12.5	67.84	25.2	17.16	0.8	1.8	
25.63	13	67.74	26.2	17.17	0.8	1.3	

Table A14.1. Cross-section data for proton–proton scattering in the laboratory and center-of-mass (CM) system (arbitrary excerpt of available data at proton energies that are useful for hydrogen analysis). Data taken from original sources are marked bold; all other values were calculated with relativistic equations and the assumption of a symmetric CM cross section about $\theta_{cm} = 90°$. Additionally, the second-to-last column lists the percentage deviation of the measured data from nonlinear regression with Eq. (A14.2) using the values in Table A14.3 (continued).

Energy E_{in} (MeV)	θ (deg)	$d\sigma/d\Omega$ (mb/sr)	θ_{cm} (deg)	$d\sigma/d\Omega_{cm}$ (mb/sr)	Error (%)	Fit dev. (%)	Data set
25.63	**14**	67.95	28.2	**17.3**	**0.8**	0.6	Jeong *et al.* (1960)
25.63	**15**	68.14	30.2	**17.43**	**0.8**	0.1	(continued)
25.63	**16**	68.77	32.2	**17.68**	**0.8**	0.0	
25.63	**17**	68.86	34.2	**17.8**	**0.8**	0.5	
25.63	**18**	68.96	36.2	**17.93**	**0.8**	0.8	
25.63	**20**	69.12	40.3	**18.2**	**0.8**	0.7	
25.63	**22**	68.65	44.3	**18.33**	**0.8**	0.8	
25.63	**25**	67.73	50.3	**18.52**	**0.8**	0.5	
25.63	**30**	64.73	60.3	**18.56**	**0.8**	0.6	
25.63	**35**	61.39	70.4	**18.65**	**0.8**	0.2	
25.63	**40**	57.13	80.4	**18.6**	**0.8**	0.4	
25.63	**45**	52.58	90.4	**18.59**	**0.8**	0.5	
25.63	49.6	48.10	99.6	18.6	0.8		
25.63	54.6	42.98	109.6	18.65	0.8		
25.63	59.7	37.25	119.7	18.56	0.8		
25.63	64.7	31.39	129.7	18.52	0.8		
25.63	67.7	27.52	135.7	18.33	0.8		
25.63	69.7	24.94	139.7	18.2	0.8		
25.63	71.8	22.19	143.8	17.93	0.8		
25.63	72.8	20.84	145.8	17.8	0.8		
25.63	73.8	19.51	147.8	17.68	0.8		
25.63	74.8	18.06	149.8	17.43	0.8		
25.63	75.8	16.75	151.8	17.3	0.8		
25.63	76.8	15.46	153.8	17.17	0.8		
25.63	77.3	14.87	154.8	17.16	0.8		
25.63	77.8	14.22	155.8	17.09	0.8		
25.63	78.9	13.23	157.9	17.33	0.8		
25.63	79.9	12.49	159.9	17.98	0.8		
25.63	80.4	12.35	160.9	18.7	0.8		
25.63	80.9	12.46	161.9	19.9	0.75		
25.63	81.9	13.23	163.9	23.76	0.8		
25.63	82.9	16.19	165.9	33.2	0.9		
25.63	83.9	23.55	167.9	56.31	1.2		
25.63	84.9	38.21	169.9	109.6	1.8		
29.4	11.9	52.52	**24**	**13.23**	3	7.8	Panofsky and Fillmore (1950)
29.4	15.9	54.66	**32**	**14.02**	3	8.2	
29.4	19.9	57.72	**40**	**15.16**	3	3.5	
29.4	23.8	57.82	**48**	**15.64**	3	1.6	
29.4	27.8	59.60	**56**	**16.7**	3	4.5	
29.4	31.8	55.79	**64**	**16.3**	3	2.1	

Table A14.1. Cross-section data for proton–proton scattering in the laboratory and center-of-mass (CM) system (arbitrary excerpt of available data at proton energies that are useful for hydrogen analysis). Data taken from original sources are marked bold; all other values were calculated with relativistic equations and the assumption of a symmetric CM cross section about $\theta_{cm} = 90°$. Additionally, the second-to-last column lists the percentage deviation of the measured data from nonlinear regression with Eq. (A14.2) using the values in Table A14.3 (continued).

Energy E_{in} (MeV)	θ (deg)	$d\sigma/d\Omega$ (mb/sr)	θ_{cm} (deg)	$d\sigma/d\Omega_{cm}$ (mb/sr)	Error (%)	Fit dev. (%)	Data set
29.4	35.8	53.70	72	16.47	3	3.2	Panofsky and Fillmore (1950)
29.4	39.8	50.49	80	16.38	3	2.6	(continued)
29.4	43.4	46.50	87.33	16	3	0.3	
29.4	46.1	44.34	92.66	16	3		
29.4	49.8	42.20	100	16.38	3		
29.4	53.8	38.74	108	16.47	3		
29.4	57.8	34.51	116	16.3	3		
29.4	61.8	31.28	124	16.7	3		
29.4	65.8	25.35	132	15.64	3		
29.4	69.9	20.64	140	15.16	3		
29.4	73.9	15.37	148	14.02	3		
29.4	77.9	10.93	156	13.23	3		
30.14	5.5	101.97	11	25.22	2.68		Fillmore (1951)
30.14	7.9	58.47	16	14.54	2.68		
30.14	11.9	50.89	24	12.82	2.68	8.0	
30.14	15.9	50.99	32	13.08	2.68	12.8	
30.14	19.9	55.74	40	14.64	2.68	4.2	
30.14	23.8	56.09	48	15.17	2.68	1.8	
30.14	27.8	53.00	56	14.85	2.68	4.3	
30.14	31.8	49.70	64	14.52	2.68	6.7	
30.14	35.8	50.86	72	15.6	2.68	0.7	
30.14	39.8	47.44	80	15.39	2.68	0.7	
30.14	43.3	43.57	87	14.95	2.68	3.6	
30.14	46.3	41.30	93	14.95	2.68		
30.14	49.8	39.65	100	15.39	2.68		
30.14	53.8	36.69	108	15.6	2.68		
30.14	57.8	31.44	116	14.85	2.68		
30.14	61.8	27.81	124	14.85	2.68		
30.14	65.8	24.58	132	15.17	2.68		
30.14	69.9	19.93	140	14.64	2.68		
30.14	73.9	14.34	148	13.08	2.68		
30.14	77.9	10.59	156	12.82	2.68		
30.14	81.9	8.04	164	14.54	2.68		
30.14	84.5	9.60	169	25.22	2.68		
39.4	4	422.80	8.1	103.8	4.1		Johnston and Swenson (1958)
39.4	5	166.14	10.1	40.85	1.8		
39.4	6	83.75	12.1	20.63	1.1		
39.4	7	54.69	14.1	13.5	0.9		

Table A14.1. Cross-section data for proton–proton scattering in the laboratory and center-of-mass (CM) system (arbitrary excerpt available data at proton energies that are useful for hydrogen analysis). Data taken from original sources are marked bold; all other values were calculated with relativistic equations and the assumption of a symmetric CM cross section about $\theta_{cm} = 90°$. Additionally the second-to-last column lists the percentage deviation of the measured data from nonlinear regression with Eq. (A14.2) using the values in Table A14.3 (continued).

Energy E_{in} (MeV)	θ (deg)	$d\sigma/d\Omega$ (mb/sr)	θ_{cm} (deg)	$d\sigma/d\Omega_{cm}$ (mb/sr)	Error (%)	Fit dev. (%)	Data set
39.4	8	43.93	16.2	10.87	0.8		Johnston and Swenson (1958)
39.4	8.5	41.40	17.2	10.26	0.8		(continued)
39.4	9	40.34	18.2	10.01	0.8		
39.4	9.5	40.15	19.2	9.98	0.8		
39.4	10	39.32	20.2	9.79	0.8	0.4	
39.4	10.5	39.38	21.2	9.82	0.8	0.5	
39.4	11	39.43	22.2	9.85	0.8	1.0	
39.4	11.5	39.67	23.2	9.93	0.8	1.3	
39.4	12	39.64	24.2	9.94	0.8	2.4	
39.4	12.5	40.07	25.3	10.07	0.8	2.3	
39.4	13.5	40.69	27.3	10.27	0.8	2.6	
39.4	15	41.38	30.3	10.52	0.8	2.8	
39.4	18	41.59	36.4	10.75	0.8	3.6	
39.4	20	41.47	40.4	10.86	0.8	3.5	
39.4	22	41.33	44.4	10.98	0.8	2.8	
39.4	25	40.78	50.5	11.1	0.8	2.0	
39.4	28	39.77	56.5	11.13	0.8	1.7	
39.4	30	39.06	60.5	11.16	0.8	1.4	
39.4	32	38.27	64.5	11.18	0.8	1.2	
39.4	35	36.86	70.6	11.17	0.8	1.3	
39.4	38	35.41	76.6	11.18	0.8	1.2	
39.4	40	34.32	80.6	11.16	0.8	1.3	
39.4	45	31.56	90.6	11.16	0.8	1.3	
39.4	49.4	28.95	99.4	11.16	0.8		
39.4	51.4	27.76	103.4	11.18	0.8		
39.4	54.4	25.81	109.4	11.17	0.8		
39.4	57.5	23.84	115.5	11.18	0.8		
39.4	59.5	22.44	119.5	11.16	0.8		
39.4	61.5	21.00	123.5	11.13	0.8		
39.4	64.5	18.84	129.5	11.1	0.8		
39.4	67.6	16.50	135.6	10.98	0.8		
39.4	69.6	14.89	139.6	10.86	0.8		
39.4	71.6	13.32	143.6	10.75	0.8		
39.4	74.7	10.91	149.7	10.52	0.8		
39.4	76.2	9.60	152.7	10.27	0.8		
39.4	77.2	8.73	154.7	10.07	0.8		
39.4	77.8	8.27	155.8	9.94	0.8		
39.4	78.3	7.93	156.8	9.93	0.8		
39.4	78.8	7.52	157.8	9.85	0.8		
39.4	79.3	7.16	158.8	9.82	0.8		
39.4	79.8	6.80	159.8	9.79	0.8		

Table A14.1. Cross-section data for proton–proton scattering in the laboratory and center-of-mass (CM) system (arbitrary excerpt of available data at proton energies that are useful for hydrogen analysis). Data taken from original sources are marked bold; all other values were calculated with relativistic equations and the assumption of a symmetric CM cross section about $\theta_{cm} = 90°$. Additionally, the second-to-last column lists the percentage deviation of the measured data from nonlinear regression with Eq. (A14.2) using the values in Table A14.3 (continued).

Energy E_{in} (MeV)	θ (deg)	$d\sigma/d\Omega$ (mb/sr)	θ_{cm} (deg)	$d\sigma/d\Omega_{cm}$ (mb/sr)	Error (%)	Fit dev. (%)	Data set
39.4	80.3	6.59	160.8	9.98	0.8		Johnston and Swenson (1958)
39.4	80.8	6.27	161.8	10.01	0.8		(continued)
39.4	81.3	6.07	162.8	10.26	0.8		
39.4	81.8	6.05	163.8	10.87	0.8		
39.4	82.9	6.58	165.9	13.5	0.9		
39.4	83.9	8.63	167.9	20.63	1.1		
39.4	84.9	14.24	169.9	40.85	1.8		
39.4	85.9	28.97	171.9	103.8	4.1		
50.06	6.9	37.58	**14**	**9.225**	0.3		Berdoz et al. (1986)
50.06	7.9	31.74	**16**	**7.81**	0.3		
50.06	8.9	30.17	**18**	**7.444**	0.3		
50.06	9.9	30.18	**20**	**7.472**	0.3	2.5	
50.06	10.9	30.72	**22**	**7.632**	0.3	1.5	
50.06	12.3	31.63	**25**	**7.903**	0.3	0.5	
50.06	13.8	32.41	**28**	**8.151**	0.3	0.4	
50.06	14.8	32.83	**30**	**8.299**	0.3	0.7	
50.06	15.8	33.23	**32**	**8.442**	0.3	1.3	
50.06	17.3	33.42	**35**	**8.564**	0.3	1.7	
50.06	18.8	33.57	**38**	**8.681**	0.3	2.4	
50.06	19.8	33.47	**40**	**8.714**	0.3	2.5	
50.06	20.7	33.37	**42**	**8.747**	0.3	2.8	
50.06	22.2	33.09	**45**	**8.771**	0.3	2.9	
50.06	23.7	32.75	**48**	**8.787**	0.3	3.1	
50.06	24.7	32.57	**50**	**8.812**	0.3	3.3	
50.06	27.2	31.71	**55**	**8.779**	0.3	3.0	
50.06	29.7	30.92	**60**	**8.778**	0.3	3.1	
50.06	32.2	29.85	**65**	**8.716**	0.3	2.4	
50.06	34.6	28.96	**70**	**8.719**	0.3	2.5	
50.06	37.1	27.92	**75**	**8.695**	0.3	2.3	
50.06	39.6	26.81	**80**	**8.66**	0.3	1.9	
50.06	42.1	25.81	**85**	**8.677**	0.3	2.1	
50.06	44.6	24.66	**90**	**8.661**	0.3	1.9	
50.06	47.1	23.57	95	8.677	0.3		
50.06	49.6	22.34	100	8.66	0.3		
50.06	52.1	21.21	105	8.695	0.3		
50.06	54.6	20.00	110	8.719	0.3		
50.06	57.2	18.70	115	8.716	0.3		
50.06	59.7	17.50	120	8.778	0.3		
50.06	62.2	16.14	125	8.779	0.3		
50.06	64.7	14.81	130	8.812	0.3		

Table A14.1. Cross-section data for proton–proton scattering in the laboratory and center-of-mass (CM) system (arbitrary excerpt o available data at proton energies that are useful for hydrogen analysis). Data taken from original sources are marked bold; all othe values were calculated with relativistic equations and the assumption of a symmetric CM cross section about $\theta_{cm} = 90°$. Additionally the second-to-last column lists the percentage deviation of the measured data from nonlinear regression with Eq. (A14.2) using th values in Table A14.3 (continued).

Energy E_{in} (MeV)	θ (deg)	$d\sigma/d\Omega$ (mb/sr)	θ_{cm} (deg)	$d\sigma/d\Omega_{cm}$ (mb/sr)	Error (%)	Fit dev. (%)	Data set
50.06	65.7	14.20	132	8.787	0.3		Berdoz *et al.* (1986)
50.06	67.2	13.33	135	8.771	0.3		(continued)
50.06	68.7	12.44	138	8.747	0.3		
50.06	69.8	11.82	140	8.714	0.3		
50.06	70.8	11.20	142	8.681	0.3		
50.06	72.3	10.20	145	8.564	0.3		
50.06	73.8	9.21	148	8.442	0.3		
50.06	74.8	8.50	150	8.299	0.3		
50.06	75.8	7.80	152	8.151	0.3		
50.06	77.3	6.77	155	7.903	0.3		
50.06	78.9	5.76	158	7.632	0.3		
50.06	79.9	5.13	160	7.472	0.3		
50.06	80.9	4.60	162	7.444	0.3		
50.06	81.9	4.29	164	7.81	0.3		
50.06	82.9	4.44	166	9.225	0.3		

A14.2 SCATTERING CROSS SECTION AS A FUNCTION OF ANGLE AND ENERGY

For quantification purposes, it is useful to have the scattering cross section as a function of both scattering angle and incident kinetic energy, thus implying a two-dimensional dependence. The scattering cross-section analysis is usually done by addition of the contributions of scattering amplitudes from Coulomb scattering (i.e., Mott scattering), S- and P-wave nuclear scattering, and also vacuum polarization (see, e.g., Knecht *et al.*, 1959; Bergervoet *et al.*, 1988). For application of proton–proton scattering to the quantification of hydrogen, v derived a simple regression formula here without physic analysis of the scattering amplitude and with residuals in tl range of a few percent of the data. This might be sufficie accuracy for most quantification purposes. To this end, v analyzed the difference between the measured data and the Mot scattering cross section. This is given in the CM system fe particles with spin s, mass m [given in atomic mass units (amu) and charge Ze (where e is the electronic charge) (see, e.g Bromley *et al.*, 1961) by

$$\frac{d\sigma_{Mott}}{d\Omega_{cm}}(\theta_{cm}) = A\left[\sin^{-4}\frac{\theta_{cm}}{2} + \cos^{-4}\frac{\theta_{cm}}{2} + \frac{(-1)^{2s}}{2s+1}\frac{2}{\sin^2(\theta_{cm}/2)\cos^2(\theta_{cm}/2)}\Phi(\theta_{cm})\right].$$

(A14.1)

In this expression,

$$\Phi(\theta_{cm}) = \cos\left[\frac{Z^2e^2}{\hbar v}\ln\left(\tan^2\frac{\theta_{cm}}{2}\right)\right]$$

$$A = \left(\frac{Z^2e^2}{2m^*v^2}\right)^2 \approx (1.296 \text{ mb})\times\left(\frac{Z^2}{E_{cm}[\text{MeV}]}\right)^2, \text{ and} \qquad \approx \cos\left[0.1575\,Z^2\sqrt{\frac{m[\text{amu}]}{2E_{cm}[\text{MeV}]}}\ln\left(\tan^2\frac{\theta_{cm}}{2}\right)\right],$$

with the reduced mass $m^* = m/2$, the relative velocity v, scattering angle θ, and kinetic energy E_{cm} [Eq. (9.28), Chapter 9], each for the CM system. For protons, of course, one has $Z = 1$ and $s = \frac{1}{2}$. The approximations are given for the nonrelativistic case only!

With the restriction to scattering angles of $\theta \geq 10°$ ($\theta_{cm} \geq 20°$), then the pp-scattering cross-section data are fitted well with a simple exponential regression, so that the scattering cross section in the CM system can be calculated as

$$\frac{d\sigma}{d\Omega_{cm}}(\theta_{cm}) = \frac{d\sigma_{Mott}}{d\Omega_{cm}}(\theta_{cm}) + A\{1 - \exp[C(B + \theta_{cm})]\} , \quad (A14.2)$$

where $d\sigma_{Mott}/d\sigma_{cm}$ is a function of E_{in} and θ_{cm} from Eq. (A14.1) and A, B, and C are the least-squares fit parameters evaluated for each data set at a certain energy, E_{in}, as given in Table A14.2. This simple fit gives a good approximation; however, at small angles close to the fit cutoff at $\theta_{cm} = 20°$, the residuals reach up to a few percent of the data, although the stated errors of the data might be smaller. At a further step, we fitted the parameters A, B, and C for all data sets as a function of kinetic energy in the laboratory system, E_{in}. We suggest the following fit functions

$$A(E_{in}) = (a_0 + a_1 E_{in})^{-a_2} , \quad (A14.3)$$

$$B(E_{in}) = b_0 + a_1 e^{b_2 E_{in}} , \quad (A14.4)$$

$$C(E_{in}) = c_0 + c_1 E_{in} . \quad (A14.5)$$

These functions are plotted in Fig. A14.1, with the parameters and statistical results as reported in Table A14.3. Note that we masked the value for $E_{in} = 29.4$ MeV for the regression analysis of parameter $C(E_{in})$. Finally, this gives the pp-scattering cross section in the CM system with Eq. (A14.2) as a function of the CM scattering angle $\theta_{cm} = 20–90°$ and kinetic energy $E_{in} = 1–50$ MeV. However, note that the accuracy of this fit is limited. The deviations from the measured data are calculated in Table A14.1 (residuum as a percentage of the data value). For energies below 3 MeV, the fit does not represent the data well, but at higher energies, the deviations exceed 3% for only a few single data points, in particular, for small scattering angles. However, for hydrogen analysis, angles of $\theta_{cm} > 50°$ are commonly used. Here, the data are fitted with deviations of less than 1%, which is in the range of the reported errors and, hence, is suitable for quantification purposes. Also, note that the fit function is to be used only for $\theta_{cm} \leq 90°$, but the scattering cross section is symmetrical with respect to 90° in the CM system anyway.

251

Table A14.2. Fit parameters for nonlinear regression with Eq. (A14.2). Uncertainties are taken from standard least-squares fit routine (Origin 8). χ_{red}^2 is the reduced chi-squared value, and R_{corr}^2 is the R-squared value corrected for the degrees of freedom.

Data set	Energy E_{in} (MeV)	$A \pm \delta A$ (mb/sr)	$B \pm \delta B$ (deg)	$C \pm \delta C$ (1/deg)	χ_{red}^2	R_{corr}^2
Knecht *et al.* (1959)	1.397	153 ± 4	-38.45 ± 0.22	-0.077 ± 0.003	150.97	0.997
Worthington *et al.* (1953)	1.855	160 ± 3	-33.01 ± 0.24	-0.084 ± 0.004	136.23	0.995
Worthington *et al.* (1953)	2.425	152 ± 2	-28.71 ± 0.19	-0.087 ± 0.004	96.96	0.996
Worthington *et al.* (1953)	3.037	137.7 ± 1.4	-26.15 ± 0.25	-0.086 ± 0.003	46.46	0.995
Worthington *et al.* (1953)	3.527	126.2 ± 1.5	-24.2 ± 0.4	-0.088 ± 0.004	66.00	0.993
Worthington *et al.* (1953)	3.899	118.4 ± 1.1	-23.1 ± 0.3	-0.087 ± 0.003	32.03	0.995
Worthington *et al.* (1953)	4.203	112.8 ± 1.0	-21.9 ± 0.5	-0.083 ± 0.004	18.43	0.995
Slobodrian *et al.* (1976)	4.21	113.58 ± 0.25	-18.5 ± 0.5	-0.069 ± 0.002	0.00	0.998
Imai *et al.* (1975)	4.978	97.2 ± 0.9	-21.22 ± 0.16	-0.096 ± 0.003	19.01	0.996
Slobodrian *et al.* (1968)	6.141	82.2 ± 0.3	-18.88 ± 0.12	-0.093 ± 0.002	2.81	0.998
Imai *et al.* (1975)	6.968	72.9 ± 0.5	-18.39 ± 0.20	-0.094 ± 0.003	8.63	0.997
Imai *et al.* (1975)	8.03	63.2 ± 0.6	-18.24 ± 0.18	-0.105 ± 0.004	18.88	0.995
Slobodrian *et al.* (1968)	8.097	64.09 ± 0.19	-12.34 ± 0.22	-0.081 ± 0.002	0.87	0.998
Jarmie *et al.* (1970)	9.918	52.29 ± 0.14	-15.5 ± 0.3	-0.093 ± 0.003	2.10	0.999
Slobodrian *et al.* (1968)	9.918	51.59 ± 0.24	-15.62 ± 0.24	-0.095 ± 0.002	6.95	0.996
Jarmie *et al.* (1970)	13.6	37.43 ± 0.10	-14.72 ± 0.25	-0.097 ± 0.002	0.80	0.999
Kikuchi *et al.* (1960)	14.16	35.67 ± 0.13	-15.2 ± 0.6	-0.100 ± 0.004	0.39	0.989
Yntema and White (1954)	18.2	27.32 ± 0.05	-16.6 ± 0.9	-0.116 ± 0.006	0.35	0.996
Burkig *et al.* (1959)	19.8	24.26 ± 0.16	-15.20 ± 0.17	-0.125 ± 0.004	1.04	0.996
Jeong *et al.* (1960)	25.63	18.39 ± 0.08	-15.06 ± 0.08	-0.133 ± 0.003	1.96	0.998
Panofsky and Fillmore (1950)	29.4	16.42 ± 0.17	-13.2 ± 1.4	-0.085 ± 0.009	0.41	0.985
Fillmore (1951)	30.1	14.98 ± 0.23	-16.72 ± 0.23	-0.129 ± 0.010	1.83	0.990
Johnston and Swenson (1958)	39.4	11.08 ± 0.03	-14.79 ± 0.10	-0.142 ± 0.003	1.39	0.999
Berdoz *et al.* (1986)	50.06	8.69 ± 0.02	-14.77 ± 0.04	-0.159 ± 0.002	10.41	0.999

Table A14.3. Fit parameters for regression analysis of parameters A, B, and C with Eqs. (A14.3)–(A14.5).

Parameter	Value	χ_{red}^2	R_{corr}^2
a_0 (mb/sr)	365 ± 23		
a_1 (1/MeV)	0.37 ± 0.04	32.6	0.9988
a_2	1.267 ± 0.024		
b_0 (deg)	-14.92 ± 0.14		
b_1 (deg)	-36.6 ± 1.9	17.8	0.9838
b_2 (1/MeV)	-0.368 ± 0.020		
c_0 (1/deg)	$(-78.6 \pm 2.4) \times 10^{-3}$	8.4	0.888
c_1 [1/(deg MeV)]	$(1.68 \pm 0.13) \times 10^{-3}$		

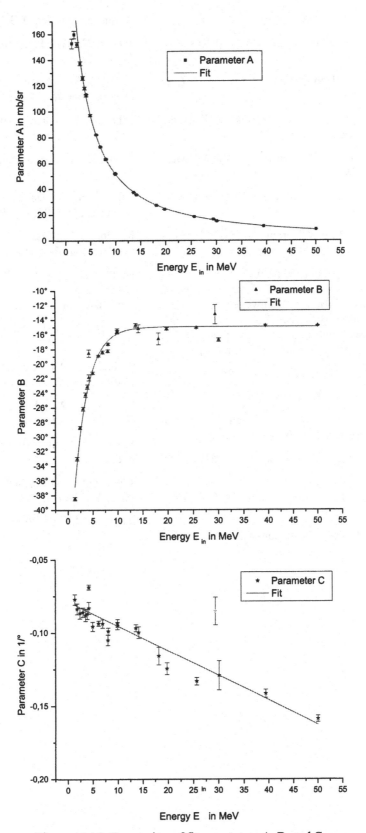

Figure A14.1. Regression of fit parameters A, B, and C.

REFERENCES

Allred, J.C., Armstrong, A.H., Bondelid, R.O., and Rosen, L. (1952), *Phys. Rev.* **88** (3), 433.

Berdoz, A., Foroughi, F., and Nussbaum, C. (1986), *J. Phys.* **G12** (6), L133.

Bergervoet, J.R., van Campen, P.C., van der Sanden, W.A., and de Swart, J.J. (1988), *Phys. Rev.* **C38** (1), 15.

Blair, J.M., Freier, G., Lampi, E.E., Sleator, W., and Williams, J.H. (1948), *Phys. Rev.* **74** (5), 553.

Bromley, D.A., Kuehner, J.A., and Almqvist, E. (1961), *Phys. Rev.* **123** (3), 878.

Burkig, J.W., Richardson, J.R., and Schrank, G.E. (1959), *Phys. Rev.* **113** (1), 290.

Bystricky, J., Carlson, P., Lechanoine, C., Lehar, F., Monnig, F., and Schubert, K.R. (1980), *Elastic and Charge Exchange Scattering of Elementary Particles* (Hellwege, K.-H., and Schopper, H., eds.), Landolt–Börnstein Tables (New Series, Group I), Numerical Data and Functional Relationships in Science and Technology, Springer, Berlin, vol. 9a, pp. 110ff.

Dayton, I.E., and Schrank, G. (1956), *Phys. Rev.* **101** (4), 1358.

Fillmore, F.L. (1951), *Phys. Rev.* **83**, 1252.

IAEA Nuclear Data Services (2009), International Atomic Energy Agency, Vienna, Austria; http://www-nds.iaea.org.

Imai, K., Nisimura, K., Tamura, N., and Sato, H. (1975), *Nucl. Phys.* **A246** (1), 76–92.

Jarmie, N., Jett, J.H., Detch, J.L., and Hutson, R.L. (1970), *Phys. Rev. Lett.* **25** (1), 34.

Jarmie, N., Jett, J.H., Detch, J.L., and Hutson, R.L. (1971), *Phys. Rev.* **C3** (1), 10.

Jarmie, N., and Jett, J.H. (1974), *Phys. Rev.* **C10** (1), 54.

Jarmie, N., Jett, J.H., and Semper, R.J. (1974), *Phys. Rev.* (5), 1748.

Jeong, T.H., Johnston, L.H., Young, D.E., and Waddel (1960), *Phys. Rev.* **118** (4), 1080.

Johnston, L.H., and Swenson, D.A. (1958), *Phys. Rev.* 212.

Johnston, L.H., and Young, D.E. (1959), *Phys. Rev.* 989.

Kikuchi, S., Sanada, J., Suwa, S., Hayashi, I., Nisimura, Fukunaga, K. (1960), *J. Phys. Soc. Jpn.* **15** (1), 9.

Knecht, D.J., Messelt, S., Berners, E.D., and Northclif (1959), *Phys. Rev.* **114** (2), 550.

NNDC Databases (2009), National Nuclear Data Brookhaven National Laboratory, Upton, http://www.nndc.bnl.gov.

Panofsky, W.K.H., and Fillmore, F.L. (1950), *Phys. Rev.* 57.

Slobodrian, R.J., Conzett, H.E., Shield, E., and Tivo (1968), *Phys. Rev.* **174** (4), 1122.

Slobodrian, R.J., Frois, B., Birchall, J., and Roy, R. *Nucl. Instrum. Methods* **136** (3), 525.

Wassmer, H., and Muhry, H. (1973), *Helv. Phys. Acta* 626.

Worthington, H.R., McGruer, J.N., and Findley, D.E. *Phys. Rev.* **90** (5), 899.

Yntema, J.L., and White, M.G. (1954), *Phys. Rev.* **95** (5),

APPENDIX
15

ACTIVATION ANALYSIS DATA

Compiled by

G. Blondiaux and J. L. Debrun

CNRS-CERI, Orleans, France

C. J. Maggiore

Los Alamos National Laboratory, Los Alamos, New Mexico, USA

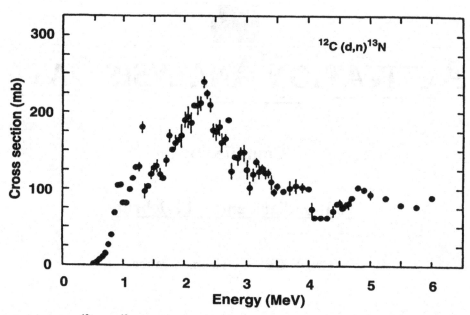

FIG. A15.1. Cross section for the ^{12}C(d,n)^{13}N reaction. From Michelmann *et al.* (1990).

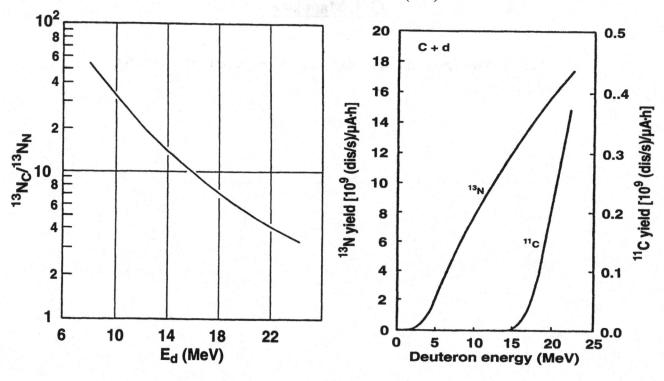

FIG. A15.2. Ratio of the ^{13}N activity from the ^{12}C(d,n)^{13}N reaction to that from the ^{14}N(d,t)^{13}N + ^{14}N(d,dn)^{13}N reactions, for equal C and N concentrations, versus the irradiation energy. From Engelmann and Marschal (1971).

FIG. A15.3. Thick-target yield for the ^{12}C(d,n)^{13}N reaction, from Krasnov *et al.* (1970). Also shown is the ^{12}C(d,t)^{11}C yield. Target is graphite.

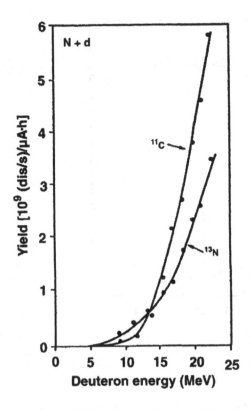

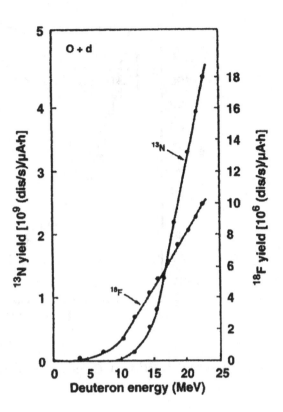

FIG. A15.4. Thick-target yield for the production of ^{13}N by the ^{14}N(d,t)^{13}N + ^{14}N(d,dn)^{13}N reactions, from Krasnov *et al.* (1970). Also shown is the ^{14}N(d,αn)^{11}C yield. Target is AlN.

FIG. A15.5. Thick-target yield for the production of ^{13}N by the ^{16}O(d,αn)^{13}N reaction, from Krasnov *et al.* (1970). Also shown is the ^{17}O(d,n)^{18}F yield. Target is Al$_2$O$_3$.

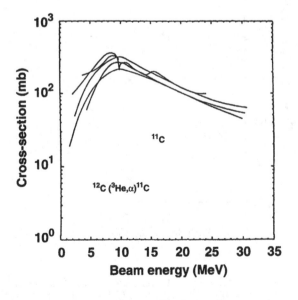

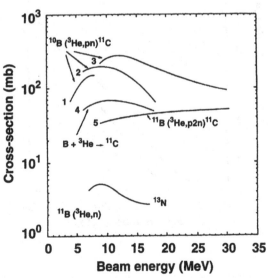

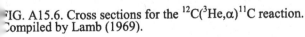

FIG. A15.6. Cross sections for the ^{12}C(^{3}He,α)^{11}C reaction. Compiled by Lamb (1969).

FIG. A15.7. Cross sections for the production of ^{11}C by irradiation of boron. Compiled by Lamb (1969).

257

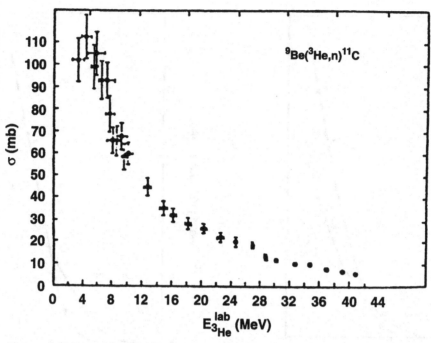

FIG. A15.8. Cross section for the ^{9}Be(^{3}He,n)^{11}C reaction. From Anders (1981).

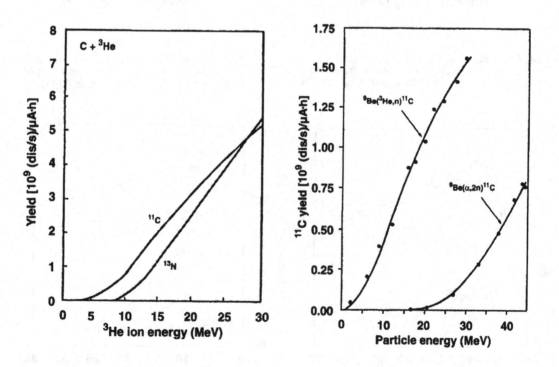

FIG. A15.9. Thick-target yield for the production of ^{11}C by the ^{12}C(^{3}He,α)^{11}C reaction, from Krasnov *et al.* (1970). Also shown is the ^{12}C(^{3}He,d)^{13}N yield. Target is graphite.

FIG. A15.10. Thick-target yield for the ^{9}Be(^{3}He,n)^{11}C reaction, from Krasnov *et al.* (1970). Also shown is the ^{9}Be(α,2n)^{11}C yield. Target is beryllium.

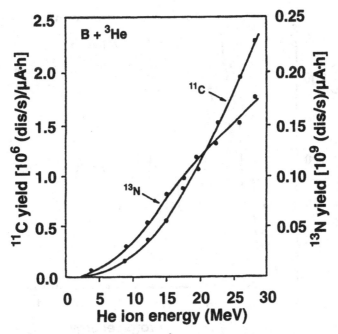

FIG. A15.11. Thick-target yield for the production of ^{11}C by ^{3}He irradiation of boron, from Krasnov *et al.* (1970). Also shown is the production of ^{13}N. Target is boron.

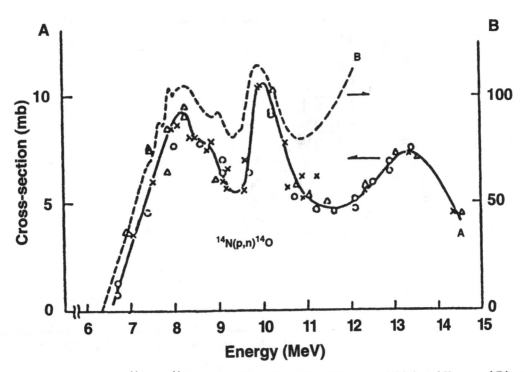

FIG. A15.12. Cross section for the ^{14}N(p,n)^{14}O reaction. From Nozaki and Iwamoto (1981) and Kuan and Risser (1964).

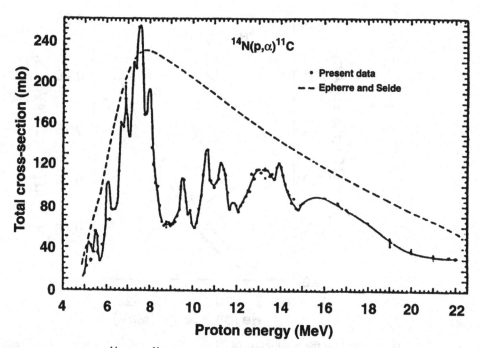

FIG. A15.13. Cross section for the $^{14}N(p,\alpha)^{11}C$ reaction. From Jacobs *et al.* (1974).

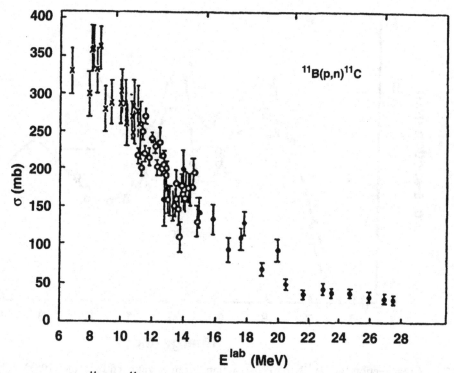

FIG. A15.14. Cross section for the $^{11}B(p,n)^{11}C$ reaction. From Anders (1981).

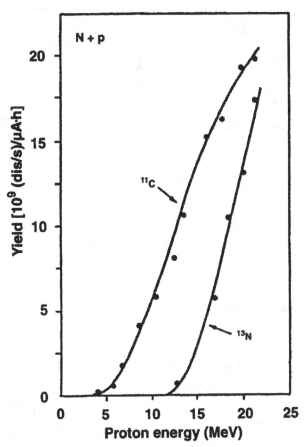

FIG. A15.15. Thick-target yield for the $^{14}N(p,\alpha)^{11}C$ reaction. Target is AlN.

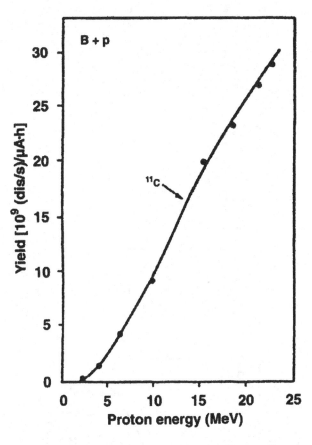

FIG. A15.16. Thick-target yield for the $^{11}B(p,n)^{11}C$ reaction.

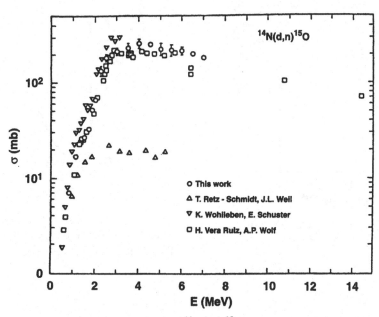

FIG. A15.17. Cross section for the $^{14}N(d,n)^{15}O$ reaction. From Kohl *et al.* (1990).

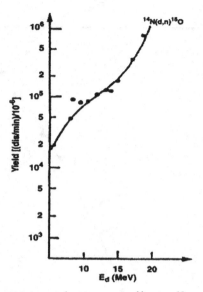

FIG. A15.18. Thick-target yield for the $^{14}N(d,n)^{15}O$ reaction, from Engelmann (1971). Target is AlN.

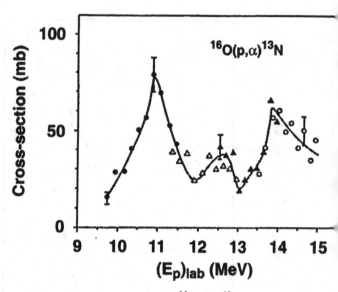

FIG. A15.19. Cross section for the $^{16}O(p,\alpha)^{13}N$ reaction. From Furukawa and Tanaka (1960).

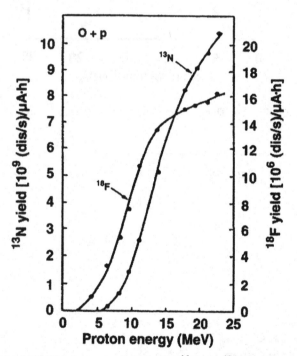

FIG. A15.20. Thick-target yield for the $^{16}O(p,\alpha)^{13}N$, from Krasnov *et al.* (1970). Also shown is the $^{18}O(p,n)^{18}F$ yield. Target is Al$_2$O$_3$.

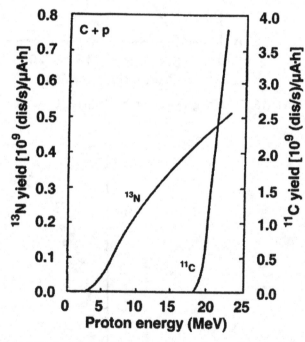

A15.21. Thick-target yield for the production of ^{13}N by irradiatio of carbon with protons, from Krasnov *et al.* (1970). Also shown the $^{12}C(p,pn)^{11}C$ yield. Target is graphite.

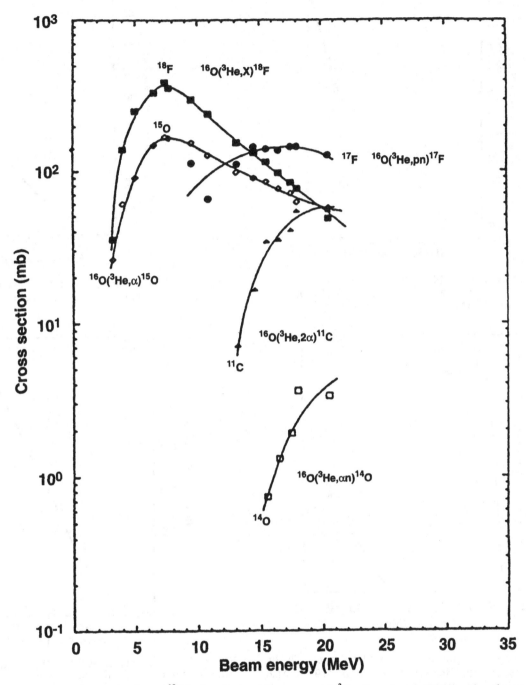

FIG. A15.22. Cross section for the production of ^{18}F by irradiation of oxygen with ^{3}He, from Lamb (1969). Also shown are the cross sections for ^{11}C, ^{14}O, ^{15}O, and ^{17}F production.

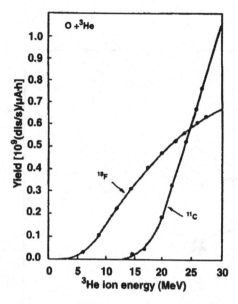

FIG. A15.23. Thick-target yield for the production of ^{18}F by irradiation of oxygen with ^{3}He, from Krasnov *et al.* (1970). Also shown is the ^{16}O(^{3}He,2α)^{11}C yield. Target is Al$_2$O$_3$.

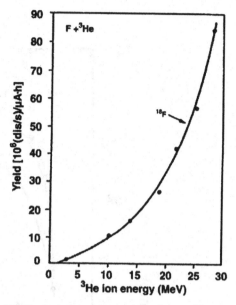

FIG. A15.24. Thick-target yield for the ^{19}F(^{3}He,α)^{18}F reaction, from Krasnov *et al.* (1970). Target is Teflon.

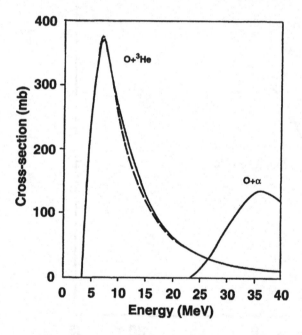

FIG. A15.25. Cross section for the production of ^{18}F by irradiation of oxygen with ^{4}He, from Nozaki *et al.* (1974). Also shown for ^{3}He.

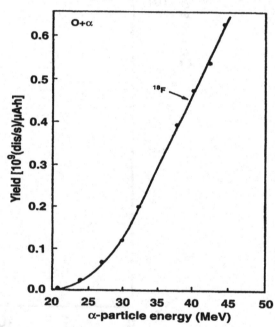

FIG. A15.26. Thick-target yield for the production of ^{18}F by irradiation of oxygen with ^{4}He.

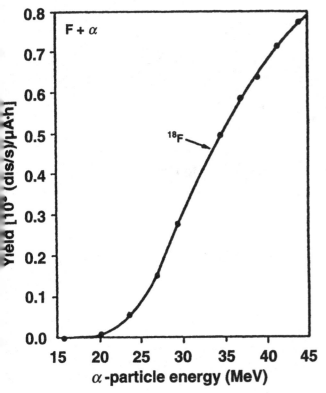

FIG. A15.27. Thick-target yield for the $^{19}F(\alpha,\alpha n)^{18}F$ reaction, from Krasnov et al. (1970). Target is Teflon.

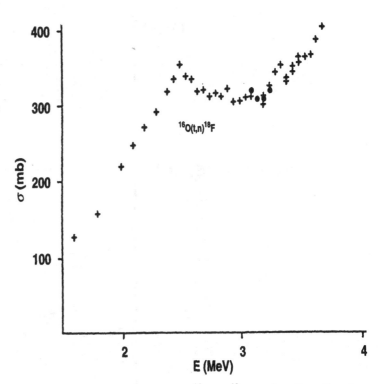

FIG. A15.28. Cross section for the $^{16}O(t,n)^{18}F$ reaction. From Revel et al. (1977).

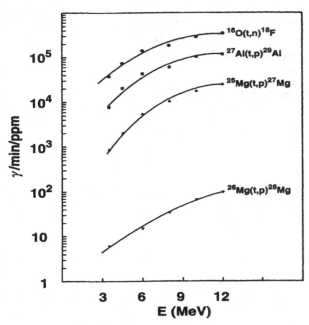

FIG. A15.29. Thick-target yield for the $^{16}O(t,n)^{18}F$ reaction, from Cordes et al. (1987). Also shown are yields for the production of ^{27}Mg, ^{28}Mg, and ^{29}Al by triton irradiation of Mg and Al. Target Al$_2$O$_3$. Yield expressed in γ/min/ppm at the end of 1 h of irradiation at 1 μA.

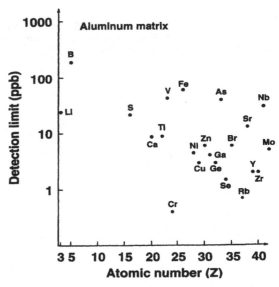

FIG. A15.30. Calculated best detection limits for 22 elements from Z = 3 to Z = 42 in an aluminum matrix. Standard irradiation conditions : 1 h, 1 μA, 10 MeV protons. From Debrun et al. (1976).

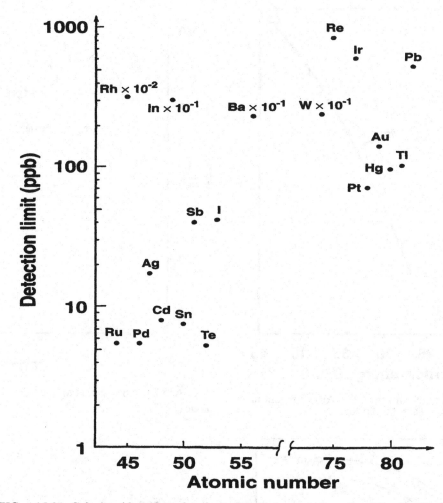

FIG. A15.31. Calculated best detection limits for 19 elements from Z = 44 to Z = 82, in an aluminum matrix. Standard irradiation conditions : 1 h, 1 μA, 10 MeV protons. From Barrandon *et al.* (1976).

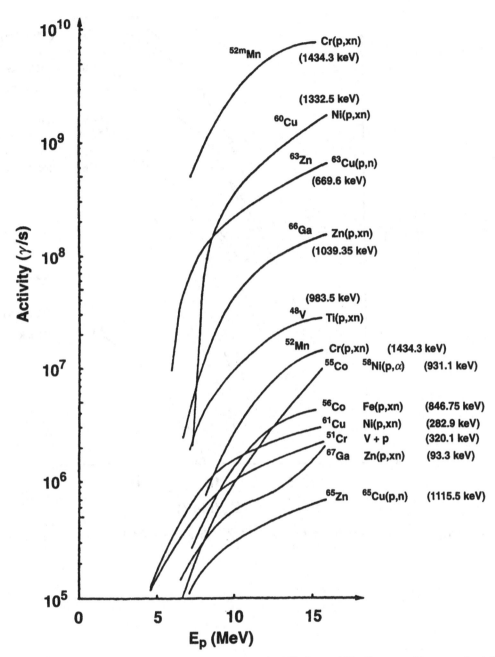

FIG. A15.32. Thick-target yields for the proton activation of Ti, V, Cr, Fe, Ni, Cu, and Zn. Expressed in γ-rays/s at the end of 1 h of irradiation at 1 μA. From Albert et al. (1987a, 1987b).

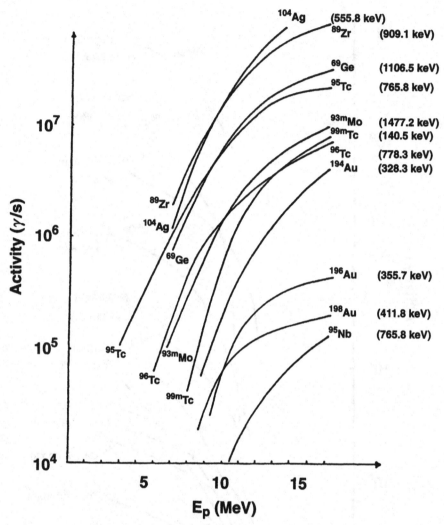

FIG. A15.33. Thick-target yields for the proton activation of Y, Zr, Nb, Mo, and Pd. Expressed in γ-rays/s at the end of 1 h of irradiation at 1 μA. From Albert et al. (1987a, 1987b).

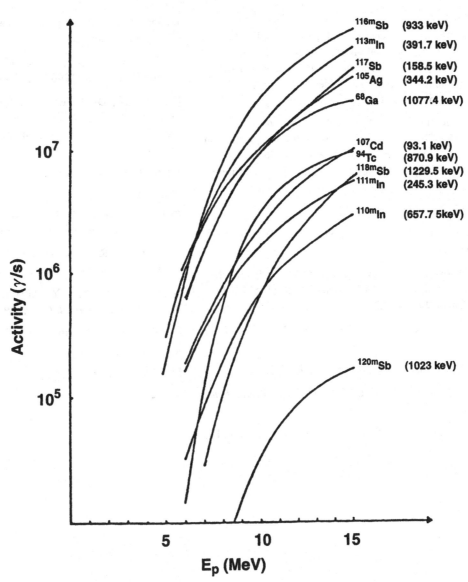

FIG. A15.34. Thick-target yields for the proton activation of Mo, Pd, Ag, Cd, and Sn. Expressed in γ-rays/s at the end of 1 h of irradiation at 1 μA. From Albert et al. (1987a, 1987b).

REFERENCES

Albert, P., Blondiaux, G., Debrun, J.L., Giovagnoli, A., and Valladon, M. (1987a), "Activation cross-sections for elements from lithium to sulphur", in *Handbook on Nuclear Activation Data*, IAEA Technical Report Series No. 273, IAEA, Vienna, Austria, sec. 3-2, pp. 479–536.

Albert, P., Blondiaux, G., Debrun, J.L., Giovagnoli, A., and Valladon, M. (1987b), "Thick target yields for the production of radioisotopes", in *Handbook on Nuclear Activation Data*, IAEA Technical Report Series No. 273, IAEA, Vienna, Austria, sec. 3-3, pp. 537–628.

Anders, B. (1981), *Z. Phys.* **A301**, 353.

Barrandon, J.N., Benaben, P., and Debrun, J.L. (1976), *Anal. Chim. Acta* **83**, 157.

Bordes, N., Blondiaux, G., Maggiore, C.J., Valladon, M., Debrun, J.L., Coquille, R., and Gauneau, M. (1987), *Nucl. Instrum. Methods* **B24–25**, 722.

Debrun, J.L., Barrandon, J.N., and Benaben, P. (1976), *Anal. Chem.* **48**,167.

Engelmann, C. (1971), *J. Radioanal. Chem.* **7**, 89.

Engelmann, C, and Marschal, A. (1971), *Radiochem. Radioanal. Lett.* **6**, 189

Furukawa, M., and Tanaka, S. (1960), *J. Phys. Soc. Jpn.* **15**, 2167.

Jacobs, W.W., Bodansky, D., Chamberlin, D., and Oberg, D.L. (1974), *Phys. Rev.* **C9**, 2134.

Kohl, F., Krauskopf, J., Misaelides, P., Michelmann, R., Wolf, G., and Bethge, K. (1990), *Nucl. Instrum. Methods* **B50**, 19.

Krasnov, N., Dmitriyev, P.P., Dmitriyeva, S.P., Konstantinov, I.O., and Molin, G.A. (1970), *Uses of Cyclotrons in Chemistry, Metallurgy and Biology* (Amphlett, C.B., ed.), Butterworths, London.

Kuan, H.M., and Risser, J.R. (1964), *Nucl. Phys.* **51**, 518.

Lamb, J.F. (1969), *Radioactivation by 3He Bombardment: A Practical Analytical System*, Report UCRL-18981, Lawrence Berkeley Laboratory, University of California, Berkeley, CA.

Michelmann, R.W., Krauskopf, J., Meyer, J.D., and Bethge, K. (1990), *Nucl Instrum. Methods* **B51**, 1

Nozaki, T., Yatsurugi, Y., Akiyama, N., Endo, Y., and Makide, Y. (1974), *J. Radioanal. Chem.* **19**, 109.

Nozaki, T.M., and Iwamoto, M. (1981), *Radiochim. Acta* **29**, 57.

Revel, G., Da Cunha Belo, M., Linck, I., and Kraus, L. (1977), *Rev. Phys. Appl* **12**, 81.

PARTICLE-INDUCED X-RAY EMISSION DATA (A SUBSET OF THE GUPIX DATABASE)

Compiled by

J. L. Campbell and S. M. Taylor

Guelph-Waterloo Physics Institute, University of Guelph, Guelph, Ontario, Canada

CONTENTS

Table A16.1. K X-ray energies and relative intensities for elements $6 \leq Z \leq 60$. If a label does not appear in the leftmost column, then that transition does not occur for any of the elements in that block; if no energy is listed for a transition, then that transition does not occur for that specific element. The column headers contain the atomic symbol and the atomic mass (g/mol) of the elements. The two columns of data for each element contain the energy of the transition line in keV and the relative intensity of each line, respectively. The intensities sum to 1.0. Doublets having very small energy separations are combined. The radiative Auger satellites are also included where pertinent.

Line	C 12.01		N 14.01		O 16.00		F 19.00		Ne 20.18	
KL_2 ($K\alpha_2$)	0.275	0.3333	0.392	0.3335	0.525	0.3239	0.677	0.3282	0.848	0.3302
KL_3 ($K\alpha_1$)	0.284	0.6667	0.391	0.6665	0.525	0.6470	0.677	0.6551	0.849	0.6580
K_{LL}		0.0		0.0	0.502	0.0291	0.643	0.0167	0.800	0.0119

Line	Na 22.99		Mg 24.31		Al 26.98		Si 28.09		P 30.97	
KL_2 ($K\alpha_2$)	1.041	0.3308	1.253	0.3313	1.486	0.3268	1.739	0.3230	2.009	0.3188
KL_3 ($K\alpha_1$)	1.041	0.6604	1.254	0.6620	1.487	0.6536	1.740	0.6418	2.010	0.6286
K_{LL}	0.978	0.0084	1.164	0.0060	1.369	0.0054	1.591	0.0066	1.820	0.0078
K_{LM}	1.040	0.0004	1.252	0.0007	1.486	0.0010	1.732	0.0012	1.994	0.0013
K_{MM}		0.0		0.0	1.548	0.0002	1.819	0.0006	2.109	0.0014

Line	S 32.06		Cl 35.45		Ar 39.95		K 39.10		Ca 40.08	
KL_2 ($K\alpha_2$)	2.307	0.3122	2.621	0.3069	2.952	0.3011	3.311	0.2989	3.688	0.2962
KL_3 ($K\alpha_1$)	2.308	0.6179	2.622	0.6073	2.954	0.5971	3.314	0.5909	3.692	0.5854
K_{LL}	2.077	0.0084	2.351	0.0055	2.626	0.0040	2.934	0.0013	3.250	0.0007
K_{LM}	2.292	0.0013	2.605	0.0014	2.925	0.0013	3.280	0.0013	3.648	0.0013
KM_2 ($K\beta_3$)	2.464	0.0192	2.816	0.0248	3.187	0.0309	3.589	0.0343	4.012	0.0369
KM_3 ($K\beta_1$)	2.464	0.0382	2.816	0.0490	3.187	0.0607	3.590	0.0686	4.013	0.0739
KM_4 ($K\beta_5^{II}$)		0.0		0.0		0.0		0.0	4.038	0.0006
KM_5 ($K\beta_5^{I}$)		0.0		0.0		0.0		0.0	4.038	0.0008
K_{MM}	2.429	0.0027	2.773	0.0051	3.136	0.0049	3.529	0.0046	3.943	0.0042

Line	Sc 44.96		Ti 47.90		V 50.94		Cr 51.99		Mn 54.94	
KL_2 ($K\alpha_2$)	4.086	0.2977	4.505	0.2972	4.945	0.2971	5.405	0.2974	5.888	0.2970
KL_3 ($K\alpha_1$)	4.091	0.5875	4.511	0.5854	4.952	0.5844	5.415	0.5840	5.899	0.5828
K_{LL}	3.586	0.0005	3.941	0.0004		0.0		0.0		0.0
K_{LM}	4.037	0.0013	4.451	0.0013	4.886	0.0013	5.341	0.0013	5.815	0.0013
KM_2 ($K\beta_3$)	4.459	0.0359	4.931	0.0369	5.425	0.0372	5.943	0.0373	6.486	0.0383
KM_3 ($K\beta_1$)	4.461	0.0713	4.934	0.0728	5.430	0.0736	5.949	0.0740	6.493	0.0755
KM_4 ($K\beta_5^{II}$)	4.486	0.0008	4.963	0.0012	5.463	0.0014	5.986	0.0014	6.536	0.0011
KM_5 ($K\beta_5^{I}$)	4.493	0.0013	4.966	0.0018	5.465	0.0021	5.987	0.0021	6.536	0.0017
K_{MM}	4.381	0.0035	4.843	0.0032	5.328	0.0029	5.837	0.0026	6.368	0.0023

Line	Fe 55.85		Co 58.93		Ni 58.71		Cu 63.54		Zn 65.37	
KL_2 ($K\alpha_2$)	6.392	0.2972	6.915	0.2973	7.461	0.2972	8.028	0.2975	8.616	0.2973
KL_3 ($K\alpha_1$)	6.405	0.5820	6.930	0.5812	7.478	0.5806	8.048	0.5797	8.639	0.5785
K_{LM}	6.312	0.0012	6.830	0.0012	7.366	0.0012	7.928	0.0012	8.503	0.0012
KM_2 ($K\beta_3$)	7.055	0.0389	7.646	0.0394	8.262	0.0398	8.904	0.0402	9.570	0.0409
KM_3 ($K\beta_1$)	7.061	0.0765	7.651	0.0772	8.263	0.0781	8.906	0.0787	9.573	0.0799
KM_4 ($K\beta_5^{II}$)	7.109	0.0009	7.706	0.0007	8.329	0.0006	8.977	0.0005	9.651	0.0003
KM_5 ($K\beta_5^{I}$)	7.110	0.0013	7.706	0.0010	8.330	0.0009	8.977	0.0008	9.651	0.0005
K_{MM}	6.925	0.0021	7.504	0.0019	8.105	0.0016	8.735	0.0015	9.389	0.0015

Table A16.1. K X-ray energies and relative intensities for elements $6 \leq Z \leq 60$. If a label does not appear in the leftmost column, then that transition does not occur for any of the elements in that block; if no energy is listed for a transition, then that transition does not occur for that specific element. The column headers contain the atomic symbol and the atomic mass (g/mol) of the elements. The two columns of data for each element contain the energy of the transition line in keV and the relative intensity of each line, respectively. The intensities sum to 1.0. Doublets having very small energy separations are combined. The radiative Auger satellites are also included where pertinent (continued).

Line	Ga 69.72		Ge 72.59		As 74.92		Se 78.96		Br 79.91	
KL$_2$ (Kα_2)	9.225	0.2971	9.855	0.2957	10.508	0.2946	11.182	0.2933	11.878	0.2926
KL$_3$ (Kα_1)	9.252	0.5775	9.886	0.5743	10.544	0.5717	11.222	0.5683	11.924	0.5647
K$_{LM}$	9.094	0.0012	9.706	0.0012	10.340	0.0011	10.990	0.0011	11.667	0.0011
KM$_2$ (Kβ_3)	10.260	0.0414	10.975	0.0421	11.720	0.0429	12.490	0.0437	13.284	0.0443
KM$_3$ (Kβ_1)	10.264	0.0810	10.982	0.0825	11.726	0.0839	12.496	0.0855	13.292	0.0865
KM$_4$ (Kβ_5^{II})	10.346	0.0002	11.074	0.0001		0.0		0.0		0.0
KM$_5$ (Kβ_5^{I})	10.351	0.0003	11.075	0.0002		0.0		0.0		0.0
K$_{MM}$	10.072	0.0013	10.769	0.0012	11.523	0.0011	12.264	0.0010	13.036	0.0010
KN$_2$ (Kβ_2^{II})		0.0	11.101	0.0009	11.864	0.0016	12.652	0.0024	13.469	0.0034
KN$_3$ (Kβ_2^{I})		0.0	11.103	0.0018	11.864	0.0031	12.652	0.0047	13.469	0.0065

Line	Kr 83.80		Rb 85.47		Sr 87.62		Y 88.90		Zr 91.22	
KL$_2$ (Kα_2)	12.598	0.2916	13.336	0.2907	14.098	0.2899	14.883	0.2893	15.691	0.2884
KL$_3$ (Kα_1)	12.651	0.5628	13.395	0.5598	14.165	0.5569	14.958	0.5544	15.775	0.5520
K$_{LM}$	12.359	0.0010	13.073	0.0010	13.807	0.0009	14.565	0.0009	15.345	0.0009
KM$_2$ (Kβ_3)	14.103	0.0442	14.952	0.0448	15.825	0.0453	16.726	0.0459	17.653	0.0465
KM$_3$ (Kβ_1)	14.111	0.0865	14.961	0.0876	15.835	0.0886	16.738	0.0898	17.667	0.0909
KM$_4$ (Kβ_5^{II})		0.0		0.0		0.0		0.0	17.815	0.0002
KM$_5$ (Kβ_5^{I})		0.0		0.0		0.0		0.0	17.818	0.0003
K$_{MM}$	13.819	0.0010	14.639	0.0010	15.478	0.0011	16.345	0.0012	17.237	0.0013
KN$_2$ (Kβ_2^{II})	14.311	0.0044	15.185	0.0052	16.084	0.0058	17.013	0.0063	17.968	0.0067
KN$_3$ (Kβ_2^{I})	14.312	0.0085	15.186	0.0100	16.085	0.0114	17.016	0.0122	17.972	0.0130

Line	Nb 92.91		Mo 95.94		Tc 97.00		Ru 101.07		Rh 102.90	
KL$_2$ (Kα_2)	16.521	0.2880	17.374	0.2874	18.251	0.2868	19.150	0.2863	20.074	0.2861
KL$_3$ (Kα_1)	16.615	0.5499	17.479	0.5480	18.367	0.5463	19.279	0.5444	20.216	0.5428
K$_{LM}$	16.147	0.0008	16.975	0.0008	17.823	0.0008	18.694	0.0007	19.589	0.0007
KM$_2$ (Kβ_3)	18.607	0.0471	19.590	0.0476	20.599	0.0481	21.634	0.0485	22.699	0.0490
KM$_3$ (Kβ_1)	18.623	0.0919	19.607	0.0929	20.619	0.0937	21.657	0.0947	22.724	0.0953
KM$_4$ (Kβ_5^{II})	18.778	0.0002	19.769	0.0002	20.788	0.0002	21.834	0.0003	22.908	0.0003
KM$_5$ (Kβ_5^{I})	18.781	0.0003	19.773	0.0003	20.791	0.0003	21.838	0.0004	22.913	0.0004
K$_{MM}$	18.154	0.0015	19.103	0.0016	20.075	0.0018	21.072	0.0019	22.097	0.0018
KN$_2$ (Kβ_2^{II})	18.950	0.0069	19.961	0.0072	21.001	0.0075	22.070	0.0078	23.168	0.0080
KN$_3$ (Kβ_2^{I})	18.956	0.0135	19.967	0.0140	21.007	0.0146	22.076	0.0151	23.174	0.0156

Table A16.1. K X-ray energies and relative intensities for elements $6 \leq Z \leq 60$. If a label does not appear in the leftmost column, then that transition does not occur for any of the elements in that block; if no energy is listed for a transition, then that transition does not occur for that specific element. The column headers contain the atomic symbol and the atomic mass (g/mol) of the elements. The two columns of data for each element contain the energy of the transition line in keV and the relative intensity of each line, respectively. The intensities sum to 1.0. Doublets having very small energy separations are combined. The radiative Auger satellites are also included where pertinent (continued).

Line	Pd 106.40		Ag 107.87		Cd 112.40		In 114.82		Sn 118.69	
KL_2 ($K\alpha_2$)	21.020	0.2860	21.990	0.2858	22.984	0.2857	24.002	0.2855	25.044	0.2856
KL_3 ($K\alpha_1$)	21.177	0.5414	22.163	0.5398	23.174	0.5380	24.210	0.5362	25.271	0.5341
K_{LM}	20.507	0.0007	21.445	0.0007	22.403	0.0006	23.384	0.0006	24.388	0.0006
KM_2 ($K\beta_3$)	23.791	0.0494	24.912	0.0498	26.060	0.0503	27.238	0.0506	28.444	0.0511
KM_3 ($K\beta_1$)	23.819	0.0962	24.943	0.0971	26.095	0.0978	27.276	0.0985	28.486	0.0991
KM_4 ($K\beta_5^{II}$)	24.010	0.0003	25.141	0.0003	26.301	0.0003	27.489	0.0004	28.707	0.0004
KM_5 ($K\beta_5^{I}$)	24.016	0.0004	25.147	0.0004	26.307	0.0005	27.497	0.0005	28.715	0.0005
K_{MM}	23.149	0.0015	24.225	0.0010	25.324	0.0007	26.450	0.0006	27.602	0.0005
KN_2 ($K\beta_2^{II}$)	24.296	0.0082	25.451	0.0085	26.640	0.0089	27.858	0.0092	29.106	0.0096
KN_3 ($K\beta_2^{I}$)	24.300	0.0159	25.458	0.0165	26.646	0.0172	27.865	0.0179	29.114	0.0186

Line	Sb 121.75		Te 127.60		I 126.90		Xe 131.30		Cs 132.90	
KL_2 ($K\alpha_2$)	26.111	0.2853	27.202	0.2853	28.317	0.2848	29.458	0.2843	30.625	0.2834
KL_3 ($K\alpha_1$)	26.359	0.5321	27.472	0.5299	28.612	0.5284	29.779	0.5266	30.973	0.5254
KM_2 ($K\beta_3$)	29.679	0.0513	30.944	0.0516	32.239	0.0518	33.562	0.0520	34.920	0.0522
KM_3 ($K\beta_1$)	29.726	0.0996	30.995	0.1001	32.295	0.1004	33.624	0.1008	34.987	0.1011
KM_4 ($K\beta_5^{II}$)	29.954	0.0004	31.231	0.0004	32.538	0.0005	33.871	0.0005	35.245	0.0005
KM_5 ($K\beta_5^{I}$)	29.964	0.0006	31.242	0.0006	32.550	0.0006	33.887	0.0007	35.259	0.0007
KN_2 ($K\beta_2^{II}$)	30.387	0.0099	31.697	0.0101	33.039	0.0104	34.397	0.0106	35.812	0.0108
KN_3 ($K\beta_2^{I}$)	30.396	0.0193	31.717	0.0198	33.050	0.0202	34.405	0.0207	35.823	0.0212
KO_{23}	30.489	0.0014	31.812	0.0021	33.166	0.0030	34.549	0.0039	35.972	0.0046

Line	Ba 137.34		La 138.91		Ce 140.12		Pr 140.91		Nd 144.24	
KL_2 ($K\alpha_2$)	31.817	0.2835	33.034	0.2835	34.279	0.2815	35.550	0.2844	36.847	0.2844
KL_3 ($K\alpha_1$)	32.194	0.5231	33.442	0.5213	34.720	0.5167	36.026	0.5187	37.361	0.5177
KM_2 ($K\beta_3$)	36.304	0.0524	37.720	0.0526	39.170	0.0525	40.653	0.0531	42.166	0.0533
KM_3 ($K\beta_1$)	36.378	0.1016	37.801	0.1019	39.258	0.1016	40.748	0.1027	42.271	0.1031
KM_4 ($K\beta_5^{II}$)	36.645	0.0005	38.076	0.0006	39.542	0.0006	41.040	0.0006	42.569	0.0007
KM_5 ($K\beta_5^{I}$)	36.660	0.0007	38.093	0.0008	39.560	0.0081	41.060	0.0009	42.591	0.0009
KN_2 ($K\beta_2^{II}$)	37.249	0.0111	38.719	0.0114	40.220	0.0114	41.754	0.0116	43.326	0.0117
KN_3 ($K\beta_2^{I}$)	37.261	0.0218	38.733	0.0223	40.236	0.0222	41.773	0.0227	43.344	0.0229
KN_{45} ($K\beta_4$)		0.0		0.0	40.332	0.0003	41.877	0.0003	43.450	0.0003
KO_{23}	37.425	0.0052	38.910	0.0057	40.423	0.0052	41.968	0.0051	43.547	0.0051

Table A16.2. Relative X-ray line intensities for the L1 subshell. From Campbell, J.L., and Wang, J.-X. (1989), *At. Data Nucl. Data Tables* **43**, 281.

Z	L1L3	L1M2	L1M3	L1N2	L1N3	L1O23	L1P23	L1M45	L1N45
18	8.559E–02	3.125E–01	6.019E–01	0	0	0	0	0	0
19	6.223E–02	3.205E–01	6.173E–01	0	0	0	0	0	0
20	4.786E–02	3.260E–01	6.261E–01	0	0	0	0	0	0
21	3.998E–02	3.295E–01	6.304E–01	0	0	0	0	8.738E–05	0
22	3.401E–02	3.329E–01	6.329E–01	0	0	0	0	2.113E–04	0
23	2.955E–02	3.350E–01	6.351E–01	0	0	0	0	3.738E–04	0
24	2.667E–02	3.365E–01	6.362E–01	0	0	0	0	6.452E–04	0
25	2.331E–02	3.390E–01	6.368E–01	0	0	0	0	8.210E–04	0
26	2.102E–02	3.403E–01	6.375E–01	0	0	0	0	1.108E–03	0
27	1.914E–02	3.419E–01	6.375E–01	0	0	0	0	1.441E–03	0
28	1.757E–02	3.437E–01	6.369E–01	0	0	0	0	1.820E–03	0
29	1.669E–02	3.452E–01	6.358E–01	0	0	0	0	2.337E–03	0
30	1.552E–02	3.465E–01	6.352E–01	0	0	0	0	2.714E–03	0
31	1.448E–02	3.479E–01	6.345E–01	0	0	0	0	3.092E–03	0
32	1.326E–02	3.411E–01	6.187E–01	8.346E–03	1.523E–02	0	0	3.372E–03	0
33	1.227E–02	3.362E–01	6.068E–01	1.454E–02	2.652E–02	0	0	3.683E–03	0
34	1.139E–02	3.298E–01	5.950E–01	2.121E–02	3.852E–02	0	0	3.982E–03	0
35	1.064E–02	3.246E–01	5.817E–01	2.807E–02	5.076E–02	0	0	4.265E–03	0
36	9.979E–03	3.189E–01	5.685E–01	3.509E–02	6.301E–02	0	0	4.528E–03	0
37	9.460E–03	3.151E–01	5.588E–01	3.995E–02	7.188E–02	0	0	4.824E–03	0
38	9.045E–03	3.129E–01	5.507E–01	4.376E–02	7.844E–02	0	0	5.128E–03	0
39	8.727E–03	3.117E–01	5.458E–01	4.611E–02	8.219E–02	0	0	5.458E–03	0
40	8.475E–03	3.113E–01	5.412E–01	4.805E–02	8.518E–02	0	0	5.791E–03	6.392E–05
41	8.304E–03	3.117E–01	5.382E–01	4.913E–02	8.644E–02	0	0	6.155E–03	1.358E–04
42	8.143E–03	3.112E–01	5.345E–01	5.074E–02	8.878E–02	0	0	6.504E–03	1.981E–04
43	8.032E–03	3.113E–01	5.304E–01	5.226E–02	9.091E–02	0	0	6.860E–03	2.707E–04
44	7.964E–03	3.112E–01	5.268E–01	5.367E–02	9.279E–02	0	0	7.222E–03	3.537E–04
45	7.920E–03	3.113E–01	5.230E–01	5.498E–02	9.474E–02	0	0	7.598E–03	4.475E–04
46	7.915E–03	3.117E–01	5.206E–01	5.582E–02	9.543E–02	0	0	7.988E–03	5.702E–04
47	7.938E–03	3.113E–01	5.164E–01	5.755E–02	9.777E–02	0	0	8.372E–03	6.687E–04
48	7.983E–03	3.108E–01	5.122E–01	5.934E–02	1.002E–01	0	0	8.753E–03	7.658E–04
49	8.062E–03	3.102E–01	5.077E–01	6.110E–02	1.029E–01	0	0	9.176E–03	8.695E–04
50	8.108E–03	3.076E–01	5.013E–01	6.241E–02	1.049E–01	5.205E–03	0	9.501E–03	9.647E–04
51	8.203E–03	3.064E–01	4.956E–01	6.377E–02	1.059E–01	9.128E–03	0	9.898E–03	1.065E–03
52	8.307E–03	3.047E–01	4.899E–01	6.490E–02	1.073E–01	1.350E–02	0	1.028E–02	1.162E–03
53	8.445E–03	3.032E–01	4.835E–01	6.608E–02	1.086E–01	1.828E–02	0	1.067E–02	1.262E–03
54	8.602E–03	3.017E–01	4.774E–01	6.703E–02	1.096E–01	2.336E–02	0	1.106E–02	1.359E–03
55	8.786E–03	3.008E–01	4.715E–01	6.814E–02	1.107E–01	2.716E–02	0	1.146E–02	1.460E–03
56	8.999E–03	3.001E–01	4.663E–01	6.919E–02	1.117E–01	3.026E–02	0	1.189E–02	1.565E–03
57	9.253E–03	2.997E–01	4.613E–01	7.047E–02	1.129E–01	3.232E–02	0	1.234E–02	1.674E–03
58	9.598E–03	3.019E–01	4.610E–01	7.088E–02	1.125E–01	2.945E–02	0	1.290E–02	1.733E–03
59	9.933E–03	3.032E–01	4.583E–01	7.160E–02	1.127E–01	2.902E–02	0	1.344E–02	1.816E–03
60	1.032E–02	3.044E–01	4.556E–01	7.234E–02	1.128E–01	2.864E–02	0	1.400E–02	1.903E–03

Table A16.2. Relative X-ray line intensities for the L1 subshell. From Campbell, J.L., and Wang, J.-X. (1989), *At. Data Nucl. Data Tables* **43**, 281 (continued).

	L1L3	L1M2	L1M3	L1N2	L1N3	L1O23	L1P23	L1M45	L1N45
	1.072E–02	3.056E–01	4.532E–01	7.296E–02	1.127E–01	2.826E–02	0	1.457E–02	1.988E–03
	1.117E–02	3.070E–01	4.504E–01	7.365E–02	1.127E–01	2.787E–02	0	1.515E–02	2.075E–03
	1.166E–02	3.085E–01	4.476E–01	7.425E–02	1.126E–01	2.748E–02	0	1.575E–02	2.162E–03
	1.216E–02	3.091E–01	4.435E–01	7.505E–02	1.126E–01	2.901E–02	0	1.632E–02	2.272E–03
	1.273E–02	3.118E–01	4.417E–01	7.551E–02	1.121E–01	2.678E–02	0	1.702E–02	2.348E–03
	1.334E–02	3.131E–01	4.390E–01	7.624E–02	1.119E–01	2.639E–02	0	1.766E–02	2.445E–03
	1.399E–02	3.147E–01	4.360E–01	7.677E–02	1.116E–01	2.603E–02	0	1.834E–02	2.542E–03
	1.467E–02	3.165E–01	4.327E–01	7.736E–02	1.114E–01	2.568E–02	0	1.903E–02	2.642E–03
	1.544E–02	3.181E–01	4.296E–01	7.813E–02	1.110E–01	2.535E–02	0	1.975E–02	2.747E–03
	1.624E–02	3.199E–01	4.264E–01	7.865E–02	1.105E–01	2.508E–02	0	2.046E–02	2.853E–03
	1.708E–02	3.211E–01	4.218E–01	7.933E–02	1.103E–01	2.629E–02	0	2.117E–02	2.979E–03
	1.797E–02	3.218E–01	4.174E–01	8.019E–02	1.101E–01	2.753E–02	0	2.189E–02	3.108E–03
	1.894E–02	3.228E–01	4.130E–01	8.088E–02	1.098E–01	2.867E–02	0	2.262E–02	3.244E–03
	1.995E–02	3.242E–01	4.082E–01	8.163E–02	1.095E–01	2.976E–02	0	2.337E–02	3.390E–03
	2.104E–02	3.253E–01	4.032E–01	8.245E–02	1.094E–01	3.085E–02	0	2.414E–02	3.539E–03
	2.221E–02	3.271E–01	3.980E–01	8.323E–02	1.091E–01	3.181E–02	0	2.491E–02	3.696E–03
	2.349E–02	3.278E–01	3.936E–01	8.427E–02	1.090E–01	3.222E–02	0	2.576E–02	3.865E–03
	2.482E–02	3.289E–01	3.885E–01	8.495E–02	1.086E–01	3.352E–02	0	2.657E–02	4.034E–03
	2.622E–02	3.306E–01	3.831E–01	8.567E–02	1.082E–01	3.452E–02	0	2.741E–02	4.208E–03
	2.772E–02	3.320E–01	3.780E–01	8.643E–02	1.077E–01	3.577E–02	0	2.801E–02	4.388E–03
	2.937E–02	3.330E–01	3.725E–01	8.714E–02	1.072E–01	3.705E–02	0	2.914E–02	4.578E–03
	3.098E–02	3.338E–01	3.665E–01	8.777E–02	1.066E–01	3.827E–02	1.348E–03	2.998E–02	4.761E–03
	3.282E–02	3.350E–01	3.602E–01	8.864E–02	1.058E–01	3.932E–02	2.384E–03	3.087E–02	4.965E–03
	3.467E–02	3.358E–01	3.544E–01	8.937E–02	1.049E–01	4.037E–02	3.566E–03	3.177E–02	5.170E–03
	3.677E–02	3.371E–01	3.479E–01	9.000E–02	1.040E–01	4.129E–02	4.853E–03	3.263E–02	5.378E–03
	3.891E–02	3.381E–01	3.414E–01	9.081E–02	1.031E–01	4.221E–02	6.237E–03	3.362E–02	5.591E–03
	4.122E–02	3.390E–01	3.352E–01	9.166E–02	1.022E–01	4.315E–02	7.208E–03	3.459E–02	5.821E–03
	4.363E–02	3.404E–01	3.288E–01	9.240E–02	1.012E–01	4.401E–02	7.974E–03	3.558E–02	6.057E–03
	4.625E–02	3.418E–01	3.223E–01	9.317E–02	1.002E–01	4.486E–02	8.554E–03	3.661E–02	6.299E–03
	4.904E–02	3.430E–01	3.158E–01	9.404E–02	9.912E–02	4.577E–02	9.082E–03	3.762E–02	6.553E–03
	5.201E–02	3.448E–01	3.098E–01	9.494E–02	9.807E–02	4.623E–02	8.658E–03	3.873E–02	6.821E–03
	5.517E–02	3.467E–01	3.031E–01	9.571E–02	9.693E–02	4.682E–02	8.692E–03	3.986E–02	7.092E–03

Table A16.3. Relative X-ray line intensities for the L2 subshell. From Campbell, J.L., and Wang, J.-X. (1989), *At. Data Nucl. Data Tables* **43**, 281.

Z	L2M1	L2M4	L2N1	L2N4	L2O1	L2O4	L2P1
18	1	0	0	0	0	0	0
19	1	0	0	0	0	0	0
20	1	0	0	0	0	0	0
21	1	0	0	0	0	0	0
22	1	0	0	0	0	0	0
23	1	0	0	0	0	0	0
24	1	0	0	0	0	0	0
25	1	0	0	0	0	0	0
26	8.246E–02	9.103E–01	7.201E–03	0	0	0	0
27	6.939E–02	9.248E–01	5.859E–03	0	0	0	0
28	5.949E–02	9.358E–01	4.745E–03	0	0	0	0
29	5.044E–02	9.481E–01	1.433E–03	0	0	0	0
30	4.609E–02	9.507E–01	3.200E–03	0	0	0	0
31	4.277E–02	9.535E–01	3.745E–03	0	0	0	0
32	4.033E–02	9.557E–01	3.967E–03	0	0	0	0
33	3.837E–02	9.575E–01	4.108E–03	0	0	0	0
34	3.684E–02	9.589E–01	4.226E–03	0	0	0	0
35	3.556E–02	9.601E–01	4.358E–03	0	0	0	0
36	3.451E–02	9.610E–01	4.489E–03	0	0	0	0
37	3.355E–02	9.617E–01	4.797E–03	0	0	0	0
38	3.271E–02	9.622E–01	5.060E–03	0	0	0	0
39	3.195E–02	9.629E–01	5.151E–03	0	0	0	0
40	3.078E–02	9.492E–01	4.977E–03	1.503E–02	0	0	0
41	2.978E–02	9.356E–01	5.005E–03	2.899E–02	5.843E–04	0	0
42	2.905E–02	9.266E–01	5.066E–03	3.901E–02	2.411E–04	0	0
43	2.829E–02	9.170E–01	5.076E–03	4.935E–02	2.353E–04	0	0
44	2.758E–02	9.073E–01	5.066E–03	5.985E–02	2.266E–04	0	0
45	2.708E–02	8.969E–01	5.067E–03	7.074E–02	2.182E–04	0	0
46	2.646E–02	8.841E–01	4.999E–03	8.418E–02	2.071E–04	0	0
47	2.606E–02	8.758E–01	5.019E–03	9.295E–02	2.144E–04	0	0
48	2.567E–02	8.681E–01	5.036E–03	1.006E–01	5.098E–04	0	0
49	2.526E–02	8.611E–01	5.063E–03	1.080E–01	5.971E–04	0	0
50	2.493E–02	8.546E–01	5.113E–03	1.147E–01	6.448E–04	0	0
51	2.455E–02	8.491E–01	5.156E–03	1.205E–01	6.840E–04	0	0
52	2.426E–02	8.439E–01	5.203E–03	1.259E–01	7.266E–04	0	0
53	2.398E–02	8.391E–01	5.240E–03	1.309E–01	7.786E–04	0	0
54	2.373E–02	8.348E–01	5.277E–03	1.353E–01	8.440E–04	0	0
55	2.349E–02	8.306E–01	5.319E–03	1.397E–01	9.086E–04	0	0
56	2.325E–02	8.262E–01	5.356E–03	1.441E–01	9.706E–04	0	1.592E–04
57	2.299E–02	8.228E–01	5.380E–03	1.477E–01	1.016E–03	0	1.768E–04
58	2.292E–02	8.241E–01	5.368E–03	1.465E–01	9.519E–04	0	1.444E–04
59	2.286E–02	8.229E–01	5.389E–03	1.477E–01	9.428E–04	0	1.382E–04
60	2.275E–02	8.222E–01	5.391E–03	1.486E–01	9.350E–04	0	1.325E–04

Table A16.3. Relative X-ray line intensities for the L2 subshell. From Campbell, J.L., and Wang, J.-X. (1989), *At. Data Nucl. Data Tables* **43**, 281 (continued).

Z	L2M1	L2M4	L2N1	L2N4	L2O1	L2O4	L2P1
61	2.271E–02	8.215E–01	5.401E–03	1.493E–01	9.261E–04	0	1.275E–04
62	2.261E–02	8.209E–01	5.399E–03	1.501E–01	9.171E–04	0	1.231E–04
63	2.253E–02	8.206E–01	5.398E–03	1.505E–01	9.091E–04	0	1.191E–04
64	2.244E–02	8.186E–01	5.421E–03	1.524E–01	9.444E–04	0	1.359E–04
65	2.241E–02	8.195E–01	5.406E–03	1.517E–01	8.949E–04	0	1.124E–04
66	2.237E–02	8.190E–01	5.422E–03	1.522E–01	8.878E–04	0	1.094E–04
67	2.237E–02	8.183E–01	5.426E–03	1.529E–01	8.822E–04	0	1.068E–04
68	2.230E–02	8.180E–01	5.426E–03	1.533E–01	8.755E–04	0	1.039E–04
69	2.230E–02	8.172E–01	5.444E–03	1.540E–01	8.697E–04	0	1.013E–04
70	2.228E–02	8.168E–01	5.447E–03	1.545E–01	8.628E–04	0	9.834E–05
71	2.223E–02	8.155E–01	5.456E–03	1.558E–01	8.908E–04	0	1.101E–04
72	2.219E–02	8.139E–01	5.484E–03	1.574E–01	9.210E–04	0	1.160E–04
73	2.207E–02	8.094E–01	5.476E–03	1.584E–01	9.466E–04	3.552E–03	1.194E–04
74	2.204E–02	8.062E–01	5.497E–03	1.599E–01	9.771E–04	5.281E–03	1.221E–04
75	2.195E–02	8.032E–01	5.508E–03	1.610E–01	1.007E–03	7.192E–03	1.242E–04
76	2.188E–02	7.998E–01	5.525E–03	1.623E–01	1.039E–03	9.291E–03	1.258E–04
77	2.178E–02	7.953E–01	5.532E–03	1.632E–01	1.063E–03	1.313E–02	0
78	2.175E–02	7.927E–01	5.559E–03	1.640E–01	1.092E–03	1.485E–02	5.624E–05
79	2.168E–02	7.893E–01	5.581E–03	1.649E–01	1.118E–03	1.741E–02	5.656E–05
80	2.166E–02	7.867E–01	5.597E–03	1.656E–01	1.149E–03	1.915E–02	1.296E–04
81	2.165E–02	7.836E–01	5.621E–03	1.668E–01	1.179E–03	2.098E–02	1.577E–04
82	2.160E–02	7.809E–01	5.649E–03	1.678E–01	1.211E–03	2.273E–02	1.788E–04
83	2.160E–02	7.781E–01	5.672E–03	1.688E–01	1.239E–03	2.436E–02	1.967E–04
84	2.153E–02	7.756E–01	5.701E–03	1.698E–01	1.270E–03	2.590E–02	2.133E–04
85	2.154E–02	7.728E–01	5.725E–03	1.710E–01	1.301E–03	2.741E–02	2.289E–04
86	2.154E–02	7.702E–01	5.755E–03	1.721E–01	1.333E–03	2.883E–02	2.434E–04
87	2.155E–02	7.676E–01	5.789E–03	1.732E–01	1.365E–03	3.027E–02	2.645E–04
88	2.153E–02	7.651E–01	5.817E–03	1.742E–01	1.395E–03	3.168E–02	2.846E–04
89	2.155E–02	7.623E–01	5.851E–03	1.755E–01	1.425E–03	3.305E–02	3.026E–04
90	2.156E–02	7.599E–01	5.879E–03	1.765E–01	1.454E–03	3.438E–02	3.193E–04
91	2.157E–02	7.579E–01	5.911E–03	1.778E–01	1.478E–03	3.504E–02	3.151E–04
92	2.160E–02	7.558E–01	5.945E–03	1.789E–01	1.507E–03	3.592E–02	3.219E–04

Table A16.4. Relative X-ray line intensities for the L3 subshell. From Campbell, J. L., and Wang, J.-X. (1989), *At. Data Nucl. Data Tables* **43**, 281.

Z	L3M1	L3M4	L3M5	L3N1	L3N4	L3N5	L3O1	L3O45	L3P1
18	1	0	0	0	0	0	0	0	0
19	1	0	0	0	0	0	0	0	0
20	1	0	0	0	0	0	0	0	0
21	1	0	0	0	0	0	0	0	0
22	1	0	0	0	0	0	0	0	0
23	1	0	0	0	0	0	0	0	0
24	1	0	0	0	0	0	0	0	0
25	1	0	0	0	0	0	0	0	0
26	8.581E–02	9.302E–02	8.137E–01	7.437E–03	0	0	0	0	0
27	7.263E–02	9.457E–02	8.268E–01	6.042E–03	0	0	0	0	0
28	6.288E–02	9.575E–02	8.364E–01	4.982E–03	0	0	0	0	0
29	5.377E–02	9.734E–02	8.474E–01	1.535E–03	0	0	0	0	0
30	4.935E–02	9.732E–02	8.499E–01	3.474E–03	0	0	0	0	0
31	4.616E–02	9.740E–02	8.524E–01	4.072E–03	0	0	0	0	0
32	4.384E–02	9.744E–02	8.545E–01	4.251E–03	0	0	0	0	0
33	4.204E–02	9.758E–02	8.559E–01	4.523E–03	0	0	0	0	0
34	4.067E–02	9.760E–02	8.570E–01	4.720E–03	0	0	0	0	0
35	3.954E–02	9.762E–02	8.580E–01	4.879E–03	0	0	0	0	0
36	3.865E–02	9.762E–02	8.587E–01	5.003E–03	0	0	0	0	0
37	3.789E–02	9.755E–02	8.593E–01	5.304E–03	0	0	0	0	0
38	3.722E–02	9.762E–02	8.596E–01	5.569E–03	0	0	0	0	0
39	3.665E–02	9.758E–02	8.601E–01	5.714E–03	0	0	0	0	0
40	3.569E–02	9.607E–02	8.468E–01	5.723E–03	1.515E–03	1.341E–02	7.861E–04	0	0
41	3.490E–02	9.471E–02	8.356E–01	5.767E–03	2.925E–03	2.578E–02	3.226E–04	0	0
42	3.429E–02	9.381E–02	8.272E–01	5.884E–03	3.924E–03	3.461E–02	3.170E–04	0	0
43	3.372E–02	9.283E–02	8.185E–01	5.965E–03	4.954E–03	4.370E–02	3.081E–04	0	0
44	3.320E–02	9.182E–02	8.097E–01	6.039E–03	5.999E–03	5.293E–02	3.003E–04	0	0
45	3.293E–02	9.084E–02	8.001E–01	6.089E–03	7.100E–03	6.265E–02	2.909E–04	0	0
46	3.263E–02	8.961E–02	7.892E–01	6.091E–03	8.442E–03	7.400E–02	0	0	0
47	3.237E–02	8.862E–02	7.815E–01	6.163E–03	9.295E–03	8.181E–02	2.635E–04	0	0
48	3.213E–02	8.780E–02	7.746E–01	6.235E–03	1.002E–02	8.859E–02	6.283E–04	0	0
49	3.199E–02	8.704E–02	7.680E–01	6.345E–03	1.074E–02	9.513E–02	7.559E–04	0	0
50	3.183E–02	8.632E–02	7.624E–01	6.467E–03	1.137E–02	1.009E–01	8.068E–04	0	0
51	3.171E–02	8.564E–02	7.573E–01	6.576E–03	1.192E–02	1.060E–01	8.927E–04	0	0
52	3.165E–02	8.514E–02	7.525E–01	6.697E–03	1.243E–02	1.106E–01	9.687E–04	0	0
53	3.158E–02	8.457E–02	7.482E–01	6.818E–03	1.289E–02	1.149E–01	1.036E–03	0	0
54	3.158E–02	8.412E–02	7.442E–01	6.946E–03	1.331E–02	1.187E–01	1.100E–03	0	0
55	3.162E–02	8.378E–02	7.403E–01	7.065E–03	1.370E–02	1.224E–01	1.200E–03	0	0
56	3.169E–02	8.346E–02	7.362E–01	7.176E–03	1.409E–02	1.259E–01	1.301E–03	0	2.123E–04
57	3.175E–02	8.305E–02	7.326E–01	7.300E–03	1.444E–02	1.293E–01	1.381E–03	0	2.383E–04
58	3.205E–02	8.335E–02	7.335E–01	7.356E–03	1.432E–02	1.279E–01	1.304E–03	0	1.971E–04
59	3.227E–02	8.328E–02	7.324E–01	7.446E–03	1.441E–02	1.287E–01	1.305E–03	0	1.903E–04
60	3.246E–02	8.288E–02	7.322E–01	7.528E–03	1.443E–02	1.290E–01	1.301E–03	0	1.842E–04

Table A16.4. Relative X-ray line intensities for the L3 subshell. From Campbell, J. L., and Wang, J.-X. (1989), *At. Data Nucl. Data Tables* **43**, 281 (continued).

Z	L3M1	L3M4	L3M5	L3N1	L3N4	L3N5	L3O1	L3O45	L3P1
61	3.275E–02	8.280E–02	7.314E–01	7.612E–03	1.449E–02	1.295E–01	1.302E–03	0	1.794E–04
62	3.308E–02	8.293E–02	7.304E–01	7.727E–03	1.453E–02	1.299E–01	1.306E–03	0	1.753E–04
63	3.333E–02	8.261E–02	7.302E–01	7.796E–03	1.454E–02	1.301E–01	1.306E–03	0	1.710E–04
64	3.360E–02	8.259E–02	7.282E–01	7.914E–03	1.469E–02	1.315E–01	1.371E–03	0	1.970E–04
65	3.399E–02	8.263E–02	7.287E–01	7.995E–03	1.461E–02	1.306E–01	1.316E–03	0	1.650E–04
66	3.437E–02	8.256E–02	7.280E–01	8.104E–03	1.463E–02	1.309E–01	1.321E–03	0	1.624E–04
67	3.474E–02	8.259E–02	7.277E–01	8.214E–03	1.466E–02	1.306E–01	1.328E–03	0	1.602E–04
68	3.510E–02	8.238E–02	7.270E–01	8.319E–03	1.464E–02	1.311E–01	1.331E–03	0	1.576E–04
69	3.546E–02	8.250E–02	7.261E–01	8.425E–03	1.469E–02	1.313E–01	1.339E–03	0	1.554E–04
70	3.591E–02	8.226E–02	7.256E–01	8.530E–03	1.469E–02	1.315E–01	1.344E–03	0	1.528E–04
71	3.625E–02	8.222E–02	7.240E–01	8.659E–03	1.480E–02	1.325E–01	1.404E–03	0	1.737E–04
72	3.674E–02	8.201E–02	7.225E–01	8.789E–03	1.490E–02	1.334E–01	1.467E–03	0	1.851E–04
73	3.698E–02	8.156E–02	7.185E–01	8.903E–03	1.497E–02	1.341E–01	1.528E–03	3.231E–03	1.920E–04
74	3.740E–02	8.134E–02	7.155E–01	9.043E–03	1.506E–02	1.351E–01	1.594E–03	4.798E–03	1.971E–04
75	3.776E–02	8.098E–02	7.126E–01	9.163E–03	1.514E–02	1.359E–01	1.657E–03	6.525E–03	2.006E–04
76	3.810E–02	8.073E–02	7.095E–01	9.308E–03	1.522E–02	1.368E–01	1.732E–03	8.427E–03	2.038E–04
77	3.850E–02	8.026E–02	7.058E–01	9.422E–03	1.526E–02	1.373E–01	1.794E–03	1.161E–02	0
78	3.889E–02	7.995E–02	7.032E–01	9.569E–03	1.531E–02	1.379E–01	1.861E–03	1.327E–02	9.227E–05
79	3.933E–02	7.962E–02	6.999E–01	9.717E–03	1.537E–02	1.385E–01	1.924E–03	1.551E–02	9.384E–05
80	3.973E–02	7.937E–02	6.972E–01	9.873E–03	1.542E–02	1.391E–01	1.998E–03	1.712E–02	2.191E–04
81	4.017E–02	7.900E–02	6.945E–01	1.003E–02	1.547E–02	1.397E–01	2.072E–03	1.882E–02	2.725E–04
82	4.068E–02	7.870E–02	6.918E–01	1.018E–02	1.551E–02	1.402E–01	2.151E–03	2.041E–02	3.180E–04
83	4.112E–02	7.852E–02	6.891E–01	1.034E–02	1.556E–02	1.409E–01	2.236E–03	2.191E–02	3.547E–04
84	4.166E–02	7.826E–02	6.867E–01	1.051E–02	1.561E–02	1.413E–01	2.318E–03	2.325E–02	3.890E–04
85	4.216E–02	7.789E–02	6.840E–01	1.068E–02	1.566E–02	1.422E–01	2.409E–03	2.457E–02	4.208E–04
86	4.269E–02	7.765E–02	6.818E–01	1.084E–02	1.570E–02	1.426E–01	2.495E–03	2.582E–02	4.512E–04
87	4.325E–02	7.742E–02	6.791E–01	1.101E–02	1.577E–02	1.433E–01	2.582E–03	2.706E–02	4.937E–04
88	4.373E–02	7.716E–02	6.765E–01	1.121E–02	1.581E–02	1.441E–01	2.667E–03	2.827E–02	5.364E–04
89	4.435E–02	7.685E–02	6.742E–01	1.137E–02	1.585E–02	1.446E–01	2.754E–03	2.945E–02	5.760E–04
90	4.491E–02	7.659E–02	6.719E–01	1.155E–02	1.589E–02	1.451E–01	2.836E–03	3.056E–02	6.150E–04
91	4.552E–02	7.641E–02	6.700E–01	1.173E–02	1.596E–02	1.459E–01	2.904E–03	3.101E–02	6.139E–04
92	4.617E–02	7.616E–02	6.677E–01	1.191E–02	1.600E–02	1.467E–01	2.971E–03	3.176E–02	6.348E–04

Table A16.5. Binding energies in eV for the K, L, and M shells. From Sevier, K.D. (1979), *At. Data Nucl. Data Tables* **24**, 323. The energy of a particular X-ray diagram line is the difference between the binding energies of the levels involved. For example, the KL. (Kα_1) line of iron has an energy of 7113 − 708 = 6405 eV.

Z	K	L1	L2	L3	M1	M2	M3	M4	M5
1	13.598								
2	24.587								
3	54.75	5.3							
4	111.9	8							
5	188	12.6	4.7						
6	284.1	10	9						
7	400.5	15	8.9	9.7					
8	532	23.7	6.8	7.4					
9	685.4	34	8.4	8.7					
10	870.1	48.47	21.66	21.56					
11	1072.1	63.3	31.1	31	0.7				
12	1305	89.4	51.5	51.3	2.1				
13	1559.6	117.7	73.15	72.72	0.7	0.5			
14	1838.9	148.7	99.5	98.9	7.6	3			
15	2145.5	189.3	136.2	135.3	16.2	9.6	10.1		
16	2472	229.2	165.4	164.2	15.8	7.8	8.2		
17	2822.4	270.2	201.6	200	17.5	6.7	6.7		
18	3202.9	326	250.55	248.5	29.24	15.93	15.76		
19	3607.4	377.1	296.3	293.6	33.9	18.1	17.8	0	0
20	4038.1	437.8	350	346.4	43.7	25.8	25.5	0	0
21	4492.8	500.4	406.7	402.2	53.8	33.8	31.5	6.6	0
22	4966.4	563.7	461.5	455.5	60.3	35.6	32.2	3.7	0
23	5465.1	628.2	520.5	512.9	66.5	40	35	2.2	0
24	5989.2	694.6	583.7	574.5	74.1	45.9	39.9	2.9	2.2
25	6539	769	651.4	640.3	83.9	53.1	46.4	3.5	2.7
26	7113	846.1	721.1	708.1	92.9	58.1	52	3.9	3.1
27	7708.9	925.6	793.6	778.6	100.7	63.2	57.7	2.7	3.3
28	8332.8	1008.1	871.9	854.7	111.8	71.2	69.7	3.9	3.3
29	8978.9	1096.1	951	931.1	119.8	75.3	72.8	1.8	1.5
30	9658.6	1193.6	1042.8	1019.7	135.9	88.6	85.6	7.9	8
31	10367.1	1297.7	1142.3	1115.4	158.1	106.8	102.9	20.7	15.7
32	11103.1	1414.3	1247.8	1216.7	180	127.9	120.8	29.2	28.5
33	11866.7	1526.5	1358.6	1323.1	203.5	146.4	140.5	41.7	40.9
34	12657.8	1653.9	1476.2	1435.8	231.5	168.2	161.9	57.4	56.4
35	13473.7	1782	1596	1549.9	256.5	189.3	181.5	70.1	69
36	14325.6	1921	1727.2	1674.9	292.1	222.1	214.4	95.04	93.82
37	15199.7	2065.1	1863.9	1804.4	322.1	247.4	238.5	111.8	110.3
38	16104.6	2216.3	2006.8	1939.6	357.5	279.8	269.1	135	133.1
39	17038.4	2372.5	2155.5	2080	393.6	312.4	300.3	159.6	157.4
40	17997.6	2531.6	2306.7	2222.3	430.3	344.2	330.5	182.4	180

Table A16.5. Binding energies in eV for the K, L, and M shells. From Sevier, K.D. (1979), *At. Data Nucl. Data Tables* **24**, 323. The energy of a particular X-ray diagram line is the difference between the binding energies of the levels involved. For example, the KL3 (Kα₁) line of iron has an energy of 7113 − 708 = 6405 eV (continued).

Z	K	L1	L2	L3	M1	M2	M3	M4	M5
41	18985.6	2697.7	2464.7	2370.5	468.4	378.4	363	207.4	204.6
42	19999.5	2865.5	2625.1	2520.2	504.6	409.7	392.3	230.3	227
43	21044	3042.5	2793.2	2676.9	544	444.9	425	256.4	252.9
44	22117.2	3224	2966.9	2837.9	585	482.8	460.6	283.6	279.4
45	23219.9	3411.9	3146.1	3003.8	627.1	521	496.2	311.7	307
46	24350.3	3604.3	3330.3	3173.3	669.9	559.1	531.5	340	334.7
47	25514	3805.8	3523.7	3351.1	717.5	602.4	571.4	372.8	366.7
48	26711.2	4018	3727	3537.5	770.2	650.7	616.5	410.5	403.7
49	27939.9	4237.5	3938	3730.1	825.6	702.2	664.3	450.8	443.1
50	29200.1	4464.7	4156.1	3928.8	883.8	756.4	714.4	493.3	484.8
51	30491.2	4698.3	4380.4	4132.3	943.7	811.9	765.6	536.9	527.5
52	31813.8	4939.2	4612	4341.4	1006	869.7	818.7	582.5	572.1
53	33169.4	5188.1	4852.1	4557.1	1072.1	930.5	874.6	631.3	619.4
54	34561.4	5452.8	5103.7	4782.2	1148.4	999	937	690.6	674.7
55	35984.6	5714.3	5359.4	5011.9	1217.1	1065	997.6	739.5	725.5
56	37440.6	5988.8	5623.6	5247	1292.8	1136.7	1062.2	796.1	780.7
57	38924.6	6266.3	5890.6	5482.5	1361.3	1204.4	1123.4	848.5	831.7
58	40443	6548.8	6164.2	5723.4	1434.6	1272.8	1185.4	901.3	883.3
59	41990.6	6834.8	6440.4	5964.3	1511	1337.4	1242.2	951.1	931
60	43568.9	7126	6721.5	6207.9	1575.3	1402.8	1297.4	999.9	977.7
61	45184	7427.9	7012.8	6459.3	1648.6	1471.4	1356.9	1051.5	1026.9
62	46834.2	7736.8	7311.8	6716.2	1722.8	1540.7	1419.8	1106	1080.2
63	48519	8052	7617.1	6976.9	1800	1613.9	1480.6	1160.6	1130.9
64	50239.1	8375.6	7930.3	7242.8	1880.8	1688.3	1544	1217.2	1185.2
65	51995.7	8708	8251.6	7514	1967.5	1767.7	1611.3	1275	1241.2
66	53788.5	9045.8	8580.8	7790.1	2046.8	1841.8	1675.6	1332.5	1294.9
67	55617.7	9394.2	8917.8	8071.1	2128.3	1922.8	1741.2	1391.5	1351.4
68	57485.5	9751.3	9264.3	8357.9	2216.7	2005.8	1811.8	1453.3	1409.3
69	59389.6	10115.7	9616.9	8648	2306.8	2089.8	1884.5	1514.6	1467.7
70	61332.3	10486.4	9978.2	8943.6	2398.1	2173	1949.8	1576.3	1527.8
71	63313.8	10870.4	10348.6	9244.1	2491.2	2263.5	2023.6	1639.4	1588.5
72	65350.8	11270.7	10739.4	9560.7	2600.9	2365.4	2107.6	1716.4	1661.7
73	67416.4	11681.5	11136.1	9881.1	2708	2468.7	2194	1793.2	1735.1
74	69525	12099.8	11544	10206.8	2819.6	2574.9	2281	1871.6	1809.2
75	71676.4	12526.7	11958.7	10535.3	2931.7	2681.6	2367.3	1948.9	1882.9
76	73870.8	12968	12385	10870.2	3048.5	2792.2	2457.2	2030.8	1960.1
77	76111	13418.5	12824.1	11215.2	3173.7	2908.7	2550.7	2116.1	2040.4
78	78394.8	13880.1	13272.6	11563.7	3297.2	3026.7	2645.7	2201.7	2121.4
79	80724.9	14352.8	13733.6	11918.7	3424.9	3147.8	2743	2291.1	2205.7
80	83102.3	14839.3	14208.7	12283.9	3561.6	3278.5	2847.1	2384.9	2294.9
81	85530.4	15346.7	14697.9	12657.5	3704.1	3415.7	2956.6	2485.1	2389.3
82	88004.5	15860.8	15200	13035.2	3850.7	3554.2	3066.4	2585.6	2484

Table A16.5. Binding energies in eV for the K, L, and M shells. From Sevier, K.D. (1979), *At. Data Nucl. Data Tables* **24**, 323. Th energy of a particular X-ray diagram line is the difference between the binding energies of the levels involved. For example, the KL (Kα₁) line of iron has an energy of 7113 − 708 = 6405 eV (continued).

Z	K	L1	L2	L3	M1	M2	M3	M4	M5
83	90525.9	16387.5	15711.1	13418.6	3999.1	3696.3	3176.9	2687.6	2579.6
84	93099.9	16927.9	16238	13810.6	4153.5	3844.3	3293.4	2793.6	2679.2
85	95724	17481.5	16777.3	14208	4311.7	3995.8	3410.5	2901.8	2780.7
86	98397.2	18048.7	17329.7	14611.4	4474.3	4151.5	3530.5	3012.3	2884.2
87	101129.9	18634.1	17900.5	15025.6	4645.7	4316	3657.3	3129.7	2994.9
88	103916.2	19236.7	18484.3	15444.4	4822	4485	3786.6	3248.4	3104.9
89	106756.3	19845.9	19083	15871.2	5000.6	4656.8	3916.7	3370.1	3219.7
90	109649.1	20472.1	19693.2	16300.3	5182.3	4830.4	4046.1	3490.8	3332
91	112596.1	21111.4	20313.7	16729.1	5366.9	5002.7	4173.8	3606.4	3439.4
92	115600.6	21757.4	20947.6	17166.3	5548	5182.2	4303.4	3727.6	3551.7

Table A16.6. Binding energies in eV for the N shell. From Sevier, K.D. (1979), *At. Data Nucl. Data Tables* **24**, 323.

Z	N1	N2	N3	N4	N5	N6	N7
19	1						
20	1.8						
21	1.7						
22	1.6						
23	1.7						
24	1						
25	1.9						
26	2.1						
27	1.9						
28	2.2						
29	1.2						
30	1.3						
31	5.6	0.8					
32	9	2.3					
33	12.5	2.5	2.5				
34	16.2	5.6	5.6				
35	27.3	5.2	4.6				
36	27.52	14.66	14				
37	29.3	14.8	14	0	0	0	0
38	37.7	20.7	19.5	0	0	0	0
39	45.4	25.1	22.8	2.4	0	0	0
40	51.3	29.3	25.7	3	0	0	0
41	58.1	35.6	29.6	3.2	0	0	0
42	61.8	38.3	32.3	1.9	1.2	0	0
43	68.8	42.8	36.9	2	1.2	0	0
44	74.9	47	41.2	2.4	1.8	0	0
45	81	51.9	46.3	2.8	2.2	0	0
46	86.4	54.4	50	1.7	1.3	0	0
47	95.2	62.6	55.9	3.6	3.1	0	0
48	107.6	70.8	65	9.7	9	0	0
49	121.9	81.9	75.1	16.8	15.8	0	0
50	136.5	93.9	86	24.6	23.4	0	0
51	152	104.3	95.4	32.2	30.8	0	0
52	168.3	116.8	96.9	40.8	39.2	0	0
53	186.4	130.1	119	50.7	48.9	0	0
54	217.7	163.9	156.5	69.52	67.55	0	0
55	230.8	172.3	161.6	78.8	76.5	0	0
56	253	191.8	179.7	92.5	89.9	0	0
57	270.4	205.8	191.4	100.7	97.7	0	0
58	289.6	223.3	207.2	113.6	107.6	0.1	0
59	304.5	236.3	217.6	117.9	110.1	2	0
60	315.2	243.3	224.6	123.4	113.5	1.5	0

Table A16.6. Binding energies in eV for the N shell. From Sevier, K.D. (1979), *At. Data Nucl. Data Tables* **24**, 323 (continued).

Z	N1	N2	N3	N4	N5	N6	N7
61	331.4	254.7	236.2	127.6	115.6	3.5	0
62	345.7	265.6	247.4	137.5	123.3	5.5	0
63	360.2	283.9	256.6	141.4	127.7	1.5	0
64	375.8	288.5	270.9	149.5	134.5	2	0
65	397.9	310.2	285	154.5	142	4	1.6
66	416.3	331.8	292.9	161.4	149.4	5.5	3.3
67	435.7	343.5	306.6	167.8	156.5	4.8	2.8
68	449.1	366.2	320	176.7	167.6	5.3	3.6
69	471.7	385.9	336.6	185.5	175.7	6.2	4.7
70	487.2	396.7	343.5	198.1	184.9	7	5.8
71	506.2	410.1	359.3	204.8	195	7.8	6.2
72	538.1	437	380.4	223.8	213.7	18.2	16.3
73	565.5	464.8	404.5	241.3	229.3	27.5	25.6
74	595	491.6	425.3	258.8	245.4	37.4	35.1
75	625	517.9	444.4	273.7	260.2	48.1	45.7
76	654.3	546.5	468.2	289.4	272.8	53.8	51
77	690.1	577.1	494.3	311.4	294.9	63.8	60.8
78	722.8	608.4	519	330.7	313.4	74.3	70.9
79	758.8	643.7	545.4	352	333.9	87.3	83.7
80	800.3	676.9	571	378.3	359.8	103.3	99.4
81	845.5	721.3	609	406.6	386.2	123	118.7
82	893.6	763.9	644.5	435.2	412.9	141.8	136.9
83	938.2	805.3	678.9	463.6	440	162.3	157.2
84	987.5	850.9	715.2	495.7	469.9	184.6	178.9
85	1038.2	897.7	753.7	527.6	500.1	207	200.8
86	1090.5	946.2	791.2	560.4	531.1	230.1	223.6
87	1149	1000.7	835.1	598.7	567.5	258.6	251.6
88	1208.4	1057.6	879.1	635.9	602.7	287.9	280.4
89	1269.4	1112.8	924.3	673.9	641.1	316.4	308.4
90	1329.5	1168.2	967.2	713.7	676.6	344.4	335
91	1387.1	1224.3	1006.7	743.4	708.2	371.2	359.5
92	1440.8	1271.8	1044.9	780.2	737.7	390.7	379.9

Table A16.7. Binding energies in eV for the O and P shells. From Sevier, K.D. (1979), *At. Data Nucl. Data Tables* **24**, 323.

Z	O1	O2	O3	O4	O5	P1	P2	P3
37	0.1							
38	0.1							
39	0.1							
40	0.1							
41	0.1							
42	0.1							
43	0.1							
44	0.1							
45	0.1							
46	0.1							
47	0.1							
48	0.1							
49	0.1	0.8						
50	0.9	1.1						
51	6.7	2.2	2					
52	11.6	2.6	2					
53	13.6	3.8	2.9					
54	23.39	13.43	12.13					
55	22.7	13.1	11.4	0	0	1		
56	29.1	16.6	14.6	0	0	1		
57	32.3	16.6	13.3	0	0	1		
58	37.8	21.8	18.8	0	0	1		
59	37.4	24.6	21.2	0	0	1		
60	37.5	23.6	19.8	0	0	1		
61	36	24.5	20.1	0	0	1		
62	37.4	23.6	18.9	0	0	1		
63	31.8	25.2	20.4	0	0	1		
64	36.1	24.3	18.3	0	0	1		
65	39	26.3	21.3	0	0	1		
66	62.9	28.2	22.9	0	0	1		
67	51.2	24.9	19.5	0	0	1		
68	59.8	27.9	22.3	0	0	1		
69	53.2	36.2	30.4	0	0	1		
70	54.1	27.4	21.4	0	0	1		
71	56.8	33	25.5	4.6	0	1		
72	64.9	38.2	29	6.6	2	1		
73	71.1	43.7	34.7	5.7	2	1		
74	77.1	46.7	36.5	6.1	2	1		
75	82.8	48.4	36.8	3.8	2.5	1		
76	83.7	58	45.4	0.4	2	1		
77	95.2	63	49.6	4.2	3.2	1		
78	101.7	65.3	51.6	2.8	1.4	1		
79	107.8	71.7	56.9	3.3	1.8	1		
80	120.3	80.5	61.8	7.5	5.7	1.5		

Table A16.7. Binding energies in eV for the O and P shells. From Sevier, K.D. (1979), *At. Data Nucl. Data Tables* **24**, 32 (continued).

Z	O1	O2	O3	O4	O5	P1	P2	P3
81	136.3	99.6	74.5	15.3	13.1	0.8		
82	147.3	104.8	84.5	21.8	19.2	3.1	0.7	
83	159.3	116.8	92.9	26.5	24.4	7.5	1.2	0.2
84	177.5	131.8	103.7	33.8	30.6	11	3.2	1.4
85	193.4	145.6	113.6	40.9	37.4	15	5.7	2.8
86	209.6	159.5	123.9	48	44.2	18.7	7.6	4.1
87	230.9	178.7	138.7	60	55.6	26.3	13.2	8.8
88	254.4	200.4	152.8	69.4	63.8	35.5	19.2	13.7
89	273.5	216.9	167.8	83.3	77.7	39.8	24.1	17
90	290.2	232	180.8	94.1	87.3	41.4	25.8	17.3
91	309.6	244.6	186.3	97.3	89.2	46.7	28.1	18.9
92	323.3	259.3	195.9	104.4	95.2	49.5	30.8	18.6

Table A16.8. K- and L-subshell fluorescence yields and Coster–Kronig probabilities. The K fluorescence yields are from a semiempirical fit by W. Bambyneks to selected experimental data reported in Hubbell, J. H., Trehan, P. N., Singh, N., Chand, B., Mehta, D., Garg, M. L., Garg, R. R., Singh, S., and Puri, S. (1994), *J. Phys. Chem. Ref. Data* **23**, 339. The **bold** L-shell quantities are from Krause, M. O. (1979), *J. Phys. Chem. Ref. Data* **8**, 307. The remainder of the L-shell quantities are from Campbell, J. L. (2003), *At. Data Nucl. Data Tables* **85**, 291 and Campbell, J. L. (2009) *At. Data Nucl. Data Tables* **95**, 115.

Z	ω_K	ω_{L1}	ω_{L2}	ω_{L3}	f_{12}	f_{13}	f_{23}
3	0.000293						
4	0.000693						
5	0.00141						
6	0.00258						
7	0.00435						
8	0.00691						
9	0.0104						
10	0.0152						
11	0.0213						
12	0.0291						
13	0.0387						
14	0.0504						
15	0.0642						
16	0.0804						
17	0.0989						
18	0.1199						
19	0.1432						
20	0.1687						
21	0.1962						
22	0.2256						
23	0.2564						
24	0.2885						
25	0.3213	**0.00084**	**0.005**	**0.005**	**0.3**	**0.58**	
26	0.3546	**0.001**	**0.0063**	**0.0063**	**0.3**	**0.57**	
27	0.3880	**0.0012**	**0.0077**	**0.0077**	**0.3**	**0.56**	
28	0.4212	**0.0014**	**0.0086**	**0.0093**	**0.3**	**0.55**	**0.028**
29	0.4538	**0.0016**	**0.01**	**0.011**	**0.3**	**0.54**	**0.028**
30	0.4857	**0.0018**	**0.011**	**0.012**	**0.29**	**0.54**	**0.026**
31	0.5166	**0.021**	**0.012**	**0.013**	**0.29**	**0.53**	**0.032**
32	0.5464	**0.0024**	**0.013**	**0.015**	**0.28**	**0.53**	**0.05**
33	0.5748	**0.0028**	**0.014**	**0.016**	**0.28**	**0.53**	**0.063**
34	0.6019	**0.0032**	**0.016**	**0.018**	**0.28**	**0.52**	**0.076**
35	0.6275	**0.0036**	**0.018**	**0.02**	**0.28**	**0.52**	**0.088**
36	0.6517	**0.0041**	**0.02**	**0.022**	**0.27**	**0.52**	**0.073**
37	0.6744	**0.0046**	**0.022**	**0.024**	**0.27**	**0.52**	**0.08**
38	0.6956	**0.0051**	**0.024**	**0.026**	**0.27**	**0.52**	**0.087**
39	0.7155	**0.0059**	**0.026**	**0.028**	**0.26**	**0.52**	**0.094**
40	0.7340	**0.0068**	**0.028**	**0.031**	**0.26**	**0.52**	**0.1**

Table A16.8. K- and L-subshell fluorescence yields and Coster–Kronig probabilities. The K fluorescence yields are from a semiempirical fit by W. Bambyneks to selected experimental data reported in Hubbell, J. H., Trehan, P. N., Singh, N., Chand, B. Mehta, D., Garg, M. L., Garg, R. R., Singh, S., and Puri, S. (1994), *J. Phys. Chem. Ref. Data* **23**, 339. The **bold** L-shell quantities are from Krause, M. O. (1979), *J. Phys. Chem. Ref. Data* **8**, 307. The remainder of the L-shell quantities are from Campbell, J. L. (2003) *At. Data Nucl. Data Tables* **85**, 291 and Campbell, J. L. (2009) *At. Data Nucl. Data Tables* **95**, 115 (continued).

Z	ω_K	ω_{L1}	ω_{L2}	ω_{L3}	f_{12}	f_{13}	f_{23}
41	0.7512	**0.0094**	**0.031**	**0.034**	**0.1**	**0.61**	0.106
42	0.7672	**0.01**	**0.034**	**0.037**	**0.1**	**0.61**	0.112
43	0.7821	**0.011**	**0.037**	**0.04**	**0.1**	**0.61**	0.118
44	0.7958	**0.012**	**0.04**	**0.043**	**0.1**	**0.61**	0.124
45	0.8086	**0.013**	**0.043**	**0.046**	**0.1**	**0.6**	0.13
46	0.8204	**0.014**	**0.047**	**0.049**	**0.1**	**0.6**	0.138
47	0.8313	**0.016**	**0.051**	**0.052**	**0.1**	**0.59**	0.141
48	0.8415	**0.018**	**0.056**	**0.056**	**0.1**	**0.59**	0.143
49	0.8508	**0.02**	**0.061**	**0.06**	**0.1**	**0.59**	0.146
50	0.8595	**0.037**	**0.065**	**0.064**	**0.17**	**0.59**	0.148
51	0.8676	**0.039**	**0.069**	**0.069**	**0.17**	**0.27**	0.151
52	0.8750	**0.041**	**0.074**	**0.074**	**0.18**	**0.28**	0.153
53	0.8819	**0.044**	**0.079**	**0.079**	**0.18**	**0.28**	0.156
54	0.8883	**0.046**	**0.083**	**0.085**	**0.19**	**0.28**	0.159
55	0.8942	**0.049**	**0.09**	**0.091**	**0.19**	**0.28**	0.159
56	0.8997	**0.052**	**0.096**	**0.097**	**0.19**	**0.28**	0.159
57	0.9049	**0.055**	**0.103**	**0.104**	**0.19**	**0.29**	0.159
58	0.9096	**0.058**	**0.11**	**0.111**	**0.19**	**0.29**	0.158
59	0.9140	**0.061**	**0.117**	**0.118**	**0.19**	**0.29**	0.158
60	0.9181	**0.064**	**0.136**	**0.134**	**0.19**	**0.3**	0.158
61	0.9220	**0.066**	0.145	0.142	**0.19**	**0.3**	0.156
62	0.9255	**0.071**	0.155	0.15	**0.19**	**0.3**	0.154
63	0.9289	**0.075**	0.164	0.158	**0.19**	**0.3**	0.152
64	0.9320	**0.102**	0.175	0.167	**0.19**	**0.279**	0.150
65	0.9349	0.107	0.186	0.175	0.182	0.285	0.148
66	0.9376	0.111	0.197	0.184	0.174	0.29	0.146
67	0.9401	0.116	0.208	0.193	0.166	0.296	0.144
68	0.9425	0.121	0.219	0.203	0.158	0.301	0.143
69	0.9447	0.131	0.231	0.212	0.15	0.306	0.141
70	0.9467	0.134	0.243	0.222	0.142	0.312	0.140
71	0.9487	0.138	0.256	0.231	0.134	0.317	0.138
72	0.9505	0.141	0.268	0.241	0.126	0.322	0.136
73	0.9522	0.144	0.28	0.251	0.118	0.328	0.134
74	0.9538	0.148	0.291	0.261	0.11	0.333	0.132
75	0.9553		0.304	0.271		0.482	0.131
76	0.9567		0.318	0.282		0.482	0.130
77	0.9580	0.145	0.331	0.292	0.076	0.482	0.128
78	0.9592	0.114	0.344	0.303	0.075	0.545	0.126
79	0.9604	0.117	0.358	0.313	0.074	0.615	0.125
80	0.9615	0.121	0.37	0.322	0.072	0.615	0.123

Table A16.8. K- and L-subshell fluorescence yields and Coster–Kronig probabilities. The K fluorescence yields are from a semiempirical fit by W. Bambyneks to selected experimental data reported in Hubbell, J. H., Trehan, P. N., Singh, N., Chand, B., Mehta, D., Garg, M. L., Garg, R. R., Singh, S., and Puri, S. (1994), *J. Phys. Chem. Ref. Data* **23**, 339. The **bold** L-shell quantities are from Krause, M. O. (1979), *J. Phys. Chem. Ref. Data* **8**, 307. The remainder of the L-shell quantities are from Campbell, J. L. (2003), *At. Data Nucl. Data Tables* **85**, 291 and Campbell, J. L. (2009) *At. Data Nucl. Data Tables* **95**, 115 (continued).

Z	ω_K	ω_{L1}	ω_{L2}	ω_{L3}	f_{12}	f_{13}	f_{23}
81	0.9625	0.124	0.384	0.332	0.069	0.615	0.121
82	0.9634	0.128	0.397	0.343	0.066	0.62	0.119
83	0.9643	0.132	0.411	0.353	0.063	0.62	0.117
84	0.9652	0.135	0.424	0.363	0.06	0.62	0.115
85	0.9659	0.138	0.438	0.374	0.057	0.62	0.113
86	0.9667	0.142	0.451	0.384	0.053	0.62	0.111
87	0.9674	0.146	0.464	0.394	0.05	0.62	0.109
88	0.9680	0.15	0.476	0.404	0.047	0.62	0.107
89	0.9686	0.154	0.49	0.414	0.044	0.62	0.105
90	0.9691	0.159	0.503	0.424	0.04	0.62	0.103
91	0.9696	0.164	0.495	0.434	0.038	0.62	0.141
92	0.9701	0.168	0.506	0.444	0.035	0.62	0.140

Table A16.9. Cross sections (barns) for K-shell ionization by protons as a function of atomic number Z and energy (MeV). From Chen, M.-H., and Crasemann, B. (1985), *At. Data Nucl. Data Tables* **33**, 217, and Chen, M.-H., and Crasemann, B. (1989), *At. Data Nucl. Data Tables* **41**, 257.

E (MeV)	22	26	29	30	32
0.10	4.440E–02	2.953E–03	3.884E–04	1.928E–04	4.598E–05
0.20	1.960E+00	1.416E–01	3.234E–02	1.990E–02	7.614E–03
0.30	5.771E+00	8.120E–01	2.149E–01	1.392E–01	6.018E–02
0.40	1.577+01	2.440E+00	6.944E–01	4.631E–01	2.107E–01
0.50	3.258E+01	5.392E+00	1.604E+00	1.088E+00	5.114E–01
0.60	5.671E+01	9.905E+00	3.063E+00	2.095E+00	1.012E+00
0.70	8.852E+01	1.614E+01	5.141E+00	3.561E+00	1.746E+00
0.80	1.273E+02	2.426E+01	7.900E+00	5.517E+00	2.747E+00
0.90	1.731E+02	3.416E+01	1.141E+01	7.999E+00	4.037E+00
1.00	2.245E+02	4.597E+01	1.563E+01	1.106E+01	5.648E+00
1.25	3.742E+02	8.254E+01	2.945E+01	2.106E+01	1.104E+01
1.50	5.445E+02	1.283E+02	4.746E+01	3.445E+01	1.845E+01
1.75	7.252E+02	1.811E+02	6.935E+01	5.072E+01	2.769E+01
2.00	9.068E+02	2.383E+02	9.434E+01	6.967E+01	3.870E+01
2.25	1.088E+03	2.993E+02	1.215E+02	9.048E+01	5.106E+01
2.50	1.260E+03	3.612E+02	1.504E+02	1.132E+02	6.478E+01
2.75	1.427E+03	4.244E+02	1.809E+02	1.367E+02	7.932E+01
3.00	1.580E+03	4.862E+02	2.117E+02	1.615E+02	
3.25	1.731E+03	5.484E+02	2.434E+02	1.864E+02	1.107E+02
3.50	1.864E+03	6.076E+02	2.746E+02	2.114E+02	1.270E+02
3.75	1.992E+03	6.645E+02	3.054E+02	2.370E+02	1.438E+02
4.00	2.105E+03	7.208E+02	3.366E+02	2.619E+02	1.604E+02
4.25	2.208E+03	7.732E+02	3.664E+02	2.864E+02	1.769E+02
4.50	2.307E+03	8.230E+02	3.954E+02	3.112E+02	1.934E+02
4.75	2.393E+03	8.723E+02	4.246E+02	3.347E+02	2.102E+02
5.00	2.471E+03	9.171E+02	4.519E+02	3.576E+02	2.262E+02

Table A16.9. Cross sections (barns) for K-shell ionization by protons as a function of atomic number Z and energy (MeV). From Chen, M.-H., and Crasemann, B. (1985), *At. Data Nucl. Data Tables* **33**, 217, and Chen, M.-H., and Crasemann, B. (1989), *At. Data Nucl. Data Tables* **41**, 257 (continued).

E (MeV)	36	40	43	45	47
0.10					
0.20	1.155E–03	1.724E–04	4.058E–05	1.502E–05	5.386E–06
0.30	1.206E–02	2.566E–03	8.332E–04	3.949E–04	1.872E–04
0.40	4.776E–02	1.178E–02	4.355E–03	2.276E–03	1.202E–03
0.50	1.248E–01	3.336E–02	1.317E–02	7.244E–03	4.043E–03
0.60	2.582E–01	7.268E–02	2.994E–02	1.698E–02	9.800E–03
0.70	4.619E–01	1.349E–01	5.741E–02	3.325E–02	1.956E–02
0.80	7.480E–01	2.248E–01	9.777E–02	5.749E–02	3.446E–02
0.90	1.130E+00	3.466E–01	1.533E–01	9.123E–02	5.535E–02
1.00	1.611E+00	5.054E–01	2.261E–01	1.358E–01	8.318E–02
1.25	3.292E+00	1.075E+00	4.944E–01	3.023E–01	1.885E–01
1.50	5.714E+00	1.928E+00	9.055E–01	5.610E–01	3.531E–01
1.75	8.863E+00	3.073E+00	1.469E+00	9.204E–01	5.870E–01
2.00	1.276E+01	4.521E+00	2.195E+00	1.388E+00	8.929E–01
2.25	1.729E+01	6.283E+00	3.094E+00	1.972E+00	1.275E+00
2.50	2.243E+01	8.320E+00	4.149E+00	2.665E+00	1.739E+00
2.75	2.818E+01	1.062E+01	5.361E+00	3.468E+00	2.278E+00
3.00	3.434E+01	1.321E+01	6.742E+00	4.378E+00	2.893E+00
3.25	4.090E+01	1.600E+01	8.252E+00	5.405E+00	3.582E+00
3.50	4.790E+01	1.899E+01	9.892E+00	6.518E+00	4.355E+00
3.75	5.508E+01	2.216E+01	1.165E+01	7.722E+00	5.185E+00
4.00	6.246E+01	2.556E+01	1.356E+01	9.009E+00	6.079E+00
4.25	7.018E+01	2.904E+01	1.553E+01	1.040E+01	7.033E+00
4.50	7.786E+01	3.263E+01	1.760E+01	1.184E+01	8.042E+00
4.75	8.561E+01	3.631E+01	1.974E+01	1.335E+01	9.127E+00
5.00	9.340E+01	4.018E+01	2.201E+01	1.491E+01	1.024E+01

Table A16.9. Cross sections (barns) for K-shell ionization by protons as a function of atomic number Z and energy (MeV). From Chen, M.-H., and Crasemann, B. (1985), *At. Data Nucl. Data Tables* **33**, 217, and Chen, M.-H., and Crasemann, B. (1989), *At. Data Nucl. Data Tables* **41**, 257 (continued).

E (MeV)	48	50	54	56	60
0.10					
0.20	3.192E–06	1.076E–06	1.052E–07		
0.30	1.283E–04	6.002E–05	1.268E–05	5.658E–06	1.059E–06
0.40	8.733E–04	4.632E–04	1.310E–04	6.927E–05	1.917E–05
0.50	3.036E–03	1.716E–03	5.601E–04	3.220E–04	1.080E–04
0.60	7.489E–03	4.396E–03	1.562E–03	9.428E–04	3.529E–04
0.70	1.512E–02	9.097E–03	3.426E–03	2.126E–03	8.535E–04
0.80	2.678E–02	1.639E–02	6.416E–03	4.079E–03	1.712E–03
0.90	4.327E–02	2.684E–02	1.080E–02	6.974E–03	3.028E–03
1.00	6.535E–02	4.094E–02	1.683E–02	1.100E–02	4.897E–03
1.25	1.498E–01	9.527E–02	4.065E–02	2.709E–02	1.262E–02
1.50	2.825E–01	1.824E–01	7.952E–02	5.365E–02	2.564E–02
1.75	4.706E–01	3.067E–01	1.366E–01	9.269E–02	4.509E–02
2.00	7.186E–01	4.721E–01	2.133E–01	1.462E–01	7.187E–02
2.25	1.033E+00	6.813E–01	3.117E–01	2.149E–01	1.068E–01
2.50	1.410E+00	9.384E–01	4.328E–01	2.999E–01	1.503E–01
2.75	1.852E+00	1.240E+00	5.775E–01	4.019E–01	2.037E–01
3.00	2.358E+00	1.587E+00		5.214E–01	2.661E–01
3.25	2.936E+00	1.980E+00	9.418E–01	6.606E–01	3.384E–01
3.50	3.569E+00	2.425E+00	1.159E+00	8.160E–01	4.205E–01
3.75	4.259E+00	2.907E+00	1.400E+00	9.890E–01	5.125E–01
4.00	5.004E+00	3.430E+00	1.665E+00	1.179E+00	6.144E–01
4.25	5.817E+00	3.993E+00	1.951E+00	1.387E+00	7.261E–01
4.50	6.667E+00	4.594E+00	2.267E+00	1.611E+00	8.500E–01
4.75	7.562E+00	5.244E+00	2.597E+00	1.851E+00	9.815E–01
5.00	8.499E+00	5.915E+00	2.948E+00	2.113E+00	1.122E+00

Table A16.10. Cross sections (barns) for L1-subshell ionization by protons as a function of atomic number Z and energy (MeV). From Chen, M.-H., and Crasemann, B. (1985), *At. Data Nucl. Data Tables* **33**, 217, and Chen, M.-H., and Crasemann, B. (1989), *At. Data Nucl. Data Tables* **41**, 257.

E (MeV)	22	26	29	30	32
0.15	3.195E+04	4.736E+03	1.277E+03	7.660E+02	2.830E+02
0.25	8.545E+04	1.847E+04	6.356E+03	4.159E+03	1.754E+03
0.50	1.826E+05	5.700E+04	2.546E+04	1.865E+04	9.940E+03
0.75	2.208E+05	8.124E+04	4.122E+04	3.171E+04	1.856E+04
1.00	2.346E+05	9.611E+04	5.169E+04	4.094E+04	2.561E+04
1.25	2.316E+05	1.025E+05	5.788E+04	4.688E+04	3.055E+04
1.50	2.232E+05	1.044E+05	6.126E+04	5.039E+04	3.384E+04
1.75	2.138E+05	1.038E+05	6.300E+04	5.230E+04	3.597E+04
2.00	2.024E+05	1.019E+05	6.338E+04	5.312E+04	3.724E+04
2.25	1.922E+05	9.931E+04	6.283E+04	5.323E+04	3.792E+04
2.50	1.827E+05	9.602E+04	6.200E+04	5.270E+04	3.816E+04
2.75	1.731E+05	9.299E+04	6.088E+04	5.204E+04	3.810E+04
3.00	1.642E+05	8.957E+04	5.938E+04	5.118E+04	3.772E+04
3.25	1.571E+05	8.665E+04	5.802E+04	5.003E+04	3.731E+04
3.50	1.497E+05	8.346E+04	5.640E+04	4.899E+04	3.668E+04
3.75	1.430E+05	8.082E+04	5.502E+04	4.775E+04	3.611E+04
4.00	1.369E+05	7.795E+04	5.344E+04	4.669E+04	3.538E+04
4.25	1.313E+05	7.522E+04	5.212E+04	4.546E+04	3.475E+04
4.50	1.261E+05	7.305E+04	5.064E+04	4.425E+04	3.399E+04
4.75	1.214E+05	7.065E+04	4.920E+04	4.328E+04	3.323E+04
5.00	1.170E+05	6.838E+04	4.782E+04	4.215E+04	3.261E+04

E (MeV)	36	40	43	45	47
0.15	6.564E+01	3.171E+01	2.081E+01	1.549E+01	1.137E+01
0.25	3.243E+02	6.530E+01	2.943E+01	2.067E+01	1.584E+01
0.50	2.849E+03	7.507E+02	2.840E+02	1.489E+02	7.891E+01
0.75	6.558E+03	2.205E+03	9.791E+02	5.665E+02	3.259E+02
1.00	1.016E+04	3.892E+03	1.920E+03	1.191E+03	7.385E+02
1.25	1.315E+04	5.500E+03	2.910E+03	1.893E+03	1.229E+03
1.50	1.549E+04	6.931E+03	3.844E+03	2.587E+03	1.743E+03
1.75	1.726E+04	8.132E+03	4.684E+03	3.242E+03	2.240E+03
2.00	1.858E+04	9.122E+03	5.418E+03	3.829E+03	2.703E+03
2.25	1.954E+04	9.923E+03	6.045E+03	4.339E+03	3.123E+03
2.50	2.020E+04	1.054E+04	6.575E+03	4.789E+03	3.491E+03
2.75	2.059E+04	1.104E+04	7.016E+03	5.176E+03	3.822E+03
3.00	2.086E+04	1.143E+04	7.365E+03	5.504E+03	4.110E+03
3.25	2.100E+04	1.169E+04	7.662E+03	5.771E+03	4.351E+03
3.50	2.097E+04	1.191E+04	7.884E+03	6.004E+03	4.567E+03
3.75	2.092E+04	1.203E+04	8.077E+03	6.185E+03	4.751E+03
4.00	2.076E+04	1.213E+04	8.207E+03	6.344E+03	4.899E+03
4.25	2.060E+04	1.215E+04	8.323E+03	6.460E+03	5.020E+03
4.50	2.036E+04	1.218E+04	8.388E+03	6.564E+03	5.132E+03
4.75	2.008E+04	1.214E+04	8.448E+03	6.631E+03	5.212E+03
5.00	1.986E+04	1.207E+04	8.466E+03	6.677E+03	5.288E+03

Table A16.10. Cross sections (barns) for L1-subshell ionization by protons as a function of atomic number Z and energy (MeV). From Chen, M.-H., and Crasemann, B. (1985), *At. Data Nucl. Data Tables* **33**, 217, and Chen, M.-H., and Crasemann, B. (1989), *At. Data Nucl. Data Tables* **41**, 257 (continued).

E (MeV)	48	50	54	56	60
0.15	9.648E+00	6.893E+00	3.442E+00	2.401E+00	1.227E+00
0.25	1.401E+01	1.106E+01	6.797E+00	5.218E+00	3.089E+00
0.50	5.734E+01	3.127E+01	1.183E+01	8.359E+00	5.250E+00
0.75	2.443E+02	1.392E+02	4.466E+01	2.559E+01	1.003E+01
1.00	5.753E+02	3.467E+02	1.243E+02	7.417E+01	2.792E+01
1.25	9.846E+02	6.262E+02	2.499E+02	1.548E+02	6.244E+01
1.50	1.420E+03	9.385E+02	4.058E+02	2.625E+02	1.130E+02
1.75	1.851E+03	1.259E+03	5.797E+02	3.867E+02	1.768E+02
2.00	2.258E+03	1.573E+03	7.594E+02	5.199E+02	2.498E+02
2.25	2.634E+03	1.869E+03	9.367E+02	6.561E+02	3.283E+02
2.50	2.973E+03	2.144E+03	1.110E+03	7.898E+02	4.096E+02
2.75	3.270E+03	2.394E+03	1.275E+03	9.210E+02	4.915E+02
3.00	3.537E+03	2.616E+03	1.431E+03	1.047E+03	5.714E+02
3.25	3.772E+03	2.819E+03	1.572E+03	1.163E+03	6.504E+02
3.50	3.969E+03	2.994E+03	1.705E+03	1.274E+03	7.250E+02
3.75	4.146E+03	3.155E+03	1.823E+03	1.375E+03	7.977E+02
4.00	4.291E+03	3.289E+03	1.935E+03	1.471E+03	8.649E+02
4.25	4.422E+03	3.413E+03	2.032E+03	1.556E+03	9.301E+02
4.50	4.524E+03	3.514E+03	2.123E+03	1.637E+03	9.893E+02
4.75	4.608E+03	3.608E+03	2.201E+03	1.708E+03	1.047E+03
5.00	4.688E+03	3.682E+03	2.270E+03	1.776E+03	1.098E+03

E (MeV)	62	64	66	67	70
0.15	8.865E–01	6.414E–01		4.062E–01	2.610E–01
0.25	2.365E+00	1.806E+00		1.217E+00	8.239E–01
0.50	4.361E+00	3.644E+00		2.807E+00	2.143E+00
0.75	6.856E+00	4.988E+00		3.490E+00	2.642E+00
1.00	1.742E+01	1.112E+01	7.520E+00	6.128E+00	3.763E+00
1.25	4.004E+01	2.536E+01	1.648E+01	1.334E+01	7.265E+00
1.50	7.376E+01	4.854E+01	3.224E+01	2.623E+01	1.418E+01
1.75	1.194E+02	7.970E+01	5.388E+01	4.417E+01	2.472E+01
2.00	1.727E+02	1.186E+02	8.198E+01	6.788E+01	3.847E+01
2.25	2.317E+02	1.625E+02	1.147E+02	9.601E+01	5.606E+01
2.50	2.940E+02	2.095E+02	1.505E+02	1.274E+02	7.644E+01
2.75	3.581E+02	2.591E+02	1.889E+02	1.607E+02	9.892E+01
3.00	4.216E+02	3.097E+02	2.287E+02	1.959E+02	1.227E+02
3.25	4.852E+02	3.598E+02	2.686E+02	2.314E+02	1.479E+02
3.50	5.473E+02	4.101E+02	3.092E+02	2.678E+02	1.734E+02
3.75	6.063E+02	4.584E+02	3.486E+02	3.039E+02	1.996E+02
4.00	6.640E+02	5.062E+02	3.881E+02	3.390E+02	2.253E+02
4.25	7.177E+02	5.512E+02	4.256E+02	3.732E+02	2.514E+02
4.50	7.700E+02	5.954E+02	4.629E+02	4.072E+02	2.770E+02
4.75	8.180E+02	6.365E+02	4.978E+02	4.393E+02	3.024E+02
5.00	8.629E+02	6.765E+02	5.321E+02	4.709E+02	3.269E+02

Table A16.10. Cross sections (barns) for L1-subshell ionization by protons as a function of atomic number Z and energy (MeV). From Chen, M.-H., and Crasemann, B. (1985), *At. Data Nucl. Data Tables* **33**, 217, and Chen, M.-H., and Crasemann, B. (1989), *At. Data Nucl. Data Tables* **41**, 257 (continued).

E (MeV)	74	78	79	80
0.15	1.464E–01	8.533E–02	7.495E–02	6.606E–02
0.25	4.889E–01	2.965E–01	2.632E–01	2.330E–01
0.50	1.468E+00	9.946E–01	9.017E–01	8.185E–01
0.75	1.920E+00	1.418E+00	1.313E+00	1.214E+00
1.00	2.301E+00	1.617E+00	1.501E+00	1.394E+00
1.25	3.518E+00	2.040E+00	1.822E+00	1.647E+00
1.50	6.329E+00	3.134E+00	2.674E+00	2.312E+00
1.75	1.114E+01	5.225E+00	4.352E+00	3.661E+00
2.00	1.806E+01	8.487E+00	7.068E+00	5.858E+00
2.25	2.683E+01	1.297E+01	1.082E+01	8.978E+00
2.50	3.760E+01	1.858E+01	1.557E+01	1.303E+01
2.75	5.016E+01	2.524E+01	2.125E+01	1.792E+01
3.00	6.418E+01	3.314E+01	2.800E+01	2.366E+01
3.25	7.910E+01	4.189E+01	3.562E+01	3.029E+01
3.50	9.501E+01	5.134E+01	4.401E+01	3.755E+01
3.75	1.113E+02	6.150E+01	5.288E+01	4.534E+01
4.00	1.282E+02	7.198E+01	6.217E+01	5.367E+01
4.25	1.450E+02	8.294E+01	7.194E+01	6.226E+01
4.50	1.626E+02	9.415E+01	8.198E+01	7.123E+01
4.75	1.800E+02	1.057E+02	9.235E+01	8.060E+01
5.00	1.969E+02	1.172E+02	1.029E+02	9.002E+01

E (MeV)	83	86	90	92
0.15	4.557E–02	3.213E–02	2.076E–02	1.706E–02
0.25	1.637E–01	1.168E–01	7.616E–02	6.269E–02
0.50	6.080E–01	4.543E–01	3.098E–01	2.588E–01
0.75	9.593E–01	7.530E–01	5.437E–01	4.641E–01
1.00	1.130E+00	9.192E–01	6.972E–01	6.071E–01
1.25	1.264E+00	1.015E+00	7.804E–01	6.897E–01
1.50	1.577E+00	1.165E+00	8.524E–01	7.479E–01
1.75	2.257E+00	1.498E+00	9.808E–01	8.317E–01
2.00	3.445E+00	2.132E+00	1.240E+00	9.984E–01
2.25	5.223E+00	3.124E+00	1.676E+00	1.286E+00
2.50	7.626E+00	4.524E+00	2.330E+00	1.731E+00
2.75	1.071E+01	6.352E+00	3.246E+00	2.372E+00
3.00	1.430E+01	8.648E+00	4.405E+00	3.196E+00
3.25	1.841E+01	1.128E+01	5.823E+00	4.247E+00
3.50	2.321E+01	1.426E+01	7.531E+00	5.483E+00
3.75	2.847E+01	1.775E+01	9.418E+00	6.917E+00
4.00	3.412E+01	2.157E+01	1.154E+01	8.545E+00
4.25	4.009E+01	2.568E+01	1.393E+01	1.029E+01
4.50	4.650E+01	3.009E+01	1.666E+01	1.245E+01
4.75	5.330E+01	3.497E+01	1.966E+01	1.477E+01
5.00	6.009E+01	3.982E+01	2.267E+01	1.712E+01

Table A16.11. Cross sections (barns) for L2-subshell ionization by protons as a function of atomic number Z and energy (MeV). From Chen, M.-H., and Crasemann, B. (1985), *At. Data Nucl. Data Tables* **33**, 217, and Chen, M.-H., and Crasemann, B. (1989), *At. Data Nucl. Data Tables* **41**, 257.

E (MeV)	22	26	29	30	32
0.15	7.350E+04	1.593E+04	5.988E+03	4.136E+03	1.993E+03
0.25	1.355E+05	3.418E+04	1.405E+04	1.009E+04	5.238E+03
0.50	2.364E+05	7.301E+04	3.381E+04	2.546E+04	1.445E+04
0.75	2.821E+05	9.800E+04	4.866E+04	3.763E+04	2.241E+04
1.00	3.036E+05	1.134E+05	5.902E+04	4.650E+04	2.895E+04
1.25	3.075E+05	1.220E+05	6.614E+04	5.268E+04	3.371E+04
1.50	3.041E+05	1.265E+05	7.045E+04	5.694E+04	3.724E+04
1.75	2.971E+05	1.284E+05	7.331E+04	5.979E+04	3.982E+04
2.00	2.871E+05	1.286E+05	7.498E+04	6.141E+04	4.167E+04
2.25	2.779E+05	1.272E+05	7.580E+04	6.250E+04	4.283E+04
2.50	2.673E+05	1.256E+05	7.576E+04	6.305E+04	4.370E+04
2.75	2.569E+05	1.231E+05	7.556E+04	6.298E+04	4.409E+04
3.00	2.470E+05	1.208E+05	7.479E+04	6.284E+04	4.438E+04
3.25	2.390E+05	1.179E+05	7.379E+04	6.226E+04	4.433E+04
3.50	2.302E+05	1.150E+05	7.296E+04	6.152E+04	4.412E+04
3.75	2.220E+05	1.120E+05	7.173E+04	6.092E+04	4.396E+04
4.00	2.142E+05	1.097E+05	7.042E+04	6.000E+04	4.355E+04
4.25	2.070E+05	1.069E+05	6.942E+04	5.903E+04	4.307E+04
4.50	2.003E+05	1.042E+05	6.808E+04	5.802E+04	4.272E+04
4.75	1.940E+05	1.015E+05	6.675E+04	5.726E+04	4.216E+04
5.00	1.881E+05	9.898E+04	6.542E+04	5.624E+04	4.158E+04

E (MeV)	30	40	43	45	47
0.15	5.022E+02	1.287E+02	5.038E+01	2.746E+01	1.525E+01
0.25	1.525E+03	4.518E+02	1.956E+02	1.132E+02	6.664E+01
0.50	4.996E+03	1.771E+03	8.609E+02	5.399E+02	3.430E+02
0.75	8.543E+03	3.334E+03	1.720E+03	1.121E+03	7.420E+02
1.00	1.178E+04	4.858E+03	2.622E+03	1.758E+03	1.188E+03
1.25	1.443E+04	6.247E+03	3.481E+03	2.382E+03	1.642E+03
1.50	1.656E+04	7.474E+03	4.274E+03	2.965E+03	2.081E+03
1.75	1.831E+04	8.540E+03	4.990E+03	3.511E+03	2.496E+03
2.00	1.972E+04	9.430E+03	5.615E+03	4.007E+03	2.874E+03
2.25	2.078E+04	1.021E+04	6.179E+03	4.444E+03	3.228E+03
2.50	2.167E+04	1.084E+04	6.658E+03	4.846E+03	3.541E+03
2.75	2.229E+04	1.140E+04	7.091E+03	5.190E+03	3.834E+03
3.00	2.282E+04	1.184E+04	7.450E+03	5.510E+03	4.088E+03
3.25	2.315E+04	1.223E+04	7.780E+03	5.777E+03	4.327E+03
3.50	2.345E+04	1.252E+04	8.044E+03	6.027E+03	4.529E+03
3.75	2.357E+04	1.275E+04	8.267E+03	6.232E+03	4.722E+03
4.00	2.362E+04	1.298E+04	8.481E+03	6.408E+03	4.881E+03
4.25	2.370E+04	1.312E+04	8.639E+03	6.579E+03	5.022E+03
4.50	2.365E+04	1.322E+04	8.769E+03	6.711E+03	5.160E+03
4.75	2.355E+04	1.333E+04	8.903E+03	6.822E+03	5.268E+03
5.00	2.351E+04	1.338E+04	8.989E+03	6.915E+03	5.362E+03

Table A16.11. Cross sections (barns) for L2-subshell ionization by protons as a function of atomic number Z and energy (MeV). From Chen, M.-H., and Crasemann, B. (1985), *At. Data Nucl. Data Tables* **33**, 217, and Chen, M.-H., and Crasemann, B. (1989), *At. Data Nucl. Data Tables* **41**, 257 (continued).

E (MeV)	48	50	54	56	60
0.15	1.130E+01	6.253E+00	2.006E+00	1.141E+00	4.141E–01
0.25	5.094E+01	2.976E+01	1.056E+01	6.286E+00	2.453E+00
0.50	2.717E+02	1.718E+02	7.029E+01	4.485E+01	1.965E+01
0.75	5.992E+02	3.948E+02	1.751E+02	1.166E+02	5.476E+01
1.00	9.767E+02	6.591E+02	3.079E+02	2.106E+02	1.038E+02
1.25	1.363E+03	9.381E+02	4.553E+02	3.176E+02	1.623E+02
1.50	1.737E+03	1.218E+03	6.091E+02	4.303E+02	2.260E+02
1.75	2.098E+03	1.489E+03	7.638E+02	5.465E+02	2.935E+02
2.00	2.435E+03	1.744E+03	9.138E+02	6.624E+02	3.625E+02
2.25	2.741E+03	1.988E+03	1.061E+03	7.743E+02	4.306E+02
2.50	3.029E+03	2.210E+03	1.199E+03	8.846E+02	4.990E+02
2.75	3.284E+03	2.422E+03	1.334E+03	9.882E+02	5.648E+02
3.00	3.523E+03	2.610E+03	1.457E+03	1.090E+03	6.299E+02
3.25	3.732E+03	2.790E+03	1.572E+03	1.183E+03	6.914E+02
3.50	3.929E+03	2.947E+03	1.685E+03	1.271E+03	7.503E+02
3.75	4.097E+03	3.090E+03	1.785E+03	1.358E+03	8.086E+02
4.00	4.246E+03	3.228E+03	1.883E+03	1.436E+03	8.623E+02
4.25	4.391E+03	3.346E+03	1.970E+03	1.508E+03	9.133E+02
4.50	4.510E+03	3.451E+03	2.050E+03	1.580E+03	9.616E+02
4.75	4.614E+03	3.556E+03	2.123E+03	1.644E+03	1.010E+03
5.00	4.706E+03	3.642E+03	2.197E+03	1.703E+03	1.054E+03

E (MeV)	62	64	66	67	70
0.15	2.557E–01	1.585E–01		8.144E–02	4.286E–02
0.25	1.557E+00	9.936E–01		5.275E–01	2.847E–01
0.50	1.304E+01	8.800E+00		4.968E+00	2.821E+00
0.75	3.785E+01	2.622E+01		1.547E+01	9.154E+00
1.00	7.349E+01	5.211E+01	3.744E+01	3.179E+01	1.942E+01
1.25	1.169E+02	8.436E+01	6.164E+01	5.263E+01	3.304E+01
1.50	1.654E+02	1.208E+02	8.962E+01	7.704E+01	4.932E+01
1.75	2.172E+02	1.603E+02	1.199E+02	1.039E+02	6.743E+01
2.00	2.700E+02	2.017E+02	1.522E+02	1.321E+02	8.712E+01
2.25	3.241E+02	2.434E+02	1.850E+02	1.616E+02	1.075E+02
2.50	3.772E+02	2.859E+02	2.188E+02	1.913E+02	1.288E+02
2.75	4.306E+02	3.276E+02	2.520E+02	2.215E+02	1.501E+02
3.00	4.818E+02	3.685E+02	2.857E+02	2.511E+02	1.714E+02
3.25	5.314E+02	4.095E+02	3.182E+02	2.803E+02	1.932E+02
3.50	5.808E+02	4.485E+02	3.499E+02	3.098E+02	2.143E+02
3.75	6.270E+02	4.863E+02	3.819E+02	3.379E+02	2.350E+02
4.00	6.713E+02	5.241E+02	4.121E+02	3.652E+02	2.561E+02
4.25	7.156E+02	5.593E+02	4.412E+02	3.928E+02	2.761E+02
4.50	7.563E+02	5.931E+02	4.707E+02	4.185E+02	2.955E+02
4.75	7.949E+02	6.271E+02	4.979E+02	4.434E+02	3.153E+02
5.00	8.316E+02	6.583E+02	5.240E+02	4.673E+02	3.337E+02

Table A16.11. Cross sections (barns) for L2-subshell ionization by protons as a function of atomic number Z and energy (MeV). From Chen, M.-H., and Crasemann, B. (1985), *At. Data Nucl. Data Tables* **33**, 217, and Chen, M.-H., and Crasemann, B. (1989), *At. Data Nucl. Data Tables* **41**, 257 (continued).

E (MeV)	74	78	79	80
0.15	1.860E–02	8.532E–03	7.098E–03	5.897E–03
0.25	1.263E–01	5.877E–02	4.880E–02	4.065E–02
0.50	1.332E+00	6.448E–01	5.395E–01	4.512E–01
0.75	4.545E+00	2.293E+00	1.936E+00	1.633E+00
1.00	1.002E+01	5.240E+00	4.474E+00	3.805E+00
1.25	1.762E+01	9.508E+00	8.155E+00	7.002E+00
1.50	2.702E+01	1.496E+01	1.291E+01	1.113E+01
1.75	3.778E+01	2.138E+01	1.855E+01	1.611E+01
2.00	4.971E+01	2.866E+01	2.498E+01	2.175E+01
2.25	6.234E+01	3.651E+01	3.195E+01	2.800E+01
2.50	7.570E+01	4.494E+01	3.946E+01	3.462E+01
2.75	8.972E+01	5.363E+01	4.724E+01	4.156E+01
3.00	1.030E+02	6.257E+01	5.525E+01	4.887E+01
3.25	1.172E+02	7.184E+01	6.359E+01	5.624E+01
3.50	1.311E+02	8.105E+01	7.190E+01	6.372E+01
3.75	1.450E+02	9.029E+01	8.024E+01	7.144E+01
4.00	1.591E+02	9.976E+01	8.882E+01	7.902E+01
4.25	1.726E+02	1.090E+02	9.717E+01	8.659E+01
4.50	1.860E+02	1.181E+02	1.055E+02	9.434E+01
4.75	1.995E+02	1.271E+02	1.139E+02	1.018E+02
5.00	2.124E+02	1.363E+02	1.221E+02	1.092E+02

E (MeV)	83	86	90	92
0.15	3.462E–03	2.111E–03	1.163E–03	8.999E–04
0.25	2.373E–02	1.425E–02	7.547E–03	5.684E–03
0.50	2.678E–01	1.608E–01	8.405E–02	6.215E–02
0.75	9.884E–01	6.049E–01	3.191E–01	2.362E–01
1.00	2.354E+00	1.468E+00	7.895E–01	5.864E–01
1.25	4.411E+00	2.802E+00	1.538E+00	1.155E+00
1.50	7.158E+00	4.612E+00	2.583E+00	1.956E+00
1.75	1.050E+01	6.888E+00	3.929E+00	2.991E+00
2.00	1.441E+01	9.550E+00	5.541E+00	4.260E+00
2.25	1.872E+01	1.259E+01	7.416E+00	5.727E+00
2.50	2.345E+01	1.590E+01	9.492E+00	7.376E+00
2.75	2.842E+01	1.944E+01	1.175E+01	9.208E+00
3.00	3.361E+01	2.324E+01	1.420E+01	1.116E+01
3.25	3.905E+01	2.714E+01	1.675E+01	1.322E+01
3.50	4.454E+01	3.115E+01	1.940E+01	1.542E+01
3.75	5.011E+01	3.535E+01	2.218E+01	1.766E+01
4.00	5.586E+01	3.952E+01	2.499E+01	1.996E+01
4.25	6.151E+01	4.373E+01	2.784E+01	2.231E+01
4.50	6.717E+01	4.808E+01	3.073E+01	2.476E+01
4.75	7.280E+01	5.233E+01	3.372E+01	2.718E+01
5.00	7.857E+01	5.658E+01	3.665E+01	2.961E+01

Table A16.12. Cross sections (barns) for L3-subshell ionization by protons as a function of atomic number Z and energy (MeV). From Chen, M.-H., and Crasemann, B. (1985), *At. Data Nucl. Data Tables* **33**, 217, and Chen, M.-H., and Crasemann, B. (1989), *At. Data Nucl. Data Tables* **41**, 257.

E (MeV)	22	26	29	30	30
0.15	1.535E+05	3.397E+04	1.286E+04	8.971E+03	4.399E+03
0.25	2.824E+05	7.252E+04	3.001E+04	2.169E+04	1.139E+04
0.50	4.908E+05	1.542E+05	7.196E+04	5.444E+04	3.117E+04
0.75	5.840E+05	2.063E+05	1.033E+05	8.030E+04	4.811E+04
1.00	6.271E+05	2.382E+05	1.252E+05	9.902E+04	6.221E+04
1.25	6.340E+05	2.556E+05	1.395E+05	1.120E+05	7.229E+04
1.50	6.261E+05	2.645E+05	1.488E+05	1.208E+05	7.970E+04
1.75	6.083E+05	2.680E+05	1.546E+05	1.266E+05	8.505E+04
2.00	5.900E+05	2.679E+05	1.579E+05	1.299E+05	8.859E+04
2.25	5.705E+05	2.647E+05	1.588E+05	1.320E+05	9.120E+04
2.50	5.484E+05	2.611E+05	1.591E+05	1.325E+05	9.261E+04
2.75	5.269E+05	2.556E+05	1.579E+05	1.327E+05	9.364E+04
3.00	5.063E+05	2.496E+05	1.568E+05	1.317E+05	9.381E+04
3.25	4.869E+05	2.446E+05	1.546E+05	1.309E+05	9.392E+04
3.50	4.716E+05	2.383E+05	1.520E+05	1.292E+05	9.338E+04
3.75	4.545E+05	2.321E+05	1.500E+05	1.273E+05	9.295E+04
4.00	4.386E+05	2.260E+05	1.472E+05	1.258E+05	9.201E+04
4.25	4.238E+05	2.212E+05	1.443E+05	1.237E+05	9.093E+04
4.50	4.099E+05	2.155E+05	1.422E+05	1.215E+05	8.975E+04
4.75	3.969E+05	2.100E+05	1.393E+05	1.193E+05	8.888E+04
5.00	3.847E+05	2.046E+05	1.365E+05	1.177E+05	8.760E+04

E (MeV)	36	40	43	45	47
0.15	1.154E+03	3.115E+02	1.274E+02	7.168E+01	4.099E+01
0.25	3.414E+03	1.054E+03	4.741E+02	2.823E+02	1.712E+02
0.50	1.099E+04	4.000E+03	1.993E+03	1.274E+03	8.249E+02
0.75	1.872E+04	7.470E+03	3.929E+03	2.598E+03	1.748E+03
1.00	2.572E+04	1.085E+04	5.964E+03	4.040E+03	2.776E+03
1.25	3.145E+04	1.393E+04	7.901E+03	5.458E+03	3.824E+03
1.50	3.611E+04	1.664E+04	9.683E+03	6.799E+03	4.833E+03
1.75	3.986E+04	1.893E+04	1.126E+04	8.037E+03	5.786E+03
2.00	4.285E+04	2.093E+04	1.268E+04	9.160E+03	6.654E+03
2.25	4.508E+04	2.262E+04	1.389E+04	1.014E+04	7.460E+03
2.50	4.693E+04	2.398E+04	1.499E+04	1.104E+04	8.173E+03
2.75	4.820E+04	2.518E+04	1.589E+04	1.181E+04	8.835E+03
3.00	4.928E+04	2.610E+04	1.672E+04	1.251E+04	9.405E+03
3.25	4.992E+04	2.684E+04	1.738E+04	1.310E+04	9.911E+03
3.50	5.049E+04	2.753E+04	1.799E+04	1.361E+04	1.039E+04
3.75	5.071E+04	2.800E+04	1.846E+04	1.409E+04	1.078E+04
4.00	5.077E+04	2.835E+04	1.886E+04	1.447E+04	1.116E+04
4.25	5.088E+04	2.872E+04	1.924E+04	1.479E+04	1.147E+04
4.50	5.071E+04	2.891E+04	1.950E+04	1.511E+04	1.173E+04
4.75	5.046E+04	2.902E+04	1.971E+04	1.534E+04	1.196E+04
5.00	5.013E+04	2.919E+04	1.994E+04	1.553E+04	1.219E+04

Table A16.12. Cross sections (barns) for L3-subshell ionization by protons as a function of atomic number Z and energy (MeV). From Chen, M.-H., and Crasemann, B. (1985), *At. Data Nucl. Data Tables* **33**, 217, and Chen, M.-H., and Crasemann, B. (1989), *At. Data Nucl. Data Tables* **41**, 257 (continued).

E (MeV)	48	50	54	56	60
0.15	3.095E+01	1.772E+01	6.076E+00	3.587E+00	1.399E+00
0.25	1.327E+02	8.034E+01	3.052E+01	1.890E+01	7.962E+00
0.50	6.628E+02	4.290E+02	1.866E+02	1.230E+02	5.790E+01
0.75	1.424E+03	9.568E+02	4.442E+02	3.030E+02	1.518E+02
1.00	2.293E+03	1.576E+03	7.644E+02	5.343E+02	2.773E+02
1.25	3.184E+03	2.229E+03	1.118E+03	7.923E+02	4.230E+02
1.50	4.059E+03	2.884E+03	1.487E+03	1.068E+03	5.833E+02
1.75	4.893E+03	3.512E+03	1.853E+03	1.350E+03	7.517E+02
2.00	5.658E+03	4.117E+03	2.218E+03	1.628E+03	9.217E+02
2.25	6.376E+03	4.686E+03	2.565E+03	1.904E+03	1.095E+03
2.50	7.035E+03	5.203E+03	2.903E+03	2.166E+03	1.263E+03
2.75	7.616E+03	5.691E+03	3.215E+03	2.423E+03	1.430E+03
3.00	8.137E+03	6.126E+03	3.518E+03	2.662E+03	1.589E+03
3.25	8.628E+03	6.521E+03	3.792E+03	2.886E+03	1.742E+03
3.50	9.046E+03	6.897E+03	4.045E+03	3.106E+03	1.893E+03
3.75	9.443E+03	7.221E+03	4.293E+03	3.305E+03	2.033E+03
4.00	9.774E+03	7.512E+03	4.511E+03	3.489E+03	2.165E+03
4.25	1.006E+04	7.795E+03	4.711E+03	6.372E+03	2.297E+03
4.50	1.032E+04	8.030E+03	4.909E+03	3.832E+03	2.416E+03
4.75	1.058E+04	8.239E+03	5.079E+03	3.980E+03	2.527E+03
5.00	1.077E+04	8.449E+03	5.233E+03	4.128E+03	2.640E+03

E (MeV)	62	64	66	67	70
0.15	8.985E–01	5.785E–01		3.157E–01	1.764E–01
0.25	5.284E+00	3.507E+00		1.989E+00	1.148E+00
0.50	3.996E+01	2.815E+01		1.699E+01	1.038E+01
0.75	1.087E+02	7.812E+01		4.886E+01	3.095E+01
1.00	2.022E+02	1.480E+02	1.103E+02	9.501E+01	6.170E+01
1.25	3.134E+02	2.320E+02	1.754E+02	1.520E+02	1.005E+02
1.50	4.367E+02	3.267E+02	2.487E+02	2.172E+02	1.453E+02
1.75	5.664E+02	4.285E+02	3.288E+02	2.875E+02	1.951E+02
2.00	7.017E+02	5.335E+02	4.133E+02	3.627E+02	2.478E+02
2.25	8.369E+02	6.423E+02	4.993E+02	4.394E+02	3.037E+02
2.50	9.739E+02	7.501E+02	5.862E+02	5.188E+02	3.605E+02
2.75	1.107E+03	8.594E+02	6.751E+02	5.974E+02	4.191E+02
3.00	1.236E+03	9.652E+02	7.617E+02	6.755E+02	4.771E+02
3.25	1.366E+03	1.068E+03	8.490E+02	7.547E+02	5.349E+02
3.50	1.487E+03	1.172E+03	9.326E+02	8.307E+02	5.937E+02
3.75	1.608E+03	1.269E+03	1.014E+03	9.047E+02	6.503E+02
4.00	1.720E+03	1.363E+03	1.096E+03	9.796E+02	7.058E+02
4.25	1.826E+03	1.457E+03	1.172E+03	1.050E+03	7.598E+02
4.50	1.928E+03	1.543E+03	1.245E+03	1.117E+03	8.150E+02
4.75	2.029E+03	1.626E+03	1.316E+03	1.186E+03	8.664E+02
5.00	2.121E+03	1.704E+03	1.388E+03	1.249E+03	9.161E+02

Table A16.12. Cross sections (barns) for L3-subshell ionization by protons as a function of atomic number Z and energy (MeV). From Chen, M.-H., and Crasemann, B. (1985), *At. Data Nucl. Data Tables* **33**, 217, and Chen, M.-H., and Crasemann, B. (1989), *At. Data Nucl. Data Tables* **41**, 257 (continued).

E (MeV)	74	78	79	80
0.15	8.232E–02	4.061E–02	3.424E–02	2.888E–02
0.25	5.564E–01	2.815E–01	2.394E–01	2.030E–01
0.50	5.382E+00	2.878E+00	2.470E+00	2.119E+00
0.75	1.682E+01	9.371E+00	8.118E+00	7.026E+00
1.00	3.467E+01	1.992E+01	1.738E+01	1.515E+01
1.25	5.791E+01	3.397E+01	2.989E+01	2.621E+01
1.50	8.533E+01	5.112E+01	4.506E+01	3.978E+01
1.75	1.164E+02	7.076E+01	6.261E+01	5.535E+01
2.00	1.498E+02	9.224E+01	8.186E+01	7.277E+01
2.25	1.857E+02	1.156E+02	1.028E+02	9.145E+01
2.50	2.226E+02	1.399E+02	1.248E+02	1.112E+02
2.75	2.613E+02	1.652E+02	1.480E+02	1.322E+02
3.00	3.000E+02	1.917E+02	1.716E+02	1.536E+02
3.25	3.391E+02	2.182E+02	1.958E+02	1.755E+02
3.50	3.792E+02	2.451E+02	2.208E+02	1.982E+02
3.75	4.183E+02	2.728E+02	2.455E+02	2.208E+02
4.00	4.570E+02	3.000E+02	2.703E+02	2.434E+02
4.25	4.952E+02	3.270E+02	2.950E+02	2.667E+02
4.50	5.343E+02	3.537E+02	3.204E+02	2.894E+02
4.75	5.713E+02	3.812E+02	3.449E+02	3.119E+02
5.00	6.075E+02	4.075E+02	3.691E+02	3.341E+02

E (MeV)	83	86	90	92
0.15	1.762E–02	1.103E–02	6.099E–03	4.666E–03
0.25	1.256E–01	7.939E–02	4.423E–02	3.379E–02
0.50	1.351E+00	8.752E–01	4.992E–01	3.839E–01
0.75	4.596E+00	3.044E+00	1.778E+00	1.380E+00
1.00	1.012E+01	6.809E+00	4.068E+00	3.194E+00
1.25	1.779E+01	1.216E+01	7.408E+00	5.849E+00
1.50	2.731E+01	1.896E+01	1.171E+01	9.336E+00
1.75	3.851E+01	2.695E+01	1.692E+01	1.354E+01
2.00	5.098E+01	3.610E+01	2.287E+01	1.845E+01
2.25	6.472E+01	4.607E+01	2.956E+01	2.390E+01
2.50	7.927E+01	5.680E+01	3.677E+01	2.985E+01
2.75	9.457E+01	6.835E+01	4.448E+01	3.633E+01
3.00	1.108E+02	8.029E+01	5.275E+01	4.312E+01
3.25	1.272E+02	9.265E+01	6.127E+01	5.002E+01
3.50	1.440E+02	1.057E+02	7.009E+01	5.777E+01
3.75	1.615E+02	1.187E+02	7.938E+01	6.542E+01
4.00	1.789E+02	1.320E+02	8.870E+01	7.327E+01
4.25	1.963E+02	1.454E+02	9.819E+01	8.128E+01
4.50	2.138E+02	1.593E+02	1.078E+02	8.965E+01
4.75	2.318E+02	1.729E+02	1.178E+02	9.793E+01
5.00	2.492E+02	1.865E+02	1.276E+02	1.063E+02

Table A16.13. Fitted empirical reference cross sections for K-shell ionization by alpha particles. From Paul, H., and Bolik, O. (1993), At. Data and Nucl. Data Tables **54**, 75.

E (MeV)	C	N	O	F	Ne	Na	Mg
0.10	2.84E+03	2.87E+02	6.35E+01	1.74E+01	5.23E+00	1.58E+00	4.85E–01
0.20	8.95E+04	7.69E+03	1.46E+03	3.70E+02	1.14E+02	3.75E+01	1.36E+01
0.40	1.14E+06	1.40E+05	2.87E+04	7.25E+03	2.14E+03	6.86E+02	2.43E+02
0.60	2.83E+06	4.88E+05	1.19E+05	3.26E+04	1.03E+04	3.37E+03	1.20E+03
0.80	4.19E+06	9.33E+05	2.62E+05	8.05E+04	2.69E+04	9.29E+03	3.45E+03
1.0	5.11E+06	1.35E+06	4.31E+05	1.44E+05	5.20E+04	1.89E+04	7.23E+03
1.2	5.61E+06	1.68E+06	5.93E+05	2.17E+05	8.28E+04	3.19E+04	1.27E+04
1.4	5.89E+06	1.95E+06	7.39E+05	2.89E+05	1.17E+05	4.72E+04	1.95E+04
1.6	6.00E+06	2.14E+06	8.60E+05	3.57E+05	1.52E+05	6.41E+04	2.76E+04
1.8	6.01E+06	2.27E+06	9.65E+05	4.18E+05	1.86E+05	8.16E+04	3.63E+04
2.0	5.93E+06	2.37E+06	1.05E+06	4.71E+05	2.18E+05	9.99E+04	4.56E+04
2.3	5.73E+06	2.46E+06	1.14E+06	5.41E+05	2.62E+05	1.25E+05	5.98E+04
2.6	5.49E+06	2.49E+06	1.21E+06	5.97E+05	2.99E+05	1.49E+05	7.42E+04
3.0	5.17E+06	2.48E+06	1.27E+06	6.50E+05	3.41E+05	1.76E+05	9.19E+04
3.5	4.79E+06	2.42E+06	1.30E+06	6.98E+05	3.82E+05	2.06E+05	1.12E+05
4.0	4.45E+06	2.33E+06	1.31E+06	7.28E+05	4.10E+05	2.30E+05	1.28E+05
4.5	4.14E+06	2.24E+06	1.29E+06	7.43E+05	4.31E+05	2.48E+05	1.42E+05
5.0	3.88E+06	2.14E+06	1.27E+06	7.51E+05	4.45E+05	2.62E+05	1.54E+05
5.5	3.64E+06	2.05E+06	1.24E+06	7.49E+05	4.55E+05	2.73E+05	1.64E+05
6.0	3.43E+06	1.98E+06	1.20E+06	7.43E+05	4.61E+05	2.82E+05	1.71E+05
7.0	3.08E+06	1.80E+06	1.14E+06	7.22E+05	4.62E+05	2.92E+05	1.83E+05
8.0	2.80E+06	1.66E+06	1.07E+06	6.96E+05	4.56E+05	2.97E+05	1.91E+05
9.0	2.56E+06	1.54E+06	1.01E+06	6.68E+05	4.46E+05	2.96E+05	1.96E+05
10.0	2.36E+06	1.44E+06	9.54E+05	6.40E+05	4.34E+05	2.93E+05	1.97E+05

E (MeV)	Al	Si	P	S	Cl	Ar	K
0.10	1.53E–01	4.77E–02	1.46E–02	4.54E–03	1.38E–03		
0.20	5.25E+00	2.11E+00	8.68E–01	3.78E–01	1.67E–01	7.28E–02	3.18E–02
0.40	9.57E+01	4.06E+01	1.82E+01	8.82E+00	4.40E+00	2.22E+00	1.15E+00
0.60	4.70E+02	1.99E+02	8.86E+01	4.35E+01	2.21E+01	1.16E+01	6.25E+00
0.80	1.37E+03	5.83E+02	2.62E+02	1.28E+02	6.54E+01	3.43E+01	1.87E+01
1.0	2.99E+03	1.29E+03	5.83E+02	2.88E+02	1.48E+02	7.76E+01	4.26E+01
1.2	5.34E+03	2.39E+03	1.10E+03	5.45E+02	2.81E+02	1.48E+02	8.18E+01
1.4	8.48E+03	3.84E+03	1.82E+03	9.16E+02	4.75E+02	2.52E+02	1.40E+02
1.6	1.23E+04	5.71E+03	2.76E+03	1.41E+03	7.38E+02	3.94E+02	2.20E+02
1.8	1.68E+04	7.93E+03	3.85E+03	2.02E+03	1.07E+03	5.78E+02	3.24E+02
2.0	2.16E+04	1.04E+04	5.17E+03	2.72E+03	1.48E+03	8.04E+02	4.54E+02
2.3	2.93E+04	1.47E+04	7.45E+03	4.00E+03	2.19E+03	1.22E+03	7.00E+02
2.6	3.74E+04	1.93E+04	1.01E+04	5.49E+03	3.05E+03	1.71E+03	1.00E+03
3.0	4.84E+04	2.56E+04	1.38E+04	7.75E+03	4.39E+03	2.51E+03	1.48E+03
3.5	6.11E+04	3.39E+04	1.87E+04	1.08E+04	6.30E+03	3.68E+03	2.21E+03
4.0	7.27E+04	4.15E+04	2.38E+04	1.40E+04	8.35E+03	5.00E+03	3.05E+03
4.5	8.25E+04	4.86E+04	2.86E+04	1.72E+04	1.05E+04	6.39E+03	3.98E+03
5.0	9.15E+04	5.51E+04	3.31E+04	2.05E+04	1.26E+04	7.86E+03	4.95E+03
5.5	9.94E+04	6.06E+04	3.73E+04	2.35E+04	1.48E+04	9.29E+03	5.95E+03
6.0	1.06E+05	6.58E+04	4.12E+04	2.63E+04	1.69E+04	1.07E+04	6.97E+03
7.0	1.16E+05	7.44E+04	4.76E+04	3.14E+04	2.06E+04	1.36E+04	9.04E+03
8.0	1.24E+05	8.11E+04	5.31E+04	3.55E+04	2.40E+04	1.61E+04	1.09E+04
9.0	1.29E+05	8.58E+04	5.75E+04	3.92E+04	2.67E+04	1.84E+04	1.27E+04
10	1.32E+05	8.96E+04	6.07E+04	4.22E+04	2.92E+04	2.03E+04	1.43E+04

Table A16.13. Fitted empirical reference cross sections for K-shell ionization by alpha particles. From Paul, H., and Bolik, O. (1993 *At. Data and Nucl. Data Tables* **54**, 75 (continued).

E (MeV)	Ca	Sc	Ti	V	Cr	Mn	Fe
0.10							
0.20	1.38E–02	6.13E–03	2.66E–03	1.13E–03	4.68E–04		
0.40	6.08E–01	3.30E–01	1.81E–01	1.01E–01	5.60E–02	3.16E–02	1.64E–02
0.60	3.46E+00	1.98E+00	1.15E+00	6.77E–01	4.04E–01	2.45E–01	1.41E–01
0.80	1.06E+01	6.18E+00	3.67E+00	2.22E+00	1.37E+00	8.57E–01	5.16E–01
1.0	2.41E+01	1.41E+01	8.54E+00	5.25E+00	3.28E+00	2.09E+00	1.29E+00
1.2	4.63E+01	2.73E+01	1.65E+01	1.02E+01	6.46E+00	4.15E+00	2.62E+00
1.4	7.97E+01	4.71E+01	2.84E+01	1.77E+01	1.12E+01	7.22E+00	4.66E+00
1.6	1.26E+02	7.46E+01	4.53E+01	2.81E+01	1.78E+01	1.16E+01	7.54E+00
1.8	1.86E+02	1.11E+02	6.75E+01	4.21E+01	2.67E+01	1.74E+01	1.14E+01
2.0	2.62E+02	1.57E+02	9.59E+01	6.00E+01	3.82E+01	2.48E+01	1.64E+01
2.3	4.08E+02	2.45E+02	1.51E+02	9.48E+01	6.07E+01	3.97E+01	2.63E+01
2.6	5.93E+02	3.59E+02	2.22E+02	1.40E+02	9.01E+01	5.92E+01	3.95E+01
3.0	8.96E+02	5.49E+02	3.43E+02	2.18E+02	1.41E+02	9.29E+01	6.25E+01
3.5	1.34E+03	8.47E+02	5.35E+02	3.44E+02	2.24E+02	1.49E+02	1.01E+02
4.0	1.88E+03	1.19E+03	7.61E+02	5.01E+02	3.30E+02	2.21E+02	1.50E+02
4.5	2.49E+03	1.59E+03	1.03E+03	6.77E+02	4.56E+02	3.07E+02	2.11E+02
5.0	3.16E+03	2.04E+03	1.33E+03	8.84E+02	5.93E+02	4.08E+02	2.82E+02
5.5	3.84E+03	2.52E+03	1.66E+03	1.11E+03	7.53E+02	5.16E+02	3.64E+02
6.0	4.55E+03	3.02E+03	2.01E+03	1.36E+03	9.27E+02	6.40E+02	4.51E+02
7.0	5.99E+03	4.05E+03	2.76E+03	1.90E+03	1.31E+03	9.17E+02	6.54E+02
8.0	7.46E+03	5.09E+03	3.53E+03	2.46E+03	1.73E+03	1.22E+03	8.82E+02
9.0	8.79E+03	6.15E+03	4.29E+03	3.04E+03	2.16E+03	1.55E+03	1.13E+03
10	1.00E+04	7.12E+03	5.08E+03	3.61E+03	2.60E+03	1.88E+03	1.39E+03

E (MeV)	Co	Ni	Cu	Zn	Ga	Ge	As
0.10							
0.20							
0.40	9.22E–03	5.16E–03	2.89E–03	1.59E–03	8.69E–04	4.66E–04	
0.60	8.66E–02	5.36E–02	3.34E–02	2.07E–02	1.29E–02	8.03E–03	4.98E–03
0.80	3.28E–01	2.12E–01	1.38E–01	8.95E–02	5.87E–02	3.85E–02	2.54E–02
1.0	8.39E–01	5.52E–01	3.67E–01	2.45E–01	1.65E–01	1.11E–01	7.56E–02
1.2	1.72E+00	1.15E+00	7.74E–01	5.24E–01	3.58E–01	2.46E–01	1.70E–01
1.4	3.08E+00	2.07E+00	1.41E+00	9.64E–01	6.66E–01	4.62E–01	3.24E–01
1.6	5.02E+00	3.40E+00	2.32E+00	1.60E+00	1.11E+00	7.79E–01	5.50E–01
1.8	7.62E+00	5.17E+00	3.56E+00	2.46E+00	1.72E+00	1.21E+00	8.63E–01
2.0	1.10E+01	7.50E+00	5.17E+00	3.59E+00	2.52E+00	1.78E+00	1.28E+00
2.3	1.77E+01	1.21E+01	8.38E+00	5.84E+00	4.12E+00	2.93E+00	2.10E+00
2.6	2.66E+01	1.82E+01	1.27E+01	8.86E+00	6.27E+00	4.47E+00	3.22E+00
3.0	4.23E+01	2.91E+01	2.03E+01	1.42E+01	1.01E+01	7.24E+00	5.24E+00
3.5	6.87E+01	4.75E+01	3.32E+01	2.34E+01	1.67E+01	1.20E+01	8.69E+00
4.0	1.03E+02	7.16E+01	5.03E+01	3.56E+01	2.54E+01	1.83E+01	1.33E+01
4.5	1.45E+02	1.01E+02	7.17E+01	5.09E+01	3.65E+01	2.64E+01	1.93E+01
5.0	1.96E+02	1.37E+02	9.73E+01	6.95E+01	5.01E+01	3.63E+01	2.66E+01
5.5	2.54E+02	1.79E+02	1.28E+02	9.13E+01	6.60E+01	4.81E+01	3.54E+01
6.0	3.19E+02	2.27E+02	1.62E+02	1.17E+02	8.46E+01	6.17E+01	4.56E+01
7.0	4.65E+02	3.34E+02	2.43E+02	1.76E+02	1.29E+02	9.49E+01	7.03E+01
8.0	6.34E+02	4.60E+02	3.35E+02	2.46E+02	1.83E+02	1.35E+02	1.01E+02
9.0	8.18E+02	5.99E+02	4.41E+02	3.26E+02	2.42E+02	1.82E+02	1.37E+02
10	1.02E+03	7.49E+02	5.56E+02	4.13E+02	3.09E+02	2.32E+02	1.76E+02

Table A16.13. Fitted empirical reference cross sections for K-shell ionization by alpha particles. From Paul, H., and Bolik, O. (1993), *At. Data and Nucl. Data Tables* **54**, 75 (continued).

E (MeV)	Se	Br	Kr	Rb	Sr	Y	Zr
0.10							
0.20							
0.40							
0.60	3.08E–03	1.90E–03	1.48E–03	9.50E–04	6.09E–04	3.89E–04	
0.80	1.68E–02	1.11E–02	7.83E–03	5.35E–03	3.66E–03	2.50E–03	1.71E–03
1.0	5.17E–02	3.55E–02	2.39E–02	1.69E–02	1.19E–02	8.44E–03	6.01E–03
1.2	1.19E–01	8.32E–02	5.51E–02	3.94E–02	2.84E–02	2.05E–02	1.49E–02
1.4	2.28E–01	1.62E–01	1.07E–01	7.72E–02	5.62E–02	4.12E–02	3.03E–02
1.6	3.92E–01	2.81E–01	1.84E–01	1.34E–01	9.86E–02	7.28E–02	5.41E–02
1.8	6.18E–01	4.47E–01	2.94E–01	2.15E–01	1.59E–01	1.18E–01	8.82E–02
2.0	9.18E–01	6.67E–01	4.41E–01	3.25E–01	2.40E–01	1.79E–01	1.35E–01
2.3	1.52E+00	1.11E+00	7.44E–01	5.50E–01	4.09E–01	3.07E–01	2.31E–01
2.6	2.34E+00	1.72E+00	1.16E+00	8.63E–01	6.44E–01	4.84E–01	3.67E–01
3.0	3.82E+00	2.81E+00	1.94E+00	1.44E+00	1.08E+00	8.13E–01	6.18E–01
3.5	6.37E+00	4.71E+00	3.33E+00	2.46E+00	1.85E+00	1.40E+00	1.07E+00
4.0	9.78E+00	7.26E+00	5.17E+00	3.87E+00	2.92E+00	2.21E+00	1.69E+00
4.5	1.42E+01	1.05E+01	7.61E+00	5.71E+00	4.31E+00	3.28E+00	2.51E+00
5.0	1.96E+01	1.46E+01	1.06E+01	7.97E+00	6.07E+00	4.63E+00	3.55E+00
5.5	2.62E+01	1.95E+01	1.43E+01	1.08E+01	8.18E+00	6.28E+00	4.83E+00
6.0	3.38E+01	2.53E+01	1.87E+01	1.41E+01	1.07E+01	8.22E+00	6.33E+00
7.0	5.24E+01	3.95E+01	2.97E+01	2.25E+01	1.72E+01	1.32E+01	1.02E+01
8.0	7.57E+01	5.72E+01	4.34E+01	3.32E+01	2.54E+01	1.96E+01	1.52E+01
9.0	1.03E+02	7.84E+01	6.00E+01	4.60E+01	3.54E+01	2.75E+01	2.14E+01
10	1.34E+02	1.03E+02	7.92E+01	6.10E+01	4.72E+01	3.67E+01	2.87E+01

E (MeV)	Nb	Mo	Tc	Ru	Rh	Pd	Ag
0.10							
0.20							
0.40							
0.60							
0.80	1.18E–03	8.07E–04	5.53E–04	3.77E–04	2.57E–04		
1.0	4.30E–03	3.07E–03	2.20E–03	1.58E–03	1.13E–03	7.96E–04	5.69E–04
1.2	1.09E–02	8.01E–03	5.90E–03	4.35E–03	3.21E–03	2.34E–03	1.73E–03
1.4	2.25E–02	1.68E–02	1.26E–02	9.43E–03	7.10E–03	5.28E–03	3.99E–03
1.6	4.06E–02	3.05E–02	2.31E–02	1.75E–02	1.34E–02	1.01E–02	7.74E–03
1.8	6.66E–02	5.04E–02	3.84E–02	2.94E–02	2.26E–02	1.73E–02	1.34E–02
2.0	1.02E–01	7.77E–02	5.95E–02	4.58E–02	3.55E–02	2.73E–02	2.13E–02
2.3	1.76E–01	1.35E–01	1.04E–01	8.06E–02	6.27E–02	4.88E–02	3.83E–02
2.6	2.81E–01	2.15E–01	1.67E–01	1.30E–01	1.01E–01	7.95E–02	6.27E–02
3.0	4.74E–01	3.66E–01	2.84E–01	2.22E–01	1.74E–01	1.38E–01	1.09E–01
3.5	8.20E–01	6.34E–01	4.94E–01	3.87E–01	3.05E–01	2.43E–01	1.93E–01
4.0	1.30E+00	1.01E+00	7.87E–01	6.18E–01	4.88E–01	3.92E–01	3.12E–01
4.5	1.94E+00	1.51E+00	1.18E+00	9.25E–01	7.32E–01	5.91E–01	4.71E–01
5.0	2.75E+00	2.14E+00	1.68E+00	1.32E+00	1.05E+00	8.46E–01	6.76E–01
5.5	3.75E+00	2.92E+00	2.29E+00	1.81E+00	1.43E+00	1.16E+00	9.31E–01
6.0	4.95E+00	3.86E+00	3.03E+00	2.40E+00	1.90E+00	1.55E+00	1.24E+00
7.0	7.95E+00	6.22E+00	4.92E+00	3.90E+00	3.11E+00	2.54E+00	2.04E+00
8.0	1.19E+01	9.33E+00	7.37E+00	5.84E+00	4.66E+00	3.85E+00	3.10E+00
9.0	1.68E+01	1.32E+01	1.05E+01	8.32E+00	6.65E+00	5.49E+00	4.43E+00
10	2.26E+01	1.79E+01	1.42E+01	1.13E+01	9.06E+00	7.49E+00	6.05E+00

Table A16.13. Fitted empirical reference cross sections for K-shell ionization by alpha particles. From Paul, H., and Bolik, O. (1993) *At. Data and Nucl. Data Tables* **54**, 75 (continued).

E (MeV)	Cd	In	Sn	Sb	Te	I	Xe
1.0	4.07E–04	2.89E–04					
1.2	1.28E–03	9.46E–04	6.99E–04	5.16E–04	3.81E–04	2.79E–04	
1.4	3.02E–03	2.29E–03	1.74E–03	1.32E–03	1.00E–03	7.57E–04	5.73E–04
1.6	5.95E–03	4.58E–03	3.54E–03	2.74E–03	2.12E–03	1.64E–03	1.27E–03
1.8	1.04E–02	8.10E–03	6.33E–03	4.96E–03	3.89E–03	3.06E–03	2.40E–03
2.0	1.67E–02	1.31E–02	1.03E–02	8.15E–03	6.47E–03	5.13E–03	4.08E–03
2.3	3.02E–02	2.39E–02	1.90E–02	1.52E–02	1.22E–02	9.75E–03	7.84E–03
2.6	4.97E–02	3.95E–02	3.16E–02	2.54E–02	2.04E–02	1.65E–02	1.34E–02
3.0	8.67E–02	6.93E–02	5.56E–02	4.49E–02	3.64E–02	2.96E–02	2.41E–02
3.5	1.54E–01	1.24E–01	9.98E–02	8.08E–02	6.58E–02	5.37E–02	4.40E–02
4.0	2.50E–01	2.01E–01	1.62E–01	1.32E–01	1.08E–01	8.82E–02	7.25E–02
4.5	3.78E–01	3.05E–01	2.47E–01	2.01E–01	1.64E–01	1.35E–01	1.11E–01
5.0	5.43E–01	4.38E–01	3.55E–01	2.90E–01	2.37E–01	1.95E–01	1.61E–01
5.5	7.49E–01	6.05E–01	4.91E–01	4.01E–01	3.28E–01	2.70E–01	2.23E–01
6.0	9.99E–01	8.08E–01	6.57E–01	5.36E–01	4.40E–01	3.62E–01	2.99E–01
7.0	1.65E+00	1.33E+00	1.09E+00	8.87E–01	7.29E–01	6.01E–01	4.97E–01
8.0	2.51E+00	2.04E+00	1.66E+00	1.36E+00	1.12E+00	9.22E–01	7.63E–01
9.0	3.60E+00	2.92E+00	2.39E+00	1.96E+00	1.61E+00	1.33E+00	1.11E+00
10	4.91E+00	4.00E+00	3.28E+00	2.70E+00	2.22E+00	1.84E+00	1.53E+00

E (MeV)	Cs	Ba	La	Ce	Pr	Nd
1.4	4.33E–04	3.03E–04	2.26E–04	1.69E–04		
1.6	9.82E–04	7.20E–04	5.53E–04	4.24E–04	3.25E–04	2.49E–04
1.8	1.89E–03	1.44E–03	1.13E–03	8.83E–04	6.91E–04	5.42E–04
2.0	3.26E–03	2.54E–03	2.02E–03	1.61E–03	1.28E–03	1.02E–03
2.3	6.33E–03	5.08E–03	4.10E–03	3.32E–03	2.69E–03	2.19E–03
2.6	1.09E–02	8.91E–03	7.27E–03	5.95E–03	4.88E–03	4.01E–03
3.0	1.98E–02	1.64E–02	1.35E–02	1.12E–02	9.26E–03	7.70E–03
3.5	3.63E–02	3.05E–02	2.53E–02	2.10E–02	1.76E–02	1.47E–02
4.0	5.99E–02	5.08E–02	4.23E–02	3.53E–02	2.96E–02	2.49E–02
4.5	9.18E–02	7.82E–02	6.52E–02	5.46E–02	4.60E–02	3.88E–02
5.0	1.33E–01	1.14E–01	9.50E–02	7.97E–02	6.72E–02	5.69E–02
5.5	1.85E–01	1.58E–01	1.33E–01	1.11E–01	9.39E–02	7.96E–02
6.0	2.49E–01	2.13E–01	1.78E–01	1.50E–01	1.27E–01	1.07E–01
7.0	4.14E–01	3.54E–01	2.97E–01	2.50E–01	2.12E–01	1.80E–01
8.0	6.36E–01	5.44E–01	4.57E–01	3.85E–01	3.26E–01	2.77E–01
9.0	9.21E–01	7.86E–01	6.61E–01	5.58E–01	4.73E–01	4.02E–01
10	1.27E+00	1.09E+00	9.14E–01	7.72E–01	6.54E–01	5.57E–01

Table A16.14. Attenuation coefficients at selected X-ray energies (keV) in various materials. These values were derived from XCOM, code created by M.J. Berger and J.H. Hubbell [(1987) NBSIR Report 87-3597, National Bureau of Standards, Washington,DC]. XCOM is also accessible on the National Institute of Standards and Technology Web site (http://physics.nist.gov/PhysRefData/Xcom/Text/XCOM.html). Units are all cm^2/g.

Material	X-ray energy (keV)								
	0.5	1	2	5	10	15	20	30	40
Be	4920	604	74.7	4.37	0.647	0.307	0.225	0.179	0.164
C	16300	2210	303	19.10	2.370	0.807	0.442	0.256	0.208
Al	7550	1190	2260	193	26.2	7.960	3.440	1.130	5.680
Si	9900	157	2780	245	33.9	10.3	4.46	1.44	0.701
SiO_2	20900	3180	1670	140	19.0	5.81	2.55	0.873	0.465
Si_3N_4	15200	2270	1860	160	21.9	6.70	2.93	0.985	0.512
Ti	34200	5870	986	684	111	35.9	15.9	4.97	2.21
Fe	48800	9080	1630	140	171	57.1	25.7	8.18	3.63
Fe_2O_3	43300	7730	1350	112	121	40.5	18.2	5.83	2.62
Ni	58800	9860	2050	179	209	70.8	32.2	10.3	4.60
Cu	69300	10600	2150	190	216	74.1	33.8	10.9	4.86
Ag	34300	7040	1400	738	119	40.0	18.4	36.7	17.2
Sn	38400	8160	1670	847	138	46.6	21.5	41.2	19.4
Ta	14000	3510	3770	533	238	134	63.3	21.9	10.3
W	14700	3680	3920	553	96.9	139	65.7	22.7	10.7
Au	18400	4650	1140	666	118	164	78.8	27.5	13.0
Pb	20300	5210	1290	730	131	112	86.4	30.3	14.4
Mylar	20300	2910	421	27.9	3.48	1.13	0.580	0.301	0.230
Kapton	19300	2730	390	25.5	3.18	1.04	0.542	0.288	0.223

APPENDIX

17

ION CHANNELING DATA

Compiled by

M. L. Swanson

University of North Carolina, Chapel Hill, North Carolina, USA

Lin Shao

Texas A&M University, College Station, Texas, USA

CONTENTS

A17.1 NOTATION FOR CHANNELING

a = Thomas–Fermi screening length

a_0 = Bohr radius = 5.292×10^{-11} m

d = atomic spacing along axial direction

d_0 = lattice constant

d_p = interplanar spacing

e^2 = electronic charge squared = 1.44×10^{-13} cm MeV

E = ion energy

dE/dx = energy loss along the ion beam path

F_{RS} = square root of the Molière string potential

F_{PS} = square root of the planar potential

H = number of counts per unit energy loss in an RBS spectrum

M_1, M_2 = atomic masses of ions and target atoms

N = atomic density per unit volume

N_d = defect concentration per unit volume

T = crystal temperature

u_1 = one-dimensional vibrational amplitude

u_2 = two-dimensional vibrational amplitude (= $1.414 u_1$)

x = distance along ion beam path

Z_1, Z_2 = atomic numbers of ions and target atoms

χ = normalized yield = (aligned yield)/(random yield)

χ_h = normalized yield from host atoms

χ_s = normalized yield from solute atoms

$\chi_h^{\langle uvw \rangle}$, $\chi_s^{\langle uvw \rangle}$ = normalized yields from host and solute atoms, respectively, for alignment along $\langle uvw \rangle$ axial channels

$\chi_h^{\{hkl\}}$, $\chi_s^{\{hkl\}}$ = normalized yields for host and solute atoms, respectively, for alignment along $\{hkl\}$ planar channels

χ_{hD} = dechanneled fraction of ions

ϕ = crystal azimuthal angle

ϕ_D = the Debye function

λ_p = dechanneling length caused by dislocations

θ = crystal tilt angle

θ_D = Debye temperature (K)

σ_d = cross section for dechanneling by defects

σ_{th} = cross section for dechanneling by thermal vibrations

ψ_1 = characteristic angle for channeling

$\psi_{1/2}$ = half-width of the channeling dip

A17.2 EQUATIONS FOR CHANNELING

(A17.1) $a = 0.04685(Z_1^{1/2} + Z_2^{1/2})^{-2/3} \approx 0.04685 Z_2^{-1/3}$

(units of nm) (see Table A17.3)

(A17.2) $u_1 = 12.1\{[\phi_D(x'')/x'' + 1/4](M_2\theta_D)^{-1}\}^{1/2}$

(10^{-8} cm)

where $x'' = \theta_D/T$ and the Debye function ϕ_D is given in Fig. A17.1

(A17.3) $\psi_1 = (2Z_1Z_2e^2/Ed)^{1/2}$ (radians)

$\psi_1 = 0.307(Z_1Z_2/Ed)^{1/2}$

(where d is in units of 10^{-8} cm and the ion energy E is in MeV)

(A17.4) $\psi_{1/2} = 0.8F_{RS}(x')\psi_1$ for **axial channels**, if ψ_1 (rad) $< a/d$, i.e., for MeV light ions or very high energy heavy ions (Barrett, 1971), where $x' = 0.85u_2/a$ and $F_{RS}(x') = [f_M(x')]^{1/2}$ is the square root of the Molière string potential [see Eq. (12.6) and Fig. A17.2]. Experimental and calculated values are compared in Table A17.5.

(A17.5) $\psi_{1/2} = 0.757(a\psi_1/d)^{1/2}$ (degrees) for **axial channels**, if ψ_1 (rad) $> a/d$, i.e., for heavy ions in the keV to MeV range; e.g., for 50 keV As on $\langle 110 \rangle$ Si, $\psi_{1/2} = 4.85°$. Thermal vibrations are neglected.

(A17.6) $\psi_{1/2} = 0.72F_{PS}(x', y')\psi_a$ for **planar channels** (Barrett, 1971), where $x' = 1.6u_1/a$, $y' = d_p/a$, F_{PS} is the square root of the adimensional planar potential using Molière's screening function (see Fig. A17.3), and $\psi_a = 0.545(Z_1Z_2Nd_pa/E)^{1/2}$ (degrees).

(A17.7) $\chi_h^{\langle uvw \rangle} = Nd\pi(2u_1^2 + a^2)$ [**axial channel** approximation sometimes attributed to Linhard].

(A17.8) $\chi_h^{\langle uvw \rangle} = 18.8Ndu_1^2(1 + \xi^{-2})^{1/2}$ for **axial channels**. This is Barrett's (1971) Monte Carlo result, where $\xi = 126u_1/(\psi_{1/2}d)$, with $\psi_{1/2}$ given in degrees. See Fig. A17.4 for a fitting with somewhat different parameters.

At high energy where $\psi^{1/2} \ll u_1/d$, Eq. (A17.8) becomes Eq. (A17.9);

(A17.9) $\chi_h^{\langle uvw \rangle} = 18.8Ndu_1^2$

Example: For 0.5 MeV He at 293 K on Ge $\langle 110 \rangle$ $\chi_{Ge}^{\langle 110 \rangle} = 0.020$ from Eq. (A17.7), 0.0268 from Eq. (A17.8), and 0.0240 from Eq. (A17.9).

(A17.10) $\chi_h^{\{hkl\}} = 2a/d_p$ for **planar channels**, as estimated after Linhard (1965). This approximation is independent of energy and temperature; e.g., for He on {110} Ge, $d_p = 0.200$ nm from Tables A17.1 and A17.3, a = 0.0148 nm from Table A17.3, and thus $\chi_{Ge}^{\{110\}} = 0.148$.

(A17.11) $\chi_h = P(\theta_c)$ for the normalized yield of a crystal covered by an amorphous overlayer, where the reduced critical angle, θ_c, is given by A17.12.

(A17.12) $\theta_c = aE\psi_{1/2}/(2Z_1Z_2e^2) = 1.5 \times 10^2 a[E/(Z_1Z_2d)]^{1/2}F_{RS}$ with a and d in units of 10^{-8} cm and E in MeV. Values of $P(\theta_c)$ are shown in Fig. A17.5 for values of the reduced thickness m given by m $= \pi a^2 Nt$. Values of m from 0.2 to 20 correspond to Si thicknesses from 34 $\times$ 10^{-8} cm to 3400 $\times 10^{-8}$ cm, respectively (Mayer and Rimini, 1977; Lugujjo and Mayer, 1973).

Table A17.1 Values by which the lattice constant must be multiplied to compute the interatomic spacings d in axial directions and the interplanar spacings d_p for planar directions in the simplest (monatomic) cubic structures (Gemmell, 1974).

Structure	Atoms per unit cell	Axis			Plane		
		$\langle 100 \rangle$	$\langle 110 \rangle$	$\langle 111 \rangle$	$\{100\}$	$\{110\}$	$\{111\}$
fcc	4	1	$1/\sqrt{2}$	$\sqrt{3}$	1/2	$1/2\sqrt{2}$	$1/\sqrt{3}$
bcc	2	1	$\sqrt{2}$	$\sqrt{3}/2$	1/2	$1/\sqrt{2}$	$1/2\sqrt{3}$
diamond cubic	8	1	$1/\sqrt{2}$	$\sqrt{3}/4, 3\sqrt{3}/4$	1/4	$1/2\sqrt{2}$	$1/4\sqrt{3}, \sqrt{3}/4$

Table A17.2. Values by which the lattice constant must be multiplied to computer the interatomic spacings d in axial directions and the interplanar spacing d_p for planar directions in the most common simple diatomic compounds (atoms labeled "A" and "B") having cubic structures* (Gemmell, 1974).

Structure	Atoms per unit cell	Axis			Plane		
		$\langle 100 \rangle$	$\langle 110 \rangle$	$\langle 111 \rangle$	$\{100\}$	$\{110\}$	$\{111\}$
Rocksalt (like NaCl)	4A + 4B	ABAB ... 1/2	pure $1/\sqrt{2}$	ABAB ... $\sqrt{3}/2$	mixed 1/2	mixed $1/2\sqrt{2}$	pure ABAB .. $1/2\sqrt{3}$
Fluorite (like CaF$_2$)	4A + 8B	pure 1 (A), 1/2(B)	pure $1/\sqrt{2}$	BAB..BAB $\sqrt{3}/4$, $\sqrt{3}/2$	pure ABAB 1/4	mixed $1/2\sqrt{2}$	pure BAB..BAB.. $1/4\sqrt{3}$, $\sqrt{3}/4$
Zinc Blende (like ZnS)	4A + 4B	pure 1	pure $1/\sqrt{2}$	AB...AB $\sqrt{3}/4$, $3\sqrt{3}/4$	pure ABAB 1/4	mixed $1/2\sqrt{2}$	pure AB..AB.. $1/4\sqrt{3}$, $\sqrt{3}/4$

*For the axial case, the term "pure" indicates that each row contains only one atomic species. For rows containing both species, the ordering in the row is given. For the planar case, the term "pure" indicates that each sheet of atoms contains only one atomic species and the way in which the sheets are ordered is shown. The term "mixed" refers to cases where each planar sheet of atoms contains both atomic species.

Table A17.3. Crystal parameters at room temperature (Gemmell, 1974). (θ_D values are approximate – from specific heat data).

Z_2	M_2	Element	Structure	Screening length $a=0.04685Z_2^{-1/3}$ (nm)	Debye temp.(θ_D) (K)	Vibrational amplitude u_1(293K) (nm)	Lattice constant d_0 (nm)
6	12.01	C	diamond cubic	0.0258	2000	0.004	0.3567
13	26.98	Al	f. c. c.	0.0199	390	0.0105	0.4050
14	28.09	Si	diamond cubic	0.0194	543	0.0075	0.5431
23	50.94	V	b. c. c.	0.0165	360	0.0082	0.3024
24	52.00	Cr	b. c. c.	0.0162	485	0.0061	0.2884
26	55.85	Fe	b. c. c.	0.0158	420	0.0068	0.2867
28	58.71	Ni	f. c. c.	0.0154	425	0.0065	0.3524
29	63.54	Cu	f. c. c.	0.0152	315	0.0084	0.3615
32	72.59	Ge	diamond cubic	0.0148	290	0.0085	0.5657
41	92.91	Nb	b. c. c.	0.0136	275	0.0079	0.3300
42	95.94	Mo	b. c. c.	0.0135	380	0.0057	0.3147
45	102.91	Rh	f. c. c.	0.0132	340	0.0061	0.3803
46	106.40	Pd	f. c. c.	0.0131	275	0.0074	0.3890
47	107.87	Ag	f. c. c.	0.0130	215	0.0093	0.4086
73	180.95	Ta	b. c. c.	0.0112	245	0.0064	0.3306
74	183.85	W	b. c. c.	0.0112	310	0.0050	0.3165
78	195.09	Pt	f. c. c.	0.0110	225	0.0066	0.3923
79	196.97	Au	f. c. c.	0.0109	170	0.0087	0.4078
82	207.19	Pb	f. c. c.	0.0108	88	0.0164	0.4951

Table A17.4. Commonly studied diatomic compounds having cubic lattice structures at room temperature (Gemmell, 1974).

Compound	Structure	Lattice constant (nm)
AlSb	ZnS	0.6135
BaF_2	CaF_2	0.6200
CaF_2	CaF_2	0.5463
CsI	CsCl	0.4567
GaP	ZnS	0.5451
GaAs	ZnS	0.5654
GaSb	ZnS	0.6118
InAs	ZnS	0.6036
InSb	ZnS	0.6478
KCl	NaCl	0.6293
KBr	NaCl	0.6600
KI	NaCl	0.7066
LiF	NaCl	0.4017
MgO	NaCl	0.4211
NaF	NaCl	0.4620
NaCl	NaCl	0.5640
NaI	NaCl	0.6473
PbS	NaCl	0.5936
RbBr	NaCl	0.6854
SrF_2	CaF_2	0.5800
ThO_2	CaF_2	0.5600
UC	NaCl	0.4959
UO_2	CaF_2	0.5468

Table A17.5. Comparison of calculated and measured* value of $\psi_{1/2}$ for axial channeling, using Eq. (A17.4) (Barrett, 1971).

Target	Direction	Ion	Energy (MeV)	$\psi_{1/2}$ (degrees) Calculated	Measured
C (diamond)	$\langle 011 \rangle$	H	1.0	0.53	0.54
		He	1.0	0.75	0.75
Al	$\langle 011 \rangle$	H	0.4	0.84	0.90
	$\langle 011 \rangle$		1.4	0.45	0.42
Si	$\langle 011 \rangle$	H	0.25	1.03	1.02
			0.5	0.73	0.68
			1.0	0.51	0.53
			2.0	0.36	0.36
			3.0	0.30	0.26
		He	0.5	1.03	1.10
			1.0	0.73	0.75
			2.0	0.51	0.55
Ge	$\langle 011 \rangle$	He	1.0	0.93	0.95
W	$\langle 111 \rangle$	H	3.0	0.83	0.85
		He	6.0	0.83	0.80
	$\langle 001 \rangle$	H	2.0	0.95	1.00
		He	2.0	1.34	1.39
		C	10.0	0.98	1.10
		O	10.0	1.13	1.23
		Cl	10.0	1.60	1.82
Au	$\langle 011 \rangle$	Cl	20.0	0.84	1.10
		I	60.0	0.80	1.00

*Estimated error for measurements about ± 0.07°

Table A17.6. Comparison of calculated and measured* values of $\psi_{1/2}$ for planar channeling (Barret, 1971).

Target	Plane	Ion	Energy (MeV)	$\psi_{1/2}$ (degrees) Calculated	Measured
Si	{100}	H	3.0	0.07	0.07
	{110}	H	3.0	0.10	0.09
Ge	{110}	He	1.0	0.31	0.30
	{100}	He	1.0	0.23	0.18
	{211}	He	1.0	0.18	0.20
W	{100}	H	2.0	0.25	0.22
	{110}	H	2.0	0.32	0.26
	{100}	He	2.0	0.35	0.27
	{110}	He	2.0	0.45	0.38
	{100}	C	10.0	0.25	0.20
	{110}	C	10.0	0.32	0.28
	{110}	O	10.0	0.36	0.33
	{110}	Cl	10.0	0.50	0.42
Au	{100}	Cl	20.0	0.27	0.31
	{110}	Cl	20.0	0.21	0.24
	{111}	Cl	20.0	0.30	0.32

*Estimated error for measurements about ± 0.03°

Table A17.7. Measured critical angles in diamond type lattice for monatomic and diatomic compounds (S.T. Picraux, 1969).

Crystal	Z_1	E (MeV)	$\Psi_{1/2}$ axial		$\Psi_{1/2}$ planar				
			<111>	<110>	{110}	{111}	{001}	{112}	{113}
Si	H+	0.25		1.02	0.32	0.33	0.25	0.20	0.16
$\alpha=1.12_5$		0.50		0.68	0.24	0.26	0.17	0.16	0.13
		1.00		0.53	0.16	0.20	0.12	0.11	0.10
		2.00		0.36	0.12	0.13	0.08	0.08	0.08
	He+	0.50	0.98	1.10	0.30	0.32	0.23	0.19	0.16
		1.00	0.69	0.75	0.22	0.26	0.18	0.13	0.16
		2.00	0.46	0.55	0.17	0.16	0.13	0.09	0.09
Ge	He+	0.50	1.13		0.40				
		1.00	0.80	0.95	0.30	0.23	0.18	0.20	
$\alpha=1.00$		1.90	0.54		0.23				
C (diamond)	H+	1.00	0.46	0.54	0.16				
$\alpha=1.44$	He+	1.00	0.58	0.75					
GaAs	He+	0.50	1.07		0.33				
$\alpha=1.03$		1.00	0.81		0.24				
		1.90	0.48		0.14				
GaP	He+	0.50	1.03	0.38					
$\alpha=1.14$		1.00	0.74	0.99	0.25	0.26	0.18	0.18	
		1.90	0.59		0.18				
GaSb	He+	0.50	1.16						
$\alpha=0.94$		1.00	0.88	1.10	0.28	0.25	0.17		

Estimated error for axial measurements ±0.06°
Estimated error for planar measurments ±0.03°
Calculated values are given in A17.5 and A17.6.

Table A17.8. Angles between planes in cubic crystals.

HKL	hkl	Angles between HKL and hkl planes					
100	100	0.00	90.00				
	110	45.00	90.00				
	111	54.75					
	210	26.56	63.43	90.00			
	211	35.26	65.90				
	221	48.19	70.53				
	310	18.43	71.56	90.00			
	311	25.24	72.45				
	320	33.69	56.31	90.00			
	321	36.70	57.69	74.50			
	322	43.31	60.98				
	331	46.51	76.74				
	332	50.24	64.76				
	410	14.04	75.96	90.00			
	411	19.47	76.37				
110	110	0.00	60.00	90.00			
	111	35.26	90.00				
	210	18.43	50.77	71.56			
	211	30.00	54.74	73.22	90.00		
	221	19.47	45.00	76.37	90.00		
	310	26.56	47.87	63.43	77.08		
	311	31.48	64.76	90.00			
	320	11.31	53.96	66.91	78.69		
	321	19.11	40.89	55.46	67.79	79.11	
	322	30.96	46.69	80.12	90.00		
	331	13.26	49.54	71.07	90.00		
	332	25.24	41.08	81.33	90.00		
	410	80.96	46.69	59.04	80.12		
	411	33.56	60.00	70.53	90.00		
111	111	0.00	70.53				
	210	39.23	75.04				
	211	19.47	61.87	90.00			
	221	15.79	54.74	78.90			
	310	43.09	68.58				
	311	29.50	58.52	79.98			
	320	36.81	80.76				
	321	22.21	51.89	72.02	90.00		
	322	11.42	65.16	81.95			
	331	22.00	48.53	82.39			
	332	10.02	60.50	75.75			
	410	45.56	65.16				
	411	35.26	57.02	74.21			
210	210	0.00	36.87	53.13	66.42	78.46	90.00
	211	24.09	43.09	56.79	79.48	90.00	
	221	26.56	41.81	53.40	63.43	72.65	90.00
	310	8.13	31.95	45.00	64.90	73.57	81.87
	311	19.29	47.61	66.14	82.25		
	320	7.12	29.74	41.91	60.25	68.15	75.64
		82.87					
	321	17.02	33.21	53.30	61.44	68.99	83.14
		90.00					

Table A17.9. Parameters in Eq. (12.25) for describing stopping powers of channeled ions (Niemann *et al.*, 1996; Azevedo *et al.*, 2002). The stopping powers are given in eV/Å for ion energies quoted in MeV.

Eq. (12.25):
$$\frac{dE}{dx} = \frac{E^{1/2}\ln(2.71828+\beta E)}{\alpha_0 + \alpha_1 E^{1/4} + \alpha_2 E^{1/2} + \alpha_3 E + \alpha_4 E^{3/2}}$$

Ion	Direction	α_0	α_1	α_2	α_3	α_4	β
^{4}He	Random	2.83×10^{-2}	-6.59×10^{-2}	5.37×10^{-2}	-2.41×10^{-2}	1.88×10^{-2}	3.92
	$\langle 111 \rangle$	7.178×10^{-2}	-4.446×10^{-2}	-7.886×10^{-2}	6.648×10^{-2}	4.861×10^{-2}	1.557
	$\langle 110 \rangle$	8.179×10^{-2}	-4.989×10^{-2}	-8.941×10^{-2}	7.522×10^{-2}	1.095×10^{-2}	6.418
^{7}Li	Random	6.092×10^{-2}	-1.051×10^{-2}	7.331×10^{-2}	-2.157×10^{-2}	6.57×10^{-2}	15.7
	$\langle 111 \rangle$	1.106×10^{-1}	-2.79×10^{-2}	-7.381×10^{-2}	5.317×10^{-2}	2.09×10^{-2}	31.42
	$\langle 110 \rangle$	1.129×10^{-1}	-4.119×10^{-2}	-9.218×10^{-2}	4.922×10^{-2}	2.54×10^{-2}	5.734
	$\langle 100 \rangle$	1.216×10^{-1}	-8.130×10^{-2}	-6.576×10^{-2}	1.305×10^{-2}	5.317×10^{-2}	192.86

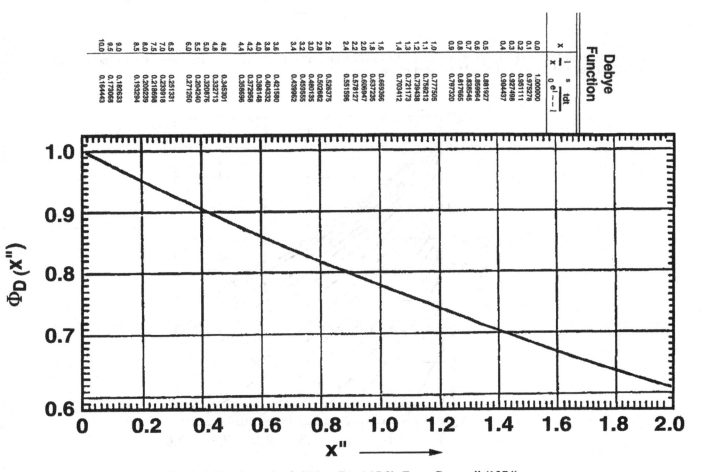

FIG. A17.1. The Debye function $\phi_D(x'')$, where $x'' = \theta_D/T$ (see Eq. A17.2). From Gemmell (1974).

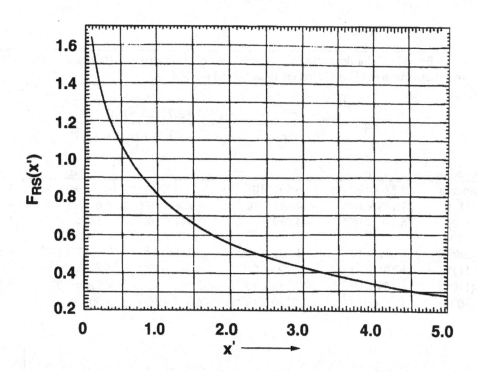

FIG. A17.2. A plot of $F_{RS}(x')$ versus x', where $F_{RS}(x')$ is the square root of the Molière string potential f_M and $x' = 1.2u_1/a$ [see Eqs. (12.6) and (A17.4)]. From Mayer and Rimini (1977).

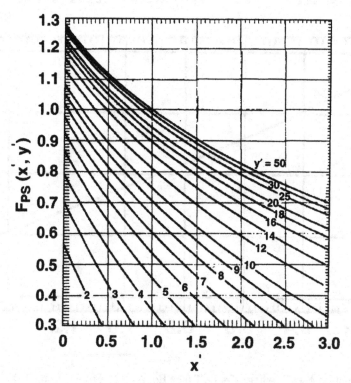

FIG. A17.3. A plot of $F_{PS}(x',y')$ versus x', for different values of y', where $F_{PS}(x')$ is the square root of the adimensional planar potential, $x' = 1.6u_1/a$, and $y' = d_p/a$ (see Eq. A17.6). From Barrett (1971).

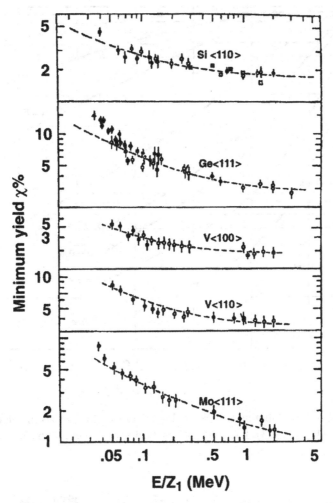

FIG. A17.4. Axial minimum yields χ for several ion–target combinations are plotted as a function of ratio E/Z_1, where E and Z_1 are the energy and the atomic number, respectively, of the incident ion.

The dashed lines represent a fit of experimental data using the formula (Barrett, 1971)

$$\chi = 2\pi Ndcu_1^2\left(1 + \frac{\psi_{1/2}^2 d^2}{k^2 u_1^2}\right),$$

where the constants c and k have the following values:

		c	k
Si	$\langle 110\rangle$	2.7	1.2
Ge	$\langle 111\rangle$	2.8	1.0
V	$\langle 100\rangle$	2.6	1.2
V	$\langle 110\rangle$	2.6	1.1
Mo	$\langle 111\rangle$	2.8	0.8

Good average values for c and k are 2.7 and 1.1, respectively. All experimental data shown in Figure A17.4 are well fitted by the following equations [the constants are somewhat different from those in Eq. (A17.8)]

$$\chi = 17Ndu_1^2\left(1 + \xi^{-2}\right)^{1/2}$$

$$\xi = 63u_1/(\psi_{1/2}d)\qquad,$$

where N is the atomic density [atoms/$(nm)^3$], d is the atomic spacing along an axial direction (nm), u_1 is the one-dimensional rms vibration amplitude (nm), and $\Psi_{1/2}$ is the axial half-angle (degrees). From Mayer and Rimini (1977).

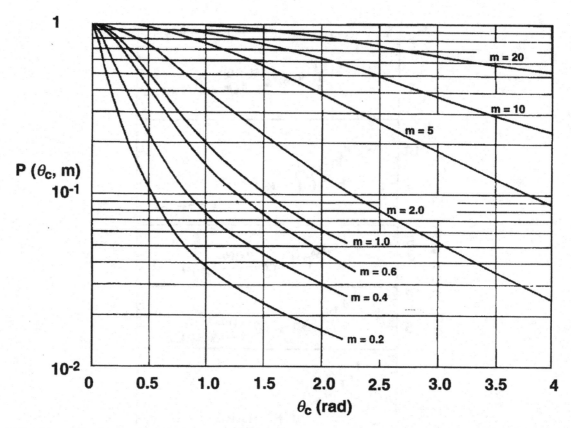

FIG. A17.5. Normalized integrated distribution P versus reduced angle θ_C for various m values [see Eqs. (A17.11) and (A17.12)]. From Lugujjo and Mayer (1973). The function P is used as a measure of the normalized yield χ for crystals covered with an amorphous layer.

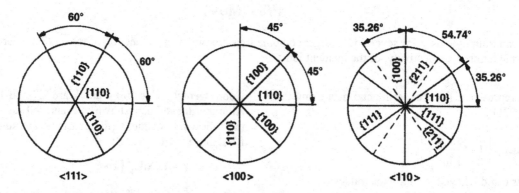

FIG. A17.6. Low-index planes around the $\langle 111 \rangle$, $\langle 100 \rangle$, and $\langle 110 \rangle$ axes in a cubic structure. To assist in orienting crystals, the major axes and planes are listed below in order of increasing spacing along the rows and planes; thus, from left to right, the channeling dips progress from strong to weak; e.g., $\langle 110 \rangle$ channels give the lowest minimum yields for the fcc and diamond structures.

Structure	Axes \<uvw\>	Planes {hkl}
fcc	110, 100, 111	111, 100, 110
bcc	111, 100, 110	110, 100, 211
diamond	110, 111, 100	110, 111, 100

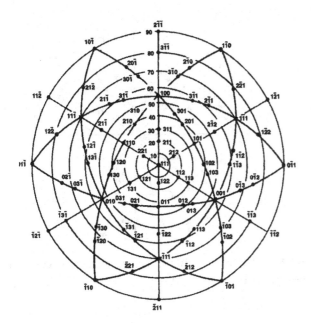

FIG. A17.7. Standard ⟨111⟩ stereographic projection for a cubic crystal.

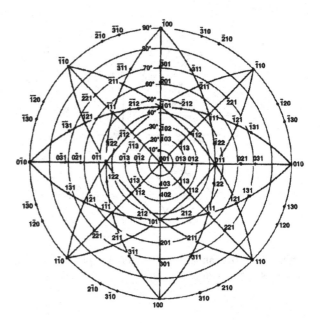

FIG. A17.8. Standard ⟨100⟩ stereographic projection for a cubic crystal.

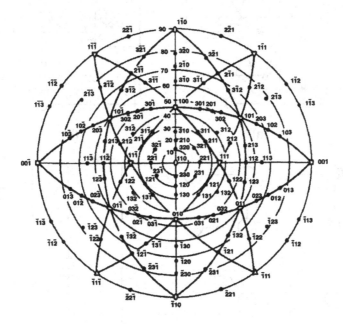

FIG. A17.9. Standard ⟨110⟩ stereographic projection for a cubic crystal.

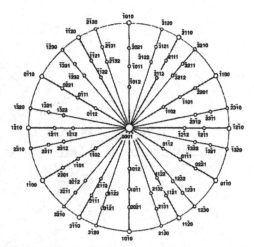

FIG. A17.10. Standard ⟨0001⟩ stereographic projection for a close-packed hexagonal crystal.

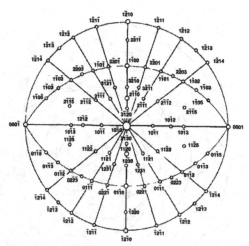

FIG. A17.11. Standard ⟨10$\overline{1}$0⟩ stereographic projection for a close-packed hexagonal crystal.

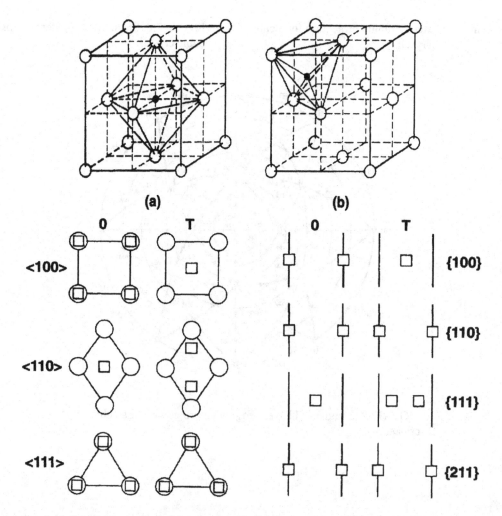

FIG. A17.12. Top: (a) Octahedral and (b) tetrahedral interstitial sites in an fcc crystal. Bottom: Projections of octahedral and tetrahedral interstitial sites onto the major axial and planar channels of an fcc structure (from Carstanjen, 1980).

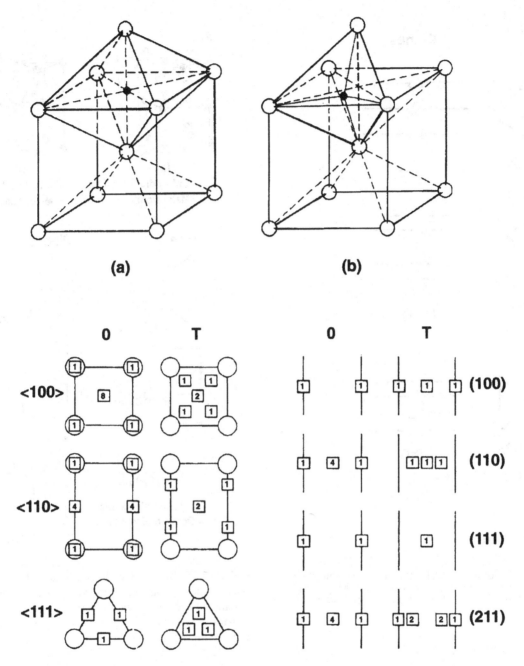

FIG. A17.13. Top: (a) Octahedral and (b) tetrahedral interstitial sites in a bcc crystal. Bottom: Projections of octahedral and tetrahedral interstitial sites onto the major axial and planar channels of a bcc structure (from Carstanjen, 1980).

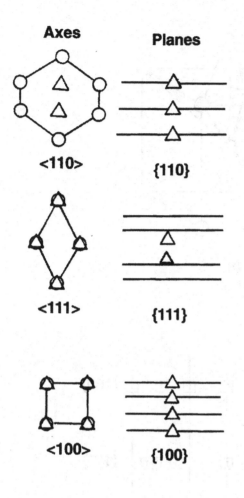

Axes

<110>

<111>

<100>

Planes

{110}

{111}

{100}

Δ Tetrahedral site

a/√2

a

(a)
○ Octahedral interstices
● Metal atoms

a√3/2√2

(b)
○ Tetrahedral interstices
● Metal atoms

Axes

<0001> ĉ

<10Ī0> â'

<11Ī0> â

x = Tetrahedral interstice
○ = Octahedral interstice

Planes

x ○ x

(0001) Basal

x x ○ x

(10Ī0) Prism

(11Ī0)

○ x ○x x

(10Ī1) Pyramidal

IG. A17.14. Projection of tetrahedral interstitial sites onto the major xial and planar channels of a diamond crystal (from Mayer and imini, 1977). See also Morita (1976) for other interstitial positions.

FIG. A17.15. Top: (a) Octahedral (a) and (b) tetrahedral interstiti; sites in a hexagonal close-packed crystal. Bottom: Projections of tetrahedral and octahedral interstitial sites onto the major axial an planar channels of a hexagonal close-packed structure (from May and Rimini, 1977; Howe).

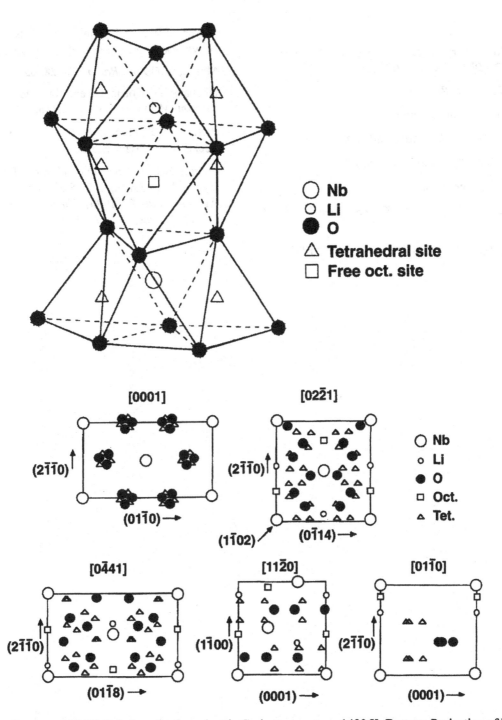

FIG. A17.16. Top: Structure of LiNbO₃ below the ferroelectric Curie temperature, 1480 K. Bottom: Projection of Li, Nb, and O sites and tetrahedral and octahedral interstitial sites onto the major axial and planar channels of trigonal LiNbO₃ (from Soares).

REFERENCES

Azevedo, G. de M., Behar, M., Dias, J.F., Grande, P.L., and da Silva, D. L. (2002), *Phys. Rev.* **B65**, 075203.

Barrett, J.H. (1971), *Phys. Rev.* **B3**, 1527.

Boerma, D.O. Private communication.

Carstanjen, H.-D. (1980), *Phys. Status Solidi* **A59**, 11.

Gemmell, D.S. (1974), *Rev. Mod. Phys.* **46**, 129.

Howe, L.M. Private communication.

Lindhard, J. (1965), *Mat.-Fys. Medd. K. Dan. Vidensk. Selsk.* **34** (14).

Lugujjo, E., and Mayer, J.W. (1973), *Phys. Rev.* **B7**, 1782.

Mayer, J.W., and Rimini, E. (eds.). (1977), *Ion Beam Handbook for Material Analysis*, Academic Press, New York.

Morita, K. (1976), *Radiat. Eff.* **28**, 65.

Niemann, D., Konac, G., and Kalbitzer, S. (1996), *Nucl. Instrum. Methods Phys. Res.* **B118**, 11.

Picraux, S.T. (1969), Ph.D. Thesis, California Institute of Technology, Pasadena, CA.

Soares, J.C. Private communication. [See also: Rebouta, L., Soares, J.C, daSilva, M.F., Sanz-Garcia, J.A., Dieguez, E., and Agullo-Lopez, F. (1990), *Nucl. Instrum. Methods* **B50**, 428 (1992), *J. Mater. Res.* **7**, 130.]

APPENDIX
18

THIN-FILM MATERIALS AND PREPARATION

Compiled by

R. A. Weller

Vanderbilt University, Nashville, Tennessee, USA

CONTENTS

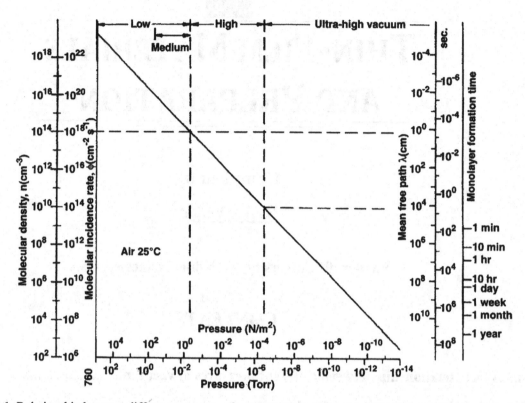

FIG. A18.1. Relationship between different measures of vacuum. From Roth, A., 1982, *Vacuum Technology,* 2nd Ed., North Holland, Amsterdam. Used with permission.

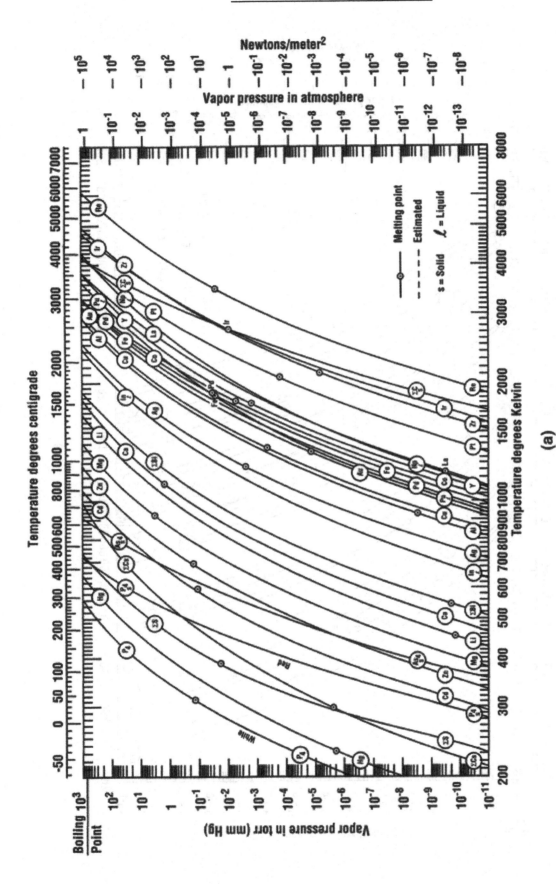

FIG. A18.2. Vapor pressures of the elements. From Honig, R.E., and Kramer, D.A. (1969), *RCA Rev.* **30**, 285. Used with permission.

(a)

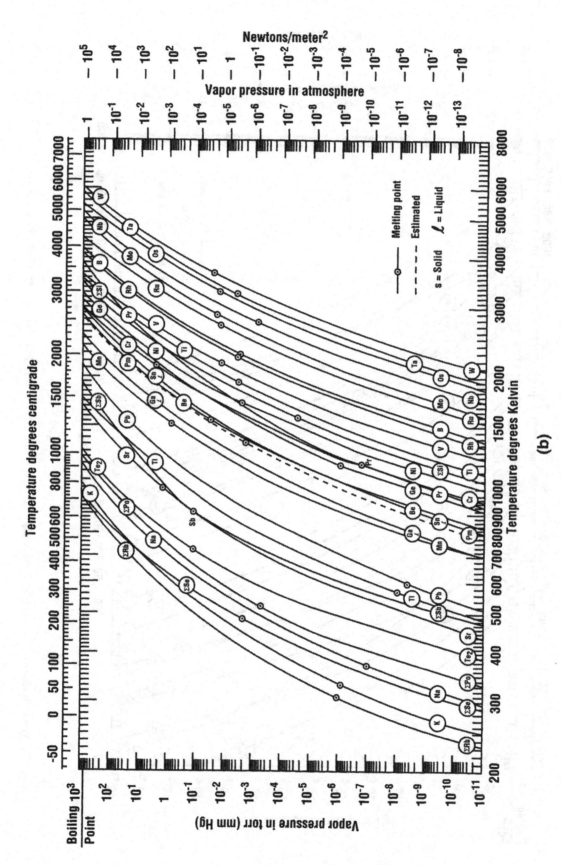

FIG. A18.2. (continued).

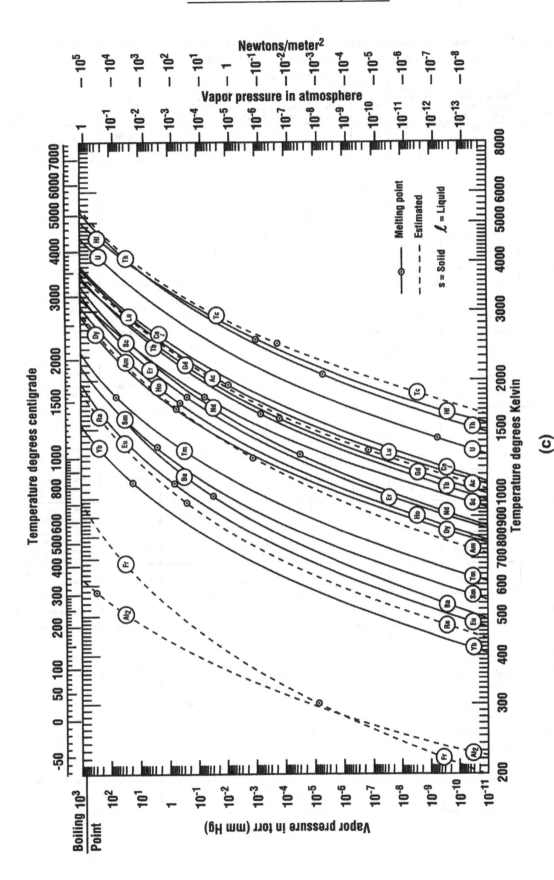

FIG. A18.2. (continued).

(c)

Table A18.1. Lebow thin film databook. Compiled by Edward Graper, Lebow Corporation. Copyright 1990. Used by permission. A reproduction for any purpose without the written permission of Lebow Company is prohibited.

Name	Symbol	Melting point (°C)	Bulk density (g/cm³)	Acoustic impedance ratio (z)	Temperature (°C) vap. press.		Evaporation techniques			Index of refraction @ microns	Remarks
					(10^{-8} torr)	(10^{-4} torr)	Electron beam	Resistance			
								Source	Material		
Aluminum	Al	660	2.70	1.05	677	1010	Xlnt.	Coil Bar	W TiB$_2$-BN	1.0 @ .85 33 @ 12	Alloys and wets tungsten; stranded superior.
Aluminum Antimonide	AlSb	1080	4.3	--	--	--	--	--	--	3.62	--
Aluminum Arsenide	AlAs	1600	3.7	--	--	1300	--	--	--	--	--
Aluminum Bromide	AlBr$_3$	97	2.64	--	--	50	--	Boat	Mo	--	--
Aluminum Carbide	Al$_4$C$_3$	2100	2.36	--	--	800	Fair	--	--	2.75 @ .70	Sublimes
Aluminum 2% Copper	Al2%Cu	640	2.82	--	--	--	--	Bar	TiB$_2$-BN	--	Wire feed and flash, difficult from dual-sources.
Aluminum Fluoride	AlF$_3$	2191	2.88	--	410	700	Poor	Boat	Mo	1.38 @ .55	Sublimes.
Aluminum Nitride	AlN	2200	3.26	--	--	1750	Fair	--	--	--	Decomposes. Reactive evap Al in 10^{-3} N$_2$ with glow discharge.
Aluminum Oxide (α) (alumina)	Al$_2$O$_3$	2045	3.96	.36	--	1550	Xlnt.	--	--	1.59 @ .60 1.56 @ 1.6	Sapphire xlnt. in EB, forms smooth hard films.
Aluminum Phosphide	AlP	2000	2.42	--	--	--	--	--	--	--	--
Aluminum 2% Silicon	Al2%Si	640	2.69	--	--	1010	--	Bar	TiB$_2$-BN	--	Wire feed and flash, difficult from dual source.
Antimony	Sb	631	6.62	.59	278	428	Fair	Boat	Mo	3.4 @ 1.0 5.1 @ 11	Toxic. Film structure is rate dependent Use Mo E.B. liner.
Antimony Telluride	Sb$_2$Te$_3$	629	6.50	--	--	600	--	Boat	Mo	--	Toxic. Decomposes over 750°C.
Antimony Oxide	Sb$_2$O$_3$	656	5.82	--	--	300	Good	Boat	Mo	2.10 @ .50	Toxic, sublimes. Decomposes on W. Use low rate. Z. Physik 165,202 (1961).
Antimony Selenide	Sb$_2$Se$_3$	629	6.50	1.87	--	--	--	Crucible	C	--	Toxic. Stoichiometry variable.
Antimony Sulfide	Sb$_2$S$_3$	550	4.12	--	--	200	Good	Boat	Mo	3.01 @ .55	Toxic. No decomposition.
Arsenic	As	814	5.73	--	107	210	Poor	Crucible Boat	C Mo	--	Toxic. Sublimes rapidly at low temperature.
Arsenic Selenide	As$_2$Se$_3$	360	4.75	--	--	--	--	Boat	Mo	3.03 @ .82 2.9 @ 92	Toxic. JVST 10, 748 (1973).
Arsenic Sulfide	As$_2$S$_3$	300	3.43	--	--	400	Fair	Boat	Mo	2.69 @ .56 2.84 @ 88	Toxic. JVST 10, 748 (1973).
Arsenic Telluride	As$_2$Te$_3$	362	5.0	--	--	--	--	Flash	W	--	Toxic. JVST 10, 748 (1975).
Barium	Ba	725	3.51	--	545	735	Fair	Boat	Mo	.85 @ .50	Toxic. Wets w/o alloying, reacts with ceramics.
Barium Chloride	BaCl$_2$	961	3.86	--	--	650	Good	Boat	Mo	1.74 @ .58	Use gentle preheat to outgas.

Table A18.1. (continued).

Name	Symbol	Melting point (°C)	Bulk density (g/cm³)	Acoustic impedance ratio (z)	Temperature (°C) vap. press. (10⁻⁸ torr)	(10⁻⁴ torr)	Electron beam	Resistance Source	Resistance Material	Index of refraction @ microns	Remarks
Barium Fluoride	BaF_2	1280	4.89	.90	--	480	Good	Boat	Mo	1.51 @ .27 1.40 @ 10.3	Sublimes. Density rate dependent.
Barium Oxide	BaO	1923	5.72	--	--	1300	Poor	Boat	Pt	1.98 @ .59	Decomposes slightly.
Barium Sulfide	BaS	1200	4.25	--	--	1100	--	Boat	Mo	2.16 @ .59	Sublimes.
Barium Titanate	$BaTiO_3$	1620	5.85	.32	decomposes ...		--	--	--	2.4 @ .8	Decomposes, yields free Ba; sputter or co-evaporate.
Beryllium	Be	1283	1.85	.55	710	1000	Xlnt.	Boat	W Ta	2.5 @ .5	Metal powder and oxides very toxic. Wets W/Mo/Ta.
Beryllium Chloride	$BeCl_2$	440	1.90	--	--	150	--	Boat	Mo	--	Very toxic.
Beryllium Fluoride	BeF_2	800	1.99	--	--	480	Good	Boat	Mo	1.33 @ .59	Very toxic, sublimes.
Beryllium Oxide	BeO	2575	3.01	--	--	1900	Good	Boat	Ta	1.82 @ .19 1.72 @ .55	Powders very toxic. No decomposition from EB guns.
Bismuth	Bi	271	9.80	.81	330	520	Xlnt.	Boat	Mo	.82 @ .35 4.5 @ 1.0	Vapors are toxic. High resistivity.
Bismuth Fluoride	BiF_3	727	5.32	--	--	300	--	Crucible	C	1.74 @ 1.0 1.64 @ 10	Toxic, sublimes. App. Opt. 18,105 (1979).
Bismuth Oxide	Bi_2O_3	811	8.9	--	--	1390	Poor	Boat	W	2.48 @ .58	Vapors are toxic. JVST 12, 63 (1975).
Bismuth Selenide	Bi_2Se_3	710	7.66	--	--	650	Good	--	--	--	Toxic. Sputter or co-evaporate.
Bismuth Telluride	Bi_2Te_3	585	6.82	--	--	600	--	Boat	Mo	--	Toxic. Sputter or co-evaporate.
Bismuth Titanate	$Bi_2Ti_2O_7$	--	--	--	decomposes ...		--	--	--	--	Toxic. Decomposes. Sputter or co-evaporate in 10^{-2} O_2.
Bismuth Sulfide	Bi_2S_3	685	7.39	--	--	--	--	--	--	1.5	Toxic.
Boron	B	2100	2.34	.45	1278	1797	Poor	Crucible	C	--	Material explodes with rapid cooling. Forms carbide with container.
Boron Carbide	B_4C	2350	2.52	--	2500	2650	Xlnt.	--	--	--	Similar to chromium.
Boron Nitride	BN	2300	2.25	--	--	1600	Poor	--	--	--	Sputtering pref. Decomposes. JVST A5(4),2696 (1987).
Boron Oxide	B_2O_3	460	2.46	--	--	1400	Good	Boat	Mo	1.46	--
Boron Sulfide	B_2S_3	310	1.55	--	--	800	--	--	--	--	--
Cadmium	Cd	321	8.65	.6	64	180	Fair	Boat	Mo	1.13 @ .6	Poisons vacuum systems, low sticking coefficient. Use Mo E.B. liner.
Cadmium Antimonide	CdSb	456	6.92	--	--	--	--	Boat	Mo	--	--

Inficon Z-Ratio° = acoustic impedance ratio, z.
Z-Ratio° Leybold Inficon

Maxtec Inc. Acoustic Impedance (A. I.) = 8.83 ÷ z

Table A18.1. (continued).

Name	Symbol	Melting point (°C)	Bulk density (g/cm³)	Acoustic impedance ratio (z)	Temperature (°C) vap. press. (10^{-8} torr)	Temperature (°C) vap. press. (10^{-4} torr)	Electron beam	Resistance Source	Resistance Material	Index of refraction @ microns	Remarks
Cadmium Arsenide	Cd_3As_2	721	6.21	–	–	–	–	–	–	–	Toxic.
Cadmium Bromide	$CdBr_2$	567	5.19	–	–	300	–	Boat	Mo	–	Sublimes.
Cadmium Chloride	$CdCl_2$	960	4.05	–	–	400	–	–	–	–	Sublimes.
Cadmium Fluoride	CdF_2	1100	6.64	–	–	600	–	–	–	1.56 @ .58	–
Cadmium Iodide	CdI_2	387	5.67	–	–	250	–	–	–	–	–
Cadmium Oxide	CdO	1430	8.15	–	–	530	–	–	–	2.49 @ .67	Disproportionates.
Cadmium Selenide	CdSe	1351	5.79	–	–	580	Good	Box	Mo	2.4 @ .58	Toxic, sublimes.
Cadmium Silicide	$CdSiO_2$	–	–	–	–	600	–	–	–	1.69	–
Cadmium Sulfide	CdS	1750	4.82	1.02	–	550	Good	Box Crucible	Mo Quartz	2.43 @ .67 2.31 @ 1.4 2.27 @ 7.0	Sublimes. Sticking coeff. affected by sub temp. Comp. variable JVST 12,188 (1975)
Cadmium Telluride	CdTe	1041	6.20	.98	–	450	–	Box Boat	Mo Mo	2.68 @ 4.0 2.51 @ 32	Toxic. Stoichiometry depends on substrate temp. JVST 8,412 (1971).
Calcium	Ca	845	1.55	2.36	272	459	Poor	Boat	Mo	.29 @ .58	Flammable, sublimes. Corrodes in air. Optic 18,59 (1961).
Calcium Fluoride	CaF_2	1360	3.18	.85	–	1100	Xlnt.	Boat	Mo	1.47 @ .24 1.32 @ 9.4	Rate control important. Use gentle preheat to Outgas.
Calcium Oxide	CaO	2580	3.35-3.38	–	–	1700	–	Boat	W	1.84 @ .59	Forms volatile oxides with W and Mo.
Calcium Silicate	$CaO \cdot SiO_2$	1540	2.90	–	–	–	Good	–	–	1.61	–
Calcium Sulfide	CaS	subl.	2.5	–	–	1100	–	Box	Mo	2.14 @ .59	Decomposes.
Calcium Titanate	$CaTiO_3$	1975	4.10	–	1490	1690	Poor	–	–	2.34 @ .59	Disproportionates except in sputtering.
Calcium Tungstate	$CaWO_4$	1620	6.06	–	–	–	Good	Boat	W	1.92 @ .59	–
Carbon (Diamond)	C	3727	3.52	.22	–	–	–	–	–	2.94 @ .19 2.42 @ .66	Deposit by CVD.
Carbon (Graphite)	C	subl.	2.26	4.33	1657	2137	Fair	–	–	1.47	Sublimes. EB preferred, Arc evaporat. Poor film adhesion.
Cerium	Ce	795	6.67	.86	970	1380	Good	Boat	W	1.91 @ .59	Films oxidize easily.
Cerium (III) Fluoride	CeF_3	1460	6.16	–	–	900	Good	Box Boat	Mo W	1.63 @ .55 1.55 @ 12	Use gentle preheat to outgas.
Cerium (IV) Oxide	CeO_2	2600	7.13	–	1890	2310	Good	Boat	W	2.18 @ .55	Sublimes. Use 250°C sub. temperature. Reacts with W. J Opt Soc Am 48,324 (1958).

Table A18.1. (continued).

Name	Symbol	Melting point (°C)	Bulk density (g/cm³)	Acoustic impedance ratio (z)	Temperature (°C) vap. press.		Evaporation techniques			Index of refraction @ microns	Remarks
					$(10^{-8}$ torr)	$(10^{-4}$ torr)	Electron beam	Resistance			
								Source	Material		
Cerium Oxide	Ce_2O_3	1691	6.89	.41	–	–	Fair	Boat	.02 W	2.18 @ .58	Alloys with source. J.Opt.Soc.Am 48,324 (1958).
Cesium	Cs	29	1.89	–	–16	30	–	Boat	Mo	–	Flammable.
Cesium Bromide	CsBr	636	4.44	1.72	–	400	–	Boat	Mo	1.79 @ .36 1.56 @ 39	–
Cesium Chloride	CsCl	646	3.99	–	–	500	–	Boat	Mo	1.64	Hygroscopic.
Cesium Fluoride	CsF	682	4.11	–	–	500	–	Boat	Mo	–	–
Cesium Hydroxide	CsOH	272	3.67	–	–	550	–	Boat	Pt	–	–
Cesium Iodide	CsI	621	4.51	2.95	–	500	–	Boat	W	1.99 @ .23 1.62 @ 53	
Chiolote	$Na_5Al_3F_{14}$	–	2.9	–	–	800	–	Boat	Mo	1.33	–
Chromium	Cr	1875	7.19	.31	837	1157	Good	plated rod or basket	W	.83 @ .13 3.19 @ .63	Sublimes. Films very adherent. High rates possible.
Chromium Boride	CrB	2760	6.17	–	–	–	–	–	–	–	–
Chromium Bromide	$CrBr_2$	842	4.36	–	–	550	–	Boat	Mo	–	–
Chromium Carbide	Cr_3C_2	1890	6.68	–	–	2000	Fair	Boat	W	–	–
Chromium Chloride	$CrCl_2$	824	2.88	–	–	550	–	Box	Mo	–	Sublimes easily.
Chromium Oxide	Cr_2O_3	2435	5.21	–	–	2000	Good	Box	Mo	2.5 @ .59	Disproportionates to lower oxides, reoxidizes @ 600°C in air.
Chromium Silicide	Cr_3Si_2	1710	6.51	–	–	–	–	–	–	–	–
Chromium-Silicon Monoxide	Cr-SiO	influenced by composition					Good	Bar	W	–	Flash.
Cobalt	Co	1495	8.92	.34	850	1200	Xlnt.	Boat Coil	.02 W W	1.10 @ .23 5.65 @ 2.2	Alloys with refractory metals.
Cobalt Bromide	$CoBr_2$	678	4.91	–	–	400		Box	Mo	–	Sublimes.
Cobalt Chloride	$CoCl_2$	724	3.37	–	–	472		Box	Mo	1.51 @ .63	Sublimes.
Cobalt Oxide	CoO	1935	6.45	–	–	–	–	–	–	–	Sputtering preferred.
Copper	Cu	1083	8.94	.43	727	1017	Xlnt.	Boat Coil	Mo W	.87 @ .45 15.5 @ 12	Films do not adhere well. Use intermediate Cr layer, O_2 free Cu req'd.
Copper Chloride	CuCl	431	4.19	–	–	580	–	–	–	1.93	–

Inficon Z-Ratio° = acoustic impedance ratio, z.   Maxtec Inc. Acoustic Impedance (A. I.) = 8.83 ÷ z
       Z-Ratio° Leybold Inficon

Table A18.1. (continued).

Name	Symbol	Melting point (°C)	Bulk density (g/cm³)	Acoustic impedance ratio (z)	Temperature (°C) vap. press.		Evaporation techniques			Index of refraction @ microns	Remarks
					(10⁻⁸ torr)	(10⁻⁴ torr)	Electron beam	Resistance			
								Source	Material		
Copper Oxide	Cu₂O	1235	6.0	--	--	600	Good	Box	Ta	2.70 @ .59	Sublimes. Evaporate in 10⁻² to 10⁻⁴ of O₂; J.Electrochem. Soc. 110,119 (1967).
Copper (I) Sulfide	Cu₂S	1100	5.6	.68	--	--	--	Boat	Mo	--	--
Copper (II) Sulfide	CuS	1113	6.75	.82	--	500	--	--	--	1.45	Sublimes.
Cryolite	Na₃AlF₆	1000	2.9	--	1020	1480	Xlnt.	Boat	Mo	2.34 @ .63	Large chunks reduce spitting. Little decomposition. App.Opt.15, 1969 (1976).
Dysprosium	Dy	1407	8.54	.60	625	900	Good	Boat	Ta	--	Flammable.
Dysprosium Fluoride	DyF₃	1360	--	--	--	800	Good	Box	Ta	--	Sublimes.
Dysprosium Oxide	Dy₂O₃	2340	7.81	--	--	1400	--	Boat	W	--	Loses oxygen.
Erbium	Er	1461	9.09	.74	680	980	Good	Boat	Ta	--	Sublimes.
Erbium Fluoride	ErF₃	1350	7.81	--	--	750	--	Boat	Mo	--	JVST A3(6),2320.
Erbium Oxide	Er₂O₃	2400	8.64	--	--	1600	--	Boat	W	--	Loses oxygen.
Europium	Eu	826	5.26	1.62	280	480	Fair	Boat	Ta	--	Flammable, sublimes. Low tantalum solubility.
Europium Fluoride	EuF₂	1390	6.5	--	--	950	--	Box	Mo	--	--
Europium Oxide	Eu₂O₃	2056	7.42	--	--	1600	Good	Boat	W	--	Loses oxygen; films clear and hard.
Europium Sulfide	EuS	--	5.75	--	--	--	Good	--	--	--	--
Gadolinium	Gd	1312	7.89	.67	760	1175	Xlnt.	Boat	Ta	--	High Ta solubility. Flammable.
Gadolinium Oxide	Gd₂O₃	2310	7.41	--	--	--	Fair	Box	W	1.8 @ .55	Loses oxygen.
Gallium	Ga	30	5.91	.59	619	907	Good	--	--	--	Alloys with refractory metals. Use EB gun.
Gallium Antimonide	GaSb	710	5.6	--	--	--	Fair	Boat	W Ta	3.80 @ 2.2	Flash evaporate.
Gallium Arsenide	GaAs	1238	5.31	1.59	--	--	Good	Boat	W	3.34 @ .78 2.12 @ 23	Flash evaporate.
Gallium Nitride	GaN	800	6.1	--	--	200	--	--	--	--	Sublimes. Evaporate Ga in 10⁻³N₂.
Gallium Oxide(ß)	Ga₂O₃	1900	5.88	--	--	--	--	Boat	W	--	Loses oxygen.
Gallium Phosphide	GaP	1348	4.1	--	520	920	--	Boat	W	3 @ 2.15	Decomposes. Vapor mostly P.

Table A18.1. (continued).

Name	Symbol	Melting point (°C)	Bulk density (g/cm³)	Acoustic impedance ratio (z)	Temperature (°C) vap. press.		Evaporation techniques			Index of refraction @ microns	Remarks
					(10⁻⁸ torr)	(10⁻⁴ torr)	Electron beam	Resistance			
								Source	Material		
Germanium	Ge	937	5.32	.51	812	1167	Xlnt.	Boat Crucible	.02-.04 W C	4.20 @ 2.1 4.00 @ 20	–
Germanium Nitride	Ge₃N₂	450	5.25	–	–	650	–	–	–	–	Sublimes. Sputtering preferred.
Germanium Oxide	GeO₂	1086	6.24	–	–	625	Good	Box	Mo	1.61 @ .59	Similar to SiO, film predominately GeO.
Germanium Telluride	GeTe	725	6.20	–	–	381	–	Box	Mo	–	–
Glass, Schott 8329	–	–	2.20	–	–	–	Xlnt.	–	–	1.47	Evaporable alkali glass. Melt in air before evaporating.
Gold	Au	1962	19.32	.39	807	1132	Xlnt.	Boat Coil	W W	1.50 @ .13 32 @ 12	Films soft, not very adherent. JVST 12,704 (1975).
Hafnium	Hf	2222	13.09	.34	2160	3090	Good	–	–	–	–
Hafnium Boride	HfB₂	3250	10.5	–	–	–	–	–	–	–	–
Hafnium Carbide	HfC	3890	12.20	–	–	2600	–	–	–	–	Sublimes.
Hafnium Nitride	HfN	3305	13.8	–	–	–	–	–	–	–	–
Hafnium Oxide	HfO₂	2811	9.69	–	–	2500	Fair	Boat	W	2.08 @ .48	Film HfO. App. Opt. Apr. 1977.
Hafnium Silicide	HfSi₂	1680	8.02	–	–	–	–	–	–	–	–
Holnium	Ho	1461	8.80	.58	650	950	Good	Boat	W	–	Sublimes.
Holnium Fluoride	HoF₃	1143	7.64	–	–	800	–	–	–	–	–
Holnium Oxide	Ho₂O₃	2360	8.36	–	–	–	–	Boat	W	–	Loses oxygen. App. Opt. 16,439
Inconel	Ni/Cr/Fe	1425	8.5	.33	–	–	Good	Boat Coil	.02 W W	–	Use fine wire pre-wrapped on W. Low rate required for smooth films.
Indium	In	157	7.31	.84	487	742	Xlnt.	Boat	Mo	–	Wets W and Cu; use Mo liner in guns.
Indium Antimonide	InSb	535	5.76	.77	500	400	–	Boat	W	1.00 @ .55 4.0 @ 7.9 3.8 @ 22	Toxic. Decomposes; sputter preferred; or co-evaporate from 2 sources; flash.
Indium Arsenide	InAs	943	5.7	–	780	970	–	Boat	W	3.3 @ 10	Toxic. Sputtering preferred; or co-evaporate frome 2 sources; flash.
Indium Oxide	In₂O₃	1565	7.18	–	–	1200	Good	Boat	W	–	Sublimes. Film In₂O; transparent conductor. JVST 12,99 (1975).
Indium Phosphide	InP	1071	4.9	–	580	730	–	Boat	W	3.3 @ 8.8	Deposits P rich. Flash evaporate.
Indium Selenide	In₂Se₃	890	5.67	–	–	–	–	–	–	–	Sputter, co-evaporate or flash.

Inficon Z-Ratio° = acoustic impedance ratio, z.
Z-Ratio° Leybold Inficon

Maxtec Inc. Acoustic Impedance (A. I.) = 8.83 ÷ z

Table A18.1. (continued).

Name	Symbol	Melting point (°C)	Bulk density (g/cm³)	Acoustic impedance ratio (z)	Temperature (°C) vap. press. (10^{-8} torr)	Temperature (°C) vap. press. (10^{-4} torr)	Evaporation techniques Electron beam	Evaporation techniques Resistance Source	Evaporation techniques Resistance Material	Index of refraction @ microns	Remarks
Indium (III) Sulfide	In_2S_3	1050	4.45	–	–	850	–	–	–	–	Sublimes. Film In_2S.
Indium (I) Sulfide	In_2S	653	5.87	–	–	650	–	–	–	–	–
Indium Telluride	In_2Te_3	667	5.8	–	–	–	–	–	–	–	Sputter, co-evaporate, or flash.
Iridium	Ir	2454	22.45	.13	1850	2380	Fair	–	–	–	–
Iron	Fe	1536	7.87	.35	858	1180	Good	Boat	W	2.0 @ .58	Attacks W. Films hard, smooth. Use gentle preheat to outgas.
Iron Bromide	$FeBr_2$	684	4.64	–	–	561	–	–	–	–	–
Iron Chloride	$FeCl_2$	674	3.16	–	–	300	–	–	–	1.57 @ .59	Sublimes.
Iron Iodide	FeI_2	592	5.31	–	–	400	–	–	–	–	–
Iron (II) Oxide	FeO	1420	5.7	–	–	–	Poor	–	–	2.32 @ .59	Decomposes; sputtering preferred.
Iron (III) Oxide	Fe_2O_3	1538	5.18	–	–	–	Good	Boat	W	3.0	Disproportionates to Fe_3O_4 at 1530°C.
Iron Sulfide	FeS	1195	4.74	–	–	–	–	–	–	–	Decomposes.
Kanthal	FeCrAl	1500	7.1	–	–	1150	–	Boat Coil	.02 W W	1.74 @ .58	JVST 7, 739 (1980).
Lanthanum	La	920	6.1	.93	990	1368	Xint.	Boat	W	–	Films will burn in air if scraped.
Lanthanum Boride	LaB_6	2210	2.61	–	–	–	Good	–	–	–	Toxic.
Lanthanum Fluoride	LaF_3	1491	5.99	–	–	900	Good	Boat	Mo	1.40 @ .30 1.20 @ 8.8	Sublimes. No decomposition. Heat substrate over 300°C.
Lanthanum Oxide	La_2O_3	2315	6.51	–	–	1400	Good	Boat	W	1.95 @ .55 1.89 @ 2.0	Loses oxygen.
Lead	Pb	327	11.34	1.10	342	497	Xint.	Boat	Mo	2.6 @ .58	Toxic. Use Mo liner in E.B. gun.
Lead Bromide	$PbBr_2$	373	6.68	–	–	300	–	Boat	Mo	–	Toxic.
Lead Chloride	$PbCl_2$	5015.85	–	–	325		–	Boat	Mo	2.3 @ .55 2.0 @ 10	Toxic. Little decomposition.
Lead Fluoride	PbF_2	855	8.24	–	–	400	–	Boat	Mo	1.92 @ .30 1.60 @ 11	Toxic, sublimes. Z.Physic 159,117 (1959).
Lead Iodide	PbI_2	502	6.16	–	–	320	–	Crucible	Quartz	2.7	Toxic. J. Opt. Soc. 65,914.
Lead Oxide	PbO	888	9.53	–	–	550	–	Boat	Pt Mo	2.51 @ .59	Toxic. No decomposition. J.Opt.Soc.Am. 52,161 (1962).

Table A18.1. (continued).

Name	Symbol	Melting point (°C)	Bulk density (g/cm³)	Acoustic impedance ratio (z)	Temperature (°C) vap. press.		Evaporation techniques			Index of refraction @ microns	Remarks
					(10⁻⁸ torr)	(10⁻⁴ torr)	Electron beam	Resistance			
								Source	Material		
Lead Stannate	PbSnO₃	1115	8.1	–	670	905	Poor	Boat	Pt Mo	–	Toxic. Disproportionates.
Lead Selenide	PbSe	1065	8.10	–	–	500	–	Box	Mo	3.5 @ 1.0	Toxic, sublimes.
Lead Sulfide	PbS	1114	7.5	.56	–	550	–	Box	Mo	3.9 @ .5	Toxic, sublimes. Little decomposition.
Lead Telluride	PbTe	917	8.16	–	780	1050	–	Boat	Ta	5.6 @ 5 3.4 @ 30	Vapors toxic. Deposits Te rich. Sputter or co-evaporate.
Lead Titanate	PbTiO₃	–	7.52	–	–	–	–	Boat	Ta	–	Toxic.
Lithium	Li	180	0.53	5.95	227	407	Good	Boat	Mo	–	Metal reacts rapidly in air.
Lithium Bromide	LiBr	550	3.46	–	–	500	–	Boat	Mo	1.78 @ .59	–
Lithium Chloride	LiCl	614	2.07	–	–	400	–	Boat	Mo	1.66 @ .59	Use gently preheat for outgas.
Lithium Fluoride	LiF	841	2.59	.78	878	1180	Good	Boat	Ta	1.44 @ .19 1.36 @ 3.5	Rate control important. Use preheat for outgas. App.11,2245 (1972).
Lithium Iodide	LiI	450	3.49	–	–	400	–	Boat	Mo	1.96 @ .59	–
Lithium Oxide	Li₂O	1427	2.01	–	–	850	–	Boat	W	1.64 @ .59	–
Lutetium	Lu	1652	9.84	.48	–	1300	Xlnt.	Boat	Ta	–	Ta impurity a problem.
Lutetium Oxide	Lu₂O₃	2487	9.41	–	–	1400	–	Boat	W	–	Decomposes.
Magnesium	Mg	650	1.74	1.61	180	327	Good	Coil Boat	W Mo	.52 @ .40	Flammable, sublimes. Extremely high rates possible.
Magnesium Aluminate	MgAl₂O₄	2135	3.6	–	–	–	Good	–	–	–	Natural spinel.
Magnesium Bromide	MgBr₂	700	3.72	–	–	450	–	Boat	Mo	–	Decomposes.
Magnesium Chloride	MgCl₂	714	2.32	–	–	400	–	Boat	Mo	1.6	Decomposes.
Magnesium Fluoride	MgF₂	1248	3.0	.68	–	–	Xlnt.	Boat	Mo	1.52 @ .20 1.36 @ 2.0	Rate control & sub. heat required for optical films. App.11, 2245 (1972).
Magnesium Iodide	MgI₂	700	4.24	–	–	200	–	Boat	Mo	–	–
Magnesium Oxide	MgO	2800	3.58	.38	–	1300	Good	–	–	1.77 @ .36 1.63 @ 5.1	W produces volatile oxides. App.11, 2243 (1972).
Manganese	Mn	1241	7.39	.43	508	648	Good	Basket Boat	W Mo	2.59 @ .59	Flammable, sublimes.
Manganese Bromide	MnBr₂	695	4.38	–	–	500	–	Boat	Mo	–	–

Inficon Z-Ratio° = acoustic impedance ratio, z.
Z-Ratio° Leybold Inficon

Maxtec Inc. Acoustic Impedance (A. I.) = 8.83 ÷ z

Table A18.1. (continued).

Name	Symbol	Melting point (°C)	Bulk density (g/cm³)	Acoustic impedance ratio (z)	Temperature (°C) vap. press.		Evaporation techniques			Index of refraction @ microns	Remarks
					(10⁻⁶ torr)	(10⁻⁴ torr)	Electron beam	Resistance			
								Source	Material		
Manganese Chloride	MnCl₂	650	2.98	–	–	450	–	Boat	Mo	–	–
Manganese Oxide	Mn₃O₄	1705	4.86	–	–	–	–	Boat	W	1.73	–
Manganese Sulfide	MnS	1615	3.58	.94	–	1300	–	Boat	Mo	2.7	Decomposes.
Mercury	Hg	–39	13.55	.74	–68	–6	–	–	–	–	Toxic.
Mercury Sulfide	HgS	583	8.10	–	–	250	–	–	–	–	Toxic, decomposes.
Molybdenum	Mo	2610	10.22	.27	1592	2117	Xlnt.	–	–	3.65 @ .59	Films smooth, hard. Careful degas req'd.
Molybdenum Boride	Mo₂B₃	2200	7.48	–	–	–	Poor	–	–	–	–
Molybdenum Carbide	Mo₂C	2687	9.18	–	–	–	Fair	–	–	–	Evaporation of Mo(CO)₆ yields Mo₆C.
Molybdenum Silicide	MoSi₂	2050	6.31	–	–	–	–	Boat	Mo	1.9	Slight O₂ loss.
Molybdenum Sulfide	MoS₂	1185	4.80	–	–	50	–	Boat	W	–	Decomposes.
Molybdenum Oxide	MoO₃	795	4.69	–	–	900	–	–	–	–	–
Neodynium	Nd	1024	7.00	.84	731	1062	Xlnt.	Boat	Ta	.89 @ .39 .30 @ .88	Flammable. Low Ta solubility.
Neodynium Fluoride	NdF₃	1410	6.51	–	–	900	Good	Boat	Mo	1.61 @ .55 1.58 @ 2.0	Very little decomposition.
Neodynium Oxide	Nd₂O₃	1900	7.24	–	–	1400	Good	Boat	W	2.0 @ .55 1.95 @ 2.0	Loses O₂, films clear, EB preferred. Hygroscopic. n varies with substrate temp.
Nichrome IV	Ni/Cr	1395	8.50	–	847	1217	Xlnt.	Coil Boat	W .02-.04 W	3.74 @ 8.8 10.2 @ 12.5	Alloys with refractory metals.
Nickel	Ni	1453	8.91	.33	927	1262	Fair	Coil Boat	W .02-.04 W	3.74 @ 8.8 10.2 @ 12.5	Alloys with refractory metals. Forms smooth adherent films.
Nickel Bromide	NiBr₂	963	4.64	–	–	362	–	Boat	Mo	–	Sublimes.
Nickel Chloride	NiCl₂	1001	3.55	–	–	444	–	Boat	Mo	–	Sublimes.
Nickel Oxide	NiO	1990	6.69	–	–	1480	–	–	–	2.18 @ .48	Dissociates upon heating.
Niobium (Columbium)	Nb	2468	8.57	.47	1728	2287	Xlnt.	Coil	W	1.80 @ .58	Attacks W source.
Niobium Boride	NbB₂	3050	6.97	–	–	–	–	–	–	–	–
Niobium Carbide	NbC	3500	7.82	–	–	–	Fair	–	–	–	–

Table A18.1. (continued).

Name	Symbol	Melting point (°C)	Bulk density (g/cm³)	Acoustic impedance ratio (z)	Temperature (°C) vap. press.		Evaporation techniques			Index of refraction @ microns	Remarks
					(10⁻⁸ torr)	(10⁻⁴ torr)	Electron beam	Resistance			
								Source	Material		
Niobium Nitride	NbN	2573	8.4	--	--	--	--	--	--	--	Reactive, evaporate Nb in 10^{-3} N_2.
Niobium Oxide	NbO	--	7.30	--	--	1100	--	Boat	W	--	--
Niobium Oxide (V)	Nb_2O_5	1520	4.47	--	--	--	--	Boat	.02 W	2.3	--
Niobium Telluride	$NbTe_5$	--	7.6	--	--	--	--	--	--	--	Composition variable.
Niobium-Tin	Nb_3Sn	--	--	--	--	--	Xlnt.	--	--	--	Co-evaporate from 2 sources.
Osmium	Os	3045	22.6	.13	2170	2760	Fair	--	--	--	Toxic.
Palladium	Pd	1552	12.02	.38	842	1192	Xlnt.	Boat	.02 W	1.5 @ .30 2.3 @ .54	Alloys with refractory metals; rapid evaporation suggested. Spits in EB.
Palladium Oxide	PdO	870	8.70	--	--	575	--	--	--	--	Decomposes.
Parylene-N	C_6H_8	300-400	1.1	--	--	--	--	--	--	--	Vapor depositable plastic. (Union Carbide).
Permalloy	Ni/Fe	1395	8.7	--	947	1307	Good	Boat	.02 W	--	Film low in Ni content. Use 84% Ni source. JVST 7(6),573 (1970).
Phosphorus	P	44.2	1.82	--	327	402	--	Boat	Mo	--	Metal reacts violently in air.
Platinum	Pt	1769	21.45	.24	1292	1747	Xlnt.	--	--	3.42 @ 1.0	Alloys, E.B. req'd. Films soft. Poor adhesion.
Platinum Oxide	PtO_2	450	10.2	--	--	--	--	--	--	--	--
Plutonium	Pu	635	19	--	--	--	--	Boat	W	--	Toxic, radioactive.
Polonium	Po	254	9.4	--	117	244	--	Boat	Mo	--	Radioactive.
Potassium	K	64	0.86	--	23	125	--	Boat	Mo	.74 @ .25	Metal reacts violently in air. Use gentle preheat to outgas.
Potassium Bromide	KBr	731	2.79	--	--	450	--	Boat	Mo	1.56 @ .48 1.47 @ 24	Use gentle preheat to outgas.
Potassium Chloride	KCl	776	2.51	2.05	--	510	Good	Boat	Mo	1.72 @ .20 1.25 @ 24	Melt in air to outgas.
Potassium Fluoride	KF	846	2.48	--	--	500	Poor	Boat	Mo	1.35 @ 1.4	Melt in air to outgas.
Potassium Hydroxide	KOH	360	2.04	--	--	400	--	Boat	Mo	--	Melt in air to outgas. Hygroscopic.
Potassium Iodide	KI	686	3.13	2.0	--	500	--	Boat	Mo	1.92 @ .27 1.56 @ 28	Melt in air to outgas.
Praseo-dymium	Pr	936	6.77	--	800	1150	Good	Boat	Ta	--	Flammable.

Inficon Z-Ratio° = acoustic impedance ratio, z.
Z-Ratio° Leybold Inficon

Maxtec Inc. Acoustic Impedance (A. I.) = 8.83 ÷ z

Table A18.1. (continued).

Name	Symbol	Melting point (°C)	Bulk density (g/cm³)	Acoustic impedance ratio (z)	Temperature (°C) vap. press. (10⁻⁸ torr)	(10⁻⁴ torr)	Evaporation techniques Electron beam	Resistance Source	Material	Index of refraction @ microns	Remarks
Praseo-dymium Chloride	PrCl₃	786	4.02	–	–	500	–	Boat	Mo	1.86	–
Praseo-dymium Oxide	Pr₂O₃	2125	6.88	–	–	1400	Good	Boat	W	1.92 @ .27 / 1.83 @ 2.0	Loses oxygen.
Radium	Ra	700	5.0	–	246	416	–	Boat	Mo	–	–
Rhenium	Re	3180	21.04	.14	1928	2571	Poor	–	–	3.18 @ .59	Fine wire will self-evaporate.
Rhenium Oxide	Re₂O₇	297	6.10	–	–	100	–	Boat	Mo	–	–
Rhodium	Rh	1966	12.41	.24	1272	1707	Good	Coil	W	1.62 @ .55	EB gun preferred.
Rubidium	Rb	38	1.53	2.54	–3	111	–	Boat	Mo	1.03 @ .25	–
Rubidium Chloride	RbCl	715	2.80	–	–	550	–	Boat	Mo	1.49	–
Rubidium Iodide	RbI	641	3.59	–	–	400	–	Boat	Mo	1.68 @ .58	–
Ruthenium	Ru	2500	12.45	.20	1780	2260	Poor	Boat	W	–	Spits violently in EB. Requires long degas.
Samarium	Sm	1072	7.54	.91	373	573	Good	Boat	Ta	–	–
Samarium Oxide	Sm₂O₃	2350	8.35	–	–	–	Good	Boat	Ir	1.97	Loses O₂. Films smooth, clear.
Samarium Sulfide	Sm₂S₃	1900	5.72	–	–	–	Good	–	–	–	A.IP Conf.Proc. on Mag.& Mag. Mat.B, 5,860 (1971).
Scandium	Sc	1539	2.99	.91	714	1002	Xlnt.	Boat	W	–	Alloys with Ta. Flammable.
Scandium Fluoride	ScF₃	1550	2.50	–	–	1400	Good	Boat	.02 W	–	–
Scandium Oxide	Sc₂O₃	2300	3.86	–	–	400	Fair	Boat	Ta	1.88 @ .55	Loses O₂.
Selenium	Se	217	4.79	.87	89	170	Xlnt.	Boat	Mo	1.88 @ .24 / 2.43 @ 2.36	Very toxic. Poisons vacuum systems. JVST 9, 387 (1972) JVST 12, 573 & 807 (1975)
Silicon	Si	1410	2.33	.88	992	1337	Fair	Boat	.04 W	3.49 @ 1.4 / 3.42 @ 32	Alloys with W; Some SiO produced above 4x10⁻⁶ Torr. App.Opt. 15,2348 (1976).
Silicon Boride	SiB₄	1870	2.47	–	–	–	Poor	–	–	–	–
Silicon Carbide	SiC	2700	3.22	–	–	1000	–	–	–	2.62 @ .69 / 6.86 @ 13	Sputtering preferred.
Silicon Dioxide	SiO₂	1610 –1710	2.20 –2.70	1.00 influenced by composition	–	1025	Xlnt.	–	–	1.47 @ .30 / 1.45 @ 2.8	Quartz xlnt. in EB.
Silicon (II) Monoxide	SiO	1702	2.1	.50	–	850	Poor	Box	Ta	2.10 @ .10 / 1.67 @ 6 / 2.75 @ 11	Sublimes. Baffle box source best.

Table A18.1. (continued).

Name	Symbol	Melting point (°C)	Bulk density (g/cm³)	Acoustic impedance ratio (z)	Temperature (°C) vap. press. (10⁻⁸ torr)	Temperature (°C) vap. press. (10⁻⁴ torr)	Evaporation techniques Electron beam	Resistance Source	Resistance Material	Index of refraction @ microns	Remarks
Silicon Nitride	Si₃N₄	1900	3.44	--	--	800	--	Boat	Mo	2.0 @ .12 2.05 @ 4	Sublimes.
Silicon Selenide	SiSe	--	--	--	--	550	--	Boat	Mo	--	Toxic.
Silicon Sulfide	SiS	subl.	1.85	--	--	450	--	Boat	Mo	--	--
Silicon Telluride	SiTe₂	--	4.39	--	--	550	--	Boat	Mo	--	Toxic.
Silver	Ag	961	10.49	.50	847	1105	Xlnt.	Coil Boat	W Mo	1.2 @ .30 14.5 @ 12	Evaporates well from any source.
Silver Bromide	AgBr	431	6.49	1.18	--	380	--	Boat	Mo	2.28 @ .58	--
Silver Chloride	AgCl	455	5.56	1.41	--	520	--	Boat	Mo	2.13 @ .43 1.91 @ 19	--
Silver Iodide	AgI	558	6.01	--	--	500	--	Boat	Ta	2.02 @ .59	--
Sodium	Na	97	0.97	4.8	74	192	--	Boat	Mo	.03 @ .59	Metal reacts violently in air.
Sodium Bromide	NaBr	755	3.20	--	--	400	--	Boat	Mo	2.12 @ .21 1.64 @ .59	Use gentle preheat to outgas.
Sodium Chloride	NaCl	801	2.16	1.57	--	530	Good	Boat	Mo	1.79 @ .20 1.20 @ 27	Little decomposition. Use gentle preheat to outgas. Hygroscopic.
Sodium Cyanide	NaCN	563	--	--	--	550	--	Boat	Mo	1.45 @ .59	Toxic. Use gentle preheat to outgas.
Sodium Fluoride	NaF	988	2.56	--	--	700	Good	Boat	Mo	1.39 @ .19 1.25 @ 23	Use gentle preheat to outgas.
Sodium Hydroxide	NaOH	318	2.13	--	--	470	--	Boat	Pt	1.36	Melt in air to outgas. Deliquescent.
Sodium Iodide	NaI	651	3.67	--	--	700	--	Boat	Mo	1.80 @ .49 1.76 @ .66	--
Spinel	MgO₃• 5Al₂O₃	--	8.0	--	--	--	Good	Boat	Ta	1.72 @ .66	--
Strontium	Sr	769	2.6	--	239	403	Poor	Boat	Mo	.61 @ .58	Toxic. Wets but does not alloy with refractory metal May react violently in air.
Strontium Fluoride	SrF₂	1450	4.24	--	--	1000	--	Boat	Ta	1.44 @ 1.4	--
Strontium Oxide	SrO	2461	4.9	--	--	1500	--	Boat	Ta	1.88 @ .58	Sublimes. Reacts with Mo and W.
Strontium Sulfide	SrS	>2000	3.70	--	--	--	--	--	--	2.11 @ .59	Decomposes.
Sulfur	S	115	2.07	2.29	13	57	Poor	Boat	Mo	--	Toxic. Poisons vacuum system.
Supermalloy	Ni/Fe/Mo	1410	8.9	--	--	--	Good	--	--	--	Sputtering preferred; or co-evap. from 2 sources: Permalloy and Mo.

Inficon Z-Ratio° = acoustic impedance ratio, z.
Z-Ratio° Leybold Inficon

Maxtec Inc. Acoustic Impedance (A. I.) = 8.83 ÷ z

Table A18.1. (continued).

Name	Symbol	Melting point (°C)	Bulk density (g/cm³)	Acoustic impedance ratio (z)	Temperature (°C) vap. press.		Evaporation techniques			Index of refraction @ microns	Remarks
					$(10^{-8}$ torr)	$(10^{-4}$ torr)	Electron beam	Resistance			
								Source	Material		
Tantalum	Ta	2996	16.6	.26	1960	2590	Xlnt.	–	–	2.05 @ .58	Forms good films.
Tantalum Boride	TaB₂	3000	11.15	–	–	–	–	–	–	–	–
Tantalum Carbide	TaC	3880	14.65	–	–	2500	–	–	–	–	JVST 12, 811 (1975).
Tantalum Nitride	TaN	3360	16.30	–	–	–	–	–	–	–	Reactive; evaporate Ta in 10^{-3} N₂.
Tantalum Oxide	Ta₂O₅	1800	8.74	.30	1550	1920	Good	Boat	Ta	2.28 @ .40 2.0 @ 1.5	Slight decomposition; evap. in 10^{-3} Torr of O₂. App. Opt. 19, 1737 (1980).
Tantalum Sulfide	TaS₂	1300	–	–	–	–	–	–	–	–	–
Technetium	Tc	2200	11.5	–	1570	2090	–	–	–	–	–
Teflon	PTFE	330	2.9	–	–	–	–	Box	Mo	–	Baffled source. Film structure doubtful.
Tellurium	Te	452	6.25	.53	157	277	Poor	Boat	Mo	4.9 @ 6.0	Toxic. Wets w/o alloying.
Terbium	Tb	1357	8.27	.64	800	1150	Xlnt.	Boat	Ta	–	–
Terbium Fluoride	TbF₃	1176	–	–	800	–	–	Boat	Mo	–	–
Terbium Oxide	Tb₂O₃	2387	7.87	–	–	1300	–	Boat	Ta	–	Partially decomposes.
Terbium Oxide	Tb₄O₇	–	–	–	–	–	–	Boat	Ta	–	Films TbO.
Thallium	Tl	301	11.89	1.58	280	480	Poor	Boat	Mo	–	Wets freely, very toxic.
Thallium Bromide	TlBr	480	7.56	1.77	–	200		Boat	Mo	2.65 @ .44 2.32 @ 24	Toxic, sublimes.
Thalium Chloride	TlCl	430	7.00	1.21	–	150		Boat	Mo	2.20 @ .75 2.6 @ 12	Toxic, sublimes.
Thallium Iodide (B)	TlI	440	7.09	–	–	200		Boat	Mo	2.78 @ .75	Toxic, sublimes.
Thallium Oxide	Tl₂O₃	717	9.65	–	–	350	–	Boat	Mo	–	Toxic. Goes to Tl₂ at 850°C.
Thorium	Th	1875	11.7	.54	1430	1925	Xlnt.	Boat	W	–	Toxic, radioactive.
Thorium Bromide	ThBr₄	–	5.67	–	–	–	–	Boat	Mo	2.47 @ 5	Radioactive, sublimes.
Thorium Carbide	ThC₂	2773	8.96	–	–	2300	–	–	–	–	Radioactive.
Thorium Fluoride	ThF₄	900	6.32	.74	–	750	Fair	Boat	Ta	1.52 @ .40 1.25 @ 12	Radioactive. Heat substrate to above 150°C. JVST 12,919 (1975).

Table A18.1. (continued).

Name	Symbol	Melting point (°C)	Bulk density (g/cm³)	Acoustic impedance ratio (z)	Temperature (°C) vap. press.		Evaporation techniques			Index of refraction @ microns	Remarks
					(10⁻⁸ torr)	(10⁻⁴ torr)	Electron beam	Resistance Source	Material		
Thorium Oxide	ThO₂	3050	9.86	--	--	2100	Good	Boat	.02 W	1.8 @ .55 1.75 @ 2.0	Radioactive.
Thorium Oxyfluoride	ThOF₂	900	9.1	--	--	--	--	Boat	Mo	1.52	Radioactive. Films often ThF₄.
Thorium Sulfide	ThS₂	1925	6.80	--	--	--	--	--	--	--	Radioactive. Sputtering preferred; or co-evaporate from 2 sources.
Thulium	Tm	1545	9.32	.52	461	680	Good	Boat	Ta	--	Sublimes.
Thulium Oxide	Tm₂O₃	--	8.90	--	--	1500	--	Boat	Ir	--	Decomposes.
Tin	Sn	232	7.29	.74	682	997	Xlnt.	Coil Boat	W Mo	1.48 @ .59	Wets Mo; use Ta liner in EB guns.
Tin Oxide	SnO₂	1131	6.99	--	--	1000	Xlnt.	Boat	W	2.08 @ .58	Films from W oxygen deficient, oxidize in air.
Tin Selenide	SnSe	861	6.18	--	--	400	Good	Boat	Mo	--	JVST 12, 110 (1975).
Tin Sulfide	SnS	882	5.22	--	--	450	--	Boat	Mo	--	--
Tin Telluride	SnTe	780	6.44	--	--	450	--	Boat	Mo	--	--
Titanium	Ti	1668	4.50	.63	1067	1453	Xlnt.	Boat	Ta	2.04 @ .45	Alloys with refractory metals; evolves gas on first heating.
Titanium Boride	TiB₂	2900	4.50	--	--	--	Poor	--	--	--	--
Titanium Carbide	TiC	3140	4.93	--	--	2300	--	--	--	--	JVST 12, 851 (1975).
Titanium Oxide (IV) (rutile)	TiO₂	1840	4.26	.40	--	1300	Fair	Boat	.02 W	2.55 @ .38 2.30 @ 1.0	Evaporate in 10⁻⁴ of O₂ onto 350°C substrates. App.Opt.15,2986 (1976).
Titanium (II) Oxide	TiO	1700	4.95	--	--	1500	Good	Boat	.02 W	2.2	Use gentle preheat to outgas. Films TiO₂ if evaporated like TiO₂.
Titanium Nitride	TiN	2930	5.43	--	--	--	Good	Boat	Mo	--	Sputter preferred. Decomposes with thermal evaporation.
Titanium Silicide	TiSi₂	1540	4.39	--	--	--	--	Boat	W	--	--
Tungsten	W	3387	19.3	.16	2117	2757	Good	--	--	2.76 @ .58	Forms volatile oxides. Films hard and adherent.
Tungsten Boride	WB₂	2900	12.75	--	--	--	Poor	--	--	--	--
Tungsten Carbide	W₂C	267	15.6	.18	1480	2120	Xlnt.	--	--	--	--
Tungsten Oxide	WO₃	1473	7.16	--	--	980	Good	Boat	W	2.0 @ .5 2.0 @ 2.0	Sublimes Preheat to outgas. W reduces oxide slightly. App. OPT 28, 1497.
Tungsten Selenide	WSe₂	2150	9.0	--	--	--	--	--	--	--	--

Inficon Z-Ratio° = acoustic impedance ratio, z.
Z-Ratio° Leybold Inficon

Maxtec Inc. Acoustic Impedance (A. I.) = 8.83 ÷ z

Table A18.1. (continued).

Name	Symbol	Melting point (°C)	Bulk density (g/cm³)	Acoustic impedance ratio (z)	Temperature (°C) vap. press.		Evaporation techniques			Index of refraction @ microns	Remarks
					(10⁻⁸ torr)	(10⁻⁴ torr)	Electron beam	Resistance			
								Source	Material		
Tungsten Silicide	WSi₂	2165	9.4	–	–	–	–	–	–	–	–
Tungsten Sulfide	WS₂	1250	7.51	–	–	–	–	–	–	–	–
Tungsten Telluride	WTe₃	–	9.49	–	–	–	–	–	–	–	–
Uranium	U	1132	19.07	.24	1132	1582	Good	Boat	.02 W	–	Films oxidize.
Uranium Carbide	UC₂	2260	11.28	–	–	2100	–	–	–	–	Decomposes.
Uranium Fluoride	UF₄	1000	–	–	–	300	–	Boat	Mo	–	–
Uranium (IV) Oxide	UO₂	2500	10.9	–	–	–	–	Boat	W	–	Ta causes decomposition.
Uranium Oxide	U₃O₃	dec	8.30	–	–	–	–	Boat	W	–	Decomposes at 1300°C to UO₂.
Uranium Phosphide	UP₂	–	8.57	–	–	1200	–	Boat	Ta	–	Decomposes.
Uranium Sulfide	U₂S₃	–	–	–	–	1400	–	Boat	W	–	Slight decomposition.
Vanadium	V	1890	6.11	.53	1890	1547	Xlnt.	Boat	.02 W	3.03 @ .58	Wets Mo. EB evaporated films preferred.
Vanadium Boride	VB₂	2400	5.10	–	–	–	–	–	–	–	–
Vanadium Carbide	VC	2810	5.77	–	–	1800	–	–	–	–	–
Vanadium Nitride	VN	2320	6.13	–	–	–	–	–	–	–	–
Vanadium (IV) Oxide	VO₂	1967	4.34	–	–	575	–	Boat	W	2.51 @ .63 2.76 @ 3.4	Sublimes. Deposit V metal @ 7x10³ O₂ JVST A2(2),301 (1984) & A7(3),1310 (1989).
Vanadium (V) Oxide	V₂O₅	690	3.36	–	–	500	–	–	–	–	–
Vanadium Silicide	VSi₂	1700	4.42	–	–	–	–	–	–	–	–
Ytterbium	Yb	824	6.98	1.27	520	690	Good	Boat	Ta	–	Sublimes.
Ytterbium Fluoride	YbF₃	1161	8.19	–	–	780	–	Boat	Mo	1.48 @ 2.2 1.32 @ 14	–
Ytterbium Oxide	Yb₂O₃	2227	9.17	–	–	1500	–	–	–	–	Sublimes. Loses oxygen.
Yttrium	Y	1509	4.47	.82	830	1157	Xlnt.	Boat	W	–	High Ta solubility.
Yttrium Aluminum Oxide	Y₃Al₅O₁₂	1990	–	–	–	–	Good	–	–	–	Films not ferroelectric.

Table A18.1. (continued).

Name	Symbol	Melting point (°C)	Bulk density (g/cm³)	Acoustic impedance ratio (z)	Temperature (°C) vap. press.		Evaporation techniques			Index of refraction @ microns	Remarks
					$(10^{-8}$ torr)	$(10^{-4}$ torr)	Electron beam	Resistance			
								Source	Material		
Yttrium Fluoride	YF₃	1152	4.01	–	–	–	–	–	–	1.46 @ 2.5 1.42 @ 10	–
Yttrium Oxide	Y₂O₃	2410	5.01	–	–	2000	Good	Boat	.02 W	1.79 @ 1	Sublimes. Loses oxygen, films smooth and clear.
Zinc	Zn	419	7.14	.50	127	250	Xlnt.	Boat Coil	Mo W	2.62 @ .69	Evaporates well under wide range of conditions. Use Mo E.B. liner.
Zinc Antimonide	Zn₃Sb₂	570	6.33	–	–	–	–	–	–	–	–
Zinc Bromide	ZnBr₂	391	4.99	–	–	300	–	Boat	Mo	1.58 @ .58	Decomposes.
Zinc Fluoride	ZnF₂	872	4.95	–	–	800	–	Boat	Ta	–	–
Zinc Nitride	Zn₃N₂	–	6.22	–	–	–	–	Boat	Mo	–	Decomposes.
Zinc Oxide	ZnO	1975	5.60	.55	–	1800	Fair	–	–	2.02 @ .59	Anneal in air at 450°C to reoxidize. JVST 12,879 (1975).
Zinc Selenide	ZnSe	1526	5.42	.72	–	660	–	Boat	Mo	2.61 @ .40 2.43 @ 18	Toxic. Use gentle preheat to outgas. Sublimes well. Z.Angew.Phys.19,392 (1965).
Zinc Sulfide	ZnS	1700	4.10	.77	–	800	Good	Boat	Mo	2.35 @ .55 2.60 @ 4.0 2.13 @ 13	Sublimes Gentle preheat rqd Sticking coeff varies with sub temp. JVST 6,433 (1969)
Zinc Telluride	ZnTe	1240	6.34	–	–	600	–	Boat	Mo	3.56 @ .59 2.80 @ 8.0	Toxic, sublimes. Use gentle preheat to outgas.
Zircon	ZrSiO₄	2550	4.56	–	–	–	–	–	–	1.96 @ .59	–
Zirconium	Zr	1851	6.39	.58	1748	1987	Xlnt.	Boat	.02 W	–	Flammable. Alloys with W. Films oxidize readily.
Zirconium Bromide	ZrB₂	3000	6.09	–	–	–	Good	–	–	–	–
Zirconium Carbide	ZrC	3540	6.73	–	–	2500	–	–	–	–	–
Zirconium Nitride	ZrN	2980	7.09	–	–	–	–	–	–	–	Reactively evaporate in 10⁻³ N₂ atmosphere.
Zirconium Oxide	ZrO₂	2715	5.49	–	–	2200	Good	Boat	W	2.05 @ .50 2.0 @ 2	Films oxygen deficient, clear and hard.
Zirconium Silicide	ZrSi₂	1790	4.88	–	–	–	–	–	–	–	–

Inficon Z-Ratio° = acoustic impedance ratio, z.
Z-Ratio° Leybold Inficon

Maxtec Inc. Acoustic Impedance (A. I.) = 8.83 ÷ z

APPENDIX
19

ACCELERATOR ENERGY CALIBRATION AND STABILITY

Compiled by

J. R. Tesmer

Los Alamos National Laboratory, Los Alamos, New Mexico, USA

CONTENTS

A19.1 INTRODUCTION

The precision needed in energy calibration for ion beam materials analysis is generally not as high as that needed in low-energy nuclear physics (± a few kiloelectronvolts). In fact, it was stated by the noted materials scientist B. Manfred Ullrich in his Second Law (Ullrich, 1988) that "No one cares about the beam energy!" This observation is substantiated by the large number of relative measurements made, as well as by the slow energy variation of stopping powers and Coulomb cross sections. For measurements in the Coulomb regime, energies within several tens of kiloelectronvolts of the actual value are usually acceptable. What is expected, however, is that the energy be at the same value for each sample measured. This is not always easy to accomplish. Some accelerator systems cannot maintain day-to-day calibrations because of their compact size and lack of high-resolution energy analyzing systems.

In practice, both absolute energy and beam energy spread can be important for many measurements. They are particularly important for many resonance reactions. Most of today's accelerators have adequate energy spread; that is, the beam energy spread is certainly much less than the detector resolution, and in most cases, it approaches that of the narrower resonances. Hence, the problems are (1) calibration of the beam energy, (2) day-to-day consistency of that energy, and (3) measurement of the energy spread.

A19.2 ABSOLUTE ENERGY CALIBRATION

For absolute energy calibration, the reactions in Table A19.1 are well known and extensively used. There are, however, other reactions that are not listed that could be useful in specific cases. The reader is referred to J.W. Butler's section (Section 4.9.5) in Mayer and Rimini (1977), Appendix 12 of this Handbook, or Section 15.3.1 of Chapter 15 in this Handbook for more guidance to other reactions. Where possible, the references cite compilations of data that combine several measurements to produce a best value. For very accurate energy calibrations, it is usually not sufficient to calibrate the energy at just one point. The nonlinearity in the analyzing magnet and the variability of other accelerator components can make it necessary to calibrate over a range of energies and possibly over a range of ions that are used (see Overley et al., 1968). Hysteresis effects in the analyzing

magnet make the procedure for setting of its field highly critical if great precision is needed. There are other effects as well hence, the article by Marion (1966) is a good starting point for those interested in the details of energy calibration.

A19.3 ENERGY CALIBRATION STANDARDS

A19.3.1 Gamma resonances

Either NaI(Tl) or bismuth germanate (BGO) scintillation detectors are the detectors of choice for high-energy gamma rays that are produced by (p,γ) or (α,γ) reactions (see Chapter 1 for a discussion of these detectors). Normally, the detectors used are as large as can be afforded and are placed very close to the target. Generally, a thick target is much easier to find and is sufficient for most reactions. The gamma yield is measured as a function of energy (the resulting curve can be fitted by a step function). The energy of the resonance corresponds to the midpoint on the yield curve halfway between the 12% and 88% height of the net yield. The energy difference between these points is the energy spread of the beam, assuming that the resonance width is much narrower than the energy spread and that the resonance and energy spread of the beam are Gaussian. In many cases, the wider resonances have Lorentzian (Breit-Wigner) shapes (see Chapter 7, Section 7.2.1.1). The convolution of Gaussians with Lorentzians is discussed in Chapters 7 and 8.

A19.3.2 Neutron thresholds

For neutron-threshold reactions, a neutron detector such as a long counter (a BF_3 proportional counter surrounded by a neutron moderator) or a plastic scintillator is used. However, as the energies of the thresholds increase, there is an increase in backgrounds so that, at the higher energies, it is common to measure the positrons produced by the reactions in the targets (or, rather, their annihilation radiation) instead of the neutrons (for example, see Overley et al., 1968). The net neutron yield (neutrons minus background) raised to the 2/3 power is plotted as a function of energy, and a linear fit to the net yield is extrapolated to zero yield, which is the threshold energy. It is usually the practice to stay within several kiloelectronvolts of the threshold to obtain the best straight-line fit (see Overley et al., 1968; Roush et al., 1970).

Table A19.1 Energy calibrating reactions.

Energy (keV)	Detected radiation [energy (MeV)]	Width (keV)	Reaction	Reference
Protons				
340.46 ± 0.04	γ (7.12, 6.92, 6.13), α	2.34 ± 0.04	$^{19}F(p,\alpha\gamma)^{16}O$	Aj, 1983; Uhrmacher et al., 1985; Marion, 1966
872.11 ± 0.20	γ (7.12, 6.92, 6.13)	4.7 ± 0.2	$^{19}F(p,\alpha\gamma)^{16}O$	Marion, 1966
991.88 ± 0.04	γ (10.78, 7.93, 1.77)	0.1 ± 0.02	$^{27}Al(p,\gamma)^{28}Si$	Roush et al., 1970; Marion, 1966
1747.6 ± 0.9	γ (9.17, 6.43, 2.74)	0.077 ± 0.012	$^{13}C(p,\gamma)^{14}N$	Marion, 1966
1880.44 ± 0.02	n, threshold	—	$^{7}Li(p,n)^{7}Be$	White et al., 1985
3235.7 ± 0.07	n, threshold	—	$^{13}C(p,n)^{13}N$	Marion, 1966
4234.3 ± 0.80	n, threshold	—	$^{19}F(p,n)^{19}Ne$	Marion, 1966
5803.3 ± 0.26	n, threshold	—	$^{27}Al(p,n)^{27}Si$	Naylor and White, 1977
14230.75 ± 0.02	p	1.2	$^{12}C(p,p,)^{12}C$	Huenges et al., 1973
Deuterons				
1829.2 ± 0.6	n, threshold	—	$^{15}O(d,n)^{17}F$	Bondelid et al., 1960
³He				
1437.9 ± 0.6	n, threshold	—	$^{12}C(^{3}He,n)^{14}O$	Roush et al., 1970
Alphas				
2435.1 ± 0.3	γ (12.1, 10.3)	<0.25	$^{24}Mg(\alpha,\gamma)^{28}Si$	Endt and Van der Leun, 1978; Maas et al., 1978
2865.8 ± 0.3	γ (12.4)	<1	$^{24}Mg(\alpha,\gamma)^{28}Si$	Endt and Van der Leun, 1978; Maas et al., 1978
3035.9 ± 2.3	α	8.1 ± 0.3	$^{16}O(\alpha,\alpha)$	MacArthur et al., 1980; Leavitt et al., 1990
3198.3 ± 0.3	γ (12.7, 10.9)	0.76 ± 0.17	$^{24}Mg(\alpha,\gamma)^{28}Si$	Endt and Van der Leun, 1978; Maas et al., 1978
3363 ± 5	α	5	$^{16}O(\alpha,\alpha)$	Leavitt et al., 1990
3576 ± 4	α	<4	$^{14}N(\alpha,\alpha)$	Herring, 1958
3877 ± 5	α	2	$^{16}O(\alpha,\alpha)$	Leavitt et al., 1990
4265 ± 5	α	27 ± 3	$^{12}C(\alpha,\alpha)$	Kettner et al., 1982; Aj, 1986; Leavitt et al., 1989
5058 ± 3	α	0.11 ± .02	$^{16}O(\alpha,\alpha)$	Häusser et al., 1972; Aj, 1983

A19.3.3 Elastic backscattering

Commonly, backscattering is used more than other analysis methods, and the resonances for the (p,p) and (α,α) reactions listed in Table A19.1 can be easily used for quick, fairly accurate energy calibrations. Thin films of carbon and oxygen can be readily found on most samples.

A19.3.4 Geometry

Care must be taken when detecting either neutrons or gammas to place the detector at an angle where the particles of interest will be emitted. For neutron-threshold reactions, the detector should be at 0° scattering angle and as close as possible to the target. For gammas, 0°

scattering angle will also work if the solid angle of the detector approaches 2π. For smaller solid angles, let the experimenter beware, the distribution of the emitted gamma must be taken into account. Angles that will always work (but might not be optimum for a given reaction) are 55° and 125° (see Chapter 7, Section 7.2.1.4). A great deal of time can be wasted by placing your detector at the wrong angle! Check the literature to determine the exact angular distribution of the gamma rays for the chosen reaction if sensitivity is a problem.

The detector angle for the resonances in particle-in/particle-out reactions vary, but backscattering angles are generally used.

Table A19.2. Alpha particle energy standards, including half-lives, energies, and relative intensities. Uncertainties in the la[st] digit of the energies are shown in parentheses. Very thin [212]Bi and [212]Po sources can be prepared by collecting [220]Rn on a samp[le] surface in the vicinity of a thorium source. The controlling half-live is 10.64 h, that of [212]Pb from which [212]Bi forms by β deca[y]. Sources: Rytz (1979), Lederer and Shirley (1978), Tuli (1985).

Isotope	Half-life	E_α (MeV)	I_α (%)	Isotope	Half-life	E_α (MeV)	I_α (%)
[146]Sm	1.03×10^8 y	2.470 (6) (calculated)	100	[240]Pu	6570 y	5.1682 (2) 5.1237 (2)	73.5 26.4
[150]Gd	1.8×10^6 y	2.719 (8) (calculated)	100	[243]Am	7380 y	5.350 (1) 5.2753 (5) 5.2330 (5)	0.22 87.4 11.0
[148]Gd	75 ± 30 y	3.18271 (2)	100	[210]Po	138.376 d	5.3044 (1)	99+
[232]Th	1.41×10^{10} y	4.013 (3) 3.954 (8)	77 23	[228]Th	1.9131 y	5.4232 (2) 5.3403 (2)	72.7 26.7
[238]U	4.468×10^9 y	4.197 (5) 4.150 (5)	77 23	[241]Am	432.2 y	5.544 (1) 5.4856 (1) 5.4429 (1)	0.34 85.2 13.1
[235]U	7.038×10^8 y	4.599 (2) 4.400 (2) 4.374 (4) 4.368 (2) 4.218 (2)	1.2 57 6.1 12.3 6.2	[238]Pu	87.74 y	5.4991 (2) 5.4563 (2)	71.5 28.5
[236]U	2.342×10^7 y	4.494 (3) 4.445 (5)	74 26	[244]Cm	18.10 y	5.80482 (5) 5.76270 (5)	76.4 23.6
[230]Th	7.54×10^4 y	4.688 (2) 4.621 (2)	75.5 24.5	[243]Cm	28.5 y	6.066 (1) 6.058 (1) 5.992 (1) 5.7850 (5) 5.7411 (5)	1.0 5.0 5.8 73.54 10.65
[234]U	2.45×10^5 y	4.775 (1) 4.723 (1)	72.5 27.5	[242]Cm	162.8 d	6.1128 (1) 6.0694 (1)	73.8 26.2
[209]Po	102 y	4.881 (2)	99.52	[254]Es	275.5 d	6.429 (1)	93.2
[231]Pa	3.28×10^4 y	5.059 (1) 5.028 (1) 5.014 (1) 4.952 (1) 4.736 (1)	11.0 20.0 25.4 22.8 8.4	[253]Es	20.47 d	6.63257 (5) 6.5914 (5)	90.0 6.6
[208]Po	2.898 y	5.115 (2)	100	[212]Bi	60.55 m	6.08994 (4) 6.05083 (4)	27.2 69.9
[239]Pu	2.412×10^4 y	5.1566 (4) 5.144 (1) 5.105 (1)	73.3 15.1 11.5	[212]Po	0.298 μs	8.7842 (1)	100

A19.3.5 Other calibration methods

Another method is to place a long-lived alpha source in the chamber and have the alphas from this source appear in the spectrum along with your data. An extensive list of alpha sources is given in Table A19.2. Determining the positions of this peak and two other known points will give not only the detector system gain but, in theory, also the energy of the beam if the scattering geometry is known. In fact, one of the popular analysis programs, RUMP (Doolittle, 1985, 1986, 1990), has this procedure built in. Care must be taken if this method is used to accurately calibrate the beam energy because of several problems relating to sources and detector windows. See Chapter 15 and Lennard et al. (1990). Corrections for the dead layer, the pulse-height defect, and the nonlinearity of the surface-barrier detector must be taken into account, as well as surface contamination and aging of the source for accurate beam energy determination. In some cases, an exothermic reaction has bee[n] used to replace the alpha source (Scott and Paine, 1983).

Another method that uses a backscattering or "crossover" technique relies on kinematics and the determination of the angle where particles scattered from different elements or differen[t] reactions in the target have the same energy. Normally, this i[s] applied to beam energies above a few megaelectronvolts (se[e] Birattari et al., 1992).

A19.4 CONSISTENCY

The problem of achieving day-to-day consistency can b[e] solved in several ways. For backscattering analysis, one ver[y] useful reaction for both energy calibration and a consistency eck is the $^{16}O(\alpha,\alpha)^{16}O$ (E_R = 3.036 MeV). This has the lowest ergy of any easily obtained alpha reactions, and therefore, it is eful for day-to-day checks of the beam energy. The resonance ergy is sufficiently well known and has a large cross section.

thin oxide layer can be found in almost any scattering amber for use in quickly calibrating the beam energy. This sonance is not Gaussian, and the peak height of the resonance ould be used to obtain the calibration energy; that is, for thick rgets, the energy that produces the largest height (not tegrated area) determines the resonance energy. This is one of e preferred reactions for day-to-day calibrations at normal BS energies.

Another very useful technique to assure consistency is to epare a calibration standard such as a thin layer (<1 nm) of ld on silicon dioxide. One then checks to make sure that the articles scattered from the gold and the Si (edge) appear in the me channels as in previous measurements before new data are corded. This method should be used with caution because the fect of the detector-system gain shift or zero-energy offset can e overcome by a change in beam energy.

If depth profiling with (p,γ) resonances is the analysis chnique, then consistency is easily obtained by observing the ergy at which the resonance of interest is first generated in the mple.

EFERENCES

jzenberg-Selove, F. (1983), *Nucl. Phys.* **A392**, 1.

jzenberg-Selove, F. (1986), *Nucl. Phys.* **A460**, 1.

irattari, C., Castiglioni, M., and Silari, M. (1992), *Nucl. strum. Methods* **A320**, 413.

ondelid, R.O., Butler, J.W., and Kennedy, C.A. (1960), *Phys. ev.* **120**, 889.

oolittle, L.R. (1985), *Nucl. Instrum. Methods* **B9**, 344.

oolittle, L.R. (1986), *Nucl. Instrum. Methods* **B15**, 227.

oolittle, L.R. (1990), in *High Energy and Heavy Ion Beams in Materials Analysis* (Tesmer, J.R., Maggiore, C.J., Nastasi, M., arbour, J.C., and Mayer, J.W., eds.), Materials Research ociety, Pittsburgh, PA, p. 175.

ndt, P.M., and Van der Leun, C. (1978), *Nucl. Phys.* **A310**, 1.

äusser, O., Ferguson, A.J., McDonald, A.B., Szoghy, I.M., lexander, T.K., and Disdier, D.L. (1972), *Nucl. Phys.* **A179**, 65.

Herring, D.F. (1958), *Phys. Rev.* **112**, 1217.

Huenges, E., Rosier, H., and Vonach, H. (1973), *Phys. Lett.* **B46**, 361.

Kettner, K.U., Becker, H.W., Buchmann, L., Gorres, J., Krawinkel, H., Rolfs, C. Schmalbrock, P., Trautvetter, H.P., and Vlieks, A. (1982), *Z. Phys.* **A308**, 73.

Lederer, CM., and Shirley, V.S. (eds.) (1978), *Table of Isotopes*, 7th Ed., Wiley, New York.

Lennard, W.N., Tong, S.Y., Massoumi, G.R., and Wong, L. (1990), *Nucl. Instrum. Methods* **B45**, 281.

Leavitt, J.A., McIntyre Jr., L.C., Stoss, P., Oder, J.G., Ashbaugh, M.D., Dezfouly-Arjomandy, B., Yang, Z.M., and Lin, Z. (1989), *Nucl. Instrum. Methods* **B40/41** 776.

Leavitt, J.A., McIntyre Jr., L.C., Asbaugh, M.D., Oder, J.G., Lin, Z., and Dezfouly-Arjomandy, B. (1990), *Nucl. Instrum. Methods* **B44**, 260.

Maas, J.W., Somorai, E., Graber, H.D., Van den Wijngaart, C.A., Van der Leun, C., and Endt, P.M. (1978), *Nucl. Phys.* **A301**, 213.

Marion, J.B. (1966), *Rev. Mod. Phys.* **38**, 660.

Mayer, J.W., and Rimini, E. (eds.) (1977), *Ion Beam Handbook for Materials Analysis*, Academic Press, New York.

Naylor, H., and White, R.E. (1977), *Nucl. Instrum. Methods* **144**, 331.

Overley, J.C., Parker, P.D., and Bromley, D.A. (1968), *Nucl. Instrum. Methods* **68**, 61.

Roush, M.L., West, L.A., and Marion, J.B. (1970), *Nucl. Phys.* **A147**, 235.

Rytz, A. (1979), *At. Data Nucl. Data Tables* **23**, 507.

Scott, D.M., and Paine, B.M. (1983), *Nucl. Instrum. Methods* **218**, 154.

Tuli, J.K. (1985), *Nuclear Wallet Cards*, National Nuclear Data Center, Brookhaven National Laboratory, Upton, NY.

Uhrmacher, M., Pampus, K., Bergmeister, F.J., Purschke, D., and Lieb, K.P. (1985), *Nucl. Instrum. Methods* **B9**, 234.

White, R.E., Barker, P.H., and Lovelock, D.M.J. (1985), *Metrologia* **21**, 193.

RADIATION HAZARDS OF (p,n), (p,y), AND (d,n) REACTIONS

Compiled by

P. Rossi

University of Padua, Padua, Italy, and Sandia National Laboratories, Albuquerque, New Mexico, USA

CONTENTS

A20.1 INTRODUCTION

All ions hitting materials can produce neutrons and gamma rays. To this end, aside from the alpha particles considered in Appendix 21, protons and deuterons are the most effective bullets for beam energies of less than 15 MeV, because of their high energy per nucleon. To a lesser extent, the same conclusion applies to light nuclei, such as tritium, lithium, and beryllium, which are, however, far less common and will thus not be addressed here. We present instead tables of proton- and deuteron-induced neutron and gamma yields for beam energies of 3, 5, and 15 MeV. The first two energies are representative of the most widely used energy range in ion beam analysis (IBA). The last value is usually an upper limit in IBA facilities of the kind and might indicate the worst possible scenario. There is not single compilation of these data, as for alpha yields, but the yields are evaluated here starting from the known cross sections and material properties. In performing these calculations, we employed the EXFOR database, the Nuclear National Data Center Web site (http://www.nndc.bnl.gov), and the program SRIM (SRIM, 2008; Ziegler *et al.*, 1985) on ion interactions with matter. Finally, the conversion factors from flux to equivalent dose, the relevant quantity in the assessment of radiation hazards, come from the SCALE code (SCALE, 2009; Jordan and Bowman, 1998).

A20.2 YIELD CALCULATIONS

The rate of n and γ production from a thick homogeneous target is given by the exact formula

$$\frac{dn}{dt} = IC\langle\sigma\rangle r_{eff}, \qquad (A3.1)$$

where n is the number of generated units of radiation, either neutrons or gamma rays, t is the time, I is the ion beam current (number of ions/s), C is the target-atom concentration (number of atoms/cm^3), r_{eff} is the effective range [i.e., the range (cm) for which the ion retains an energy greater than the reaction threshold (E_{th})], and $<\sigma>$ is

$$\langle\sigma\rangle = \frac{1}{r_{eff}} \int_{E_{th}}^{E(ion)} \frac{\sigma(E)}{LET(E)} \, dE. \qquad (A3.2)$$

An approximate value can be obtained by assuming that LET(E) is constant in the energy range of interest . In this case,

$$\langle\sigma\rangle_{app} = \frac{1}{r_{eff}} \frac{1}{LET} \int_{E_{th}}^{E(ion)} \sigma(E) \, dE$$

$$= \frac{1}{E(ion) - E_{th}} \int_{E_{th}}^{E(ion)} \sigma(E) \, dE. \qquad (A3.3)$$

This approximation was used for the values listed in Table A20.3 through A20.7, as this compilation is intended only to assess safety concerns and not to obtain high accuracy. Nevertheless, relative errors in the yields are expected to typically be less than 10%.

A20.3 DOSE CALCULATIONS

We evaluated the neutron and gamma dose rates at a distance of 3 m from the radiation source, either target, collimators, or beam pipes, associated with (p,n), (p,γ), and (d,n) reactions on typical materials in a ion beam laboratory. Although the maximum allowable radiation dosage recommended by the National Council on Radiation Protection and Measurement (NCRP) is 5 rem/year (NCRP, 1993), we assume as suitable maximum of one-tenth of this value, that is, 0.5 rem/year, and report the number of exposure hours per week to exceed this limit.

The dose calculations were based on the "response factor" R (Morrison *et al.*, 1972), a coefficient that transforms the particle flux [number of radiation particles/(cm^2 s)] into the dose rate (rem/h). The units of the two quantities are chosen according to established tradition. R depends on the neutron or gamma energy. We follow the energy subdivision of Morrison *et al.* (1972) as reported in Table A20.1. Values were obtained using the SCALE code (SCALE, 2009; Jordan and Bowman, 1998).

Table A20.1. Response factors, R.

Neutrons upper energy (MeV)	R [(rem h^{-1})/(s^{-1} cm^{-2})]	Photons lower energy (MeV)	R [(rem h^{-1})/(s^{-1} cm^{-2})]
1.49180E+01	1.94480E–04	8.00E+00	8.77160E–06
1.22140E+01	1.59710E–04	6.50E+00	7.47849E–06
1.00000E+01	1.47060E–04	5.00E+00	6.37479E–06
8.18730E+00	1.47720E–04	4.00E+00	5.41360E–06
6.36000E+00	1.53390E–04	3.00E+00	4.62209E–06
4.96590E+00	1.50620E–04	2.50E+00	3.95960E–06
4.06570E+00	1.38920E–04	2.00E+00	3.46860E–06
3.01190E+00	1.28430E–04	1.66E+00	3.01920E–06
2.46600E+00	1.25270E–04	1.33E+00	2.62759E–06
2.35000E+00	1.26320E–04	1.00E+00	2.20510E–06
1.82680E+00	1.28940E–04	8.00E–01	1.83260E–06
1.10900E+00	1.16870E–04	6.00E–01	1.52290E–06
5.50230E–01	6.52780E–05	4.00E–01	1.17250E–06
1.10900E–01	9.18970E–06	3.00E–01	8.75940E–07
3.35460E–03	3.71340E–06	2.00E–01	6.30610E–07
5.82950E–04	4.00860E–06	1.00E–01	3.83380E–07
1.01300E–04	4.29459E–06	5.00E–02	2.66930E–07
2.90230E–05	4.47609E–06	1.00E–02	9.34770E–07
1.06000E–05	4.56730E–06		
3.05900E–06	4.53549E–06		
5.31590E–07	4.37009E–06		
4.14000E–07	3.71469E–06		

Table A20.2 reports R and the dose-rate/particle-rate ratio for neutrons (N) and gammas (γ) at a distance of 300 cm from the radiation source. Two beam ions, hydrogen (p) and deuterium (d); three beam energies, 3, 5, and 15 MeV; and a beam current of 1 nA were considered. The energy of the emitted radiation was set equal to a rough average of Q values of the reactions on steel. However, R depends on energy only slightly. As the proton or deuteron energy is increased from 3 MeV to 15 MeV, no new important channels open, so the values in Table A20.2 apply to beams of all energies. Note that, because the (d,γ) reaction gives negligible yields at our energies, it is not considered here. The values in the last column of Table A20.2 were calculated using the equation

$$\frac{dD}{dt} \ (\text{mrem h}^{-1}) \Big/ \frac{dn}{dt} \ (\text{s}^{-1}) = \frac{1000R \ [\text{rem h}^{-1}/(\text{s}^{-1} \ \text{cm}^{-1})]}{4\pi(300)^2 \ (\text{cm}^2)}, \quad (A3.4)$$

where D represents dose.

Table A20.2. Dose-rate/particle-rate coefficients for the considered cases.

Ion	Type of radiation	⟨E(n)⟩ (MeV)	R [(rem s^{-1})/(s^{-1} cm^{-2})]	(dD/dt)/(dn/dt) [(mrem h^{-1})/s^{-1}]
p	N	5	1.53390E–04	1.36E–07
d	N	6	1.53390E–04	1.36E–07
p	γ	3	3.95960E–06	4.79E–09

Assuming a beam current of 1 nA, we performed calculations to determine the number of hours per week a person would have to be exposed to a given level of radiation to exceed a cumulative exposure of 0.5 rem/year. The results of these calculations are listed in the column labeled h/w (i.e., hours/week) in Tables A20.3–A20.7. A dash (–) indicates that the required number of hours per week exceeds 168, the total number of hours in a week, indicating that the radiation dosage is negligible. The cases where the number of hours per week is less than 4 (length of a typical ion beam experimental session) are highlighted in bold, indicating that extra caution is required.

Finally, Figure A20.1 describes the neutron emission due to the D(D,n) reaction of a deuterium beam (1 nA) impinging on a previously deuterium-implanted target. Two cases of implantation on C and Cu have been considered, with a "saturation" D surface concentration. The dose rate per unit surface area is simply proportional to the average D(D,n) cross section and can be scaled to the actual D concentration.

It should be noted that the radiation dosage is directly proportional to the beam current and inversely proportional to the square of the distance. Thus, the values presented can be adjusted for any beam current and distance.

Symbols used in the subsequent tables are as follows: Iso.Abs. = isotopic abundance; E_{Th} = approximate value of the threshold energy of the highest-yield reaction of that type; r_{eff} = effective range, the distance covered by the ion until its energy is above the threshold; $\langle CS \rangle$ = average cross section for an energy ranging from the E_{Th} up to the energy of the incident beam; DR = dose rate; I beam current intensity; h/w = maximum hours per week of operation; – = no limit.

Table A20.3. Proton-induced neutron emission (5 and 15 MeV).

			5 MeV					15 MeV				
Target	Iso.Abs.	E_{Th} (MeV)	r_{eff} (cm)	$\langle CS \rangle$ (mbarn)	Rate (s^{-1})	DR/I [mrem/(h nA)]	h/w (h)	r_{eff} (cm)	$\langle CS \rangle$ (mbarn)	Rate (s^{-1})	DR/I [mrem/(h nA)]	h/w (h)
^{14}N (air)	1.0E+00	6.4	0.0E+00	0.0	0.0E+00			1.8E+02	47.4	2.2E+06		
^{15}N (air)	3.7E–03	3.7	1.4E+01	59.4	7.4E+02			2.1E+02	112.0	2.2E+04		
^{17}O (air)	3.8E–04	3.6	1.4E+01	65.0	2.0E+01			2.1E+02	65.0	2.9E+02		
^{18}O (air)	2.0E–03	2.5	2.4E+01	147.0	3.6E+02	1.5E–04	–	2.2E+02	193.0	4.5E+03	3.1E–01	34
^{9}Be	1.0E+00	2.3	1.7E–02	255.0	3.4E+06	4.7E–01	22	1.6E–01	255.0	3.2E+07	4.3E+00	2
^{16}O (MgO)	1.0E+00		0.0E+00	0.0	0.0E+00		–	0.0E+00	0.0	0.0E+00		
^{25}Mg (MgO)	1.0E–01	5.1	0.0E+00	0.0	0.0E+00		–	7.5E–02	102.3	2.5E+05		
^{26}Mg (MgO)	1.1E–01	5.0	0.0E+00	0.1	0.0E+00	0.0E+00	–	7.5E–02	63.8	1.6E+05	5.6E–02	–
^{27}Al	1.0E+00	5.6	0.0E+00	0.0	0.0E+00	0.0E+00	–	1.1E–01	78.0	3.1E+06	4.2E–01	25
^{16}O (Al$_2$O$_3$)	1.0E+00		0.0E+00	0.0	0.0E+00			0.0E+00	0.0	0.0E+00		
^{27}Al (Al$_2$O$_3$)	1.0E+00	5.6	0.0E+00	0.0	0.0E+00			1.1E–01	78.0	2.4E+06	3.3E–01	32
^{13}C (steel)	1.1E–02	2.0	6.1E–03	45.4	3.5E+01			4.9E–02	62.7	3.9E+02		
^{12}C (steel)	1.0E+00	12.0	0.0E+00	0.0	0.0E+00			1.6E–02	45.2	9.2E+03		
^{56}Fe (steel)	9.2E–01	4.8	5.0E–04	0.0	0.0E+00			4.3E–02	318.0	6.6E+06		
^{57}Fe (steel)	2.2E–02	1.7	6.5E–03	59.4	4.4E+03			4.9E–02	336.0	1.9E+05		
^{58}Fe (steel)	2.8E–03	3.1	4.4E–03	97.5	6.1E+02	6.8E–04	–	4.7E–02	495.0	3.3E+04	9.3E–01	11
^{63}Cu	6.9E–01	3.9	2.6E–03	41.3	4.0E+04			4.3E–02	337.0	5.4E+06		
^{65}Cu	3.1E–01	1.9	6.1E–03	55.6	5.4E+04	1.3E–02	–	4.7E–02	432.0	3.2E+06	1.2E+00	9
^{206}Pb (lead)	2.4E–01	12.7	0.0E+00	0.0	0.0E+00			1.2E–02	295.0	1.7E+05		
^{207}Pb (lead)	2.2E–01	13.0	0.0E+00	0.0	0.0E+00			1.2E–02	220.0	1.2E+05		
^{208}Pb (lead)	5.2E–01	7.6	0.0E+00	0.0	0.0E+00	0.0E+00	–	3.9E–02	253.0	1.0E+06	1.8E–01	57

Table A20.4. Proton-induced neutron emission (3 MeV).

Target	Iso.Abs.	E_{Th} (MeV)	r_{eff} (cm)	\<CS\> (mbarn)	3 MeV Rate (s^{-1})	DR/I [mrem/(h nA)]	h/w (h)
^{14}N (air)	1.0E+00	6.4	0.0E+00	0.0	0.0E+00		
^{15}N (air)	3.7E–03	3.7	0.0E+00	0.0	0.0E+00		
^{17}O (air)	3.8E–04	3.6	0.0E+00	0.0	0.0E+00		
^{18}O (air)	2.0E–03	2.5	4.1E+00	23.6	1.0E+01	1.4E–06	–
^{9}Be	1.0E+00	2.3	1.7E–02	113.0	1.5E+06	2.1E–01	50
^{13}C (steel)	1.1E–02	2.0	1.6E–03	2.4	5.0E–01		
^{12}C (steel)	1.0E+00	12.0	0.0E+00	0.0	0.0E+00		
^{56}Fe (steel)	9.2E–01	4.8	0.0E+00	0.0	0.0E+00		
^{57}Fe (steel)	2.2E–02	1.7	2.0E–03	4.0	9.2E+01		
^{58}Fe (steel)	2.8E–03	3.1	0.0E+00	0.0	0.0E+00	1.3E–05	–
^{63}Cu	6.9E–01	3.9	0.0E+00	0.0	0.0E+00		
^{65}Cu	3.1E–01	1.9	1.8E–03	3.2	9.0E+02	1.2E–04	–

Table A20.5. Proton-induced gamma emission (5 and 15 MeV).

Target	Iso.Abs.	E_{Th} (MeV)	5 MeV r_{eff} (cm)	\<CS\> (mbarn)	Rate (s^{-1})	DR/I [mrem/(h nA)]	h/w (h)	15 MeV r_{eff} (cm)	\<CS\> (mbarn)	Rate (s^{-1})	DR/I [mrem/(h nA)]	h/w (h)
^{12}C (steel)	9.9E–01	0.8	7.8E–03	0.9	8.7E+01			5.1E–02	1.6	1.0E+03		
^{54}Fe (steel)	5.8E–02	3.0	6.1E–03	1.0	1.9E+02			4.9E–02	0.9	1.4E+03		
^{56}Fe (steel)	9.2E–01	3.0	4.0E–03	0.8	1.5E+03			4.7E–02	5.2	1.2E+05		
^{58}Fe (steel)	2.8E–03	0.8	4.4E–03	0.8	5.0E+00	6.4E–06	–	4.7E–02	0.8	5.4E+01	4.2E–04	–
^{63}Cu	6.9E–01	1.1	2.4E–03	0.9	8.0E+02			4.3E–02	0.8	1.3E+04		
^{65}Cu	3.1E–01	1.0	5.2E–03	0.9	7.5E+02	5.4E–06	–	4.6E–02	0.9	6.6E+03	6.8E–05	–

Table A20.6. Deuterium-induced neutron emission (5 and 15 MeV).

Target	Iso.Abs.	E_{Th} (MeV)	5 MeV r_{eff} (cm)	\<CS\> (mbarn)	Rate (s^{-1})	DR/I [mrem/(h nA)]	h/w (h)	15 MeV r_{eff} (cm)	\<CS\> (mbarn)	Rate (s^{-1})	DR/I [mrem/(h nA)]	h/w (h)
^{14}N (air)	1.0E+00	0.6	2.0E+01	136.0	7.1E+05			1.4E+02	131	4.6E+06		
^{16}O (air)	1.0E+00	1.9	1.6E+01	443.0	4.3E+05	1.5E–01	68	1.3E+02	335	2.6E+06	9.8E–01	11
^{16}O (MgO)	1.0E+00	1.9	6.4E–03	443.0	9.4E+05			5.0E–02	335	5.6E+06		
Mg (MgO)	1.0E–01		0.0E+00	0.0	0.0E+00	1.3E–01	81	0.0E+00	0	0.0E+00	7.6E–01	14
^{16}O (Al$_2$O$_3$)	1.0E+00	1.9	5.8E–03	443.0	1.1E+06			4.6E–02	335	6.7E+06		
^{27}Al (Al$_2$O$_3$)	0.0E+00		0.0E+00	0.0	0.0E+00	1.5E–01		0.0E+00	0	0.0E+00	9.2E–01	11
^{16}O (SiO$_2$)	1.0E+00	1.9	8.5E–03	443.0	1.2E+06			6.8E–02	335	7.5E+06		
Si (SiO$_2$)	0.0E+00		0.0E+00	0.0	0.0E+00	1.7E–01		0.0E+00	0	0.0E+00	1.0E–00	10
^{12}C (steel)	9.9E–01	0.4	4.9E–03	107.0	6.4E+03			3.1E–02	65	2.4E+04		
^{13}C (steel)	1.1E–02	0.2	5.0E–03	271.0	1.7E+02			3.1E–02	271	1.0E+03		
^{54}Fe (steel)	5.8E–02	2.4	3.5E–03	31.0	3.4E+03			2.9E–02	84	7.8E+04		
^{56}Fe (steel)	9.2E–01	6.6	3.5E–03	0.0	0.0E+00	1.4E–03	–	2.9E–02	134	1.9E+06	2.7E–01	38
^{63}Cu	6.9E–01	3.6	2.1E–03	0.3	2.3E+02			2.7E–02	115	1.2E+06		
^{65}Cu	3.1E–01	5	0.0E+00	0.0	0.0E+00	3.1E–05	–	2.5E–02	558	2.2E+06	4.6E–01	23

Table A20.7. Deuterium-induced neutron emission (3 MeV).

Target	Iso.Abs.	E_{Th} (MeV)	r_{eff} (cm)	\<CS\> (mbarn)	3 MeV Rate (s^{-1})	DR/I [mrem/(h nA)]	h/w (h)
^{14}N (air)	1.0E+00	0.6	8.1E+00	69.4	1.5E+05		
^{16}O (air)	1.0E+00	1.9	4.6E+00	283.0	7.7E+04	3.0E–02	–
^{16}O (MgO)	1.0E+00	1.9	1.9E–03	283.0	1.7E+05		
Mg (MgO)	1.0E–01		3.7E–03	0.0	0.0E+00	2.4E–02	–
^{16}O (Al$_2$O$_3$)	1.0E+00	1.9	1.6E–03	283.0	2.0E+05		
^{27}Al (Al$_2$O$_3$)	0.0E+00		3.3E–03	0.0	0.0E+00	2.8E–02	–
^{16}O (SiO$_2$)	1.0E+00	1.9	2.5E–03	283.0	2.3E+05		
Si (SiO$_2$)	0.0E+00		4.9E–03	0.0	0.0E+00	3.1E–02	–
^{12}C (steel)	9.9E–01	0.4	2.1E–03	108.0	2.7E+03		
^{13}C (steel)	1.1E–02	0.2	2.1E–03	271.0	7.2E+01		
^{54}Fe (steel)	5.8E–02	2.4	6.4E–04	5.4	1.1E+02		
^{56}Fe (steel)	9.2E–01	6.6	0.0E+00	0.0	0.0E+00	3.9E–04	
^{63}Cu	6.9E–01	3.6	0.0E+00	0.0	0.0E+00		
^{65}Cu	3.1E–01	5.0	0.0E+00	0.0	0.0E+00	0.0E+00	–

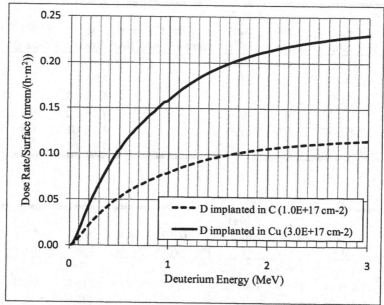

FIG. A20.1. Deuterium-induced neutron emission from a deuterium-implanted target.

REFERENCES

Jordan, W.C., and Bowman, S.M. (1998), SCALE Cross-Section Libraries, NUREG/CR-0200, rev. 6, vol. 3, sec. M4, Report ORNL/NUREG/CSD-2/V3/R6, Oak Ridge National Laboratory, Oak Ridge, TN.

Morrison, G.W., Straker, E.A., and Odegaard, R.H. (1972), *Trans. Am. Nucl. Soc.* **15** (1), 535.

NCRP (1993), *Limitation of Exposure to Ionizing Radiation*, Report No. 116, National Council on Radiation Protection and Measurements, Bethesda, MD.

SCALE (2009), *A Modular Code System for Performing Standardized Computer Analyses for Licensing Evaluation*, Report ORNL/TM-2005/39, version 6.0, vols. I–III (January 2009), available from Radiation Safety Information Computational Center at Oak Ridge National Laboratory, Oak Ridge, TN, as document C00750.

SRIM (2008), http://www.SRIM.org.

Ziegler, J.F., Biersack, J.P., and Littmark, U. (1985), "The Stopping and Range of Ions in Solids", in *The Stopping and Range of Ions in Matter* (Ziegler, J.F., ed.), Pergamon, New York, vol. 1.

APPENDIX
21
RADIATION HAZARDS OF (α,n) REACTIONS

Compiled by

S. N. Basu

Boston University, Boston, Massachusetts, USA

M. Nastasi, and J. R. Tesmer

Los Alamos National Laboratory, Los Alamos, New Mexico

CONTENTS

A21.1 INTRODUCTION

The analysis of materials using α irradiation is an established technique. However, the irradiation of low-Z elements by energetic α particles can lead to the production of neutrons (Bair and del Campo, 1979; West and Sherwood, 1982; Wilson *et al.*, 1988). These neutrons can ionize body tissues, causing health hazards (Cember, 1976). The maximum allowable radiation dosage recommended by the International Commission on Radiological Protection (ICRP) is 5 rem per year (ICRP, 1987). In this work, it is assumed that an individual would prefer to be exposed to no more than one-tenth of this maximum dose (i.e., 0.5 rem/year). This work presents the neutron radiation dosages associated with (α,n) reactions that occur when low-Z elements and some compounds are irradiated by α particles having energies up to 10 MeV. The numbers of hours of radiation exposure per week required to exceed a cumulative dosage of 0.5 rem per year are also presented.

A21.2 CALCULATIONS

The α-induced thick-target neutron yields have been measured experimentally (Bair and del Campo, 1979; West and Sherwood, 1982) and calculated theoretically (Wilson *et al.*, 1988). The thick-target neutron yields, Y, for the lower-atomic-number elements are listed in column 3 of Table A21.1 as a function of the energy of the α-particles. The neutron yield data for 10 MeV α-particles were obtained by extrapolation of data presented by West and Sherwood (1982), who reported neutron yields for α-particles having energies up to 9.9 MeV.

Assuming that the α-particles are singly charged and that the neutron generation is isotropic in space, the neutron flux, n, for a given beam current, i, at a distance, d, from the target can be calculated as

$$n = \frac{iY}{e4\pi d^2},\qquad (A21.1)$$

where e is the electronic charge. Assuming d = 10 ft ≈ 300 cm, then

$$\frac{n}{i} = 5.52 \times 10^3 \, Y \text{ neutrons/(cm}^2 \text{ s nA)}.\qquad (A21.2)$$

This flux can be converted to an α-radiation dose using the appropriate neutron-flux-to-dose-rate conversion factor. It should be noted that, for a given energy of α-particles, neutrons of various energies can be emitted from a target. However, the conversion factors for 1 MeV and 20 MeV neutrons differ by less than a factor of 2 (ANSI, 1977). Because the conversion factor is not very sensitive to the neutron energy, the value of the conversion factor for 10 MeV neutrons (2.45 × 10 neutrons/cm² = 1 rem) was used for all calculations. See Chapter 17 for the actual values. The neutron radiation dose in rem/(nA h), calculated at a distance of 10 feet from the target, is listed in column 4 of Table A21.1.

The maximum allowable ionizing radiation dose is 5 rem/year. It is assumed that an individual would prefer not to be exposed to more than one-tenth of this maximum dose. Assuming a beam current of 100 nA, we performed calculations to determine the number of hours per week a person would have to be exposed to exceed a cumulative exposure of 0.5 rem/year. The results of these calculations are listed in the fifth column of Table A21.1, labeled h/w (i.e., hours/week). A dash (–) in this column indicates that the required number of hours per week exceeds 40, indicating that the radiation dosage associated such experiments is not very significant. The cases where the number of hours per week is less than 4 (length of a typical ion beam experimental session) are highlighted in bold, indicating that extra caution is required.

It should be noted that the radiation dosage is directly proportional to the beam current and inversely proportional to the square of the distance. Thus, the values presented can be adjusted for any beam current and distance. Also, if the α-particles are doubly charged, the radiation dosage is cut by a factor of 2.

A21.3 RADIATION YIELDS FROM COMPOUNDS

Up to now, all of the data presented have been for elemental targets. The neutron yields of compounds can be calculated if the individual yields of the constitutive elements are known (West, 1979). The total thick-target yield, Y, of a compound is given as

$$Y = \frac{\sum_j a_j A_j Y_j C_{j1}}{\sum_j a_j A_j C_{j1}},\qquad (A21.3)$$

Table A21.1. Netron radiation dose for pure element targets.

Target	MeV	Neutrons/α	Rem/na-hr	Number of hours/week[1]	Reference
F	3.5	2.16e-7	1.75e-7	·	Wilson et al., 1988
	4.0	8.79e-7	7.13e-7	·	Bair and del Campo, 1979
	4.5	2.16e-6	1.75e-6	·	Bair and del Campo, 1979
	5.0	4.39e-6	3.56e-6	28	Bair and del Campo, 1979
	5.5	7.75e-6	6.29e-6	16	Bair and del Campo, 1979
	6.0	1.23e-5	9.98e-6	10	Bair and del Campo, 1979
	6.5	1.79e-5	1.45e-5	7	Bair and del Campo, 1979
Ne	3.0	2.55e-7	2.07e-7	·	Wilson et al., 1988
	3.5	5.20e-7	4.20e-7	·	Wilson et al., 1988
	4.0	9.07e-7	7.36e-7	·	Wilson et al., 1988
	4.5	1.42e-6	1.15e-6	·	Wilson et al., 1988
	5.0	2.03e-6	1.65e-6	·	Wilson et al., 1988
	5.5	2.74e-6	2.22e-6	·	Wilson et al., 1988
	6.0	3.58e-6	2.90e-6	34	Wilson et al., 1988
	6.5	4.55e-6	3.69e-6	27	Wilson et al., 1988
Na	4.0	4.30e-8	3.48e-8	·	Wilson et al., 1988
	4.5	4.34e-7	3.50e-7	·	Wilson et al., 1988
	5.0	1.46e-6	1.18e-6	·	Wilson et al., 1988
	5.5	2.53e-6	2.05e-6	·	Wilson et al., 1988
	6.0	6.53e-6	5.30e-6	19	Wilson et al., 1988
	6.5	1.11e-5	9.00e-6	11	Wilson et al., 1988
Mg	4.0	7.70e-8	6.24e-8	·	Bair and del Campo, 1979
	4.5	2.63e-7	2.13e-7	·	Bair and del Campo, 1979
	5.0	6.44e-7	5.22e-7	·	Bair and del Campo, 1979
	5.5	1.26e-6	1.02e-6	·	Bair and del Campo, 1979
	6.0	2.14e-6	1.74e-6	·	Bair and del Campo, 1979
	6.5	3.25e-6	2.64e-6	38	Bair and del Campo, 1979
	7.0	5.01e-6	4.06e-6	25	West and Sherwood, 1982
	7.5	6.92e-6	5.61e-6	18	West and Sherwood, 1982
	8.0	9.15e-6	7.42e-6	13	West and Sherwood, 1982
	8.5	1.16e-5	9.41e-6	11	West and Sherwood, 1982
	9.0	1.45e-5	1.18e-5	8	West and Sherwood, 1982
	9.5	1.75e-5	1.42e-5	7	West and Sherwood, 1982
	10.0	2.09e-5*	1.69e-5	6	West and Sherwood, 1982
Al	4.0	1.69e-8	1.37e-8	·	Bair and del Campo, 1979
	4.5	8.02e-8	6.51e-8	·	Bair and del Campo, 1979
	5.0	2.64e-7	2.14e-7	·	Bair and del Campo, 1979
	5.5	6.97e-7	5.65e-7	·	Bair and del Campo, 1979
	6.0	1.44e-6	1.17e-6	·	Bair and del Campo, 1979
	6.5	2.78e-6	2.25e-6	·	Bair and del Campo, 1979
	7.0	5.00e-6	4.06e-6	25	West and Sherwood, 1982
	7.5	7.72e-6	6.26e-6	16	West and Sherwood, 1982
	8.0	1.10e-5	8.92e-6	11	West and Sherwood, 1982
	8.5	1.50e-5	1.22e-5	8	West and Sherwood, 1982
	9.0	1.96e-5	1.59e-5	6	West and Sherwood, 1982
	9.5	2.50e-5	2.03e-5	5	West and Sherwood, 1982
	10.0	**3.13e-5***	**2.54e-5**	**4**	West and Sherwood, 1982

[1] Values represent the number of hours of exposure each week for a year, under the stated conditions (see text), that result in a dose equivalent to 0.5 rem. Bold numbers (<4) indicate that extra caution is required.

*Neutrons/α data at 10 MeV are an extrapolation of data presented in West and Sherwood (1982).

Table A21.1. Neutron radiation dose for pure element targets (continued).

Target	MeV	Neutrons/α	Rem/na-hr	Number of hours/week[1]	Reference
Si	4.0	3.97e-9	3.22e-9	-	West and Sherwood, 1982
	4.5	1.60e-8	1.30e-8	-	Bair and del Campo, 1979
	5.0.	5.20e-8	4.22e-8	-	Bair and del Campo, 1979
	5.5	1.14e-7	9.25e-8	-	Bair and del Campo, 1979
	6.0	2.31e-7	1.87e-7	-	Bair and del Campo, 1979
	6.5	3.85e-7	3.12e-7	-	Bair and del Campo, 1979
	7.0	6.51e-7	5.28e-7	-	West and Sherwood, 1982
	7.5	9.44e-7	7.66e-7	-	West and Sherwood, 1982
	8.0	1.34e-6	1.09e-6	-	West and Sherwood, 1982
	8.5	1.81e-6	1.47e-6	-	West and Sherwood, 1982
	9.0	2.37e-6	1.92e-6	-	West and Sherwood, 1982
	9.5	2.95e-6	2.39e-6	-	West and Sherwood, 1982
	10.0	3.69e-6*	2.99e-6	33	West and Sherwood, 1982
Fe	5.0	2.05e-10	1.66e-10	-	West and Sherwood, 1982
	5.5	3.35e-10	2.72e-10	-	West and Sherwood, 1982
	6.0	4.22e-9	3.42e-9	-	West and Sherwood, 1982
	6.5	2.07e-8	1.68e-8	-	West and Sherwood, 1982
	7.0	9.00e-8	7.30e-8	-	West and Sherwood, 1982
	7.5	3.10e-7	2.51e-7	-	West and Sherwood, 1982
	8.0	7.63e-7	6.19e-7	-	West and Sherwood, 1982
	8.5	1.61e-6	1.31e-6	-	West and Sherwood, 1982
	9.0	3.14e-6	2.55e-6	39	West and Sherwood, 1982
	9.5	5.72e-6	4.64e-6	22	West and Sherwood, 1982
	10.0	9.19e-6*	7.45e-6	13	West and Sherwood, 1982

1. Values represent the number of hours of exposure each week for a year, under the stated conditions (see text), that result in a dose equivalent to 0.5 rem. Bold numbers (<4) indicate that extra caution is required.

* Neutrons/α data at 10 MeV are an extrapolation of data presented in West and Sherwood (1982).

where Y_i is the thick-target yield of the elemental constituent of atomic weight A_i that constitutes a_i mole fraction of the compound. The symbol C_{j1} is a constant, defined as

$$C_{j1} = \frac{\left(\dfrac{1}{\rho}\dfrac{dE}{dx}\right)_{j,E}}{\left(\dfrac{1}{\rho}\dfrac{dE}{dx}\right)_{1,E}}, \qquad (A21.4)$$

where subscript 1 refers to some chosen reference element among the constituents. The term

$$\left(\frac{1}{\rho}\frac{dE}{dx}\right)_{j,E}$$

is the stopping cross section for α-particles of energy E in element and can be calculated using the program TRIM (Ziegler et al. 1985). The radiation yields of some compounds are listed in Table A21.2.

Table 21.2. Neutron radiation dose for compound targets.

Target	MeV	Neutrons/α	Rem/na-hr	Number of hours/week[1]	Reference
BN	4.5	4.23e-6	3.43e-6	29	West and Sherwood, 1982
	5.0	6.26e-6	5.08e-6	20	West and Sherwood, 1982
	5.5	9.00e-6	7.30e-6	14	West and Sherwood, 1982
	6.0	1.18e-5	9.57e-6	10	West and Sherwood, 1982
	6.5	1.42e-5	1.15e-5	9	West and Sherwood, 1982
316SS	5.0	7.03e-10	5.70e-10	-	West and Sherwood, 1982
	5.5	1.85e-9	1.50e-9	-	West and Sherwood, 1982
	6.0	8.19e-9	6.64e-9	-	West and Sherwood, 1982
	6.5	3.18e-8	2.58e-8	-	West and Sherwood, 1982
	7.0	1.17e-7	9.49e-8	-	West and Sherwood, 1982
	7.5	3.51e-7	2.85e-7	-	West and Sherwood, 1982
	8.0	8.09e-7	6.56e-7	-	West and Sherwood, 1982
	8.5	1.71e-6	1.39e-6	-	West and Sherwood, 1982
	9.0	3.29e-6	2.67e-6	37	West and Sherwood, 1982
	9.5	5.72e-6	4.64e-6	22	West and Sherwood, 1982
	10.0	9.08e-6*	7.36e-6	14	West and Sherwood, 1982

1. Values represent the number of hours of exposure each week for a year, under the stated conditions (see text), that result in a dose equivalent to 0.5 rem. Bold numbers (<4) indicate that extra caution is required.

* Neutrons/α data at 10 MeV are an extrapolation of data presented in West and Sherwood (1982).

ACKNOWLEDGEMENTS

The authors thank Larry Andrews and Richard Olsher for providing the flux-to-dose-rate conversion factors, as well as M. Bozoian for providing the article on calculated thick-target neutron yields.

REFERENCES

ANSI/ANS (1977), *American National Standard Neutron and Gamma-Ray Flux-to-Dose-Rate Factors*, Report ANSI/ANS-6.1.1, American Nuclear Society, La Grange Park, IL.

Bair, J.K., and del Campo, J.G. (1979), *Nucl. Sci. Eng.* **71**, 18.

Cember, H. (1976), *Introduction to Health Physics*, Pergamon, New York.

ICRP (International Commission on Radiological Protection) (1987), *Radiation—A Fact of Life*, Report IAEA/PI/A9E 85-00740, American Nuclear Society, La Grange Park, IL.

West, D., and Sherwood, A.C. (1982), *Ann. Nucl. Energy* **9**, 551.

West, D. (1979), *Ann. Nucl. Energy* **6**, 549.

Wilson, W.B., Bozoian, M, and Perry, R.T. (1988), "Calculated Alpha-induced Thick Target Yields and Spectra, with Comparison to Measured Data", in *Proceedings of the International Conference on Nuclear Data for Science and Technology*, May 30–June 3, Mito, Japan, p. 1193.

Ziegler, J.F., Biersack, J.P., and Littmark, U. (1985), "The Stopping and Range of Ions in Solids", in *The Stopping and Ranges of Ions in Matter* (Ziegler, J.F., ed.), Pergamon, New York, vol. 1.

INDEX

Printed in the United States
By Bookmasters